Sedimentology

Recent developments and applied aspects

D1373648

Sedimentology
Recent developments and applied aspects

edited by

P. J. Brenchley & B. P. J. Williams*

Jane Herdman Laboratories of Geology, University
of Liverpool, Liverpool, and *Department
of Geology, University of
Bristol, Bristol

1985

Published for
The Geological Society
by Blackwell Scientific Publications
Oxford London Edinburgh
Boston Palo Alto Melbourne

Library
I.U.P.
Indiana, Pa.

551.304   Se283d
C.1

Published by

Blackwell Scientific Publications
Osney Mead, Oxford OX2 0EL
8 John Street, London WC1N 2ES
23 Ainslie Place, Edinburgh EH3 6AJ
52 Beacon Street, Boston, Massachusetts 02108, USA
744 Cowper Street, Palo Alto, California 94301, USA
107 Barry Street, Carlton, Victoria 3053, Australia

First published 1985

© 1985 The Geological Society. Authorization to
photocopy items for internal or personal use, or
the internal or personal use of specific clients,
is granted by The Geological Society for libraries
and other users registered with the Copyright
Clearance Center (CCC) Transactional Reporting
Service, provided that a base fee of $02.00 per
copy is paid directly to CCC, 21 Congress Street,
Salem, MA 01970, USA.
0305-8719/85 $02.00.

Printed in Great Britain by
Alden Press Ltd, Oxford

DISTRIBUTORS

USA and Canada
  Blackwell Scientific Publications Inc.
  PO Box 50009, Palo Alto
  California 94303
Australia
  Blackwell Scientific Book Distributors
  31 Advantage Road, Highett
  Victoria 3190

British Library Cataloguing in Publication Data

Sedimentology: recent developments and
  applied aspects.
  1. Sedimentology
  I. Brenchley, P.J.   II. Williams, B.P.J.
  551.3'04   QE471

ISBN 0-632-01192-0
ISBN 0-632-01418-0 Pbk

Contents

Preface

In December 1982 the British Sedimentological Research Group celebrated its 21st anniversary at its annual meeting, held at the University of Liverpool. To mark the occasion twelve speakers were invited to review progress in various aspects of sedimentology. This book is the collection of those reviews.

The founding meeting of the BSRG took place at Reading on 16–17 November 1962, when the Geology Department was celebrating its occupation of the new Sedimentology Research Laboratory. The occasion was attended by about ninety guests, including the laboratory's benefactors from geology-based industries and by soft-rock geologists and geo-administrators from most U.K. universities (all research students were invited) and relevant public bodies and consulting firms. Active British sedimentologists, virtually all of whom came, formed a minority, such was the state of our science at the time.

The celebrations comprised demonstrations (with demonstrators) of research in progress, an informal social evening, addresses by three distinguished visitors from overseas, and a winding-up session.

The three 'keynote speakers' (a novel buzz-word then) symbolized the major areas of sedimentology in which the new laboratory would work: Al Fischer (non-clastic and geochemical sedimentology), spoke on the Capitan reef-complex, Ake Sundborg (clastic and experimental), on fluvial deposits and Adolf Seilacher (palaeoecology), on his trace-fossil depth-zone model.

Demonstrations (concerned by request with work still in progress) were organized by research students and staff from several British universities and by workers from the IGS (now British Geological Survey), National Coal Board and major oil and quarry companies. The projects being unfinished, lively and informal discussions were generated, with free and easy participation and no sense of who was 'senior' or 'junior'.

At the closing session Perce Allen's address began with a straight question: 'Do you think this sort of meeting is a good thing'? On receiving a clear 'yes' he asked 'Oughtn't it to be repeated at intervals, say annually or biennially, at other centres?'. The response being an enthusiastic and unanimous affirmative, Perce went on to outline how this might be done. He suggested that the basic need was an organization that depended for life on its own enthusiasm (and died without it). It therefore had to have the simplest possible constitution. In his view this should be: 'There shall be no organization (or semi-permanent committee or whatever) except a convenor. A new convenor shall be elected at each meeting. He will be responsible for ensuring that the next meeting is arranged and then retire'.

Independence from other geological bodies should be preserved 'to avoid ossification (and to be free to die if enthusiasm waned)'. Emphasis should be on the younger worker and on informality. Meetings should include field demonstrations and be always mindful of the need to re-integrate theoretical, experimental and field work on the modern and ancient sediments in all their physical, chemical and biological aspects.

The BSRG was thus born in those distant 'pre-North Sea' days when our science in the U.K. went little further than sedimentary petrology and probably had fewer than a score of serious followers.

The style and general format of the annual meeting has been retained over the subsequent years in spite of the vigorous growth of the meetings. The annual meeting which, apart from the first 3 years, has been held at the Christmas period, has visited respectively the Universities of Reading, Newcastle, Belfast, Bristol, Dundee, Keele, Oxford, Aberystwyth, Cambridge, Durham, East Anglia, Reading, Leeds, Swansea, Strathclyde, Bristol and Liverpool in 1982.

Three-hundred-and-forty people attended the Liverpool meeting when there were seventy-three 10-minute talks and forty poster displays. On this occasion the usual 2-day meeting was extended to 3 days to allow time for the review lectures.

The choice of topics for review was influenced by three factors. Firstly, we wished the topics to reflect themes which have been of concern to BSRG members in past meetings of the Group. Thus we included reviews of clastic and carbonate diagenesis but excluded the geochemistry of sediments which has never figured prominently at the meetings. Secondly, we wished to reflect the influence which the development of hydrocarbons in the North Sea has had on British sedimentology, and reciprocally the role sedimentology has increasingly played in hydrocarbon exploration. Our third concern was a practical one, namely that the published collection of reviews would not overlap in content too much with other recently published books on sedimentology. Arising from this policy we confined the discussion of sedimentary environments and facies sequences to a few broad reviews in spite of innovative research in this field which has formed an important part of many recent BSRG meetings. A good deal of this work was guided by Harold Reading and was summarized in *Sedimentary Environments and Facies*[1] which he edited. A new edition will be completed in the near future and there was little to be gained by duplicating this excellent guide to the interpretation of ancient facies.

Many of the seeds of modern sedimentology had been sown before 1962, and the last two decades have seen their vigorous growth. For example, our understanding of the dynamics of sediment transport had advanced a relatively long way as early as 1941 with the publication of Bagnold's *The Physics of Blown Sand*[2]; in 1963 the recently acquired knowledge of the sea-floor was summarized in Shepard's classic *Submarine Geology*[3] and 1958 saw the publication of Bathurst's paper on 'Diagenetic fabrics in British Dinantian Limestones'[4]. However, in spite of important progress in some fields, our understanding of water movement and sediment transport in the sea was still rudimentary, diagenetic studies were still in their infancy and in spite of Walther's early insight into the relationship between modern facies distribution and the resulting vertical facies sequences, there had been little progress in developing facies models. Furthermore the significance of magnetic 'stripes' in establishing a chronology of ocean-crust formation was not yet appreciated and the deep-sea drilling programmes of JOIDES and DSDP were in the future. Consequently the exciting history of the oceans was barely suspected two decades ago.

In the four sections of the book we have tried to reflect some of the important developments of the last two decades. In the first section John Allen has traced some of the developments in our understanding of sediment transport since Bagnold's classic studies. In the second section Roger Anderton and Maurice Tucker have respectively reviewed our present understanding of clastic and carbonate facies and the value of current facies models, whilst Roger Suthren has summarized recent developments in the relatively unexplored field of volcaniclastic facies. Modern shelf environments and processes of sediment transport (Nick McCave) and deep-sea clastic facies (Dorrik Stow) are reviewed whilst Jerry Leggett uses DSDP data and evidence from submersibles to review some aspects of the history of the oceans.

Our appreciation of diagenetic fabrics and processes has largely emerged during the last two decades. Diagenetic fabrics in carbonates have been the subject of several recent reviews, so Tony Dickson has concentrated his review on the relative value of petrography, staining and cathodoluminescence as techniques of investigation and the value of the distribution coefficient, stable isotopes and fluid inclusions in our understanding of carbonate rocks. Emphasis is also placed on deep-burial diagenesis.

Studies of the petrography and chemistry of pore-filling cements in clastic rocks are mainly relatively recent and these are summarized in the review of Burley, Kantorowicz and Waugh.

The applied aspects of sedimentology are covered in the last section of the book. Harry Clemmey shows how the epigenetic theory of metallic ore formation has been substantially replaced by syngenetic models which draw on a knowledge of facies analysis, weathering and soil-forming processes, fluid migration and pore-water geochemistry. In the final two review papers the importance of sedimentology in the exploration for oil is highlighted. The role which sedimentology has played in the search for oil in the North Sea is described by Howard Johnson and David Stewart and a case history of how detailed carbonate-facies analysis can be used in modelling reservoir potential is given for the Middle East by Trevor Burchette and Selina Britton.

We have not tried to be comprehensive in this collection of reviews but we hope that the book reflects many of the concerns of sedimentologists today, that it helps to put this work into a history of development over the last two decades and may help to define the path into the future.

It was our hope that this book would be published within a year of the December 1982 BSRG Annual Meeting. However, although the first reviews were received early in 1983, others for various unforeseen reasons were not received until late in the year. This unavoidable delay slightly 'dated' those reviews which arrived early. To the authors and to the reader we offer our apologies.

The use of colour in some of the illustrations was made possible by generous grants from British Gas, the British National Oil Corporation and British Petroleum Company.

We would like to thank Hilary Davies for preparing the index, and Carol Pudsey for her careful reading of the proofs.

P. J. BRENCHLEY & B. P. J. WILLIAMS*, Jane Herdman Laboratories of Geology, University of Liverpool, Liverpool L69 3BX, and *Department of Geology, University of Bristol, Bristol BS7 9BU.

[1]READING, H. G. (ed.) 1978. *Sedimentary Environments and Facies*, 577 pp. Blackwell Scientific Publications, Oxford.
[2]BAGNOLD, R. A. 1941. *The Physics of Blown Sand and Desert Dunes*, 265 pp. Methuen, London.
[3]SHEPARD, E. P. 1963. *Submarine Geology*, 511 pp. Harper and Row, New York.
[4]BATHURST, R. G. C. 1958. Diagenetic fabrics in some British Dinantian Limestones. *Lpool Manchr geol. J.* **2**, 11–36.

FLUID DYNAMICS AND
LOOSE-BOUNDARY HYDRAULICS

Loose-boundary hydraulics and fluid mechanics: selected advances since 1961

J. R. L. Allen

SUMMARY: A major and acknowledged role in the advance of sedimentological knowledge is played by the methods of physics as expressed through fluid mechanics and loose-boundary hydraulics. Turbulence is increasingly being seen as involving orderly flow structures, and these are significant for the origin of several sedimentary structures and for suspension transport. Recent advances in sediment-transport theory are rooted in the concept of the sediment load as a downward-acting force and in the notion of the flow as a transporting machine. Laboratory experiments are increasing our understanding of bedforms due to rivers and waves, but there is a need for well-framed field studies to clarify the influence of change in natural environments. Physical explanations are being proposed for tidal bedforms. Field studies in both modern environments and the stratigraphic record suggest that tidal patterns and strengths are recognizable from the internal features of cross-bedded units formed by sand waves. Tidal patterns are complex, however, and a wider range of situations needs to be explored than has hitherto been the case. Mathematical modelling could help to define the expectable bedding patterns. Although successful mathematical models have been constructed of flow and sedimentation in channel bends, secondary flow in real bends and the evolution of those bends are not yet well understood. Turbidity currents on a geologically significant scale have not been observed directly, but valuable insights into their character and processes have come from laboratory experiments. Recent laboratory work points to the limitations of lock-exchange experiments and emphasizes the various ways in which turbidity currents mix with the ambient medium. These studies seem to have implications, yet to be worked out, for the internal features and some sole marks of turbidites.

In 1961 the British Sedimentology Research Group stood on the threshold of a period of explosive growth in sedimentological activity in the British Isles, during which the number of those claiming to be sedimentologists has grown by an order of magnitude. From a minor topic of doubtful import on the geological horizon, sedimentology has matured into a discipline with a central place in the geological curriculum and a powerful and acknowledged role in the advance of geology's principles and applications. Moreover, sedimentology has come to be recognized as multidisciplinary in the best sense, for in it we find strands from many traditional geological fields interwoven with threads from several fields of non-geological scholarship. The cloth would have appealed to the Renaissance mind.

Some of us who stood at that threshold many years ago saw that a valid contribution to sedimentology was to be made from the fields of loose-boundary hydraulics and fluid mechanics, where the experimental and theoretical methods of physics are applied to the problems of the motion of pure fluids and of fluid–solid mixtures. We hoped thereby to achieve a better understanding of sediment erosion, transport and deposition, in all their varied natural modes, and a fuller appreciation of sedimentary structures and their physical and environmental meanings. Our vision was not new, for Sorby

(1859, 1908) a century previously had demonstrated the power of a quantitative and experimental approach to sediments, but we did feel that now its time had come.

The extent to which the methods of physics have been applied to an understanding of sedimentary processes and products has increased at much the same pace as interest in sedimentology as a whole. There is now hardly a physical process of sedimentation which has not been illuminated by laboratory experiments, quantitative studies in modern environments, and mathematical explorations starting from general principles. Indeed, the wealth of data available today is so great that the reviewer with limited space faces insuperable problems of choice. The following personal selection of topics—turbulent flow, sediment transport, bedforms, flow in channel bends, and turbidity currents—is certainly not comprehensive, but it perhaps serves to illustrate the advances that have been made in several important areas. In most it is possible to see the interaction of well-framed field studies, experiment and theory.

Turbulent boundary layers

Turbulence—the apparently random motion of a fluid which is commonly also in translation— is a phenomenon of global significance. Most natural flows in the atmosphere or hydrosphere are turbulent, and turbulence is critically important to a wide range of industrial and engineering processes. Natural turbulence arises chiefly where a fluid flows and therefore shears past a stationary boundary—e.g. a river in its bed—or where the flow separates at a bluff body—e.g. a boulder projecting up into the current. However, turbulence also arises, especially in the ocean and the atmosphere, both of which are stratified, at the interface between two extensive layers of fluid in relative motion.

Early students of turbulence were impressed by its randomness, and therefore adopted an essentially statistical-dynamical approach to turbulent flows. To the considerable understanding of turbulence and its effects achieved by these means (see Townsend 1976 for review) has been added in the last 20 years a growing appreciation of the fact that turbulence is also to a considerable degree orderly (see Cantwell 1981 for review). By using various techniques, particularly flow visualization, it has been found that turbulent currents are largely formed of flow configurations of a definite and reproducible size, shape and persistence in time, i.e. structural features that are coherent. Moreover, these configurations seem to be hierarchical and their temporal and spatial scales can be related to flow properties. Turbulence is therefore not just a form of chaos, but a strongly ordered phenomenon in which stochastic and deterministic processes and effects are married together. The coherent structures of turbulence are best understood from boundary-layer flows.

The turbulent boundary layer is divisible into at least two regions: a wall region where the flow in the innermost part (viscous sublayer) is essentially laminar and where elsewhere vortical structures and turbulence begin to develop, and an outer region where the turbulence completes its development and predominates. Kline et al. (1967) used visualization and hot-wire techniques to show that the fluid in the wall region, and especially in the viscous sublayer, was organized into an array of streamwise high- and low-speed streaks whose transverse spacing depended only on the boundary shear stress and fluid properties (see also Smith & Metzler, 1983). Periodically at a fixed point, the fluid in part of a high-speed streak became lifted up and swiftly ejected on a steeply inclined path into the outer flow, a process called streak bursting. Later work (see particularly Grass 1971), stimulated by this important finding, has modified and refined the concept of streak-bursting and our notion of the coherent structure represented by a bursting streak (see reviews by Bridge 1978a;

Cantwell 1981; Allen 1982a). The work particularly of Grass (1971), Utami & Ueno (1977) and of Head & Bandyopadhyay (1981) strongly suggests that the bursting streak is a jet of low-momentum fluid ejected from near the wall into the outer part of the flow in the form of a horseshoe vortex (modification of axisymmetric jet by mean shear flow). At low Reynolds numbers these obliquely inclined vortices are loop-shaped but at large numbers are greatly stretched axially into a hairpin-like form. Even far out into the flow these vortices retain the transverse scale of the low-speed streaks from which they appear to be derived (Head & Bandyopadhyay 1981). Significantly, the frequency of streak-bursting is determined not by wall parameters like the boundary shear stress but by the overall flow properties, notably thickness and mean velocity (Rao et al. 1971). However, there is probably no universally constant non-dimensional burst frequency (Bandyopadhyay 1982).

The measurement of the motion of hydrogen bubbles generated in one flow-visualization technique (Grass 1971), and the clever analysis of hot-wire signals (Brodkey et al. 1974), have uncovered some important dynamical properties of turbulent wall regions filled with bursting streaks. Contrary to widespread belief, a marked anisotropy, particularly normal to the wall, characterizes the turbulence. The fluctuating velocity components directed perpendicularly away from the wall prove to be significantly stronger than those directed inward toward the boundary. Therefore there is a substantial net outward flow of momentum from the wall into the outer region of the turbulent boundary layer.

Early work on turbulent boundary layers and other shear flows established that their outermost edges are only intermittently turbulent and suggested that such flows comprise an array of large eddies or vortices on the scale of the boundary layer itself (Townsend 1956). More recent visualization and hot-wire studies, notably those of Kovasznay et al. (1970), Blackwelder & Kovasznay (1972), Brown & Thomas (1977), Falco (1977), Head & Bandyopadhyay (1981), and Thomas & Bull (1983) confirmed this result and showed how the horseshoe vortices representing bursting streaks could be related to these larger vortical structures, which may also be horseshoe-shaped. The large eddies, also inclined obliquely to the flow boundary but at a shallower angle, are two to three boundary-layer thicknesses long and about one boundary-layer thickness wide. They travel with the flow at approximately the mean

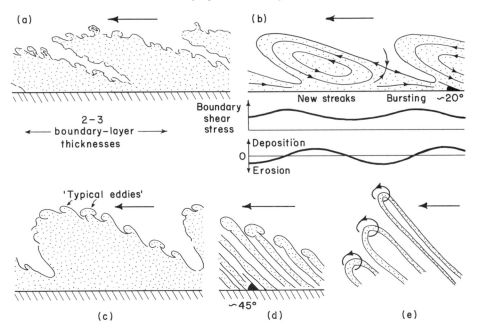

FIG. 1. Summary of experimental evidence for coherent structures in two-dimensional turbulent boundary layers on plane walls. (a) Position of the interface between marked (stippled) and unmarked fluid showing the occurrence of large-scale structures (possibly horseshoe-shaped) on the scale of the boundary layer itself (adapted from Falco 1977). (b) Flow pattern within the large structures as seen by an observer travelling approximately with the structures (adapted from Brown & Thomas 1977 and Falco 1977). The streamwise distribution of boundary shear stress is also shown, together with its implications for sediment bedload transport erosion and deposition. (c) The tips of vortices due to bursting streaks ('typical eddies' of Falco 1977) expressed on the upper surface of a large-scale structure. (d) Stack of vortices due to bursting streaks within the proximal part of a large-scale structure (adapted from Head & Bandyopadhyay 1981). (e) Form of vortices due to bursting streaks at (from left to right) small, intermediate and large Reynolds numbers (adapted from Head & Bandyopadhyay 1981).

flow speed. In a wall-related turbulent flow (Brown & Thomas 1977; Thomas & Bull 1983), the boundary shear stress is a maximum (and also strongly fluctuating) at or just upstream of the gaps between consecutive large eddies. Each large eddy appears to consist of numerous horseshoe vortices representing burst streaks, the tops of which would seem to be Falco's (1977) 'typical eddies'.

Figure 1 is an attempt using this work to portray a model of the coherent structures and related properties of turbulent boundary layers, with each part of the diagram emphasizing a particular aspect. It seems possible that streak-bursting is controlled by the movement past the bed of the large horseshoe vortices of the outer flow (Brown & Thomas 1977). The streamlines in these vortices are concave-outward near the wall, a feature suggesting that the wall streaks represent a Görtler-type instability maintained by the passage of one large eddy after another. The bursting of the streaks will clearly be

favoured by the rise in boundary shear stress observed toward the rear of each large horseshoe vortex. Streak-bursting at a fixed point will occur at about the observed frequency, if the vortices are considered to be arranged *en echelon* in rows across the flow. Nakagawa & Nezu (1981) found that the streamwise spatial separation of bursts forming at the same instant was two to three flow thicknesses.

Sedimentologists concerned with bedforms quickly seized on these important advances in fluid mechanics. Parting lineations are widely familiar structures from parallel-laminated sands and sandstones, and it is now clear that these streamwise ridges and furrows reflect boundary-layer streaks and streak-bursting (Allen 1970a; Mantz 1978) (see also reviews by Bridge 1978a; Allen 1982a). Bridge (1978a) suggests that the laminations themselves record streak-bursting. However, individual laminae, carrying multitudinous lineations, are far too extensive across the flow to be due to the

bursting of single streaks (see Head &
Bandyopadhyay, 1981), but their scale and other
characters are consistent with the downstream
convection of the large horseshoe vortices,
beneath which there is a streamwise variation in
boundary shear stress (Brown & Thomas 1977)
appropriate to a cyclical pattern of short-term
erosion and deposition (Fig. 1). Jackson (1976a)
attributed dunes in sand-bed rivers to streak-
bursting, on the basis of measurements of the
frequency with which large eddies appeared at
fixed points on the surface of the Wabash River.
These eddies are far too big to be individual
burst streaks (see Head & Bandyopadhyay,
1981) and are probably to be identified either
with the large horseshoe vortices of laboratory-
scale turbulent boundary layers, or as vortices
generated in the free-shear zones of separated
flows coupled to the dunes. Insofar as streaks
and their bursting appear to depend on the
properties and convection of the large vortical
structures (Brown & Thomas 1977), it seems
unlikely that streak-bursting of itself can be the
cause of dunes.

Sediment transport

General

There can be no detrital sedimentary record
without sediment particle erosion, transport and
deposition. Specifically, there can be no record
unless the sediment transport rate varies in space
and/or time, and in such a manner that, under
tectonic, eustatic or topographic influences,
deposition in the long term prevails over
erosion. An understanding of sediment transport, and particularly its space–time variability
on comparatively small scales, is therefore
crucial to an understanding of sedimentary
sequences.

Bagnold's physically based theory of sediment
transport is unquestionably the most significant
single contribution of recent decades to this area
of sedimentology. The theory was advanced in a
long and difficult paper (Bagnold 1956) that
failed to catch widespread attention. However,
the later more accessible and somewhat widened
version (Bagnold 1966) has profoundly influenced many areas of sedimentological
thought. Refinement of the theory by its author
continues to the present day (Bagnold 1973,
1977, 1979, 1980).

The essence of Bagnold's theory is two-fold.
Because the common rock-forming minerals are
more dense than either water or air, detritus
maintained against gravity in a state of transport
above the lower boundary of a current is a load
pressing down on the bed, no matter what the

mode of transport. This load is equal to the
immersed weight of the particles per unit bed
area or, alternatively, to the quotient of the
sediment transport rate (immersed weight basis)
and the mean grain transport velocity. But a
load pressing down on the bed must under
conditions of uniform steady transport be
supported by an equal but upward-acting force.
The ultimate source of this force can only be the
transporting agent itself. The second essence lies
in the fact that the rate of sediment transport
has the quality and dimensions of a rate of doing
work. Therefore a sediment-transporting flow
can be regarded as a machine, the transport rate
being linked through an efficiency factor to the
power of the flow, i.e. its potential ability to do
work. Hence for such as rivers, the wind, or
turbidity currents flowing over a static bed, the
transport rate must be proportional through a
non-dimensional coefficient to the product of a
characteristic boundary shear stress and a
characteristic flow velocity. Alternatively, the
transport rate may be written as proportional
through dimensional coefficients, either to the
cube of a characteristic flow velocity, or to the
3/2nd power of a characteristic boundary shear-
stress term.

The flow-derived force supporting the load is
divisible into two distinct components, each
relating to a specific dynamically defined mode
of sediment transport (Bagnold 1966) (see also
Leeder 1979a, b). The bedload under practical
conditions comprises relatively large particles in
dense array confined to near the bed, making
frequent contacts with the bed and with each
other, and which are supported above the bed by
particle impact forces. On the other hand, the
suspended load under practical conditions
comprises relatively small particles at a low
concentration distributed throughout the whole
flow and which thus seldom, if ever, touch.
These particles would seem to be buoyed up by
turbulence. Bagnold postulated that wall-related
turbulence is anistropic in a direction normal to
the boundary, so that the upward fluctuations in
a stream exceed the downward ones and thus
yield the net upward momentum flux necessary
to support the suspended grains against gravity.

Bedload

Bagnold's (1956, 1966) bedload-transport
theory rests on his study of the impact forces
arising during the bulk shearing of neutrally
buoyant uniform waxy spheres dispersed in a
fluid confined in the annular space between two
coaxial rotating drums. Curiously, few later
attempts have been made either to reproduce or

extend these rather idealized experiments. Savage (1978, 1979) and Savage & Jeffrey (1981) took some steps in this direction, partly confirming Bagnold's results, and there have been encouraging parallel developments on the theoretical side (Shen & Ackermann 1982). But there is a pressing need for further work, particularly that aimed at clarifying the role of particle size, shape and density as affecting the bedload transport of real grains.

Much more attention has been given both experimentally and theoretically to individual bedload particles travelling over a granular bed, especially when the particles are transported in the absence of moving neighbours. Early Japanese work in this area is an important source of data that has long been overlooked (Tsuchiya 1969, 1971; Tsuchiya *et al.* 1969; Tsuchiya & Aoyama 1970; Tsuchiya & Kawata 1971, 1973). Later contributions were made by Gordon *et al.* (1972), Ellwood *et al.* (1975), Luque & Van Beek (1976), White & Schulz (1977), Reizes (1978), Murphy & Hooshiari (1982) and, particularly, Abbott & Francis (1977). These studies show that a bedload grain may either slide, roll or saltate over the bed, the incidence of saltation relative to the other two modes increasing as the driving force exceeds the threshold for grain entrainment. Bedload grains in water take leaps between successive bed contacts that are of the general order of one to three grain diameters high and ten grain diameters long. The trajectories depend strongly on spin and added mass, as well as on the impact and driving forces. Grains saltating beneath the wind take trajectories generally many orders of magnitude larger than the same particles advancing in water. Forces due to spin continue to be important.

The bedload transport rate depends critically on the conditions for grain entrainment as well as on the applied force. Whereas threshold conditions are well established for practically horizontal beds under a wide range of conditions (Komar & Miller 1973, 1975a; Miller & Komar 1977; Miller *et al.* 1977; Dingler 1979), the influence of appreciable bed slope on the transport rate is virtually unknown quantitatively. Only Lysne (1969) explored slope effects experimentally, yet they appear to be crucial to the maintenance and limiting shape of many bedforms (Bagnold 1956; Fredsøe 1974; Richards 1980; Allen 1982b, c, d). A smaller entrainment force is needed for downslope than upslope transport, whence a bedform tends to decay unless a counterbalancing mechanism also operates. Bed slope has an additional effect, as yet unexplored except theoretically, namely, on

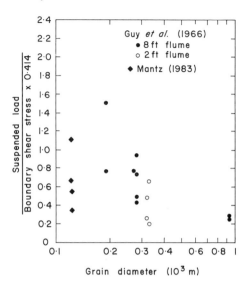

FIG. 2. Relative suspended load measured in laboratory experiments above upper-stage plane sand beds as a function of grain size (data of Guy *et al.* 1966, Tables 2–7; Mantz 1983, Tables 3, 6 and 9).

the transport rate itself (Bagnold 1956). More work is necessary for upslope than downslope transport, because in transport up a slope the load must be lifted against gravity as well as maintained above the static bed. There is an urgent need to explore these slope effects further.

Suspension

Suspension transport has traditionally been associated with fluid turbulence but treated as a diffusion phenomenon. Classically, the downward flux of suspended grains due to gravity is balanced against an opposite flux along an upward-declining concentration gradient (Rouse 1937). This analysis suffers from the weakness of being purely kinematic; it is impossible to deduce either the magnitude of the suspended load or of the suspension transport rate. Furthermore, the single function traditionally used to describe the vertical concentration of suspended sediment serves to conceal important differences in variation in different regions of the flow. In an inner region, the concentration declines according to an inverse linear relationship, but in the outer region decays logarithmically (Coleman 1969, 1970).

An interesting recent development is McTigue's (1981) demonstration that statements derived from the diffusion model can be regained using the *dynamical* theory of mixtures, in which the solid phase is allowed volume and weight. Essentially, an equilibrium can be struck between the immersed weight of the suspended particles and the fluid drag on them due to turbulence. The dynamical implications of mixture theory for suspension transport should now be explored.

Bagnold (1966) also accepted a connection between suspension transport and turbulence, but pursued an apparently very different dynamical analysis. He pointed out that turbulence could only support the suspended load if it was anisotropic, in the manner already discussed, and argued from a knowledge of shear turbulence that the load (immersed weight) should equal approximately 41% of the mean boundary shear stress. These proposals were widely discounted at the time, despite Irmay's (1960) independent evidence adduced in their favour. A major change of view has since occurred. The experiments of Grass (1971) and Brodkey et al. (1974) show that wall-related turbulence is anisotropic precisely as required. Moreover, as Leeder (1983) pointed out, experimentally determined suspended sediment loads above plane sand beds (Guy et al. 1966; Mantz 1983) are of the same order as the fraction of the mean boundary shear stress calculated by Bagnold (Fig. 2). The grain-size effect suggested by the data may either depend on the apparatus used or reflect the way the dispersed grains modify the turbulence through their drag and added mass.

We seem therefore to be approaching a period of renewed integration in sediment transport research. Suspension transport seems to depend on the anisotropy of shear turbulence, which in turn is one expression of the ordered character of a turbulent boundary layer. The experimental studies of Sumer & Ogoz (1978) and of Sumer & Deigaard (1981) draw attention to the fact that the coherent structures now recognized as occurring in all turbulent flows must take an increasingly important place in future work on the mechanisms of suspension transport. Coleman's (1969, 1970) inner and outer sediment concentration regions should have an explanation in these coherent structures and the turbulence associated with them.

Bedforms and their internal structures

Unidirectional aqueous currents

The sandy bedforms generated by one-way water streams furnish through their external shapes and internal geometry major clues to the meaning of the sedimentary record. From them can be inferred the direction and intensity of sediment transport and, when allied with other evidence, a semi-quantitative indication of flow conditions. Although Owens (1908) and Gilbert (1914) had long ago glimpsed the existence of a succession of bedforms with increasing flow strength, a reasonably detailed and comprehensive hydraulic description of this sequence has only recently arisen, largely through engineering laboratory experiments made at the Colorado State University (Fort Collins) (Guy et al. 1966). Simons et al. (1965) first drew wide attention to this work amongst geologists.

The relationship of bedforms to flow under steady uniform conditions in straight channels of laboratory scale can be expressed from the very considerable experimental data now available in stability or existence diagrams of several types. Figure 3 shows bedforms in the unit stream power–grain-size plane (Simons et al. 1965; Allen 1968, 1970, 1982a). The velocity–depth–grain-size scheme of Southard (1971) is also popular (Southard & Boguchwal 1973; Costello & Southard 1981). Bedforms found in natural environments, which are non-uniform and unsteady, and generally much deeper than laboratory flumes, are fairly satisfactorily described using the stream power–grain-size scheme (Smith 1971; Bridge & Jarvis 1976, 1982) and the velocity–depth–grain-size plot (Boothroyd & Hubbard 1974; Jackson 1976b; Dalrymple et al. 1978; Cant 1978; Rubin & McCulloch 1980). However, the comparisons are so far limited to comparatively small-scale systems. In a third and less-often-used scheme, bedforms appear in a graph of boundary shear stress (either dimensional or non-dimensional) against grain size (e.g. Leeder 1980; Allen 1982a). A fourth scheme of limited use shows bedforms in the shear-stress–grain-Reynolds-number plane (e.g. Allen & Leeder 1980).

Each scheme is restricted in applicability by the limitations of the data base (Fig. 3). Mantz (1978, 1980) has gone some way toward improving knowledge of bedforms in silts, but much is unknown about sediments of this important size class. Bedforms in gravels are also poorly understood and some workers appear to think that the larger sandy bedforms have no coarser-grade counterparts. The field occurrence of breakout-flood gravel dunes (e.g. Baker 1973) and of gravel antidunes (Shaw & Kellerhals 1977) warns against this prejudice. For all size classes, however, there is a dearth of observations at large stream powers. Do upper-stage plane beds persist to an indefinitely large stream power (assuming no free-surface effects), or can further instabilities develop? Finally, the schemes apply only to near-spherical quartz

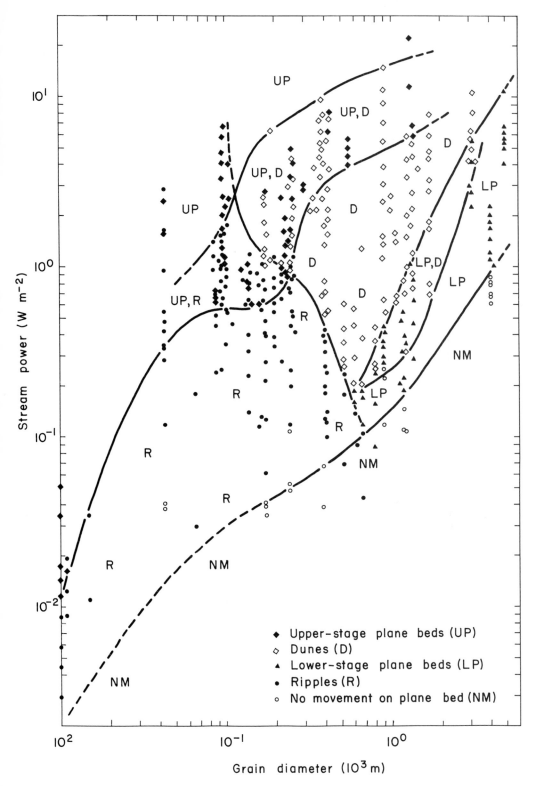

FIG. 3. Existence of bedforms in uniform-steady flow in straight laboratory channels as depicted in the stream power–grain-size plane, based on 566 observations from nineteen investigations (for data sources see Allen 1982a, p. 339). To avoid overcrowding only representative and critical data points are shown. The stream power is calculated using wall-corrected values for the boundary shear stress, and the grain diameter is adjusted to plain water at 25°C.

sands; virtually nothing is known of the flow relationships of bedforms composed of strongly non-spherical particles—e.g. shell hash.

Velocity–depth–grain-size schemes have the particular disadvantage of becoming increasingly unreliable as the natural systems to which they are applied grow larger relative to the laboratory scale. The schemes involving stream power and boundary shear stress have the limitation of showing the fields for plane beds significantly overlapped by those for ripples and dunes (Fig. 3), for which only part of the total flow force is directly involved in the sediment transport. Bridge (1981) explained the overlap between the dune and upper-stage plane bed fields, but that with lower-stage plane beds remains mysterious.

Insofar as they depend on steady-state experiments, none of the schemes is applicable without considerable caution to the sedimentary record, shaped by unsteady and non-uniform flows.

Allen (1973, 1974) and Allen & Collinson (1974) pointed out that bedforms, especially large ones, should lag in size and shape behind changes of flow, because sediment transport can occur only at a finite, flow-determined rate. There is now abundant evidence for lag, both from experiments (Simons & Richardson 1962; Gee 1975; Wijbenga & Klaasen 1983) and the field (Pretious & Blench 1951; Allen et al. 1969; Stückrath 1969; Peters 1971; Nasner 1974). Two theoretically possible mechanisms of bedform response to a changing flow (Allen 1976a) are (i) alteration of the composition of the bedform population, as new individuals better adjusted to the changed conditions replace existing forms, and (ii) change on the part of individual forms during their lifespans. The operation of these or other as yet unidentified mechanisms leads to complex and non-unique relationships between bedform and flow properties (e.g. Allen 1976b, c, 1978a), which somewhat restricts our

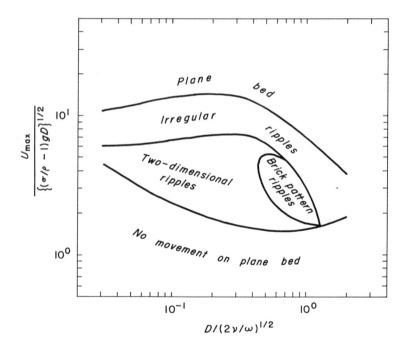

FIG. 4. Existence of bedforms related to water waves, based on experiments using monochromatic waves in a straight wave channel (after Kaneko 1980). The ordinate is a non-dimensional boundary shear stress, where U_{max} is the maximum orbital velocity of a near-bed water particle, σ and ρ respectively the solids and fluid densities, g the acceleration due to gravity, and D the sediment particle diameter. The abscissa shows a non-dimensional grain size (denominator is the thickness of the wave boundary layer), where ν is the fluid kinematic viscosity and ω the angular frequency of the waves.

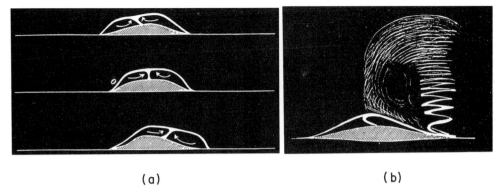

(a) (b)

FIG. 5. Reproductions of two of Darwin's (1884) original figures to illustrate the two-layer drift currents generated by waves acting on deformable granular beds. (a) The 'ink mushroom' at different stages in the wave cycle, illustrating the lower layer of vortices, the drift just above the bed being from trough to crest. (b) The 'ink tree' illustrating the upper layer of vortices, in which the drift is upward in the trough and downward above the crest (an ink mushroom appears below).

ability to interpret bedforms and their internal structures. However, lag does permit transient bedforms to be preserved—e.g. dune forms in flash-flood deposits.

Attempts by mathematical modelling to understand lag mechanisms and their effects were made by Allen (1976b, d, e, 1978a, b) and Fredsøe (1979, 1981). Allen's model relates to strongly unsteady flows, whereas Fredsøe's is more restricted in terms of the mechanisms incorporated, and is at present suitable only for gradually changing flows. Both models give results in qualitative agreement with field data. Neither model incorporates the effects of flow non-uniformity, an important limitation in comparison with natural environments, and each could be improved if more were known experimentally about lag. What exactly are the mechanisms of lag and what is their quantitative significance?

Existence diagrams such as Fig. 3 describe but do not explain sandy bedforms in one-way currents. The physically most appealing explanation of the bedforms is that they record either the stability (plane bed) or instability (wavy bed) of a deformable granular boundary in the presence of a grain-transporting flow. Kennedy (1963, 1964, 1969) made the first comprehensive stability analysis of bedforms; the approach is now a powerful and sometimes useful tool, although still with limitations. Kennedy showed mathematically that bed waves (ripples, dunes, antidunes) could grow only where there was an appropriate spatial lag between the streamwise changes of bed elevation and the streamwise changes in the rate of sediment transport over the bed. Thus a bed wave can grow in amplitude only when the transport-rate maximum lies

upstream of the wave crest. Damping occurs when the crest lies upstream. However, Kennedy's models for the flow and transport were very simple, and his lag distance was introduced as an arbitrary parameter. An intrinsic spatial lag emerges from later analyses using more realistic flow and transport models (Engelund 1970; Fredsøe 1974; Parker 1975; Nakagawa & Tsujimoto 1980; Richards 1980). The lag distance may depend on as many as six distinct effects, which in each flow combine algebraically to determine its magnitude (Allen 1983). Some reflect the way the fluid alone responds to the wavy bed, and others the response of the bedload and suspended load to the spatially changing current. These later analyses are quite good at predicting bedform existence fields and draw attention to the importance of bed slope and sediment transport mode as controls on bed stability. Because non-linear effects are ignored, they predict the quantitative aspects of bedforms less well. Puls (1981) and Fredsøe (1982) have recently and with some success calculated dune size and shape.

Currents generated by wind waves

After long neglect, attention has again been turned to the bedforms generated by wave-related currents, particularly the ripple marks. Figure 4 shows the existence field for these bedforms suggested experimentally by Kaneko (1980), others having been proposed by Allen (1967, 1970a) and Komar & Miller (1975b). Based on laboratory and field data, several models for the interpretation of ancient wave ripple marks have now been advanced (Komar 1974; Clifton 1976; Allen 1979a; Allen 1981).

Perhaps the most important development has been the rediscovery and substantial vindication through extensive theoretical and experimental studies of Darwin's (1884) original explanation for wave ripple marks. Using ink to mark the fluid, Darwin found that the oscillatory current present above a bed of the ripples (not too steep) was accompanied by slow drifts of fluid which defined a vertically stacked stationary double circulation. The drift created what he called his ink tree and ink mushroom, as reproduced in Fig. 5 from his original paper. The lower row of stationary recirculating vortices, denoted by the ink mushrooms, is very flat and the near-bed flow is from ripple troughs to crests. The opposite circulation, giving the ink trees, is observed in the upper row of vortices, which are distant from the bed and much larger. The response of the boundary to the near-bed drift was a movement of grains from ripple trough to crest and a consequent increase in the amplitude of the bed waves and further accentuation of the drift. The bed would appear to have been intrinsically unstable, which is not surprising since no natural granular boundary can be without drift-inducing waviness.

Darwin's (1884) flow pattern was not again seen experimentally until Uda & Hino (1975). Physically, it depends on force gradients that arise in an oscillatory boundary-layer flow that thickens and thins over bed waviness. Several theoretical predictions of these drift currents exist (Lyne 1971; Hall 1974; Sleath 1974, 1976; Uda & Hino 1975; Kaneko & Honji 1979; Matsunaga *et al.* 1981), supported by abundant experimental proof (Uda & Hino 1975; Hino & Fujisaki 1977; Kaneko & Honji 1979; Honji *et al.* 1980; Matsunaga & Honji 1980; Du Toit & Sleath 1981).

When wave ripple marks become sufficiently steep, effects due to flow separation seem to dominate and the forms become attributable to Bagnold's (1946) vortex type. Tunstall & Inman (1975) and Longuet-Higgins (1981) recently explored the flow over vortex ripples but confined themselves to the separation-related vortices. Darwin's drifts are present (Du Toit & Sleath 1981) but their influence in maintaining the bed features is as yet uncertain.

Tidal currents

Oscillatory tidal currents generate a variety of bedforms, the general morphology and distribution of which are now well known (Stride 1982). The mechanics of the larger of these—sand ribbons, sand waves (excluding dunes as understood from one-way flows) and

tidal current ridges—have attracted some recent attention, although there is so far little agreement as to interpretation.

Sand ribbons are longitudinal bedforms that are most commonly seen as streamwise bands of sand on a surface of contrasted texture. It has long been held that these structures reflect the action of a pattern of secondary flow in the form of paired streamwise corkscrew vortices. McLean (1981) emphasized the role of transverse roughness variations in the inducement and maintenance of such vortices, and the dependence of their scale on flow thickness. However, there are other possible causes—e.g. turbulence anisotropy and wave–wind drift interactions (Langmuir vortices). What is lacking from the field is a study designed to confirm the presence of a secondary flow in association with sand ribbons and to establish its most probable cause. There is also considerable uncertainty about the scale of sand ribbons. Their spacing as summarized by Allen (1982e) using published field data increases on the average with flow thickness, and in a manner consistent with what is theoretically expected of secondary flows composed of paired streamwise vortices. Kenyon (1970) and Belderson *et al.* (1982) believe that ribbon spacing is unrelated to flow depth.

Tidal current ridges (Off 1963) are very large linear sand banks roughly aligned with tidal currents that occur about equally spaced in groups. Numerous sand waves lie superimposed on them. Unlike sand ribbons, with a spacing typically a few times the water depth, tidal current ridges lie a distance apart approximately two orders of magnitude greater than the depth (Off 1963; Allen 1968). The ridge crests are offset by about 10° from the direction of the regional peak tidal streams (Smith 1969; McCave 1979; Kenyon *et al.* 1981), and most of the banks have markedly asymmetrical cross-sectional profiles.

An origin by the action of secondary flows similar to those responsible for sand ribbons (Off 1963; Houbolt 1968) is supported neither by the relative spacing of the ridges nor perhaps by McCave's (1979) field observations of the local currents. The large relative spacing probably also discounts their dependence implied by Kenyon *et al.* (1981) on the centrifugal instability of the tidal boundary layer. Caston (1972) concluded from field observations that the strongest tidal currents and sand transport tended on the shallower parts of a bank to turn toward the crest, maintaining the form of the bank and giving a net sand transport in the direction of the steeper face (see also

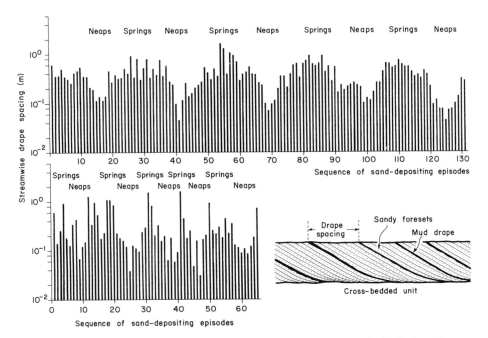

FIG. 6. Graphs illustrating spatial patterns in the arrangement of mud drapes in tidally formed cross-bedded units. The upper graph represents a late Holocene unit from the Oosterschelde (data of Visser 1980), and the lower a unit from the Folkestone Beds (Cretaceous) in the English Weald (data of Allen 1981, 1982f).

Caston & Stride 1970). In addition, there appears to be a circulation of sand around a bank (Houbolt 1968; McCave & Langhorne 1982), best revealed by the orientations of sand waves superimposed on the sand ridges. Huthnance (1973, 1982a, b) has given an important and quantitatively correct analytical justification of Caston's (1972) model, the banks being interpreted as due to an instability involving the influence of bed slope on sediment-transport rate.

Tidal current ridges are expected to comprise cross-bedded sands organized within a complex hierarchical framework of erosional bedding contacts (McCave & Langhorne 1982; Stride *et al.* 1982). Because the bank itself is apparently migrating in one direction, but sand is circulating around it, only the internal master accretion bedding would appear to denote the direction of regional net sand transport. The internally preserved cross-bedding related to dunes superimposed on sand waves, and any intermediate-order structures attributable to the sand waves themselves (Allen 1980), are those due to the oppositely acting subordinate tidal stream. Care should therefore be taken when interpreting palaeocurrent patterns from shallow-marine sandstones.

Sand waves ranging from symmetrical to asymmetrical in cross-sectional profile, and typically with dunes superimposed on them, are the largest transverse bedforms shaped by tidal currents. Their cause is as obscure now as at the time of their discovery. Some workers (Cartwright 1959; Furnes 1974) have invoked internal waves to explain them, but there are good reasons why this explanation cannot be universal. Allen (1979b) advanced the possibility that sand waves are analagous in the oscillatory tidal flows to the drift-related ripples generated by surface waves. Belderson *et al.* (1982) saw sand waves merely as modified dunes but did not satisfactorily explain what seem to be the dunes superimposed on them. Hammond & Heathershaw (1981) advanced a general wave theory which, when a beat effect is included, predicts the scale of sand waves well. The work of Davies (1982a, b) and Heathershaw (1982) on the reflection of wave energy by sea-bed topography may help to explain some sand waves. Theoretically, sand waves grow in asymmetry with increasing net sediment transport (Allen 1982c, d), and critical field studies are now needed to amplify and refine the few existing observations (Allen 1980) linking sand-wave shape with the flow and sediment-transport regimes.

Finally, we urgently need to establish by direct observation the internal structure of sand waves, and how in detail that structure is related to bedform shape, size and flow-transport regime. Models based on shallow sampling (e.g. Reineck 1963), and on considerations of form, movement and general regime (McCave 1971; Allen 1980; Langhorne 1982), will be helpful for a time, but cannot be regarded as ultimately satisfactory. Recent studies (e.g. Visser 1980) have stressed the importance of slack-water mud drapes and reactivation structures preserved within cross-bedded units as a means of unravelling tidal flow patterns, particularly the spring–neap cycle. As examples, Fig. 6 shows the downcurrent spacing of mud drapes in a subfossil cross-bedded deposit from a Dutch tidal channel (Visser 1980) and in a cross-bedded unit from the Cretaceous Folkestone Beds of the Weald (Allen 1981, 1982f). The closely spaced drapes suggest neap tides too weak to transport much sand, and those spaced further apart the more vigorous springs. The pattern of spacings in each case is roughly periodic, the length of the cycles suggesting in the one a diurnal and in the other a semidiurnal tidal regime. There are suggestions in Visser's (1980) case of the diurnal inequality widely known from European tides. Valuable as are the recent field investigations of Boersma (1969), Terwindt (1971, 1975, 1981), Visser (1980), Boersma & Terwindt (1981), Kohsiek & Terwindt (1981) and Van den Berg (1982), it should be remembered that they relate to a mesotidal to macrotidal and strongly semidiurnal regime. Tidal bedding patterns recorded from the stratigraphic record (Levell 1980; Allen 1981, 1982f; Homewood 1981; Homewood & Allen 1981) suggest that other tidal regimes should be studied before general conclusions are drawn regarding structural expressions of tidal cyclicity. Tidal regimes are complex and varied. Mathematical modelling could help to define the kinds of bedding patterns to be expected under different regimes.

Channel bends

Water flowing in an open, curved, rigid laboratory channel of moderate width/depth ratio follows a simple corkscrew path, drifting toward the outside of the bend just beneath the free surface, and inward near the bed. This type of secondary motion is due to a viscosity-related imbalance between the outward-directed centrifugal and inward-acting pressure forces on each fluid element. Sand introduced into the flow accumulates on the insides of the bends, in response to the inward-directed component of the bed shear stress (Yen 1970; Martvall & Nilsson 1972; Onishi et al. 1972; Hooke 1975;

Kikkawa et al. 1976; Zimmerman & Kennedy 1978). When the channel boundaries are altered from rigid to mobile, as in Ackers & Charlton's (1970) experiments, erosion on the outside of each bend proceeds in harmony with accretion on the inner bank, and the bend migrates sideways and downstream. Under some circumstances, a limiting bend amplitude is reached, after which the migration is downstream only.

It is not surprising that the theory and actuality of secondary flow in channel bends, and its implications for sediment transport, bend behaviour, and fluvial landscapes, should have long attracted the attention of hydraulic engineers, geomorphologists and sedimentologists. The history of research in this field is cautionary, however, for it offers a particularly clear instance of how advances can be delayed when workers operating within different traditions proceed in ignorance of each other's activities (Allen 1978c). Van Bendegom (1947), an engineer working in the Netherlands, gave the first substantial flow-based analysis of bend sedimentation, but because he published in Dutch his advance was widely overlooked. Much later, and unknown to each other, at least two groups of hydraulic engineers and a sedimentologist arrived at much the same quantitative model for deposition in channel bends.

Only the invincibly ignorant can any longer deny the relevance of secondary flow to the sedimentary processes and behaviour of curved river and tidal channels. Nonetheless, although field data overwhelmingly prove the occurrence of secondary flow (Hey & Thorne 1975; Jackson 1975; Bridge & Jarvis 1976, 1977, 1982; Bathurst et al. 1977, 1979; Hickin 1978; Dietrich et al. 1979; Thorne & Hey 1979), the flow patterns observed seldom resemble the single corkscrews generated in the laboratory. Commonly, several vortices are present, forming a sheaf whose strands loosely twist and untwist, variously tilt, and flatten and narrow as they are traced through a bend or series of bends. A good example is afforded by the River South Esk, the data of Bridge & Jarvis (1982) being re-analysed in Fig. 7. Similar vertically stacked vortices are occasionally reported from the laboratory (Prus-Chacinski 1954; Mosonyi et al. 1975). Moreover, as Bathurst et al. (1979) also observe, the number and pattern of the vortices can change substantially as the plan and cross-sectional shape of the bend vary in response to discharge fluctuations (Fig. 7). Indeed, at some river stages, the handedness of the secondary flow dominating a bend may be the opposite of that implied by the bend curvature (e.g. Jackson

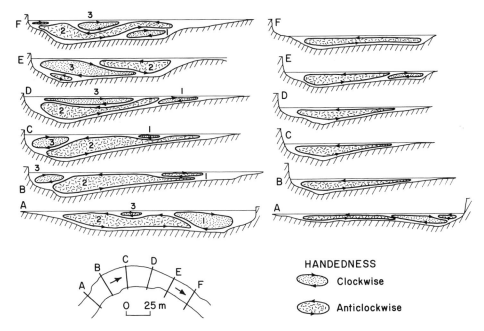

Fig. 7. Direction of rotation (observer looks downstream) of secondary flows in a single meander bend of the River South Esk, Scotland. Left-hand set of profiles for discharge of 19.6 m³ s⁻¹, and right-hand for discharge of 4.3 m³ s⁻¹. A tentative correlation of the vortices from section to section is suggested by the numbering applied at the higher discharge value. Based on a re-analysis of data originally published by Bridge & Jarvis (1982).

1975 as analysed by Allen 1982e; Hickin 1978). At which point value of the discharge is a given bend best adjusted with reference to the secondary flow developed? Are all the bends in a given stretch of a river best adjusted to the same discharge? Clearly, much more remains to be learned from the field about secondary flows and their influence. The problem in the tidal case has scarcely been touched.

In view of the complexity of secondary flow in real channels, it is perhaps surprising that attempts to model sedimentation in channel bends should have been so successful. Two kinds of mathematical model exist. The less useful (e.g. Suga 1967; Odgaard 1982) seeks to establish from dynamical considerations the channel shape and distribution of sediment, given the static stability of the bed material. The other, outlined by Van Bendegom (1947), and subsequently elaborated in two dimensions (Allen 1970a, b; Yen 1970; Ikeda 1974, 1975; Bridge 1976; Kikkawa *et al.* 1976) and three dimensions (Engelund 1974; Gottlieb 1976; Bridge 1977 1978b; Zimmermann & Kennedy 1978; Bridge & Jarvis 1982), seeks the condition for the dynamical stability of sediment in transport as bedload. In essence, dynamical stability exists when, at every point on the bed,

the upslope component of the fluid force acting on the total bedload or a single bedload particle is exactly balanced by the downslope-acting sediment weight. Combined with the use of bedform existence diagrams (e.g. Fig. 3), this type of analysis yields a remarkably accurate picture of the horizontal and vertical variation of grain size and sedimentary structures in river point bar deposits (Bridge 1978; Bridge & Jarvis 1982). But there remain some outstanding questions. Is Van Bendegom's (1947) type of force-balance equation preferable to Engelund's (1974)? Are we justified with Rozovskii (1961) in regarding the strength of secondary flow in channel bends as effectively independent of bed roughness (see Zimmermann 1977)?

Morphological and historical evidence widely suggests that river channel bends typically migrate both downstream and laterally. We are far from understanding how channel bends develop in this way, but one possible mechanism involves a spatial lag between the channel shape in plan and the spatially changing erosive potential of the secondary flow determined by that shape (Allen 1974, 1982e). On a more local scale, Begin (1981) argues that the same interaction (a lag is not however involved) explains a supposed limiting shape for channel

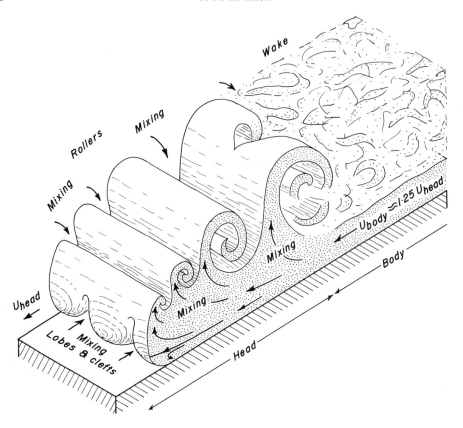

FIG. 8. Suggested structure of the leading portion of a turbidity current advancing in deep water over a nearly horizontal bottom. Based chiefly on Simpson (1969, 1972), Allen (1971), and Simpson & Britter (1979).

bends. The theory of Ikeda *et al.* (1981) and Parker *et al.* (1982) treats meandering as an instability phenomenon and attributes no effective role to either secondary flow or lag. Two instability mechanisms operating on similar characteristic wavelengths are identified; one is related to bar deposition and the other to outer-bank erosion. The theory gives good agreement with observation and is particularly appealing in that some non-linear effects observed from natural channel patterns are reproduced. The next step with all these theories must be to make numerical models capable of predicting the initiation of meanders in straight channels and their growth to realistically large amplitudes. What is the factor limiting amplitude growth in the absence of constraining bedrock walls?

Turbidity currents

Turbidity currents belong to the important class of flows called gravity currents (see Chen 1980 and Simpson 1982 for reviews). These are flows driven by a *small* difference of density relative to the neighbouring fluid or fluids. Depending on relative density, a gravity current may flow either beneath or over the surface of the ambient medium, or at some level within it when stratified. Turbidity currents are driven by an excess of density due to the presence of dispersed sediment particles.

Kuenen & Migliorini's (1950) revolutionary paper persuaded geologists that turbidity currents must be treated seriously as agents for the transport of large quantities of coarse sediment from shallow to deep water. The acceptance of the hypothesis, however, did not spark off amongst geologists quite the experimental and theoretical activity that might have been expected, although there are mitigating circumstances. Turbidity currents belong to a class of time-dependent, non-linear, non-uniform and unsteady free-boundary flows (Chen 1980), and are further complicated by containing as an essential element particles capable of sedimenting under gravity. The experiments of Middleton (1966a, b, 1967), Lovell (1971), Kersey & Hsu (1976) and Lüthi (1981a, b) provided valuable insights into the flow and transport of sediment by turbidity currents, but have the now-evident serious limitation of heavy reliance on lock-exchange techniques with few if any natural parallels. On the theoretical front, Bagnold (1962) emphasized the idea that turbidity currents might derive their energy directly from a part or the whole of the sediment dispersed in them, namely, that they are autosuspended. Middleton (1966c) and Southard & Mackintosh (1981) disputed this concept, but others have given it their support (Pantin 1979; Parker 1982). These largely untested studies highlight the fact that the key to modelling turbidity currents is how to marry the behaviour of sediment-free gravity currents with a proper treatment of the sediment; a part of this may contribute to autosuspension, another portion may be fully supported in transport by the available fluid forces, while the third and probably substantially greater part exceeds the supportable load and so is available for progressive deposition.

Probably the most important and immediately applicable advances in this difficult area have come from carefully married theoretical and experimental attempts to understand atmospheric gravity currents (Simpson 1969, 1972; Britter & Simpson 1978, 1981; Simpson & Britter 1979, 1980; Britter & Linden 1980). These suggest that turbidity currents in deep water have the general form and motion summarized in Fig. 8. The head of the current is overhanging, as has long been known, and is divided transversely into roughly periodic buttock-shaped lobes and clefts (Simpson 1969, 1972). These structures record the viscous nature of the ambient medium into which the current is advancing and, in particular, the gravitational (Rayleigh-Taylor) instability that arises where the less dense medium is being over-ridden by the heavier current (Allen 1971). The clefts carry narrow streams of ambient medium back into the head and so permit a small amount of mixing into the current (Allen 1971; Simpson 1972; Simpson & Britter 1979). Associated with the lobe and cleft structure should be a transverse variation in boundary shear stress which could explain some of the longitudinal sole markings of turbidites (Allen 1971, 1982e). Simpson (1969, 1972) further showed that spatially growing transverse billows were generated on the top of the head. These billows, representing Kelvin-Helmholtz instability, record a high rate of mixing into the medium from the current (Britter & Simpson 1978; Simpson & Britter 1979). On account of

the mixing, the body of a current in deep water flows about 25% faster than the head, so that the current consumes itself in the process of mixing into the ambient medium. Indeed, the mixing and instability of the head is the crucial factor in determining the overall motion, whether on a horizontal or sloping bed (Simpson & Britter 1979; Britter & Linden 1980).

The self-consuming nature of gravity currents emphasized in Simpson & Britter's (1979) work raises the interesting possibility that turbidity currents in a large-scale oceanic setting may develop episodically. As has often been suggested, one may envisage a typical turbidity current as generated high up on a continental margin by the mixing and dilution of slumps and slides with sea water. The current will be of finite volume and therefore length in the canyons and valleys through which at first it flows. The consumption of the last of the body by the head as the current flows along signals the completion of the creation as a kind of wake of a much larger volume of fluid which, because it contains some dispersed sediment, will be capable of further downslope flow. This process of consumption and transformation through mixing into a larger and more dilute but still finite mass might occur more than once during the flow of a turbidity current over an extensive ocean floor. Could the grain-size break at the 'bedding joints' of turbidites (e.g. Parkash 1970) express an episode in such a cascading development?

Conclusion

The advances made since 1961 in our understanding of turbulence, sediment transport, bedforms, flow in channel bends, and turbidity currents convincingly demonstrate the general value of applying the methods of physics in its broad sense to sedimentological problems. Well-designed quantitative field observations allow us some opportunity to link process and product under natural conditions, but these conditions are generally so complicated that few of the linkages can be established fully, even when many examples have been made available. Laboratory experiments allow us to test hypotheses and to explore processes under controlled and reproducible circumstances. But they are good only to the extent that we know which factors are critical under natural conditions, and suffer particularly from limitations of scale. Mathematical modelling is a powerful means for gaining insight into the controls on natural processes and especially into the relative importance of the factors involved; the value of quantitative statements about the natural process cannot be overestimated. At a more developed level, mathematical models can reproduce with considerable accuracy the behaviour and outcomes of particular natural sedimentary systems, but at present only when these are comparatively simple. There have therefore been solid gains which future generations of sedimentologists will consolidate and build upon. More power (so to speak) to them!

References

ABBOTT, J.E. & FRANCIS, J. R. D. 1977. Saltation and suspension trajectories of solids grains in water streams. *Phil. Trans. R. Soc.* **A284**, 225–54.

ACKERS, P. & CHARLTON, F. G. 1970. The geometry of small meandering streams. *Proc. Instn civ. Engrs*, Suppl. 12, 289–317.

ALLEN, G. P., DERESSEGUIER, A. & KLINGEBIEL, A. 1969. Évolution des structures sédimentaires sur un banc sableux d'estuaire en fonction de l'amplitude des marées. *C. r. hebd. Séanc. Acad. Sci., Paris*, **D269**, 2167–9.

ALLEN, J. R. L. 1967. Depth indicators of clastic sequences. *Mar. Geol.* **5**, 429–46.

—— 1968. *Current Ripples*. North-Holland, Amsterdam. 433 pp.

—— 1970a. *Physical Processes of Sedimentation*. Allen & Unwin, London.

—— 1970b. Studies in fluviatile sedimentation: a comparison of fining-upwards cyclothems, with special reference to coarse-member composition and interpretation. *J. sedim. Petrol.* **40**, 298–323.

—— 1970c. A quantitative model of grain size and sedimentary structures in lateral deposits. *Geol. J.* **7**, 129–46.

—— 1971. Mixing at turbidity current heads, and its geological implications. *J. sedim. Petrol.* **41**, 97–113.

—— 1973. Phase differences between bed configurations and flow in natural environments, and their geological relevance. *Sedimentology*, **20**, 323–9.

—— 1974. Reaction, relaxation and lag in natural sedimentary systems: general principles, examples and lessons. *Earth Sci. Rev.* **10**, 263–342.

—— 1976a. Bed forms and unsteady processes: some concepts of classification and response illustrated by common one-way types. *Earth Surf. Proc.* **1**, 361–74.

—— 1976b. Computational models for dune time-lag: general ideas, difficulties, and early results. *Sedim. Geol.* **15**, 1–53.

—— 1976c. Computational models for dune time-lag: population structures and the effects of discharge pattern and coefficient of change. *Sedim. Geol.* **16**, 99–130.

—— 1976d. Computational models for dune time-lag: an alternative boundary condition. *Sedim. Geol.* **16**, 255–79.

—— 1976e. Time-lag of dunes in unsteady flows: an analysis of Nasner's data from the R. Weser, Germany. *Sedim. Geol.* **15**, 309–321.

—— 1978a. Polymodal dune assemblages: an interpretation in terms of dune creation–destruction in periodic flows. *Sedim. Geol.* **20**, 17–28.

—— 1978b. Computational models for dune time-lag: calculations using Stein's rule for dune height. *Sedim. Geol.* **20**, 165–216.

—— 1978c. L. van Bendegom: a neglected innovator in meander studies. *Mem. Can. Soc. Petrol. Geol.* **5**, 199–209.

—— 1979a. A model for the interpretation of wave ripple-marks using their wavelength, textural composition, and shape. *J. geol. Soc. London*, **136**, 673–82.

—— 1979b. Initiation of transverse bedforms in oscillatory bottom boundary layers. *Sedimentology*, **26**, 863–765.

—— 1980. Sand waves: a model of origin and internal structure. *Sedim. Geol.* **26**, 281–328.

—— 1981. Lower Cretaceous tides revealed by cross-bedding with mud drapes. *Nature, Lond.* **289**, 579–81.

—— 1982a. *Sedimentary Structures*, Vol. 1. Elsevier, Amsterdam. 593 pp.

—— 1982b. Simple models for the shape and symmetry of tidal sand waves. I. Statically-stable equilibrium forms. *Mar. Geol.* **48**, 31–49.

—— 1982c. Simple models for the shape and symmetry of tidal sand waves. II. Dynamically-stable symmetrical equilibrium forms. *Mar. Geol.* **48**, 51–73.

—— 1982d. Simple models for the shape and symmetry of tidal sand waves. III. Dynamically-stable asymmetrical forms without flow separation. *Mar. Geol.* **48**, 321–36.

—— 1982e. *Sedimentary Structures*, Vol. 2. Elsevier, Amsterdam. 663 pp.

—— 1982f. Mud drapes in sand wave deposits: a physical model with application to the Folkestone Beds (early Cretaceous, southeast England). *Phil. Trans. R. Soc.* **A306**, 291–345.

—— 1983. River bedforms: progress and problems. *Spec. Publs int. Ass. Sediment.* **6**, 19–33.

—— & COLLINSON, J. D. 1974. The superimposition and classification of dunes formed by uni-directional aqueous flows. *Sedim. Geol.* **12**, 169–78.

—— & LEEDER, M. R. 1980. Criteria for the instability of upper-stage plane beds. *Sedimentology*, **27**, 209–17.

ALLEN, P. A. 1981. Some guidelines in reconstructing ancient sea conditions from wave ripplemarks. *Mar. Geol.* **43**, M59–67.

BAGNOLD, R. A. 1946. Motion of waves in shallow water. Interaction between waves and sand bottoms. *Proc. R. Soc. Lond.* **A187**, 1–16.

—— 1956. The flow of cohesionless grains in fluids.

Phil. Trans. R. Soc. **A249**, 235–97.

—— 1962. Autosuspension of transported sediment; turbidity currents. *Proc. R. Soc. Lond.* **A265**, 315–9.

—— 1966. An approach to the sediment transport problem from general physics. *Prof. Pap. U.S. geol. Surv.* **422-I**, 1–37.

—— 1973. The nature of saltation and of 'bed-load' transport in water. *Proc. R. Soc. Lond.* **A.332**, 473–504.

—— 1977. Bed load transport by natural rivers. *Wat. Resour. Res.* **13**, 303–12.

—— 1979. Sediment transport by wind and water. *Nordic Hydrol.* **10**, 309–22.

—— 1980. An empirical correlation of bedload transport rates in flumes and natural rivers. *Proc. R. Soc. Lond.* **A372**, 453–73.

BAKER, V. R. 1973. Paleohydrology and sedimentology of Lake Missoula flooding in eastern Washington. *Spec. Pap. geol. Soc. Am.* **144**, 1–79.

BANDYOPADHYAY, P. R. 1982. Period between bursting in turbulent boundary layers. *Physics Fluids*, **25**, 1751–4.

BATHURST, J. C., THORNE, C. R. & HEY, R. D. 1977. Direct measurement of secondary currents in river bends. *Nature, Lond.* **269**, 504–6.

——, —— & —— 1979. Secondary flow and shear stress at river bends. *J. Hydraul. Div. Am. Soc. civ. Engrs*, **105**, 1277–95.

BEGIN, Z. B. 1981. Stream curvature and bank erosion: a model based on the momentum equation. *J. Geol.* **89**, 497–504.

BELDERSON, R. H., JOHNSON, A. M. & KENYON, N. H. 1982. Bedforms. *In*: STRIDE, A. H. (ed.) *Offshore Tidal Sands*, pp. 27–57. Chapman and Hall, London.

BLACKWELDER, R. F. & KOVASZNAY, L. S. G. 1972. Large-scale motion of a turbulent boundary layer during relaminarization. *J. Fluid Mech.* **53**, 61–83.

BOERSMA, J. R. 1969. Internal structure of some tidal megaripples on a shoal in the Westerschelde Estuary, The Netherlands. Report of a preliminary investigation. *Geologie Mijnb.* **48**, 409–14.

—— & TERWINDT, J. H. J. 1981. Neap–spring tide sequences of intertidal shoal deposits in a mesotidal estuary. *Sedimentology*, **28**, 151–70.

BOOTHROYD, J. C. & HUBBARD, D. K. 1974. Bed form development and distribution patterns, Parker and Essex Estuaries, Massachusetts. *U. S. Army Corps Enginrs, Coastal Eng. Res. Cent., Misc. Pap. 1–74*, 1–39.

BRIDGE, J. S. 1976. Bed topography and grain size in open channel bends. *Sedimentology*, **23**, 407–14.

—— 1977. Flow, bed topography, grain size and sedimentary structures in open channel bends: a three-dimensional model. *Earth Surf. Proc.* **2**, 401–16.

—— 1978a. Origin of horizontal lamination under turbulent boundary layers. *Sedim. Geol.* **20**, 1–16.

—— 1978b. Palaeohydraulic interpretation using mathematical models of contemporary flow in meandering channels. *Mem. Can. Soc. Petrol. Geol.* **5**, 723–42.

—— 1981. Bed shear stress over subaqueous dunes, and the transition to upper-stage plane beds. *Sedimentology*, **28**, 33–6.

—— & JARVIS, J. 1976. Flow and sedimentary processes in the meandering River South Esk, Glen Clova, Scotland. *Earth Surf. Proc.* **1**, 303–36.

—— & —— 1977. Velocity profiles and bed shear stress over various bed configurations in a river bend. *Earth Surf. Proc.* **2**, 281–94.

—— & —— 1982. The dynamics of a river bend: a study in flow and sedimentary processes. *Sedimentology*, **29**, 499–541.

BRITTER, R. E. & LINDEN, P. F. 1980. The motion of the front of a gravity current travelling down an incline. *J. Fluid Mech.* **99**, 531–43.

—— & SIMPSON, J. E. 1978. Experiments on the dynamics of a gravity current head. *J. Fluid Mech.* **88**, 223–40.

—— & —— 1981. A note on the structure of the head of an intrusive gravity current. *J. Fluid Mech.* **112**, 459–66.

BRODKEY, R. S., WALLACE, J. M. & ECKELMAN, H. 1974. Some properties of truncated turbulence signals in bounded shear flows. *J. Fluid Mech.* **63**, 209–24.

BROWN, G. L. & THOMAS, A. S. W. 1977. Large structure in turbulent boundary layers. *Physics Fluids*, **20** (10, II), S243–52.

CANT, D. J. 1978. Bedforms and bar types in the South Saskatchewan River. *J. sedim. Petrol.* **48**, 1321–30.

CANTWELL, B. J. 1981. Organized motion in turbulent flow. *Ann. Rev. Fluid Mech.* **13**, 457–515.

CARTWRIGHT, D. E. 1959. On sand waves and tidal lee-waves. *Proc. R. Soc. Lond.* **A253**, 218–41.

CASTON, V. N. D. 1972. Linear sand banks in the southern North Sea. *Sedimentology*, **18**, 63–78.

—— & STRIDE, A. H. 1970. Tidal sand movement between some linear sand banks in the North Sea off northeast Norfolk. *Mar. Geol.* **9**, M38–42.

CHEN, J. C. 1980. Studies on gravitational spreading currents. *Rep. W. M. Keck Lab. Hydraulics Water Resources, Calif. Inst. Technol., No. KH-R-40*, 1–436.

CLIFTON, H. E. 1976. Wave-formed sedimentary structures — a conceptual model. *Spec. Publs Soc. econ. Paleont. Miner., Tulsa*, **24**, 126–48.

COLEMAN, N. L. 1969. A new examination of sediment suspension in open channels. *J. Hydraul. Res.* **7**, 67–82.

—— 1970. Flume studies of the sediment transfer coefficient. *Wat. Resour. Res.* **6**, 801–9.

COSTELLO, W. R. & SOUTHARD, J. B. 1981. Flume experiments on lower-flow-regime bed forms in coarse sand. *J. sedim. Petrol.* **51**, 849–64.

DALRYMPLE, R. W., KNIGHT, R. J. & LAMBIASE, J. J. 1978. Bedforms and their hydraulic stability relationships in a tidal environment, Bay of Fundy. *Nature, Lond.* **275**, 100–4.

DARWIN, G. H. 1884. On the formation of ripple-marks. *Proc. R. Soc. Lond.* **36**, 18–43.

DAVIES, A. G. 1982a. On the interaction between

surface waves and undulations on the sea bed. *J. Mar. Res.* **40**, 331–68.

—— 1982b. The reflection of wave energy by undulations on the sea bed. *Dyn. Atmos. Oceans*, **6**, 207–32.

DIETRICH, W. E., SMITH, J. D. & DUNNE, T. 1979. Flow and sediment transport in a sand bedded meander. *J. Geol.* **87**, 305–15.

DINGLE, J. R. 1979. The threshold of motion under oscillatory flow in a laboratory wave channel. *J. sedim. Petrol.* **49**, 287–94.

DU TOIT, C. G. & SLEATH, J. F. A. 1981. Velocity measurements close to rippled beds in oscillatory flow. *J. Fluid Mech.* **112**, 71–96.

ELLWOOD, J. M., EVANS, P. D. & WILSON, I. G. 1975. Small scale aeolian bedforms. *J. sedim. Petrol.* **45**, 554–61.

ENGELUND, F. 1970. Instability of erodable beds. *J. Fluid. Mech.* **42**, 225–44.

—— 1974. Flow and bed topography in channel bends. *J. Hydraul. Div. Am. Soc. civ. Engrs*, **100**, 1631–48.

FALCO, R. E. 1977. Coherent motions in the outer regions of turbulent boundary layers. *Physics Fluids*, **20** (10, II), S124–32.

FREDSØE, J. 1974. On the development of dunes in erodible channels. *J. Fluid Mech.* **64**, 1–16.

—— 1979. Unsteady flow in straight alluvial streams: modification of individual dunes. *J. Fluid Mech.* **91**, 427–512.

—— 1981. Unsteady flow in straight alluvial channels. II. Transition from dunes to plane beds. *J. Fluid Mech.* **102**, 431–53.

—— 1982. Shape and dimensions of stationary dunes in rivers. *J. Hydraul. Div. Am. Soc. civ. Engrs*, **108**, 932–47.

FURNES, G. K. 1974. Formation of sand waves on unconsolidated sediments. *Mar. Geol.* **16**, 145–60

GEE, D. M. 1975. Bed form response to unsteady flows. *J. Hydraul. Div. Am. Soc. civ. Engrs*, **101**, 437–49.

GILBERT, G. K. 1914. The transport of debris by running water. *Prof. Pap. U.S. geol. Surv.* **86**, 1–263.

GORDON, R., CARMICHAEL, J. B. & ISACKSON, F. J. 1972. Saltation of plastic balls in a 'one-dimensional' flume. *Wat. Resour. Res.* **8**, 444–59.

GOTTLIEB, L. 1976. Three-dimensional flow pattern and bed topography in meandering channels. *Series Paper, Inst. Hydrodyn. Hydraul. Engng, Tech. Univ. Denmark, No. 11*, 1–79.

GRASS, A. J. 1971. Structural features of turbulent flow over smooth and rough boundaries. *J. Fluid Mech.* **50**, 233–55.

GUY, H. P., SIMONS, D. B. & RICHARDSON, E. V. 1966. Summary of alluvial channel data from flume experiments, 1956–61. *Prof. Pap. U.S. geol. Surv.* **462-I**, 1–96.

HALL, P. 1974. Unsteady viscous flow in a pipe of slowly varying cross-section. *J. Fluid Mech.* **64**, 209–26.

HAMMOND, F. D. C. & HEATHERSHAW, A. D. 1981.

A wave theory for sand waves in shelf seas. *Nature, Lond.* **293**, 208–10.

HEAD, M. R. & BANDYOPADHYAY, P. 1981. New aspects of turbulent boundary-layer structure. *J. Fluid Mech.* **107**, 297–338.

HEATHERSHAW, A. D. 1982. Seabed-wave resonance and sand bar growth. *Nature, Lond.* **296**, 343–5.

HEY, R. D. & THORNE, C. R. 1975. Secondary flow in river channels. *Area*, **7**, 191–5.

HICKIN, E. J. 1978. Mean flow structure in meanders of the Squamish River, British Columbia. *Can. J. Earth Sci.* **15**, 1833–49.

HINO, M. & FUJISAKI, H. 1977. *Flow Visualization*, pp. 309–14. Hemisphere, Washington.

HOMEWOOD, P. 1981. Faciès et environnements de dépôt de la Molasse de Fribourg. *Eclog. geol. Helv.* **74**, 29–36.

—— & ALLEN, P. 1981. Wave-, tide-, and current-controlled sand bodies of Miocene Molasse, western Switzerland. *Bull. Am. soc. Petrol. Geol.* **65**, 2534–45.

HONJI, H., KANEKO, A. & MATSUNAGA, N. 1980. Flows above oscillatory ripples. *Sedimentology*, **27**, 225–9.

HOOKE, R. LeB. 1975. Distribution of sediment transport and shear stress in a meander bend. *J. Geol.* **83**, 543–65.

HOUBOLT, J. J. H. C. 1968. Recent sediments in the Southern Bight of the North Sea. *Geologie Mijnb.* **47**, 245–73.

HUTHNANCE, J. M. 1973. Tidal current asymmetries over the Norfolk sand banks. *Estuar. Coast. Mar. Sci.* **1**, 89–99.

—— 1982a. On one mechanism forming linear sand banks. *Estuar. Coast. Shelf Sci.* **14**, 79–99.

—— 1982b. On the formation of sand banks of finite extent. *Estuar. Coast. Shelf Sci.* **15**, 277–299.

IKEDA, S. 1974. On secondary flow and dynamic equilibrium of transverse bed profile in alluvial curved open channels. *Proc. Jap. Soc. civ. Engrs*, **229**, 55–65.

—— 1975. On secondary flow and bed profile in alluvial curved open channels. *Proc. 16th Conf. Int. Ass. Hydraul. Res.* **2**, 105–12.

——, PARKER, G. & SAWAI, K. 1981. Bend theory of river meanders. I. Linear development. *J. Fluid Mech.* **112**, 363–77.

IRMAY, S. 1960. Accelerations and mean trajectories in turbulent channel flow. *J. bas. Engng*, **82**, 961–8.

JACKSON, R. G. 1975. Velocity-bedform-texture patterns of meander bends in the lower Wabash River of Illinois and Indiana. *Bull. geol. Soc. Am.* **86**, 1511–22.

—— 1976a. Sedimentological and fluid-dynamic implications of the turbulent bursting phenomenon in geophysical flows. *J. Fluid Mech.* **77**, 531–60.

—— 1976b. Largescale ripples of the lower Wabash River. *Sedimentology*, **23**, 593–623.

KANEKO, A. 1980. The wavelength of oscillation sand ripples. *Rep. Res. Inst. Appl. Mech. Kyushu Univ.* **28**, 57–71.

—— & HONJI, H. 1979. Double structure of steady

Library
I.U.P.
Indiana, Pa,

551.304 Se283d
C.1   25

streaming in the oscillatory viscous flow over a wavy wall. *J. Fluid Mech.* **93**, 727–36.

KENNEDY, J. F. 1963. The mechanics of dunes and antidunes in erodible-bed channels. *J. Fluid Mech.* **16**, 521–44.

—— 1964. The formation of sediment ripples in closed rectangular conduits and in the desert. *J. geophys. Res.* **69**, 1517–24.

—— 1969. The formation of sediment ripples, dunes and antidunes. *Ann. Rev. Fluid Mech.* **1**, 147–68.

KENYON, N. H. 1970. Sand ribbons of European tidal seas. *Mar. Geol.* **9**, 25–39.

KENYON, N. H. BELDERSON, R. H., STRIDE, A. H. & JOHNSON, A. M. 1981. Offshore tidal sand banks as indicators of net sand transport and as potential deposits. *Spec. Publs int. Ass. Sediment.* **5**, 257–68.

KERSEY, D. G. & HSÜ, K. J. 1976. Energy relations of density-current flows: an experimental investigation. *Sedimentology*, **23**, 761–89.

KIKKAWA, H., IKEDA, S. & KITAGAWA, A. 1976. Flow and bed topography in curved open channels. *J. Hydraul. Div. Am. Soc. civ. Engrs*, **102**, 1327–42.

KLINE, S. J., REYNOLDS, W. C., SCHRAUB, F. A. & RUNDSTADLER, P. W. 1967. The structure of turbulent boundary layers. *J. Fluid Mech.* **30**, 741–73.

KOHSIEK, L. H. M. & TERWINDT, J. H. J. 1981. Characteristics of foreset and topset bedding in megaripples related to hydrodynamic conditions on an intertidal shoal. *Spec. Publs int. Ass. Sediment.* **5**, 27–37.

KOMAR, P. D. 1974. Oscillatory ripple marks and the evaluation of ancient wave conditions. *J. sedim. Petrol.* **44**, 169–80.

—— & MILLER, M. C. 1973. The threshold of sediment movement under oscillatory water waves. *J. sedim. Petrol.* **43**, 1101–10.

—— & —— 1975a. On the comparison between the threshold of sediment motion under waves and unidirectional currents with a discussion of the practical evaluation of the threshold. *J. sedim. Petrol.* **45**, 362–7.

—— & —— 1975b. The initiation of oscillatory ripple marks and the development of plane-bed at high shear stresses under waves. *J. sedim. Petrol.* **45**, 697–703.

KOVASZNAY, L. S. G., KIBBENS, V. & BLACKWELDER, R. F. 1970. Large scale motion in the intermittent region of a turbulent boundary layer. *J. Fluid Mech.* **41**, 283–325.

KUENEN, P. H. & MIGLIORINI, C. I. 1950. Turbidity currents as a cause of graded bedding. *J. Geol.* **58**, 91–127.

LANGHORNE, D. N. 1982. A study of the dynamics of a marine sandwave. *Sedimentology*, **29**, 571–94.

LEEDER, M. R. 1979a. 'Bedload' dynamics: grain-grain interactions in water flow. *Earth Surf. Proc.* **4**, 229–40.

—— 1979b. 'Bedload' dynamics: grain impacts, momentum transfer, and derivation of a grain Froude number. *Earth Surf. Proc.* **4**, 291–5.

—— 1980. On the stability of lower stage plane beds and the absence of current ripples in coarse sand. *J. geol. Soc. London*, **137**, 423–9.

—— 1983. On the dynamics of sediment suspension by residual Reynolds stresses—confirmation of Bagnold's theory. *Sedimentology*, **30**, 485–91.

LEVELL, B. K. 1980. A late Precambrian tidal shelf deposit, the Lower Sandfjord Formation, Finnmark, north Norway. *Sedimentology*, **27**, 539–557.

LONGUET-HIGGINS, M. S. 1981. Oscillatory flow over steep sand ripples. *J. Fluid Mech.* **107**, 1–35.

LOVELL, J. P. B. 1971. Control of slope on deposition from small-scale turbidity currents: experimental results and possible geological significance. *Sedimentology*, **17**, 81–8.

LUQUE, R. V. & VAN BEEK, R. 1976. Erosion and transport of bed-load sediment. *J. Hydraul. Res.* **14**, 127–44.

LÜTHI, S. 1981a. Experiments on non-channelized turbidity currents and their deposits. *Mar Geol.* **40**, M59–68.

—— 1981b. Some new aspects of two-dimensional turbidity currents. *Sedimentology*, **28**, 97–105.

LYNE, W. H. 1971. Unsteady viscous flow over a wavy wall. *J. Fluid Mech.* **50**, 33–48.

LYSNE, D. K. 1969. Movement of sand in tunnels. *J. Hydraul. Div. Am. Soc. civ. Engrs*, **95**, 1835–46.

MANTZ, P. A. 1978. Bedforms produced by fine, cohesionless, granular and flaky sediments under subcritical conditions. *Sedimentology*, **25**, 83–103.

—— 1980. Laboratory flume experiments on the transport of cohesionless silica silts by water streams. *Proc. Instn civ. Engrs*, **69** (2), 977–94.

—— 1983. Semi-empirical correlations for fine and coarse cohesionless sediment transport. *Proc. Instn civ. Engrs*, **75** (2), 1–33.

MARTVALL, S. & NILSSON, G. 1972. Experimental studies of meandering. *Univ. Uppsala, UNGI Rep.* **20**, 1–100.

MATSUNAGA, N. & HONJI, H. 1980. Formation of brick-pattern ripples. *Rept. Res. Inst. Appl. Mech.* **28**, 27–38.

——, KANEKO, A. & HONJI, H. 1981. A numerical study of steady streamings and oscillatory flow over a wavy wall. *J. Hydraul. Res.* **19**, 29–42.

McCAVE, I. N. 1971. Sand waves in the North Sea off the coast of Holland. *Mar. Geol.* **10**, 199–225.

—— 1979. Tidal currents at the North Hinder Light Ship, southern North Sea: flow directions and turbulence in relation to maintenance of sand banks. *Mar. Geol.* **31**, 101–14.

—— & LANGHORNE, D. N. 1982. Sand waves and sediment transport around the end of a tidal sand bank. *Sedimentology*, **29**, 95–110.

McLEAN, S. R. 1981. The role of non-uniform roughness in the formation of sand ribbons. *Mar. Geol.* **42**, 49–74.

McTIGUE, D. F. 1981. Mixture theory for suspended sediment transport. *J. Hydraul. Div. Am. Soc. civ. Engrs*, **107**, 659–73.

MIDDLETON, G. V. 1966a. Experiments on density and turbidity currents. I. Motion of the head. *Can. J. Earth Sci.* **3**, 523–46.

—— 1966b. Experiments on density and turbidity currents. II. Uniform flow of density currents. *Can. J. Earth Sci.* **3**, 627–37.

—— 1966c. Small-scale models of turbidity currents and the criterion for auto-suspension. *J. sedim. Petrol.* **36**, 202–8.

—— 1967. Experiments on density and turbidity currents. III. Deposition of sediment. *Can. J. Earth Sci.* **4**, 475–504.

MILLER, M. C. & KOMAR, P. D. 1977. The development of sediment threshold curves for unusual environments (Mars) and for inadequately studied materials (foram sands). *Sedimentology*, **24**, 709–21.

——, MCCAVE, I. N. & KOMAR, P. D. 1977. Threshold of sediment motion under unidirectional currents. *Sedimentology*, **24**, 507–27.

MOSONYI, E. MECKEL, H. & MEDER, G. 1975. Étude de developement du courant spirale dans des courbes consecutives d'un canal. *Proc. 16th Congr. Int. Ass Hydraul. Res.* **2**, 347–55.

MURPHY, P. J. & HOOSHIARI, H. 1982. Saltation in water dynamics. *J. Hydraul. Div.* **108**, 1251–67.

NAKAGAWA, H. & NEZU, I. 1981. Structure of space–time correlation of bursting phenomena. *J. Fluid Mech.* **104**, 1–43.

—— & TSUJIMOTO, T. 1980. Sand bed instability due to bed load motion. *J. Hydraul. Div. Am. Soc. civ. Engrs*, **106**, 2029–51.

NASNER, H. 1974. Über das Verhalten von Transportkorpern im Tidegebiet. *Mitt. Franzius-Inst.* **40**, 1–149.

ODGAARD, A. J. 1982. Bed characteristics in alluvial channel bends. *J. Hydraul. Div. Am. Soc. civ. Engrs*, **108**, 1268–81.

OFF, T. 1963. Rhythmic linear sand bodies caused by tidal currents. *Bull. Am. Ass. Petrol. Geol.* **47**, 324–41.

ONISHI, Y., JAIN, S. C. & KENNEDY, J. F. 1972. Effects of meandering on sediment discharges and friction factors of alluvial streams. *Rep. Iowa Inst. Hydraul. Res.* **141**, 1–150.

OWENS, J. S. 1908. Experiments on the transporting power of sea currents. *Geogr. J.* **31**, 415–25.

PANTIN, H. M. 1979. Interaction between velocity and effective density in turbidity flows: phase-plane analysis, with criteria for autosuspension. *Mar. Geol.* **31**, 59–99.

PARKASH, B. 1970. Downcurrent changes in sedimentary structures in Ordovician turbidite graywackes. *J. sedim. Petrol.* **40**, 572–90.

PARKER, G. 1975. Sediment inertia as a cause of river antidunes. *J. Hydraul. Div. Am. Soc. civ. Engrs*, **101**, 211–21.

—— 1982. Conditions for the ignition of catastrophically erosive turbidity currents. *Mar. Geol.* **46**, 307–27.

——, SAWAI, K. & IDEDA, S. 1982. Bend theory of river meanders. II. Nonlinear deformation of finite-amplitude bends. *J. Fluid Mech.* **115**, 303–14.

PETERS, J. J. 1971. *La Dynamique de la Sédimentation de la Région Divagante du Bief Maritime du Fleuve Congo.* Laboratoire de Recherches Hydrauliques à Borgerhout, Borgerhout, Belgium.

PRETIOUS, E. S. & BLENCH, T. 1951. *Final Report on Special Observations of Bed Movement in Lower Fraser River at Ladner Reach during 1950 Freshet.* National Research Council of Canada, Fraser River Model, Vancouver, Canada.

PRUS-CHACINSKI, T. M. 1954. Patterns of motion in open channel bends. *Publs Int. Ass. Sci. Hydrol.* **38**, 311–318.

PULS, W. 1981. Numerical simulation of bedform mechanics. *Mitt. Inst. Meereskunde, Hamb*, **24**, 1–147.

RAO, K. N., NARASIMHA, R. & NARAYANAN M. A. B. 1971. The 'bursting' phenomenon in a turbulent boundary layer. *J. Fluid Mech.* **48**, 339–52.

REINECK, H.-E. 1963. Sedimentgefüge im Bereich der südlichen Nordsee. *Abh. senck. naturf. Ges.* **505**, 1–138.

REIZES, J. A. 1978. Numerical study of continuous saltation. *J. Hydraul. Div. Am. Soc. civ. Engrs*, **104**, 1305–21.

RICHARDS, K. J. 1980. The formation of ripples and dunes on an erodible bed. *J. Fluid Mech.* **99**, 597–618.

ROUSE, H. 1937. Modern conceptions of the mechanics of fluid turbulence. *Trans. Am. Soc. civ. Engrs* **102**, 463–543.

ROZOVSKII, I. L. 1961. *Flow of Water in Bends of Open Channels.* Israel Program for Scientific Translations, Jerusalem.

RUBIN, D. M. & MCCULLOCH, D. S. 1980. Single and superimposed bedforms: a synthesis of San Francisco Bay and flume studies. *Sediment. Geol.* **26**, 207–231.

SAVAGE, S. B. 1978. Some studies of flow of cohesionless granular materials. *Z. angew. Math. Mech.* **58**, T313–5.

—— 1979. Gravity flow of cohesionless materials in chutes and channels. *J. Fluid Mech.* **92**, 53–96.

—— & JEFFREY, D. J. 1981. The stress tensor in granular flow at high shear rates. *J. Fluid Mech.* **110**, 255–72.

SHAW, J. & KELLERHALS, R. 1977. Palaeohydraulic interpretation of antidune bedforms with applications to antidunes in gravel. *J. sedim. Petrol.* **47**, 257–66.

SHEN, H. & ACKERMANN, N. L. 1982. Constitutive relationships for fluid-solid mixtures. *J. eng. Mech. Div. Am. Soc. civ. Engrs*, **108**, 748–63.

SIMONS, D. B. & RICHARDSON, E. V. 1962. The effect of bed roughness on depth–discharge relations in alluvial channels. *Wat.-Supply Irrig. Pap., Wash.* **1498-E**, 1–26.

——, ——, & NORDIN, C. F. 1965. Sedimentary structures generated by flow in alluvial channels. *Spec. Publs Soc. econ. Paleont. Miner., Tulsa*, **12**, 34–52.

SIMPSON, J.E. 1969. A comparison between laboratory and atmospheric density currents. *Q. Jl R. met. Soc.* **95**, 758–65.

—— 1972. Effects of the lower boundary on the head

of a gravity current. *J. Fluid Mech.* **53**, 759–68.

— 1982. Gravity currents in the laboratory, atmosphere, and ocean. *Ann. Rev. Fluid Mech.* **14**, 213–34.

— & BRITTER, R. E. 1979. The dynamics of the head of a gravity current advancing over a horizontal surface. *J. Fluid Mech.* **94**, 477–95.

— & — 1980. A laboratory model of an atmospheric mesofront. *Q. Jl R. met. Soc.* **106**, 485–500.

SLEATH, J. F. A. 1974. Mass transport on a rough bed. *J. Mar. Res.* **32**, 13–24.

— 1976. On rolling-grain ripples. *J. Hydraul. Res.* **14**, 69–81.

SMITH, C. R. & METZLER, S. P. 1983. The characteristics of low-speed streaks in the near-wall region of a turbulent boundary layer. *J. Fluid Mech.* **129**, 27–54.

SMITH, J. D. 1969. Geomorphology of a sand ridge. *J. Geol.* **77**, 39–55.

SMITH, N. D. 1971. Transverse bars and braiding in the Lower Platte River, Nebraska. *Bull. Geol. Soc. Amer.* **82**, 3407–20.

SORBY, H. C. 1859. On the structures produced by the currents present during the deposition of stratified rocks. *Geologist*, **2**, 137–47.

— 1908. On the application of quantitative methods to the study of the structure and history of rocks. *Q. Jl geol. Soc. London*, **64**, 171–233.

SOUTHARD, J. B. 1971. Representation of bed configurations in depth–velocity–size diagrams. *J. sedim. Petrol.* **41**, 903–15.

— & BOGUCHWAL, L. A. 1973. Flume experiments on the transition from ripples to lower flat bed with increasing sand size. *J. sedim. Petrol.* **43**, 1114–21.

— & MACKINTOSH, M. E. 1981. Experimental test of autosuspension. *Earth Surf. Proc. Landforms*, **6**, 103–11.

STRIDE, A. H. (ed.) 1982. *Offshore Tidal Sands.* Chapman and Hall, London.

—, BELDERSON, R. H., KENYON, N. H. & JOHNSON, M. A. 1982. Offshore tidal deposits: sand sheet and sand bank facies. *In*: STRIDE, A. H. (ed.) *Offshore Tidal Sands*, pp. 95–125. Chapman and Hall, London.

STÜCKRATH, T. 1969. Die Bewegung von Grossrippeln an der Sohle des Rio Paraná. *Mitt. Franzius-Inst.* **32**, 267–93.

SUGA, K. 1967. The stable profiles of the curved open channels. *Proc. 12th Cong. Int. Ass. Hydraul. Res.* **1**, 487–95.

SUMER, B. M. & DEIGAARD, R. 1981. Particle motions near the bottom in turbulent flow in an open channel. *J. Fluid Mech.* **109**, 311–38.

— & OGUZ, B. 1978. Particle motion near the bottom in turbulent flow in an open channel, *J. Fluid Mech.* **86**, 109–28.

TERWINDT, J. H. J. 1971. Lithofacies of inshore estuarine and tidal-inlet deposits. *Geologie Mijnb.* **50**, 515–26.

— 1975. Sequences in tidal deposits. *In*: R. N. GINSBURG (ed.) *Tidal Deposits*, pp. 85–9. Springer-Verlag, Berlin.

— 1981. Origin and sequences of sedimentary structures in inshore mesotidal deposits of the North Sea. *Spec. Publs Int. Ass. Sediment.* **5**, 4–26.

THOMAS, A. S. W. & BULL, M. K. 1983. On the role of wall-pressure fluctuations in deterministic motions in the turbulent boundary layer. *J. Fluid Mech.* **128**, 283–322.

THORNE, C. R. & HEY, F. D. 1979. Direct measurement of secondary currents at a river inflexion point. *Nature, Lond.* **280**, 226–8.

TOWNSEND, A. A. 1956. *The Structure of Turbulent Shear Flow.* Cambridge University Press, Cambridge.

— 1975. *The Structure of Turbulent Shear Flow*, 2nd edn. Cambridge University Press, Cambridge 440 pp.

TSUCHIYA, Y. 1969. Mechanics of the successive saltation of a sand particle on a granular bed in a turbulent stream. *Bull. Disaster Prevention Res. Inst. Kyoto Univ.* **19**, 31–44.

— 1971. Successive saltation of a sand grain by wind. *Proc. 12th Conf. Coast. Engng*, **2**, 1417–27.

— & AOYAMA, T. 1970. On the mechanism of saltation of a sand particle in a turbulent stream. II. On a theory of successive saltation. *Annuals Disaster Prevention Res. Inst., Kyoto Univ.* **13B**, 199–216.

— & KAWATA, Y. 1971. Mechanism of motion of sand grains by wind. II. On the characteristics of saltation of sand grains. *Annuals Disaster Prevention Res. Inst., Kyoto Univ.* **14B**, 311–25.

— & KAWATA, Y. 1973. Characteristics of saltation of sand grains by wind. *Proc. 13th Conf. Coast. Engng*, **2**, 1617–25.

—, WATADO, K. & AOYAMA, T. 1969. On the mechanism of saltation of a sand particle in a turbulent stream. I. *Annuals Disaster Prevention Res. Inst., Kyoto Univ.* **12B**, 475–90.

TUNSTALL, E. B. & INMAN, D. L. 1975. Vortex generation by oscillatory flow over rippled surfaces. *J. geophys. Res.* **80**, 3475–84.

UDA, T. & HINO, M. 1975. A solution of oscillatory viscous flow over a wavy wall. *Proc. Japan. Soc. civ. Engrs*, **237**, 27–36.

UTAMI, T. & UENO, T. 1977 Lagrangian and Eulerian measurements of large scale turbulence by flow visualization techniques. *In*: T. ASANUMA (ed.) *Flow Visualization*. Hemisphere, Washington.

VAN BENDEGOM. L. 1947. Eenige beschouwingen over riviermorphologie en rivierbetering. *De Ingenieur*, **59** (4), 1–11.

VAN DEN BERG, J. H. 1982. Migration of large-scale bedforms and preservation of cross-bedded sets in highly accretionary parts of tidal channels in the Oosterschelde, SW Netherlands. *Geologie Mijnb.* **61**, 253–63.

VISSER, M. J. 1980. Neap–spring cycles reflected in Holocene subtidal large-scale bedform deposits: a preliminary note. *Geology*, **8**, 543–6.

WHITE, B. R. & SCHULZ, J. C. 1977. Magnus effect in saltation. *J. Fluid Mech.* **81**, 497–512.

WIJBENGA, J. H. A. & KLAASEN, G. J. 1983. Changes

in bedform dimensions under unsteady flow conditions in a straight flume. *Spec. Publs int. Ass.
Sediment.* **6**, 35-48.

YEN, C.-L. 1970. Bed topography effect on flow in a
meander. *J. Hydraul. Divn Am. Soc. civ. Engrs*,
96, 57-73.

ZIMMERMAN, C. 1977. Roughness effects on the flow
direction near curved stream beds. *J. Hydraul.
Res.* **15**, 73-85.

—— & KENNEDY, J. F. 1978. Transverse bed slopes in
curved alluvial streams. *J. Hydraul. Div. Am.
Soc. civ. Engrs*, **104**, 33-48.

J. R. L. ALLEN, Department of Geology, The University, Reading RG6 2AB.

FACIES MODELS AND MODERN SEDIMENTARY ENVIRONMENTS

Clastic facies models and facies analysis

R. Anderton

SUMMARY: The term facies is used either descriptively, for a certain volume of sediment, or interpretatively for the inferred depositional environment of that sediment. Facies models are intellectual aids to the understanding of sedimentary environments and the origin of ancient sedimentary rocks. Many different models can be constructed to explain a given set of data, depending on which aspect of the facies requires the most illumination and on the types of techniques used in the analysis.

When dealing with ancient rocks, facies modelling is the last stage in the process of facies analysis which consists of the detailed description of exposed or cored sediments, their classification into objectively defined facies, the compilation of the characteristics of each facies, the deduction of the processes of deposition of each facies, the examination of the spatial relationships between facies and the recognition of facies associations, the interpretation of the overall depositional environment of the association and the detailed interpretation and modelling of individual facies. Short cuts can lead to fallacious interpretations. Reliable facies models cannot be produced without careful facies analysis. A facies model may be plausible, yet totally inapplicable to the facies from which it was supposedly derived.

Facies models are a fundamental component of the language of clastic sedimentology. Such terms as 'deltaic facies', 'washover fan facies' and even 'prograding lobe-fringe facies' are bandied about in conversation and in the literature with an air of great authority. One therefore assumes that sedimentologists, by and large, know what facies models are. However, I am less certain that the intellectual process of facies analysis, that leads from the observation of outcrop or core data to the eventual construction of a facies model, is as clearly understood.

A familiarity with published facies models is clearly essential for any practising sedimentologist. However, a knowledge of facies models, *per se*, is not as important as an understanding of the methods by which they are constructed. Facies models should not be thought of as springing from the pens of a select sedimentological elite for the rest of us to apply, if we are able. Rather, they should be constructed from first principles to meet the demands of each particular circumstance. A facies model for a particular rock unit can be no better than the analysis by which it was derived. Surprisingly therefore, there is little guidance in the literature, especially for the young researcher just starting in the field, on the subject of facies analysis, although there is a plethora of ready-prepared facies models from which to choose.

Facies analysis and the construction of facies models are skills that depend partially on the practitioner's experience and imagination. In that they are no different from other scientific techniques. Nonetheless, as with all other scientific skills, if one does not keep the

objective and logical basis of these techniques constantly in mind, it is very easy to descend into sloppy, imprecise thinking. The purpose of this paper, therefore, is to review the technique of facies analysis as it applies to clastic sediments and to show how it can be used to construct facies models. It is hoped that this will clarify the methodology of a fundamental aspect of clastic sedimentology, but an aspect which is sometimes in danger of becoming a rather intuitive and imprecise art. I make no attempt to review facies models as such. For this you can do no better than turn to the excellent recent texts edited by Reading (1978a) and Walker (1979a) and, if you have a few decades to spare, the references therein.

Definitions and philosophy

Facies

The term facies is used in many different ways in geology and can even mean different things to different people within the relatively restricted subject area of sedimentology. The meaning of the term has changed over the years (Moore 1949; Teichert 1958; Weller 1958; Krumbein & Sloss 1963; Reading 1978b) and it has been used in a somewhat different way in North America to that in Europe where the term was first used by Gressly (1838). However, today sedimentologists most commonly use the word facies to mean a certain volume of rock that can be characterized by a set of features, such as grain size, geometry and structure, that distinguish it from other rock units. This is a definition of a descriptive sedimentary facies (Table 1). Ideally,

TABLE 1. *Definitions of terms*

Descriptive sedimentary facies
A certain volume of rock that can be characterized by a set of features, such as grain size, geometry and structure, that distinguish it from other rock units.

Interpretative sedimentary facies
A label summarizing the interpretation of the processes and environments of deposition of a certain rock unit.

Facies analysis
The description and classification of any body of sediment followed by the interpretation of its processes and environments of deposition, usually in terms of a facies model.

Facies model
A description of the origin, characteristics, behaviour and evolution of the whole or part of a certain sedimentary environment, real or imaginary, in terms of a defined and realistic set of variables and boundary conditions.

---

one may then use these characteristics to infer a particular environment of deposition for the rock unit, and one may therefore refer to that rock unit as an example of, say, a fluvial facies. Now, here we have used the term facies in a rather different sense. This usage is also common even though it has incurred the wrath of some authors (e.g. Selley 1982, p. 265). I, however, see no objection to referring to a certain sandstone facies (descriptive) as an example of a fluvial facies (interpretative). The distinction should always be clear from the context, although it may not always be clear in our minds. A possible definition of an interpretative facies is: a label summarizing the interpretation of the processes and environments of deposition of a certain rock unit (Table 1).

The broad use of the term 'facies' is especially desirable when one has to consider modern environments as well as ancient rocks. One can hardly object to the application of the term tidal-flat mud facies to a sedimentary unit in which one is standing ankle-deep. In the case of modern environments, of course, such a facies label is descriptive *not* interpretative. In recent sediments an element of interpretation will usually only creep in if the term facies is used to include a depositional process which has not been directly observed, e.g. a storm-generated sand-ridge facies. A term such as crevasse splay, then, is interpretative when applied to rocks but descriptive when applied to modern environ-

ments. This distinction is so obvious as to be considered trivial. However, not keeping it clearly in mind can result in an almost subconscious transfer of the significance of modern environmental terms to ancient rock units and the acceptance, again often subconscious, of the labels of ancient facies as facts rather than mere interpretations. The use of the terms descriptive and interpretative sedimentary facies would help us to guard against this danger.

There is some confusion about the difference between the terms 'facies' and 'lithofacies'. It was Moore (1949) who urged the use of these terms to distinguish between, respectively, the purely descriptive and partially interpretative uses of the word facies. However, his definitions are no longer accepted and the term lithofacies has often been used subsequently for a type of facies based on lithological, rather than palaeontological, attributes (Krumbein & Sloss 1963). As most sedimentologists would include biological aspects as an important part of any facies description, the term lithofacies is misleading if used as a synonym for a descriptive facies. In the recent literature, facies and lithofacies are often used rather indiscriminately. There is an implicit assumption in some papers that the term lithofacies refers to rock units and facies to their environmental interpretation. However, there is seldom any consistency in this usage which is, incidentally, the reverse of that originally proposed by Moore (1949), and, accepting the broad and flexible use of the word facies discussed above, I suggest that the term lithofacies is redundant in most facies analyses. It is, however, a useful term if one wishes to draw a clear distinction with biofacies—e.g. in the description of maps based solely on lithogical data such as sand/shale ratios.

The terms subfacies and facies association are widely used and, together with facies, form a useful hierarchy of scale for describing rock units and environments. The distinction between the three terms is fairly arbitrary but they should be used in such a way as to simplify and clarify the written discussion of a rock succession. This is further discussed below. In practice, a subfacies is a more detailed subdivision of a facies and a facies association consists of several facies in a specific spatial arrangement—e.g. a vertical sequence. In many cases, it is the facies association which is the key to environmental interpretation.

Facies models

Facies is a useful term, although one has always

to be clear whether it is being using descriptively or interpretatively. What then is a facies model? Walker (1979b) refers to a facies model as a general summary of a specific sedimentary environment, written in terms that make it usable as a norm for that environment, as a framework for future observations, as a basis for hydrodynamical interpretation and as a predictive tool. I agree that it can and should be used as a framework for future observations and as a predictive tool. However, I am not so certain about the other aspects of this definition. In order to examine these, it is first necessary to make a short philosophical digression.

One's approach to facies analysis and facies modelling depends partially on one's philosophical view of the nature of the universe. If you view nature as ordered and systematic then facies analysis is an attempt to discover the character of a part of that order by classifying sedimentary rocks and modern environments into a neat system and using each to help interpret the other. An ordered view of the universe, enshrined as the principle of determinism, is supposed to be fundamental to scientific thought (e.g. Collinson & Thompson 1982, p. 3). This, of course, is a belief not a fact and should, like all beliefs, be treated with some scepticism. So if, in the light of experience, you view the universe more in terms of chaos, randomness and chance, the apparent order being largely in the mind of the observer, then facies models and facies analysis reflect an attempt to divide up the continuum into manageable but rather arbitrary units. This philosophical distinction is not so esoteric as it first appears. If you take the ordered view, then you believe that there is a finite number of environments, a finite number of descriptive sedimentary facies and a finite number of facies models by which to interpret them. You are always therefore working towards an ideal endpoint of total understanding by filling in the gaps in our framework of knowledge. On the other hand, if, like me, you have a more nihilistic view of life, the universe and everything, then you have to admit an infinite number of environments, facies and models. Every descriptive sedimentary facies is then unique and has a unique interpretation in terms of a facies model. True, many facies are similar and they may be classified in all sorts of different but essentially arbitrary ways, but lumping them together may obscure some significant features. Facies are a bit like people. It may be convenient to refer to a group of them as Welshmen, or Liberals or pigeon fanciers but that does not necessarily help you to predict how, as individuals, they will behave.

The idea that there may be a fixed number of, for example, fluvial facies models may have had some credence when BSRG started 21 years ago, but with the progressive realization of the complexity and variability of sedimentary environments it has become increasingly untenable. Just as no-one would now claim that there is only one fluvial facies model, then there is no reason to assume that there are only four, or fourteen or even 400. Each facies model for the interpretation of an ancient rock unit or a modern environment is unique. It is useful to classify these models as such classification helps to highlight the controlling variables. However, as all classifications are arbitrary, there are many equally valid and useful ways of classifying the same set of models.

So much for the digression, where does that leave us in our attempt to define a facies model? As with any type of scientific model, a facies model is a description of complex phenomena in terms of a simpler set of component parts. The description of the complex in terms of the more simple is what we mean by understanding. Clearly the choice of the number and type of component parts and the level of analysis are left to us. So even for one phenomenon there is an infinite number of possible models.

Facies models have sometimes been referred to as examples of process–response models in which the behaviour, and hence the characteristics, of a sedimentary environment are described in terms of the response of that environment to a set of process variables—e.g. the response of a beach environment to wave action. This is unsatisfactory. First, it is seldom possible to see complex natural environments in terms of simple cause and effect relationships. Demonstrating that a change in one variable can be correlated with a change in another is very far from demonstrating that one is the cause of the other. Natural systems are too complex for that, they have too many feedback loops and too many interactions with unknown variables. The best one can usually do is to describe how, not why, a system works. Secondly, and more importantly, even if one does consider natural environments in terms of linear cause and effect relationships one finds that any process is only a response to another more fundamental process —e.g. waves are a response to wind stress which in turn is a response to atmospheric pressure variations which in turn are a response to the global heat budget—and that any response can be thought of as a process which has its own consequences—e.g. the accumulation of cross-laminated sand is a consequence of the process of wave ripple formation. We can break into such a system at any point and interpret sediment behaviour, in terms of a facies model, at any of the many levels of understanding. So, although there is theoretically such a thing as a complete facies model for any sedimentary system, in practice we may consider a whole series of models, each of which describes one, or a limited number, of levels of analysis. So, for example, to interpret an ancient sandstone facies as the deposit of a certain type of fluvial channel bar is an example of 'understanding' a complex set of observations in simpler terms. However, one might then try to understand the geometry and structure of that interpretative bar in detailed terms of bedform migration and aggradation. To go to an even more fundamental level one could try to understand bedform behaviour in terms of flow conditions, flow conditions in terms of river regime and then river regime in terms of climate, weather, source-area topography and geology. Now each of these levels of analysis, or understanding, is different but equally valid in its own way and all could be used to construct different types of facies model for the same rock unit (Table 2).

The question one must ask of any facies model is 'Is it intellectually satisfying?'. The answer to this will depend more on various aspects of the questioner's intellect, training and personality than on any objectively assessable criteria. Thus, while an interpretation at a fundamental level may be rigorous to a fluid-mechanics expert, it may be nitpicking to a palaeogeographer. Conversely, what is a satisfying regional model to the plate tectonician may be hopelessly speculative and superficial to the bedform theorist.

One day it may be possible to describe natural environments completely in terms of fundamental physical laws. However, such 'explanations' will only be understandable by computers as the inherent randomness within natural environments precludes their precise description by any but the most complex mathematics. That does not mean that we should not attempt the mathematical formulation of all sedimentological behaviour, but that we have to admit that the resulting algebraic formulae, no matter how precise, may not provide as satisfying an 'understanding' of such behaviour as descriptive or geometric reasoning. Geometric thinking is fundamental to all branches of geology, whereas algebra is a tool which may, or may not, be useful. The distinction between the two was deftly illustrated by William Hamilton in 1838 who, as paraphrased by Davie (1961, p. 127) said; 'The mathematical process in algebra is like running a railroad through a tunnelled mountain: the process in geometry is like crossing the mountain on foot. The former carries us, by a short and easy transit, to our destined point, but in miasma, darkness, and torpidity, whereas the latter allows us to reach it only after time and trouble, but feasting us at each turn with glances of the earth and of the heavens, while we inhale the pleasant breeze, and gather new strength at every effort we put

TABLE 2. *Types of facies models*

A. Different levels of analysis
1. Interpretation of sandstone in terms of individual channel units.
2. Interpretation of channel units in terms of bars.
3. Interpretation of bars in terms of bedform migration and aggradation.
4. Interpretation of bedform behaviour in terms of flow conditions.
5. Interpretation of flow conditions in terms of river regime.
6. Interpretation of river regime in terms of climate, weather, source-area topography and geology.

B. Hierarchy in terms of scale

	Scale of model	Examples
1.	Environmental complex	Slope–submarine fan–basin plain
		Shelf–delta–alluvial plain
2.	Environment	Submarine fan
		Delta
3.	Sub-environment	Mid-fan lobe
		Inter-distributary bay
4.	Sedimentation sequence	Sequence of turbidite beds
		Crevasse splay sequence
5.	Sedimentation unit	Turbidite bed
		Single crevasse channel bed

forth'.

In other words, although it may ultimately be possible to explain everything in terms of fundamental physics, it is the limited capacity of the human mind that will always require us to understand complex natural phenomena at numerous levels of analysis. This is amply demonstrated by the fate of much of the excellent mathematically orientated research which remains inaccessible to and unused by the sedimentological community. It is only those algebraic analyses that can be reduced to a simple physical or geometrical concept, illustrated, for example, by a graph or diagram, that have any chance of sinking into the mass sedimentological consciousness.

The idea that different facies models must interpret sedimentary environments at different levels of understanding is also clear from a consideration of the scale of sedimentary environments. Facies models try to describe environments from the order of 10s of metres to 1000s of kilometres across, i.e. 5 orders of magnitude. Clearly, the level of detail explicit in the former cannot be approached in the latter and a whole hierarchy of models could be envisaged describing increasing levels of complexity from one sedimentation unit to an environmental complex (Table 2).

So, where has this discussion taken us in our attempt to define what is meant by a facies model? First, we have expunged the idea that it has, in any way, to describe a norm or an ideal. Secondly, we have seen that facies models can be constructed on a whole series of levels of understanding from the fundamental (i.e. couched in terms of basic physics) to the complex (i.e. couched in terms of more descriptive relationships). Thirdly, facies models can describe phenomena on a wide range of scales. May I therefore suggest that a facies model is a description of the origin, characteristics, behaviour and evolution of the whole or part of a certain sedimentary environment, real or imaginary, in terms of a defined and realistic set of variables and boundary conditions (Table 1). Such a model shows how, from an initial set of tectonic, geographic or physiographic starting conditions, a sedimentary environment will develop and evolve under the action of a defined set of variables. The environment may be real, in that it may be a modern environment in which the variables have been accurately measured, or it may be created in the imagination from fragmentary observations on ancient sedimentary rocks. A facies model may be a totally hypothetical construction designed to give insight into as yet undiscovered environments. The stipulation that the variables should be

reasonable is not meant to place too severe a constraint. Thus, it would be reasonable to construct a model to describe the aeolian environment on a distant planet with a gravity of 10 g, 500 m s^{-1} winds and an atmosphere of chlorine...or would it?

In practice, facies models may be derived from a range of data sources such as those shown in Table 3. Clearly, going from data source (a) to (e) corresponds to a decrease in the level of detail and an increase in the amount of generalization. Returning to Walker's (1979b) definition, the kind of model he is thinking of is derived from data sources (e) or possibly (d). This is the kind of generalized facies model that is very useful when introducing people to the subject. It is the type commonly described in text books and used for teaching purposes. However, I think it is a mistake to rely too heavily on such 'ideal' or 'generalized' models because, whether it is the intention of their creators or not, there is a danger that these models are taken too literally and applied out of context.

Why do we construct facies models? Firstly, because they provide us with the intellectual satisfaction of 'understanding' a phenomenon. Secondly, because we can extrapolate and interpolate the component parts of a model more easily than the whole and, thus, make various sorts of predictions. It is when we come to examine the predictive power of facies models that we can really test their mettle. They have proved to be extremely valuable in regional exploration, especially for hydrocarbons, in predicting the kind of circumstance in which economic deposits may be found and thus directing exploratory activity to specific areas and stratigraphic horizons. Unfortunately they are often less useful at the development stage because of the obvious difficulties of predicting, say, the three-dimensional geometry, porosity and permeability variations of a reservoir horizon from scant core, cutting and wireline log data derived from a few exploration wells. The problem here, that of relating vertical to lateral facies variations, is fundamental to facies modelling and its solution is still in its infancy. However, the danger now is of naively using generalized facies models to make predictions that prove to be unreliable and thus bring the technique into disrepute. The interpretation of a reservoir unit as a submarine-fan or barrier-island facies does not, in itself, allow us to make any but the most general predictions because the variations within each of these facies types are huge. The assumption that any ancient facies has the characteristics of a well-known 'ideal' facies model is unlikely to be valid. To make

TABLE 3. *Data bases for facies models*

(a) The study of a single modern environment

(b) The study of a single ancient rock unit and a single modern environment

(c) A synthesis of data from a range of similar modern environments

(d) The study of a single ancient rock unit and the synthesis of data from a range of modern environments

(e) The synthesis of data from a range of ancient rock units and modern environments

serious predictions one has to have a very detailed understanding of the facies in question and a knowledge of all the other geological variables involved—e.g. tectonic environment, source area geology and palaeoclimate. In some cases such factors may have exerted a more profound control on the characteristics of an ancient facies than the depositional environment itself.

Facies models are ephemeral. Any model is just a particular way of looking at a particular phenomenon at a particular time. As our understanding of the component parts of facies models changes so we must keep generating new models. It is also a delusion to think that successive modifications and refinements will eventually lead to a 'complete' understanding. Just as you think you are about to reach such a point, a revolution in understanding takes place, everything goes back into the melting pot and you start all over again. Science never provides *the* answer to any question but merely *an* answer.

Facies Analysis

Facies analysis is the description and classification of any body of sediment followed by the interpretation of its processes and environments of deposition, usually in the form of a facies model (Table 1). The description should be as detailed as time and money will allow, the classification should be objective. Armed with this data, the sedimentologist then deduces as much as he can about the processes responsible for sediment deposition. This deduction is based on an understanding of sedimentary processes which in turn are derived from theoretical, laboratory, and field studies plus a geometrical analysis of the rock structure itself. The sedimentologist then searches the literature for the closest modern analogue for each aspect of

the rock unit, such as its geometry, structure, and inferred processes of deposition, and also looks for examples of similar ancient facies. These analogues or comparisons will provide him with further data which he will add to that already derived and enable him to synthesize a facies model. Under exceptional circumstances he may find a ready-described facies model in the literature that fits his data exactly. This is an unlikely coincidence. Usually it is a question of pulling together and modifying observations and ideas from a variety of sources. Finally, the data, the analysis and the interpretation must be written up in a clear, intelligible and interesting way with simple diagrams explaining the ideas involved. An attempt should be made to present the interpretations at different levels or scales of understanding so that the information will be accessible to a wide audience. Thus, a mathematical analysis of the behaviour of a sand bar should also be presented in descriptive geometric terms. A discussion of the origin of a particular ancient facies should be accompanied by a map showing the palaeogeographical significance of the interpretation. This is not a question of condescendingly 'watering-down' ones high-powered science, but of making relevant aspects of it accessible to those with different skills and interests.

Facies analysis

Description

The techniques of describing the sedimentological features of sedimentary rocks in the field have been discussed recently by Collinson & Thompson (1982) and Tucker (1981). I do not want to repeat what they have said beyond giving the briefest of reviews and airing my own prejudices on the subject. However, I cannot overemphasize the importance of clear, accurate and detailed description. While one can always go back and re-interpret good data time and time again in the comfort of one's own office, re-describing a rock succession may be time-consuming and expensive if the area is remote, or may even be impossible in the case of temporary sections, subsequently infilled quarries or non-archived borehole core.

Rock successions may be examined as either essentially one-dimensional sections—i.e. borehole core, stream sections and more extensive exposures of sequences which show no significant lateral variation—or as two-dimensional sections on cliff or quarry faces which exhibit some lateral variation. (Well-exposed three-dimensional sections can be treated as a series of

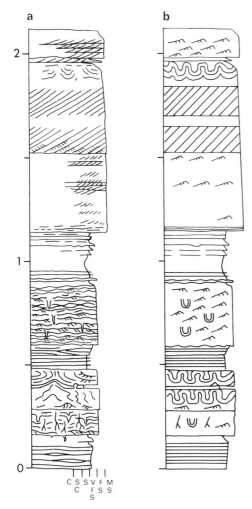

FIG. 1. Logs showing (a) realistic and (b) stylised representations of 2 m of borehole core from the Upper Limestone Group, Midland Valley of Scotland.

used for rootlets. Why not be more realistic and show the density of rootlets, their variation with depth, their thickness, geometry, nature of bifurcation and effect of host sediment structure. Clearly, all this takes time, but without this detail at this stage it will not be possible to construct a useful facies model later. In a large organization where several people, probably with different backgrounds and levels of training, may be involved in logging the same core and interpreting the resulting data, standardization of symbols is necessary. However, such standardization does result in only a small part of the available sedimentological detail being recorded on the log. Some company geologists draw a personalized detailed log, to help with their own facies interpretation, from which they abstract a standardized log for internal company consumption. This is a sensible solution which, for a small added expenditure of time, leads to a large increase in the amount of potentially useful information recorded.

Speed of logging will depend on the complexity of the facies being described as well as the required level of detail. Although several hundred metres per day may be possible in some monotonous units, a couple of metres per day would be realistic in others. The rate of progress should be linked to the scale of the detail that is observed. However, as it is often difficult to decide on a suitable scale before starting work, I suggest the use of a flexible-scale logging technique in which each bed or lamina is drawn at a scale sufficient to show any internal structure (Fig. 2). The details of each sedimentation unit are drawn to scale, but the scale may change by orders of magnitude from unit to unit. One can then delay deciding on what is the most appropriate scale until the neat version of the log is drawn up and one can extract both generalized and more detailed logs from the same field data (Fig. 2).

The recording of one-dimensional sections is fairly straightforward and well understood. The sections are logged by describing the character of each convenient unit (i.e. bed, coset, set or bundle of similar laminae), in order from bottom to top. First measure the thickness of each unit then plot its variation in grain size with height, the nature of its contacts and internal structure and finally describe any additional features that cannot be represented graphically. Two-dimensional sections, on the other hand, have first to be studied from a distance so that their lateral variability can be appreciated. The major structure of the face is then sketched so that it can be subdivided into a series of more

two-dimensional faces.) One-dimensional sections can be fully described in an annotated sedimentological log while the two-dimensional sections require detailed drawings of each face. The method and style of log drawing is a matter of personal taste. I dislike the stylized symbols beloved by oil companies, among others, although I appreciate the reasons for their use (see below). I think logs should be as detailed and realistic as the artistic ability of the drawer will allow. For example, the caterpillar symbol for cross lamination conveys little detail (Fig. 1); a more realistic version can show set thickness, foreset inclination, nature of bottomsets if any, presence of mud drapes and variability of set geometry. Similarly, a lightning symbol is often

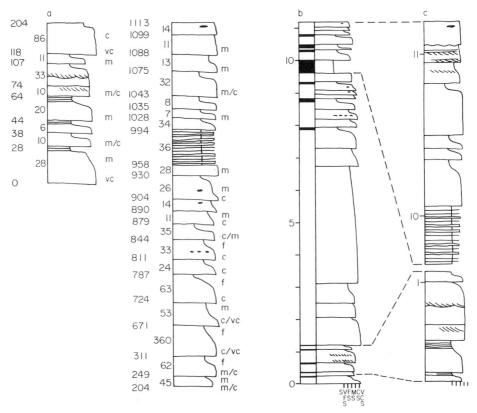

FIG. 2. Example of the use of flexible-scale logging, from an Upper Dalradian turbidite sequence, Loch Lomondside, Scotland. (a) shows a copy of the field log with each bed drawn on a scale sufficient to show any relevant detail. The thickness of each bed and the cumulative thickness from the base of the section is shown in centimetres. The notes f, m, c etc, refer to sandstone grain sizes. (b) is a true scale log, derived from this data, with some of the smaller detail generalised. Scale is in metres. (c) are more detailed true scale logs of parts of the section. (b) also has a lithology column to emphasise changes in sand/shale ratio.

homogeneous units. One can then move towards the face, make a detailed diagram of the geometry and internal structure of each unit before drawing representative logs to illustrate the variability of each unit (Fig. 3). If just one, or even a series of equally or randomly spaced logs of the whole section is drawn then the nature of the lateral relationships between units may be missed. The major structure can all too easily be missed if one is preoccupied with recording the detail. In other words, the danger with this type of section is of not seeing the wood for the trees.

As logging proceeds, all relevant palaeo-current and orientation data should be plotted on the log indicating exactly where each measurement was made. As well as the obvious palaeocurrent indicators such as sole marks, ripple crests and foreset dips, the orientation of any bedding or set surface that departs more

than a degree or two from the palaeo-horizontal should be recorded. This is especially true of two-dimensional sections of complex channel-ized sandstones. If one wants to reconstruct the history of the sandstone in terms of channel cutting, bar migration and aggradation; to work out the type, size and orientation of bars, or to deduce the relationship of smaller bedforms to bar and channel features; in short, if one wants to say anything more about a sandstone other than that it is the deposit of a fluvial or delta or submarine fan channel, one has to measure every surface, or a representative sample of every surface that can be seen. In the case of curved surfaces—e.g. channel bases or curved coset boundaries—their orientation should be measured, say, every few metres so that their original geometry can be reconstructed. All this data should be plotted on logs and diagrams of the exposed sections or faces together with

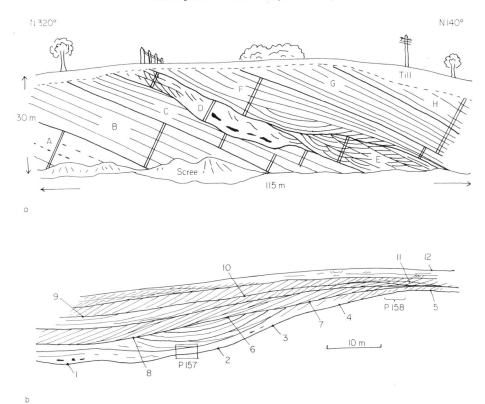

FIG. 3. (a) Hypothetical field diagram of a cliff face, either sketched in the field or drawn accurately from photographs. Face has been subdivided into units A to H and major bedding surfaces drawn. Location of logged sections shown by double lines. (b) More detailed diagram of unit F showing internal structure and numbered location of orientation measurements. Location of measured foreset orientations would also be shown. P157 etc. refers to photograph number.

sufficient structural information to enable the palaeo-orientation data to be subsequently tilted, unfolded or suitably unwound back to its original attitude.

The recording of such palaeo-surface data brings me to the discussion of the ideal way of recording long, well-exposed, complex, two-dimensional sections such as quarry faces. It is this. Photograph the whole face, allowing plenty of overlap between frames, make a photo-montage of the resulting black and white prints, trace every bit of sedimentological detail onto an overlay, photocopy the overlay and then go back to the face and plot all your orientation data and other details on the photocopies. From this information you can produce accurate scaled diagrams of large faces with sufficient detail to allow the most sophisticated facies interpretation to be made. The photo-montage is made by taking a series of overlapping photographs by moving along parallel with the

face between exposing each frame, analogous to the way an aircraft makes a stereo air photo run. Allow plenty of overlap, each point on the face should be recorded about three times and make sure that the camera film plane is both parallel to the face and at a constant distance from it. Move a measuring staff along the face so that it appears on each frame (having a field assistant will save a lot of time). Cut and match the resulting prints so that features on adjacent frames are aligned as closely as possible and fix them together with sellotape on the reverse side so that at least one major bedding surface can be traced continuously through the montage. The degree of misfit between frames increases as the face becomes more irregular and the camera–face distance decreases. However, these mis-alignments can usually be smoothed over when tracing on the overlay. In the case of real difficulty a perfect match can always be obtained by photographing the whole face from one point and panning round from frame to frame. Although this produces great scale distortion, it can sometimes be useful when trying to trace surfaces from frame to frame on the normal montage.

Of course, there are problems with these photographic methods. You need to be able to go to the field area at least twice. If this is not possible, a Polaroid-type camera can be used to make montages in the field, although this is a bit costly on film. The subtleties of lighting have to be learned and one may make numerous visits to a section before chancing on the right conditions. Diffuse light from a bright but cloudy sky is best for irregular faces but some weathered planar faces show up well under a bright oblique sun. Coastal cliffs may require near horizontal illumination from a rising or setting sun plus the use of a boat! Whatever the problem is, it is worth overcoming. Photographic techniques are sufficiently important in facies description that if one ever has to choose between taking a camera and a hammer into the field, take the camera.

Finally, as well as putting all lithological, palaeontological and structural data on your logs and two-dimensional diagrams, also indicate the positions from which any samples have been collected and the locations of small features of which you have made diagrams or taken photographs. The whole package of logs, diagrams, photographs and samples should be in such a form as to enable one to reconstruct, on paper back in the laboratory, the features of each part of the rock units under study.

Classification

Having got the description of the rocks down on paper, the next stage is to subdivide them into descriptive facies. There are two problems here, the fineness of the facies classification and the choice of facies boundaries.

If one accepts that the choice of the number of facies into which a rock succession should be divided is arbitrary, and that 'lumpers' and 'splitters' will always disagree, then the main criteria for deciding on this number is convenience and manageability. A good facies scheme is one which manages to be, at the same time, both detailed yet simple enough to be memorable. This means that the number of facies should be between two and ten, about six is ideal. Hardly anyone can read, be interested in and remember a classification with twenty facies. This means, of course, that the amount of rock and degree of complexity encompassed by one facies will depend on the amount and complexity of the rock being classified. The classification of 10 m of Carboniferous delta sediments will result in a finer facies classification than that of 10 km of Lower Palaeozoic turbidites.

Every facies classification should be objective. This means that with a given set of criteria every sedimentologist should produce the same classification for the same set of rocks. However, numerous different sets of criteria may work equally well on the same rock sequence. A lot of facies schemes come to grief because of the use of inconsistent multiple criteria for distinguishing between facies. For example, if facies A is defined as a grey mudstone with cm-thick sandstone bands and facies B as a black mudstone with mm-thick sandstone laminae, what happens when one discovers a black mudstone with cm-thick bands. Does one alter the facies definitions so that it will fit into facies A or facies B or, even worse, invent a new facies C. The answer is to use a single criterion or multiple criteria arranged in a workable hierarchy. So, for example, facies A could include all grey mudstones with sand laminae and facies B all black mudstones with sand laminae. Facies B could then be further divided into subfacies B1 and B2 on the basis of the thickness of the sand laminae. Of course, if there is a gradation between mudstones with sand laminae and sandstones with mudstone laminae, the common boundary between such facies has also to be clearly defined.

How does one go about making this objective classification? The simplest way is to look at your logs and progressively subdivide them into units using a list of characteristics such as shown in Table 4. Sometimes one characteristic—e.g. lithology—is sufficient to divide the logged sections into a handful of facies. Usually one would want to further subdivide the sandstones, for example, on the basis of internal structure

TABLE 4. *Order of priority in deciding on facies boundaries*

1. Bases of major channels or scoured surfaces
2. Tops and bases of non-clastic lithologies, (e.g. coal, limestone)
3. Abrupt changes from section composed predominantly of one grain size to another (e.g. sandstone to shale, fine sandstone to coarse)
4. Changes of internal structure within unit of similar grain size
5. . . . .

and nature of the basal surface. The type of criteria will clearly depend on the type of sediments. The more monotonous the succession, the more subtle will be the facies definitions. The most difficult successions to deal with are those with thinly interbedded units of different lithologies. It may be necessary to define a minimum thickness for a facies unit, say, 5 m, 1 m or 10 cm, as this can make a lot of difference to the way the whole facies scheme works. So, for interbedded sandstones and shales one could call the sandstones facies A and the shales facies B. Alternatively, facies A could be defined as units consisting of sandstones interbedded with shale beds less than 10 cm thick and facies B as sandstones with shales greater than 10 cm thick. Again, the facies could be defined in terms of the sand/shale ratio. Finding the right definitions is a question of trial and error. The objective is to erect a set of definitions that will subdivide all the rock under study into a series of facies such that every rock unit can be classified and that no unit can be included in more than one facies. The scheme should work for all the logged sections. If needs be, some or all of the facies may be split into subfacies. If a new section is logged subsequently, the previously recognized facies should be readily picked out. Any units that cannot be allocated to the existing scheme can then be examined to see if there is a need for additional facies to be erected. As the work progresses the facies scheme is modified and refined.

Defining a facies scheme is often the most difficult part of the whole process of facies analysis. However, having done this, one can then go on and compile a general description of the features found in the various examples of each facies. This is the facies description and it is quite distinct from the facies definition. By definition, facies A may be grey mudstone with sand laminae but the description may include such facts as that the sand laminae are carbonaceous and *usually* show parallel lamination, that the mudstones *sometimes* contain fish scales and mica flakes, and that syn-sedimentary faults are *occasionally* found. Thus, one of the objects of facies classification and description is to enable one to pool sufficient data from different examples of each facies to enable an environmental interpretation to be made. This may not be possible if each example is considered in isolation.

Interpretation

Once the hard and often tedious work of description and classification has been done, the fun of interpretation can begin. Some descriptive facies can be interpreted, at least to a first approximation, in isolation from the adjacent sediments—e.g. rooty siltstones as soil horizons or red mudstones with deep water microfossils as basinal pelagic muds. These are key facies which may strongly influence the rest of the facies interpretation. If the whole section can be interpreted in this way, there is no need for a rigorous scheme of facies analysis. However, usually one can only make direct deductions about sedimentary *processes*, using one's understanding of sedimentary structures. Thus, a sharp-based, graded, climbing ripple cross-laminated sandstone can be interpreted as the deposit of a suspension laden decelerating flow with flow power sufficient to maintain migrating current-ripples. The environment of deposition could be an alluvial floodbasin, a lake, shelf or deep basin. One treats each facies in isolation and writes down every detail that can be objectively deduced about such things as flow conditions, variations of flow with time, sediment-transport mechanisms, geometry of sand bodies, water depths, ecological significance of fauna and flora, mineralogical and geochemical inferences about water chemistry, etc. A range of possible environments can then be suggested, but it is essential to be honest and include every conceivable possibility.

The next step is to look at the relationships between facies. Vertical sequence analysis may take some statistically sophisticated form such as Markov chain analysis, but in essence it is the application of Walther's concept that if two facies are seen one above the other in a vertical sequence, they were probably deposited in adjacent sedimentary environments. Having made our process deductions about each facies, we can then examine, from the vertical sequences, the way the processes changed through time at each point on the Earth's surface represented by our logged sections. One may then recognize a simple pattern of, for example, upward shallowing and deposition from progressively more powerful flows capped by a distinctive beach facies or a soil horizon. This would be interpreted as some type of prograding shoreline sequence. However, the presence of either marine or freshwater fossils at the base of the sequence would control whether it was interpreted as a marine or lacustrine shoreline. In facies interpretation, one either interpolates the environment of non-diagnostic facies between more readily interpretable facies or uniquely interprets the depositional environment of two or several associated facies that could only be ambiguously interpreted in isolation. Thus an identical unit of thin graded sandstones in mudstone would be interpreted as an alluvial floodplain facies if interbedded with certain channel sandstones, shelf storm layers if associated with certain shelly sheet sandstones and basin plain turbidites if overlain by a submarine-fan sequence.

A vertical sequence of facies, or several facies with a restricted range of relationships between them, is known as a facies association. Thus, a repeated upward sequence of facies A to E could be referred to as facies association 1. The sequence of processes deduced could lead to its interpretation as an alluvial plain facies association, the individual facies then being interpreted as, say, swamp, lacustrine, minor lake delta, braided channel and floodplain.

The facies association is usually the key to facies interpretation. This is aided by the presence of facies containing features diagnostic of specific environments. As our understanding increases, it may seem that the number of diagnostic criteria increases. However, for every new structure that is found to be diagnostic of say, soil horizons or wave reworking, another structure is found in a wider range of environments than was previously known. Don't forget, it is not many years ago that ripple marks were thought to prove a shallow-water environment and that flutes were restricted to turbidites.

Having deduced the broad depositional environment of each facies one can then go on to work out a detailed interpretation—i.e. a facies model. It is at this stage, and not before, that one makes detailed use of modern analogues. So if one has interpreted a facies as a fluvial channel deposit, the vast fluvial literature can be searched in order to pick out the elements that explain the detailed features of the facies. However, it is even more important to squeeze every drop of information that can be wrought from the rocks themselves, and here one turns again to the detailed field logs, diagrams and descriptions. Working now with the extra insight of a hypothesis for the environment of deposition of each facies, one can search for significant points that may have been missed earlier and start to interpret the points of detail. Large two-dimensional sections are especially valuable as one can take them apart geometrically and discover the manner in which sediment accumulated. So for a complexly cross-stratified fluvial sandstone one can try to work out the sequence of erosional, aggradational and progradational events; the geometry, orientation and evolution of channels; the size, geometry and type of bars; the channel and bar pattern and its evolution with time. Once one realizes

the significance of the relevant structures, there is no end to the detail that can be extracted from good sections of such a facies. All this requires a thorough understanding of modern fluvial processes, but the literature can only provide the observer with possible lines of thought. The key to the interpretation lies in the rocks themselves.

It may seem paradoxical that, having emphasized that the processes must be deduced before one can interpret the environment of a facies association, the detailed modelling of those processes is then dependent upon the interpretation of the facies. This is because one can only go so far when interpreting processes from first principles. The added refinements depend on hypotheses about the environment of deposition that are strongly influenced by the interpretation of the whole association. So our thin graded sandstones in mudstone can be modelled in terms of the magnitude, proximity and frequency of alluvial flooding events, or the character of storm-wave and tidal currents, or the physiography and evolution of a submarine fan, depending on the overall context of the facies association in which they occur. Of course, if one's facies interpretation is wrong, it is still possible to produce a well-argued, plausible, logical and totally erroneous facies model.

One can produce a facies model to explain a facies association and/or a series of models for each facies. At this stage it may be useful to finally refine the facies classification so that there is a neater relationship between descriptive facies and interpreted environments. For example, it would be better to have the descriptive facies A, B and C as the deposits of fluvial channel, floodplain and lake environments than having the A and part of B as channel facies, part of B as floodplain and C and part of B as lacustrine facies. However, this is still only acceptable if the facies can be redefined in terms of objective, descriptive criteria.

The system of facies analysis outlined here may seem to be a rather involved and complex way of tackling a simple problem. Why not just log the sections, then look at the logs and pick out the coarsening-upwards cycles, the fining-upwards cycles, the thickening-upwards cycles and interpret these in the light of conventional wisdom without having to worry much about the processes of deposition? Well, this is a type of short-cut procedure that can work quite well for facies which behave according to well-known, generalized facies models. For facies which are a bit out of the ordinary, and are therefore likely to be new, interesting and important, this method can go catastrophically wrong because it biases the interpretation in the direction of the apparently, instantly recognizable facies.

For example, if one instantly interpreted certain sharp-based, fining-upwards cycles as being of fluvial origin, one might then be apt to overlook the fact that the floodplain facies would be better interpreted as tidal-flat sediments. In this case the more objective, step-by-step approach might lead to the sandstones being interpreted as tidal-channel deposits. A coal seam overlain by a coarsening- and thickening-upward sequence might be instantaneously recognized as a delta prograding over a subsided abandoned lobe. But what if the sequence showed no evidence for upward shallowing in spite of evidence for rapid deposition. The point is that Walther's concept does not always work, which is why it is not a law. So, perhaps the alluvial peat-bearing plain subsided so quickly in the absence of sediment deposition that it was directly overlain by a submarine fan. Or perhaps the coal is an allochthonous turbidite deposited by a rare event in a normal submarine-fan environment. Either circumstance is possible but would not be deduced by the facies-spotting technique, only by the methodical scheme for facies analysis discussed here.

Historical review

Pre-1959: facies pre-history

The word 'sedimentology' came into use in the 1930s (Wadell 1932, 1933) to describe an area of study that was expanding beyond the realms of the traditional field of sedimentary petrology. Although this label could still be referred to as an '...ugly hybrid inappropriate word...' by Goldman in 1950, sedimentology had become a recognized geological discipline by the time the International Association of Sedimentologists was founded in 1952. The basic philosophy of the subject, that one can only make detailed interpretations of sedimentary rocks if one is armed with an understanding of modern environments, experimental and theoretical work, was well established and the two-pronged attack on the modern and ancient was underway. The realization that modern sedimentary environments are of interest in their own right to non-geologists has subsequently led to the continual broadening of the subject area of sedimentology towards its present interdisciplinary nature.

The terms 'facies analysis' and 'facies models' were used in the period up to the 1950s in a rather different sense from that understood today. Facies maps were used to show lateral lithological variations of rock formations (e.g. Krumbein 1952). Facies analysis consisted of the use of such maps, together with petrographic and palaeontological data, to deduce source-area geology, location and relief, sediment-dispersal patterns and the geometry of generalized depositional basins (e.g. Pelletier 1958).

The report on a conference on facies models, held in Illinois in 1958 (Potter 1959), shows that a facies model was then thought of as a description of the three-dimensional distribution of lithologies within what we would now call a facies association. Such a model was considered to be a useful tool in geological exploration, as indeed it is, but no mention was made of sedimentary processes. Indeed sediment dispersal was considered more important than deposition and as the study of facies models was still in its infancy, it was admitted that the problems of facies models had not yet been clearly formulated.

The understanding of depositional environments was generally vague, based on traditional petrographic deductions and usually couched in such terms as 'marine', 'non-marine', 'nearshore' and 'offshore'. However, by 1950 many data on modern environments were already available, much of it from non-geological sources, and during the 1950s geological sedimentologists took an increasingly active part in studying modern environments themselves.

With one notable exception, to which I will return, the collection of data from modern environments proceeded much faster up until the late 1950s than its application to ancient rocks. Thus, while much of the literature of the period on clastic sedimentary rocks seems rather dated, many of the studies of recent sediments and environments are still interesting and relevant today. For example, the paper of Fisk *et al.* (1954) on the Mississippi delta was one of the first comprehensive descriptions of a major modern environment. It is clear, well illustrated and still exerts a great influence on the interpretation of deltaic sediments today. Extensive studies of Dutch and German tidal flats also established the sedimentological essence of this type of environment during the 1950s (e.g. Van Straaten 1954). Choosing one example from each of several other environments, one could cite Thompson's (1937) study of beach sediments, Blissenbach's (1954) work on alluvial fans and Sundborg's (1956) study of the River Klarälven to show that sedimentologists already had many of the basic data with which to formulate facies models. It should be noted that all these examples come from terrestrial to marginal marine environments which are more directly accessible than marine areas. There was extensive sampling of modern shelf sediments at this time, but much of the resulting data was essentially petrographic.

The one exception to the above discussion and, as Reading (1978b, p. 1) has pointed out, probably the first description of what we would now call a facies model was Kuenen & Migliorini's (1950) classic paper on turbidites. This combined data from a sophisticated series of laboratory experiments with deductions from Apennine graded beds to present a complete working model for turbidite formation. Even after over 30 years it still stands as an archetypal example of the sedimentological method and is remarkable for its clarity, far-sightedness and imagination. It also invalidates the often-held assumption that the interpretation of ancient sedimentary rocks is totally dependent on the detailed understanding of modern environments.

1959–1965: a trickle and then a flood

1959 was a significant year. It was the year that sedimentologists working on ancient rocks started to catch up with the work from modern environments. Within the space of a few years Kuenen & Migliorini's (1950) turbidite model, which had stood alone for almost a decade, was to be joined by facies models for a host of environments.

The first environment to receive this new wave of sedimentological interest was the delta. A flood of information on the Mississippi appeared in the late 1950s (summarized by Shepard 1960) and it was left to Scrutton (1960) to realize the significance of the prograding delta coarsening-upwards cycle to the recognition of ancient delta facies and thus provide the key to the deltaic facies model. In 1960, it was still possible for Shepard to write that 'the literature of stratigraphy and depositional environments of the past appears to have few references to ancient deltas' (p. 80). This was a situation that was not going to continue for long. Moore (1958, 1959) set the ball rolling in the Carboniferous by interpreting the Yoredale series in terms of Fisk *et al.* (1954) Mississippi data. He was able not only to draw analogies between Yoredale facies and Mississippi sub-environments but to explain the stratigraphic evolution of his sequence in terms of known deltaic process. P. Allen (1959) took this type of analysis a stage further by assembling results from a range of modern environments and weaving them into a

detailed interpretation of the environments, processes and evolution of the Wealden deltaic sediments of the Anglo-Paris basin. Although Kuenen & Migliorini's (1950) prophetic paper established the methodology for modern facies analysis, Allen's (1959) paper was the first example of a detailed facies interpretation of a specific ancient rock succession. It was the direct forerunner of all subsequent facies models.

During the 1960s, facies descriptions and models became an increasingly important part of the sedimentological literature. However, during this decade they never accounted for more than a small proportion of the published papers. The BSRG was founded and had its first meeting in 1962, the same year that saw the publication of the International Association of Sedimentologists' new journal, *Sedimentology*. The importance of the study of recent environments is reflected in five out of seventeen of its first year's papers being on the great Bahama Bank, a Florida bay, lake flats in Utah, a Californian turbidite basin and French braided streams. However, only three papers dealt with the interpretation of ancient rock sequences and of these only Duff & Walton (1962) really refer to the importance of modern analogues. The rest dealt with various examples of theoretical and experimental work, petrographic techniques and studies of sedimentary structures, a mix similar to that seen in other sedimentary journals during the 1950s.

The next important step was not a new description of an environment, but a minor technical innovation, the introduction of the Bouma, or graphical log (Bouma & Nota 1961; Bouma 1962). The simple idea of constructing a log by plotting grain size against height enabled a vast increase in the level of detail recorded by sedimentologists. For example, instead of all sandstones being recorded as dots between tramlines their great variety could start to be considered. Also, the existence of coarsening- or fining-upwards sequences became immediately apparent.

The Bouma log came out of his work on turbidites, rocks which are ideally suited to graphic representation. Turbidite studies moved forward rapidly in the 1960s, in spite of the relative lack, in comparison with other facies, of information from modern environments. Indeed, turbidites have always led the way in showing how deductions from ancient facies can lead to an understanding of modern processes and environments. Walker's (1965) paper can be cited here as a classic early example of this.

After turbidites and deltas, fluvial facies were next to come under close scrutiny. In the U.K., fluvial studies concentrated on the Old Red Sandstone, with J. R. L. Allen's (1964) paper being a milestone in this field. Progress was sufficiently rapid for Allen to be able to publish a review of fluvial sediments in 1965 containing no less than 336 references. Of these references, over half date from before 1960, demonstrating how once sedimentologists have been alerted to the importance of a particular environment, they can often find a mass of existing relevant, if dormant, literature. Allen's (1965) paper discussed four fluvial facies models; alluvial fan, braided, low sinuosity and meandering and showed how different fluvial elements could be combined into larger-scale associations.

By 1965 the subject of sedimentology in general, and clastic facies analysis in particular, had started to take on a fairly modern look. The paper by de Raaf *et al.* (1965) is one of the first good examples of the analysis of a complex succession deposited in a range of environments. In it the authors analysed a Westphalian succession in North Devon by dividing it into eleven facies. Although the logs are not 'Boumarized' there are lots of excellent photographs. Facies relationships, cyclicity and palaeocurrents are all discussed in terms of a series of basin–delta–coastal-plain cycles.

Finally, the fact that facies models had become, by 1965, part of the currency of sedimentology is shown by Visher's (1965) review of vertical sequences in facies interpretation. He recognised thirteen depositional sequences or models; four fluvial, four regressive marine, two transgressive marine, one deltaic, one bathyal-abyssal marine and one lacustrine. Although not many people would have agreed with his choice and number, it does roughly reflect the bias of the subject towards fluvial and marginal marine sequences up to this time, and the belief that sedimentary environments and facies can be neatly categorized.

1966–1979: facies come of age

This period witnessed the explosive expansion of sedimentology during which the sedimentologist became a widely recognized component of university and company exploration departments and, as a result, the amount of teaching and research in this field mushroomed. The number and size of sedimentology journals, books and meetings grew so fast that it became difficult to keep track of more than a few aspects of the subject and to be objective about what were the most important discoveries.

During this period the spotlight moved from one hitherto poorly known facies to another.

An interest in modern offshore tidal-shelf sediments, started by the pioneering work of Stride (1963) around the British Isles, and followed by important contributions such as those of Belderson & Stride (1966), Houbolt (1968) and McCave (1971), led to the widespread recognition of ancient tidal facies (e.g. Narayan 1971; Anderton 1976). Surprisingly, the understanding of wave-dominated shelves lagged behind this until the work on North American shelves (e.g. Swift 1969) led to the recognition of storm and fair-weather deposits (Hobday & Reading 1972). Studies of modern barrier coastlines and tidal flats had continued since the 1950s, but it was not until the early 1970s that they started to be widely applied to ancient facies. Fostered by the discovery of the North Sea Rotliegendes gas fields, modern and ancient desert facies came under close scrutiny (Glennie 1970). To illustrate the point that the sedimentological wheel keeps going round and round, turbidites came dramatically back into fashion in the 1970s with the application of Mutti & Ricci-Lucchi's (1972) submarine-fan model.

I don't think one should be too cynical about the widespread description of a certain facies from ancient rocks following its recognition in a modern environment. It is not entirely a question of wishful thinking. One can study a rock succession for years, without being able to interpret it satisfactorily, until work on a modern environment suddenly provides the key to a reasonable explanation. Years later an even better explanation may become possible as a result of the continual progress in the subject. It does not mean that the original interpretation was wrong, just that it was the best that could be done at the time. One must have the courage and honesty to repeatedly change and update facies models as our understanding advances. P. Allen's (1975) re-interpretation of his 1959 deltaic facies in terms of a braidplain model is an instructive example of this.

This period was concluded by the publication of Reading's (1978a) mammoth facies bible and the slimmer, but very useful, compilation by Walker (1979a). These books show that the subject has reached an advanced stage of development and that our ability to interpret ancient facies has reached what must seem to other geologists to be an almost miraculous degree of sophistication. Where do we go from here?

High technology will certainly play an increasing part in the future. I foresee the use of instant cameras, video tapes and electronic measuring devices in the field. In the laboratory the use of computer data-handling, especially the computer-graphic manipulation of logs, will become routine. The interpretation of facies will certainly become increasingly intricate. Unfortunately, this will lead to the inevitable further fragmentation of the subject so that, for example, fluvial experts and shelf experts will have increasingly little in common. However, we may be saved from this if the understanding of fundamental processes can pull the subject together. Whatever happens, we still have a tremendous amount to learn. By the next century we will be able to be as nostalgic about the quaint old 1980s, as we can now be about the 1950s.

Conclusions

There are no conclusions to this paper, only advice. Try to be objective, get your data down on paper, think about it and try to deduce as much as possible from the facies themselves

before resorting to comparisons and analogues. Don't be swayed by fashion, don't ignore the awkward detail that doesn't fit into your hypothesis and don't worry if you have to change your mind about an interpretation at a later date. And always listen to the other man's point of view, even though he is probably wrong. He may well referee your next paper.

ACKNOWLEDGMENTS: On the collective behalf of clastic sedimentologists, I would like to acknowledge the debt we all owe to many people, only some of whom are mentioned in the 'Historical Review' section. Particularly we must thank Ph. H. Kuenen and C. I. Migliorini for showing what could be done and Harold Reading and Perce Allen for, both by example and enthusiastic encouragement of others, fostering the whole subject of facies analysis and modelling in Britain. Personally, I must thank my ex-Reading colleagues, Mike Leeder and Paul Bridges, my research students over the years and all my colleagues at Strathclyde, particularly Mike Russell, for discussing, sharpening and testing my ideas.

References

ALLEN, J. R. L. 1964. Studies in fluviatile sedimentation: six cyclothems from the Lower Old Red Sandstone, Anglo-Welsh basin. *Sedimentology*, **3**, 163–98.

—— 1965. A review of the origin and characteristics of recent alluvial sediments. *Sedimentology*, **5**, 85–91.

ALLEN, P. 1959. The Wealden environment: Anglo-Paris basin. *Phil. Trans. R. Soc. London*, **242B**, 283–346.

—— 1975. Wealden of the Weald: a new model. *Proc. Geol. Ass.* **86**, 389–437.

ANDERTON, R. 1976. Tidal-shelf sedimentation: an example from the Scottish Dalradian. *Sedimentology*, **23**, 429–58.

BELDERSON, R. H. & STRIDE, A. H. 1966. Tidal current fashioning of a basal bed. *Mar. Geol.* **4**, 237–57.

BLISSENBACH, E. 1954. Geology of alluvial fans in semi-arid regions. *Bull. geol. Soc. Am.* **65**, 175–90.

BOUMA, A.H. 1962. *Sedimentology of Some Flysch Deposits: A Graphic Approach to Facies Interpretation*. Elsevier, Amsterdam. 168 pp.

—— & NOTA, D. J. G. 1961. Detailed graphic logs of sedimentary formations. *In:* SORGENFREI, T. (ed.) *Rept. int. geol. Congr. 21st Session, Norden,. 1960*, **23**, 52–74.

COLLINSON, J. D. & THOMPSON, D. B. 1982 *Sedimentary Structures*. Allen and Unwin, London 194 pp.

DAVIE, G. E. 1961. *The Democratic Intellect: Scotland and Her Universities in the Nineteenth Century*. Edinburgh University Press, Edinburgh.

DUFF, P. McL. D. & WALTON, E.K. 1962. Statistical basis for cyclothems: a quantitative study of the sedimentary succession in the East Pennine

Coalfield. *Sedimentology*, **1**, 235–55.

FISK, H. N., McFARLANE, J. R. E., KOLB, C. R. & WILBERT, J. R. L. J. 1954. Sedimentary framework of the modern Mississippi delta. *J. sedim. Petrol*. **24**, 76–99.

GLENNIE, K. W. 1970. *Desert Sedimentary Environments*. Elsevier, Amsterdam. 222 pp.

GOLDMAN, M. I. 1950. What is 'sedimentology'? *J. sedim. Petrol*. **20**, 118–9.

GRESSLY, A. 1838 Observations géologiques sur la Jura Soleurois. *Neue Denkschr. allg. schweiz. Ges. ges. Naturw*. **2**, 1–112.

HOBDAY, D. K. & READING, H. G. 1972. Fair weather *versus* storm processes in shallow marine sand bar sequences in the late Pre-Cambrian of Finmark, North Norway. *J. sedim. Petrol*. **42**, 318–24.

HOUBOLT, J. J. H. C. 1968. Recent sediments in the southern bight of the North Sea. *Geol. Mijnb*, **47**, 245–73.

KRUMBEIN, W. C. 1952. Principles of facies map interpretation. *J. sedim. Petrol*. **22**, 200–11.

—— & SLOSS, L. L. 1963. *Stratigraphy and Sedimentation*, 2nd edn. Freeman, San Francisco. 660 pp.

KUENEN, Ph. H. & MIGLIORINI, C. I. 1950. Turbidity currents as a cause of graded bedding. *J. Geol*. **58**, 91–127.

McCAVE, I. N. 1971. Sand waves in the North Sea off the coast of Holland. *Mar. Geol*. **10**, 199–225.

MOORE, D. 1958. The Yoredale Series of Upper Wensleydale and adjacent parts of North-West Yorkshire. *Proc. Yorks. geol. Soc*. **31**, 91–148.

—— 1959. Role of deltas in the formation of some British Lower Carboniferous sediments. *J. Geol*. **67**, 522–39.

MOORE, R. C. 1949. Meaning of facies. *Geol. Soc. Am. Memoir* **39**, 1–34.

MUTTI, E. & RICCI-LUCCHI, F. 1972. Le torbiditi dell

'Appennino settentrionale: introduzione all' analisi di facies. *Memorie Soc. geol. ital.* **11**, 161–99.

NARAYAN, J. 1971. Sedimentary structures in the Lower Greensand of the Weald, England, and Bas-Boulonnais, France. *Sediment. Geol.* **6**, 73–109.

PELLETIER, B. R. 1958. Pocono paleocurrents in Pennsylvania and Maryland. *Bull. geol. Soc. Am.* **69**, 1033–64.

POTTER, P. E. 1959. Facies model conference. *Science,* **129**, 1292–4.

DE RAAF, J. F. M., READING, H. G. & WALKER, R. G. 1965. Cyclic sedimentation in the Lower Westphalian of North Devon, England. *Sedimentology,* **4**, 1–52.

READING, H. G. 1978a. *Sedimentary Environments and Facies.* Blackwell Scientific Publications, Oxford. 557 pp.

—— 1978b. Introduction. *In*: READING, H. G. (ed.) *Sedimentary Environments and Facies*, pp. 1–3. Blackwell Scientific Publications, Oxford.

SCRUTON, P. C. 1960. Delta building and the deltaic sequence. *In*: SHEPARD, F. P., PHLEGER, F. B. & VAN ANDEL, T. H. (eds) *Recent Sediments, Northwest Gulf of Mexico*, pp. 82–102. American Association of Petroleum Geologists, Chicago.

SELLEY, R. C. 1982. *An Introduction to Sedimentology*, 2nd edn. Academic Press, London 417 pp.

SHEPARD, F. P. 1960. Mississippi delta: marginal environments, sediments and growth. *In*: SHEPARD, F. P., PHLEGER, F. B. & VAN ANDEL, T. H. (eds) *Recent Sediments, Northwest Gulf of Mexico*, pp. 56–81. American Association of Petroleum Geologists, Chicago.

STRIDE, A. H. 1963. Current swept floors near the southern half of Great Britain. *Q. Jl geol. Soc. London,* **119**, 175–99.

SUNDBORG, A. 1956. The River Klarälven: a study of fluvial processes. *Geogr. Annalr,* **38**, 127–36.

SWIFT, D. J. P. 1969. Inner shelf sedimentation: process and products. *In*: STANLEY, D. J. (ed.) *The New Concepts of Continental Margin Sedimentation: Application to the Geological Record, DS-4-1 to DS-4-46.* American Geological Institute, Washington.

TEICHERT, C. 1958. Concept of facies. *Bull. Am. Ass. Petrol. Geol.* **42**, 2718–44.

THOMPSON, W. O. 1937. Original structures of beaches, bars and dunes. *Bull. geol. Soc. Am.* **48**, 723–51.

TUCKER, M. E. 1981. *The Field Description of Sedimentary Rocks.* Open University Press, Milton Keynes. 112 pp.

VAN STRAATEN, L. M. J. U. 1954. Comparison and structure of recent marine sediments in the Netherlands. *Leid. geol. Meded.* **19**, 1–110.

VISHER, G. S. 1965. Use of vertical profile in environmental reconstruction. *Bull. Am. Ass. Petrol. Geol.* **49**, 41–61.

WADELL, H. 1932. Sedimentation and sedimentology. *Science,* **75**, 20.

—— 1933. Sedimentation and sedimentology. *Science,* **77**, 536–7.

WALKER, R. G. 1965. The origin and significance of the internal sedimentary structures of turbidites. *Proc. Yorks. geol. Soc.* **35**, 1–32.

—— 1979a. *Facies Models.* Geological Association of Canada, Toronto.

—— 1979b. Facies and facies models. I. General introduction. *In*: WALKER, R. G. (ed.) *Facies Models*, pp. 1–7. Geological Association of Canada, Toronto.

WELLER, J. H. 1958. Stratigraphic facies differentiation and nomenclature. *Bull. Am. Ass. Petrol. Geol.* **42**, 609–39.

R. ANDERTON, Department of Applied Geology, University of Strathclyde, 75 Montrose Street, Glasgow G1 1XJ.

Recent shelf clastic sediments

I. N. McCave

SUMMARY: Shelf sedimentation in the past 21 years has moved from an essentially descriptive study to one where theory and measurement of sediment transport play an important role in accounting for the distribution of shelf sediments. This paper looks briefly at pre-1960 works then presents a view of the major features of present research on shelf clastics. This is organized under shelf muds, tidal-current-dominated sands and wave-dominated sands. Brief consideration is given to processes of mud and sand transport and deposition prior to a discussion of sediment types and bedforms. Storms are emphasized and their capacity to emplace sand layers on the shelf by wind-driven flows is discussed.

When the British Sedimentological Research Group was founded 21 years ago the study of shelf clastic sediments had come a very great distance in the relatively short period of time since the 1920s when such studies were initiated. The position is best summarized in the second edition of F. P. Shepard's classic textbook *Submarine Geology* published in 1963. The work carried out to that date was mainly descriptive rather than dynamical and the descriptions were principally of the continental shelves around North America and NW Europe. Continental shelves have been much better mapped by hydrographers and fishermen than other areas of the seabed, and early work utilized the many notations of bottom composition as well as the relatively detailed bathymetric information available on navigational charts. The general dimensions of continental shelves, that their width averages about 75 km, that they are narrow off active continental margins and wider off passive and glaciated continental margins, and have a mean gradient of about 0.1° were all features well known and well displayed in Shepard's book. Some detailed analyses of topography had been made using the echo-sounder which came into prominence in the work of van Veen (1936) and others from the mid-1930s onwards. The boundary of the continental shelf marked by a sharp break in slope had been analysed by Dietz & Menard (1951) and the general geomorphology of continental shelves by Dietz (1952). It was recognized in these studies of morphology that a great many 'drowned' features on continental shelves dated from low stands of sea level; that there were drowned moraines in the North Sea, and drowned reefs on tropical continental shelves, for example.

Most work, however, was conducted on the grain-size characteristics and distribution of sediment size and type on shelves, and, particularly by Dutch workers, on the heavy mineral content of continental shelf sands. Extensive studies in the North Sea by the Germans and the Dutch on shelf sand size and mineralogy occurred before the war (Lüders 1939; Edelman 1939). The French also had examined sediments of the Channel and shelf off western France (Berthois 1946). In America in 1960 the results of an extremely successful investigation, the API Project 51 on the North-West Gulf of Mexico were published (Shepard *et al*. 1960). This study comprised examination of the coastal lagoons, the continental shelf and the Mississippi delta. Elsewhere the Dutch had conducted the wide-ranging investigation of the mainly clastic sediments of the continental shelves of NE South America in the Orinoco Shelf Expedition reported on in the theses of Nota (1958) and Koldewijn (1958) and the classic report of van Andel & Postma (1954) on the Gulf of Paria. K. O. Emery also increased our regional knowledge through work on the East Asian shelf, on the shelf off southern California and the Persian Gulf.

The methods employed by these workers were principally field sampling by grab and corer with use of the echo-sounder. Laboratory analysis comprised grain-size distributions and coarse fraction and sand fraction composition together with, in some cases, clay mineralogy.

Studies of deltas figured importantly in the period before 1960, particularly in the API project of Shepard and his companions on the Mississippi delta but also by the Dutch on the Rhône delta, and on the Orinoco within the study of the Gulf of Paria, and by Shepard again on the tidally influenced delta of the Fraser in British Columbia.

These studies established several important aspects of the distribution of shelf clastics. The most significant was that most sediments are not in equilibrium with the conditions existing at the present day, and that much of the sedimentary cover of the continental shelves was in fact emplaced at lower sea level. They established that the concept of seaward fining of clastic sediment was incorrect as applied to present-day continental shelves, and that in many cases outer continental-shelf sediments were coarser than those to be found on inner shelves because those outer-shelf sediments represented coarse beach and reworked river sands. They also established that there was a distinct zonal difference in the composition of sediments with those at high latitude being dominated by sands, gravels and rocky bottoms, whereas those at lower latitudes tended to comprise carbonate components and, if adjacent to a fluvial source of sediment, muds. Although this review is not about carbonate sediments their mention prompts the point that by 1961 workers on carbonates had achieved a considerable measure of the synthesis which is accepted to the present day. In particular the studies of Newell & Rigby (1957), Newell et al. (1960) and Purdy (1963) following those of Maurice Black (1933) and Leslie Illing (1954) on Grand Bahama Bank, and the work of Ginsburg (1956) in South Florida had set out the major compositional varieties of carbonate sediments and their relationship to organic and wave/current activity. Studies of this type were repeated in many other warm and pleasant parts of the world by subsequent generations of carbonate sedimentologists and it is only in the last few years that recognition has been given to the fact that high-latitude shelves away from sources of clastic sediments are also coated mainly with carbonate sediments and are objects worthy of study (Lees & Buller 1972; Lees 1975).

In the early 1960s several influential investigations and publications changed the direction of work on shelf sediments from one that was descriptive to one that became more and more based in attempting to understand the dynamics of shelf-sediment transport and sorting by waves and currents, and to relating those processes to the sedimentological textures and structures recorded in the deposits. The intention was partly to make available to geologists a better understanding of the dynamics of production of those features which would be recognizable in the geological record, and which would be the basis for the interpretation of the geological record. Another objective was to provide a better understanding of the whole shelf environment, including water movement, sediments and biota. Such studies, often conducted by geologists, are now seen as a necessary prelude to management of human activities in shelf waters.

Two new instruments became available, firstly the box corer invented by Reineck (1963b), and secondly the side-scan sonar used by Stride and his colleagues at the National Institute of Oceanography from the late 1950s onwards (Chesterman et al. 1958). In his study of the NW Gulf of Mexico, Curray (1960) had attempted a novel interpretation of the grain-size variation of the Mississippi silts in terms of transport paths of the differing size fractions across the shelf. Similarly, in the work on the sediments of the Gulf of Paria, Postma was concerned with the transport of suspended sediment and van Andel with its deposition and distribution to provide a dynamical framework for the interpretation of deposits (van Andel & Postma 1954). Reineck and Stride now provided methods for both looking at bedforms on the surface of the shelf to be interpreted in terms of shelf-sediment-transport patterns, and for collecting large enough undisturbed samples of the deposits to be able to impregnate them and examine their sedimentary structures. Study of bedforms and cross-bedding was becoming important; John Allen's paper on the subject in 1963 provoked much thought and activity, and the study of facies, bedforms and processes was gaining considerable support in other spheres of sedimentary geology and geomorphology as, for example, in the books by Potter & Pettijohn (1963) and Leopold et al. (1964).

In the remainder of this paper I shall set out what seems to me the presently accepted consensus of ideas concerning the location of sediment types and the processes controlling the transport and modification of shelf muds, sands and gravels. That synthesis has two major components, the notion that shelves are either dominated by the action of tidal currents or by wind-generated waves. Wave-dominated shelves comprise both those in so called 'high energy' regions, i.e. climatic zones where the waves are high and of long period, and 'lower energy' or sheltered regions where the waves are less effective (Davies 1980), but where other (e.g. tidal) currents are also weak. Most work has been done on shelf sands with rather less on muds, and the latter are not well integrated into a synthesis of shelf-sediment behaviour, partly because their behaviour under wave and current action is not at all well understood. We know, for example, that the critical erosion velocity for mud varies according to water content, the animals which inhabit it, its mineralogy and many other parameters, but not with any precision, whereas for sand we can get within a factor of 2 at predicting critical movement conditions and perhaps within a factor of 10 at predicting the flux of sediment. Thus the distribution of shelf muds will be taken first and slightly separately from the distribution of sands which will then be dealt with under tidal and non-tidal situations.

Muddy deposits

The conditions of formation and characteristics of muddy shelf sediments have been relatively neglected because they are more difficult to study, and because the interpretative (and economic) value of sedimentary structures in sands was thought to be higher. Sands are easier to impregnate and study in box cores, but at the larger (seismic reflection) scale only banks have revealed internal structures with any consistency, whereas cut and fill sequences in mud beds are often well displayed on seismic (3.5 kHz and Boomer) records. X-radiography is the best tool for revealing mud structures at small scale. The interpretation of muds in terms of hydraulic experiments in flumes has not been as clear as the picture obtained for sands (Dalrymple *et al.* 1978) though uncertainties remain there too. However there is a body of hydraulic experimental data but little of it also deals with internal structure.

Experimental behaviour of muds

The behaviour of muds has recently been reviewed at some length (McCave 1984). Muds are cohesive, but their degree of cohesion is not simply related to grain size. Water content (compaction), mineralogy, cation exchange capacity (CEC), temperature, organic mucus content, Bingham yield strength and salinity are all parameters with an influence. Migniot (1968) compacted all his experimental erosion data on to one curve using yield strength versus shear velocity, whereas Thorn & Parsons (1980) did the same for theirs using concentration (kg m^{-3}). Krone's (1962) hypothesis that critical erosion stress is controlled by the strength of the surface aggregates has the best physical basis of explanations put forward so far. Aggregate strengths measured by Krone (1963) go up to about 5 Pa for 'zero-order' aggregates, i.e. the smallest units. Krone (1963) estimated that during deposition, higher-order aggregates should compact to zero order under a sediment overburden of 20–30 mm. It must be remembered that if this strength represents a critical erosion stress, it is much higher than that for sand which is $\tau_c = 0.1$–0.3 Pa. The erosion rate of these muds is very complex, depending both on the compaction and the excess shear stress as well as CEC, temperature and other factors. Compacted muds would have rates of 0.1×10^{-3} kg m^{-1} s^{-1}

Pa^{-1} excess shear stress, i.e. 10^{-5} kg m^{-2} s^{-1} for $(\tau_o - \tau_c) = 0.1$ Pa. This would result in < mm day^{-1} erosion rate in most cases, but for softer muds with rates of 2.5×10^{-3} kg m^{-2} s^{-1} Pa^{-1}, this excess would result in ≥ 1 mm h^{-1} removal. Of course for higher excess shears the removal increases *pro rata*. Thus detectable scours in soft muds may be formed in less than an hour but compacted mud requires a long period of high stress to remove a millimetre.

Once eroded the suspended mud changes in size and settling velocity distribution during transport such that aggregates of 100 μm and more in diameter are common, and suspensions have median settling velocities of 0.01–2 mm s^{-1}, rising with concentration.

Deposition of fine sediment is dependent on its concentration, settling velocity and the bed shear stress τ_o. Deposition occurs below a shear stress of about 0.05–0.1 Pa (equivalent to speeds less than about 0.15 m s^{-1} at 1 m above bottom) for suspensions of low concentrations (<0.3 kg m^{-3}), but for C >1 kg m^{-3} limited deposition may occur at higher stresses up to about 1 Pa. It is well known that mud deposits can form in deeper water areas below wave influence or in places protected from wave action such as some tidal flats. However muds may also be deposited in higher-energy (wave and current) areas too. The controlling condition is that sufficient mud be deposited, while water flows slowly, to provide a layer thick enough so that at least the bottom of it is sufficiently compacted to withstand the flow speeds next time they become higher—the next tide or the next storm. Thinner layers of mud may also be compacted by sand being deposited on the mud.

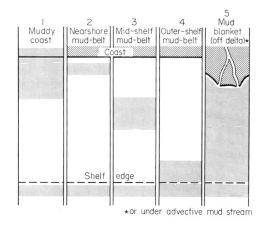

FIG. 1. Classes of shelf mud deposit location, from McCave (1972).

Location of mud deposits

Shelf muds can form as muddy coasts, near-shore, mid- and outer-shelf mud belts and as shelf mud blankets; see McCave (1972), who summarized work up to about 1970 (Fig. 1). Muddy coasts are to be found in areas adjacent to the major sediment suppliers of the world, the coastal mud flats west of the Mississippi (Morgan *et al.* 1953) on the Guyana and Surinam coast NW of the Amazon (Allersma 1971; Eisma & van de Marel 1971; Wells & Coleman 1981), on the tidal flats of Yellow River mud in the Gulf of Bohai and the western Yellow Sea (Wang 1983; Ren *et al.* 1984) and on coastal mud flats adjacent to other tropical rivers. This phenomenon appears to be a direct consequence of the very high concentration of suspended sediment in the water column, generally several hundred g m^{-3}. Note that under these conditions there is no need for a protective barrier to attenuate waves. Waves in fact do resuspend the mud but it is rapidly redeposited (Wells & Coleman 1981).

Nearshore mud belts also occur in cases where there is a nearby source of suspended matter but concentrations are lower. Examples are the Dutch coast north of the Rhine, the Belgian coast south of the Scheldt, the Italian coast south of the Po, the Chinese coast south of the Chang Jiang (Yangtze) (Wang Baoyong 1983), the eastern Texas coast of the NW Gulf of Mexico (Curray 1960), the Gulf of Gaeta (Reineck & Singh 1971), the coast off Sapelo Island (Howard & Reineck 1972), off Southern California (Howard & Reineck 1981), and the inner German Bight (Dorjes *et al.* 1970). In all these cases a transition from coastal sands to muddy sands and muds in water depths of 5–20 m is followed by a reversion to more sandy deposits farther offshore. The latter are generally relict sediments with no present supply of sand from the coast. Thus the nearshore sand and mud belt are the only parts of the system with significant active deposition at the present time. It was McCave's (1971a, 1972) contention that the seaward limit of the mud belt was due to the drop in concentration (thus deposition rate) relative to the wave or current activity (re-suspension rate and frequency). Under conditions of higher (wave) energy, it was argued, this situation is found displaced seawards as a mid-shelf mud-belt with a wider zone of wave-dominated, near-shore sands. McCave (1972) pointed to examples in the Celtic Sea and the Gulf of Gascony. More recent work has revealed their presence south of New England (Bothner *et al.* 1981), off the west coast of South Africa (Birch 1977), SE Australia (Davies 1979; Roy & Thom 1981), Nayarit in Mexico (Curray 1969), the shelf off the Congo and Gabon north of the Zaire (Congo) River (Giresse & Kouyoumontzakis 1973) and the shelf off Washington, U.S.A (Nittrouer *et al.* 1979; Nittrouer & Sternberg 1981). In all these cases there is evidence, in some cases from radiometric dating, of recent sediment accumulation in these mid-shelf mud-belts.

The outer continental shelf has an important component of relict coarse- to medium-grained sandy sediment. In most areas this has escaped blanketing by mud because the suspended sediment concentration is very low (generally < 1 g m^{-3}) and bottom current velocities are enhanced there due to tides, internal waves and tides, and motions associated with fronts (Southard & Stanley 1976; Huthnance 1981). In a few cases the outer shelf is blanketed with mud, e.g. NW Gulf of Mexico (Curray 1960; van Andel 1960) and the Oregon shelf (Kulm *et al.* 1975). In both cases there is a probable relationship to supply from the Mississippi and Columbia rivers, respectively. The northward flow of Columbia river mud appears to supply the Washington mid-shelf mud belt in winter while the southerly distribution of the plume over the outer Oregon shelf and slope in summer may be responsible for the mud cover there (Pak *et al.* 1970; McCave 1979a). Elsewhere Stanley & Wear (1978) and Stanley *et al.* (1983) define a 'mud line' on the upper slope below which mud is accumulating and above which there is erosion or non-deposition due to shelf-edge water movements.

Off major deltas, particularly those in the tropics, the whole shelf is generally blanketed with mud. Several of these deltas (the Huang Ho, Chang Jiang, Red River, Chao Phya) discharge into adjacent seas which still have an outer rim of relict sediment. Many others however (Ganges, Indus, Mississippi, Irrawaddy, Amazon, Magdalena and the rivers of eastern India) blanket open shelves with mud. In some cases the delta front merges with the continental slope, as in the case of the Mississippi, Nile and Niger, suggesting a shelf morphology and sediment distribution which has overcome most of the effects of the Holocene transgression.

Most shelf sediment distributions (sand and mud) are extremely confused, being a combination of modern supply and relict sediment (originally emplaced at lower sea levels, see Emery 1968), some of which is at present being reworked, giving it a modern appearance, but still retaining clues to its former identity (thus 'palimpsest sediments' of Swift *et al.* 1971).

Situations where the whole shelf has been blanketed with mud as a result of supply perhaps approach most closely the ideal of a graded shelf (Curray 1965; Swift 1970). It is noteworthy that in these cases the shelf is generally narrow and muddy with a shallow coastal sand belt in some cases.

Bedding characteristics of mud deposits

Few workers have embarked on comprehensive studies of sedimentary structures in marine sediments. The largest body of data is from Reineck and his associates who have investigated the inner German Bight (Dorjes *et al.* 1970; Reineck *et al.* 1967; Reineck *et al.* 1968; Gadow & Reineck 1969), Gulf of Gaeta (Italy) (Reineck & Singh 1971), shelf off Sapelo Island (Howard & Reineck 1972) and Ventura (California) shelf (Howard & Reineck 1981), summarized in English in Reineck & Singh (1980). The principal features revealed by their examination of box cores are bioturbation and storm-sand layers. The latter are dealt with later in this chapter. The former have the effect of obliterating physical structures (Moore & Scruton 1957) while providing useful ecological data. This aspect has been studied by Reineck's colleagues Dorjes and Hertweck in several of the papers cited above, and summarized in Dorjes & Hertweck (1975). The study of lebensspuren in bioturbated sediments provided a method of relative depth zonation alongside distribution of structure and texture. In the three areas studied (Gulf of Gaeta, German Bight, Sapelo Island) three zones were distinguished below low water mark; a wave-affected shoreface and upper and lower offshore zones with decreasing wave activity and sediment deposition, and increasing mud content. The most lebensspuren and maximum bioturbation are found in the upper offshore where both numbers of species and individuals peak.

Recently the study of structures in muds has gained an impetus from the joint study of X-radiographs showing internal structure and the distribution of radionuclides especially ^{137}Cs, ^{210}Pb and ^{234}Th (Nittrouer & Sternberg 1981; Kuehl *et al.* 1982). The radionuclides can be used to obtain sedimentation rates and to parameterize the bioturbation as a diffusivity D_B (dimensions L^2T^{-1}) in the manner suggested by Goldberg & Koide (1962). Guinasso & Schink (1975) proposed a parameter $G = (D_B/L_B)/R$ where L_B is the thickness of the biologically mixed layer and R is the net sedimentation rate, to indicate the relative importance of bioturbation and deposition. Nittrouer & Sternberg

(1981) suggest $G < 0.1$ yields deposits dominated by physical structures with little bioturbation and $G > 10$ will give a homogenized sediment column, corresponding to the bedded → mottled → homogeneous sequence of Moore & Scruton (1957). A non-dimensional relationship between rate of deposition and frequency of erosion was proposed by Smith (1977, p. 575) as $H = (f_e . L_e)/R$, where f_e and L_e are the frequency and depth of erosion. Nittrouer & Sternberg (1981) point out that for small H more of the fine-grained portions of sedimentary sequences should be preserved forming texturally heterogeneous deposits. Relation of G and H to sedimentary structures is best undertaken on X-radiographs because the structure is better seen there than in cores directly (see Kuehl *et al.* 1982, figs 5–7).

Shelf sands under tidal currents

Essentials of sediment transport

In the last decade many studies have been made of the dynamics of tidal current transport of sands. There is substantial agreement that the rate of bedload transport is approximately proportional to u_*^3 and that total load (including suspension) goes to some higher power, perhaps u_*^5 as in the expression of Engelund & Hansen (1967). (u_* is the shear velocity which is proportional to the flow speed.) Five of the main sediment transport equations in use for unidirectional flow have been evaluated against tidal marine data from radioactive tracers by Heathershaw (1981). He finds most satisfactory the equation of the type proposed by Bagnold (1963), simplified for use in marine situations by Gadd *et al.* (1978) as:

$$q = A(U_1 - U_{1c})^3$$

where q is the mass transport rate (kg m^{-1} s^{-1}), U_1 is the velocity at 1 m above the bed and U_{1c} is the value of U_1 at critical conditions, and A is a coefficient of proportionality equal to 7.22 for 0.18 mm sand and 1.73 for 0.45 mm sand. Predictions based on the simple $(U_1 - U_{1c})^3$ expression using tidal current data have been shown by McCave (1971b) and Heathershaw (1981) to agree well with patterns of sediment transport inferred from bedforms.

In expressions using bed shear stress the major problems are found in the estimation of the skin friction component of the total bed shear stress and its temporal variability (Smith 1977; Smith & McLean 1977a, b). The non-linear dependence of transport rate on u_* means that small errors in u_* result in large errors in q. For bedload transport there is litle lag in the system but for fine sand the time taken to diffuse up and down the water column results in a form of hysteresis (Thorn 1975). The addition of wave-induced water motion to tidal currents can greatly increase sand transport—Heathershaw (1981) estimates a factor of 5 increase due to 1 m high waves on a 0.5 m s^{-1} current in 20 m water depth—but where the erosion threshold is regularly exceeded by the tides alone the bedforms are generally dominated by currents, though this is still a matter of relative magnitude and frequency of tidal *v.* wave-induced currents. Even without storm influence, strong tides (1–1.5 m s^{-1}) can cause net sand fluxes implicit in bedform migration of the order of 100–500 tonnes m^{-1} yr^{-1}. These net tidal sand fluxes are effectively produced by small inequalities in the speed of ebb and flood tides, either because of some steady net flow or due to combinations of the semidiurnal M$_2$ tide with its M$_4$ harmonic (Howarth, 1982). These effects can be opposed, as in the southern Bight of the North Sea between the U.K. and Belgium, where the net drift transports suspended muds to the NNE but bottom stress drives the sand to the SSW. The dominance of the mean maximum bottom stress in controlling tidal sand transport and bedforms is shown by the close agreement between the patterns of peak (M$_2$ + M$_4$) stress (Pingree & Griffiths 1979) and the sand transport directions based on bedform analysis (Johnson *et al.* 1982).

This dominance occurs because there is a regular occurence of high bottom stress that moves sand. For medium and coarse sands moved as bedload, the sediment transport equations can be applied to tidal current speeds which vary relatively slowly with time. This is in distinct contrast to the situation on those shelves where tidal and wind-driven currents are too slow to entrain sediments. Movement of bed material there is dependent on waves stirring up the bed so that the drift current can move it. The shear stresses vary greatly on a 10 s time scale and the convenient assumption of steadiness cannot be used. Moreover, the stress due to both wave and steady currents must be taken into account.

This is done in the theories of Bijker (1967), Smith (1977) and Grant & Madsen (1979). The latter provide the most workable scheme. The motion and stresses under current plus waves are not simply additive but involve non-linear interactions expressed in:

$$u_{*cw} = \left(\frac{f_{cw}\alpha}{2}\right)^{1/2} |U_b|$$

and
$$\alpha = 1 + \left(\frac{|U_a|}{|U_b|}\right)^2 + 2\frac{|U_a|}{|U_b|}\cos\phi_c$$

where u_{*cw} is the shear velocity due to combined waves and currents, $|U_b|$ is the maximum bottom wave orbital velocity outside the wave boundary layer, and ϕ_c is the angle between U_a and U_b. $|U_a|$ is the steady current inside the wave boundary layer, not simply given by addition of wave and steady current vectors and so subject to determination by an iterative procedure along with ϕ_c.

Such an approach is essential to the proper evaluation of sediment transport on storm-dominated shelves. This has been done by Vincent and his associates for the shelf off New York (Vincent 1984, Vincent *et al.* 1981a, b, 1982). They use a fairly straightforward bedload sediment transport equation based on excess shear stress

$$q(t) = 0.09 \, \rho_s \, \theta'(t) \, U(t) \qquad (4)$$

Where ρ_s is the sediment density, $\theta'(t) = \theta(t) - \theta_c$ in which $\theta(t)$ is the time-dependent, non-dimensionalized shear stress $\tau_o(t)/(\rho_s - \rho)gd$ in which ρ is the water density and d the grain size, and θ_c is its critical motion value for a given sand size (Vincent *et al.* 1981a). Relatively low annual transports were estimated from a 95 day period of instrumental data, from 5 kg m^{-1} at 66 m to 10 kg m^{-1} at 32 m to 25 kg m^{-1} at 12 m depth (Vincent 1984). This transport is concentrated in storms. The rate predicted using a simple excess velocity cubed relationship (equation 1) is one to two orders of magnitude less than that obtained with equation (4) and Grant & Madsen's (1979) procedure to calculate shear stress. Nevertheless, net annual transport rates are several orders of magnitude less than under tidal currents. This makes the wave-driven transport more susceptible to dominance by relatively rare extreme events.

Bedforms under tidal currents

At the lowest speeds capable of moving sand, currents produce rippled beds, but combined with waves, gently undulating flat beds are also formed. Ripples are revealed by bottom photographs, but not by side-scan sonar, so regional mapping of them has not been conducted. Ripples are ≤40 mm high generally and of wavelength <0.60 m or ≅10^{3}d (Allen 1970a). Because they contain little sand they are quickly formed and reformed under low sediment-transport rates. This makes them unsuitable for large-scale analysis of sediment-transport directions and they have been observed more on tidal flats than the shelf. At low tidal-current velocities, ≤0.5 m s^{-1} in 30–60 m water depth, where waves start to play an important role in determining patterns, sand patches, some irregular, some elongate, occur.

In hydraulic engineers' parlance, asymmetrical transverse sandy bedforms larger than ripples are referred to as dunes. Dunes in shelf seas range in height from a fraction of a metre to 20 m, though most are 0.5–10 m. Some papers distinguish between the higher and lower ends of this range calling those 1.5–2 m megaripples and those bigger, sand waves. Advocates of a single class of large bedforms (Rubin & McCulloch 1980; Belderson *et al.* 1982) maintain there is no difference dynamically between different sizes other than that larger forms are found in deeper or faster flows or of coarser sand. For different reasons (field observations, flume experiment, theory) others consider the two classes to be different dynamically (Boothroyd & Hubbard 1975; Middleton & Southard 1977; Dalrymple *et al.* 1978; Allen 1980a). Allen (1980a) suggests that the larger tidal sand waves are a form of oscillatory ripple analogous to small-scale wave-generated symmetrical ripples. McCave (1971b) and Terwindt (1971) distinguished between migratory megaripples and sand waves mainly on the basis of superposition of the former on the latter, though Rubin & McCulloch (1980) suggest this is insufficient reason. It now seems that, if Allen is correct, the term 'sand wave' should be applied to the large slowly moving forms, but if not, there is only one class of dune bedform. In practice the movement of larger forms probably results from sand transported in the smaller superimposed ones (Reineck 1963a; Terwindt 1971).

Stride (1963, 1982) and his colleagues mapped the distribution and facing directions of the steeper slopes of sand waves and inferred sediment transport paths from them. From this emerged regions of bedload parting and convergence. In both of these regions symmetrical sand waves are found which McCave (1971b) argued represented regions of approximately zero net sediment transport. However, the location of symmetrical sand waves in a zone at the end of sand banks was suggested by McCave & Langhorne (1982) to have no net transport normal to the sand wave crests, but strong transport sub-parallel with the crests in associated smaller asymmetrical sand waves.

The dimensions of tidal sand waves generally give wavelength/height ratios in the region 20–100. Diver observations show the large forms > 2 m high to have little of their steeper slope at angle of repose (30–35°), whereas the superimposed forms are dominated by steeper slopes (S. Briggs, personal communication.) The internal structure of sand waves, a matter for speculation since there are no observations deeper

than 0.5 m below their surface, is thus unlikely to show large avalanche foresets. There may be a metre or so of avalanche at the top of a sand wave but most of the internal structure is likely to be dominated by cross-bed sets generated on small bedforms as suggested by Reineck (1963a) and Allen (1980b). The internal structure of symmetrical sand waves may also contain cross-bedding from superimposed bedforms. There has been no suggestion as to the structure of sand waves with no superimposed bedforms and no angle of repose slope, but they are important in fine sand (McCave 1971b).

In water depths of 20–45 m sand waves are the largest bedform when surface currents are in the range 0.6–0.9 m s^{-1} at their mean spring maxima. With currents above about 0.9 m s^{-1} and with plentiful sand, large banks are formed. The upper velocity limit to their formation is not known. Close to the coast the banks in many situations form ebb–flood parabolas as shown by Van Veen (1936) for the Flemish Banks and Robinson (1966) for the inner Norfolk Banks. These are up to 20 m from crest to trough, are very mobile and generally have some small associated sand waves. The channels are 1 or 2 km wide and the banks a few kilometres long (Fig. 2). Farther offshore the parabolic banks give place to linear sets of banks 3–10 km apart (Off 1963). The latter are *not* aligned parallel with the major axis of the tidal ellipse (McCave 1979b), but are mainly offset anticlockwise by about 7–15° (Kenyon *et al.* 1981). An obliquity of banks is predicted by Huthnance's (1982) theory as being optimum for bank growth. Sand is transported over the bank from both sides, but more moves on to the bank than off it. The convergent pattern of sediment movement is revealed by the orientations of bedforms shown by Houbolt (1968), Caston (1972) and McCave & Langhorne (1982).

Because of their great size, banks have attracted great interest as models for potential oil reservoirs. Recognition of ancient banks would be facilitated by some idea of their internal structure. Seismic reflection shows that some banks contain a master bedding structure with a dip of a few degrees, subparallel to the bank's steeper face (Houbolt 1968). Little is known of small-scale structures beyond the cross-bedding shown in Houbolt's short cores and the speculation of McCave & Langhorne (1982) and Stride *et al.* (1982) that the cross-bedding from the steeper face should be preserved. Most of the banks that have been examined for grain size are coarser towards the base and finer at the crest. This probably results from the fact that most banks sit on a sandy gravel pavement from which sand is winnowed and the coarser com-

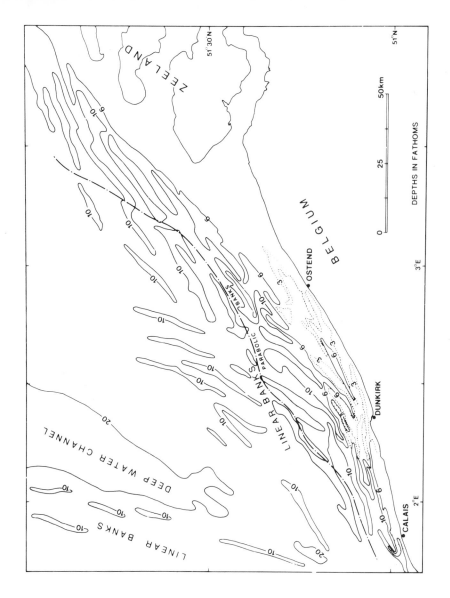

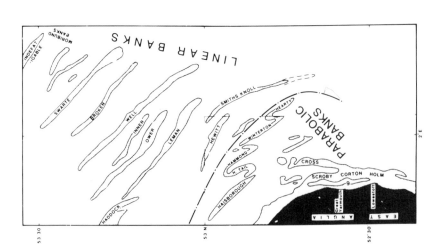

Fig. 2. Sand banks in the southern North Sea showing progression from parabolic types nearshore to linear offshore. To the north-east, off the Belgian and Dutch coasts, the current speed decreases and the banks pass into a large sand wave field.

ponents are left at the foot, while those most easily transported get to the crest. Nevertheless this is in contrast to most supposed ancient analogues of banks which coarsen upwards (Bridges 1982). However these sandstone bodies are surrounded by mudstone and presumably rose from a muddy bottom. Most modern examples do not allow us to examine this sediment-transport situation. Possible exceptions are the northernmost Flemish banks which interdigitate with the south-western end of the nearshore mud belt from the Scheldt, and the tidal sand banks found in the muddy estuary of the Gironde (Allen 1972).

Sand ribbons occur with less sand and high current speed (1–1.5 m s^{-1}). These are thin (<1 m) long patches of sand with regular spacing, aligned parallel with the current and recognized on side-scan sonar by their textural difference from the underlying gravel or other relatively immobile rough substrate. They are useful indicators of sand-transport direction but contribute little to sedimentary deposits. They are probably caused by a secondary flow on the main tidal current which probably originates from a non-uniform turbulence field generated over a bed of non-uniform roughness (McLean 1981).

At very high current velocities the gravel itself may be mobile. Belderson *et al.* (1982) indicate peak spring tidal speeds ≥ 1.5 m s^{-1} for formation of the longitudinal furrows in gravel first described by Stride *et al.* (1972). Dyer (1971) describes an area where tidal currents generally >1.5 m s^{-1} (giving a stress ≥ 10 Pa), have formed gravel waves up to 1.7 m high in 16-mm-sized material. Flemming & Stride (1967) also recorded large symmetrical ripples of 2.2 mm gravel formed under storm waves with moderate (0.9 m s^{-1}) tidal currents. The gravel in these cases is mainly of fluvial origin emplaced at low sea-level stand but with some modern carbonate components as well.

Regional distribution of tidal sand facies

Although many small coastal areas have significant tidal currents, few larger areas are tidally dominated. The principal ones are the seas around the British Isles—Southern North Sea, Channel, Celtic and Irish seas, Nantucket Shoals and Georges Bank off north-eastern U.S.A., Malacca Straits, the north-eastern Yellow Sea off North Korea and the southern end of the tongue of the ocean (TOTO) on the Bahama Banks. Only the area surrounding the British Isles has been investigated in detail, a major compilation and analysis of which has just appeared (Stride 1982).

At present we can say remarkably little about tidal sedimentary facies that is based on observations of modern sediments. Internal structures are by and large unknown because of lack of either seismic or long core data. We rely mainly on sparse grain size data, morphology and inference. In all of the areas mentioned above, with the possible exception of the Gulf of Korea, there is little modern sand transport into the tidal areas from the land. The southern North Sea and Nantucket Shoals have glacial outwash, old beach and river sands, and a little material from continuing coastal retreat. The northern Malacca Straits are supplied through the delta of the Klang whose load is 90% silt and clay (Coleman & Wright 1975). Tidal reworking of previously deposited material continues at present, and grain sizes are not in static balance with shear stresses. Sediments become finer down the tidal transport path (Stride 1963), irrespective of whether the shear stress increases or decreases along the path (Hamilton *et al.* 1980).

Parabolic banks are most commonly found nearshore off coasts of unconsolidated sediments with tidal current peak speeds up to 0.9 m s^{-1}, for example Norfolk, Belgium, Nantucket and northern Malacca Straits. Only Korea of the major bank fields is an exception. Estuary entrances with large amounts of sand also contain parabolic banks (van Veen 1950; Robinson 1960; Ludwick 1974). Off rocky coasts 'banner' (flying in the breeze) banks (Belderson *et al.* 1982) are common in tidal residual eddies adjacent to headlands or simply on the downdrift side of an obstacle (Pingree 1978; Pingree & Maddock 1979).

Further offshore in these higher velocity situations the banks give way to more linear, regularly spaced forms (Fig. 2). Swift (1975) observes that these cases resemble the shoal-retreat sand bodies created by transgression of the Atlantic shelf of the U.S.A. The outer linear banks are thus likely to be modifications of an abandoned nearshore parabolic field that has been acted on by tidal currents yielding growth of large banks at the preferred spacings suggested by Huthnance's (1982) theory. An additional aspect of this hypothesis is provided by the suggestion of Kenyon *et al.* (1981) that the outermost banks of the Norfolk group are 'moribund'. That is the bank-forming flows are no longer active (the currents are now too weak), so the banks are being lowered by wave action and their sides draped with the erosion products of that degradation (Stride *et al.* 1982). This situation is marked by absence of sand waves on banks. Other moribund banks are found in a large field in the Celtic Sea and in the

East Bank area north-west of the Dogger Bank in the North Sea. The Celtic sea banks resemble the carbonate banks of TOTO in the Bahamas in being perpendicular to the shelf edge. It is possible that here some control on spacing was originated through tidal edge-waves.

Lower velocity areas contain sand-wave fields grading down the transport path into areas of sand patches and mud deposition (Stride 1963; Belderson & Stride 1966). Only the sand-wave area off the Dutch coast is known in detail, but the internal structure of the resulting deposit is speculative. Grain-size variations over sand waves have been investigated by Ludwick & Wells (1974) but they find both sand waves with coarse sediment on the crest and fines in the trough and vice versa. The fact that most of the extensive areas of sand waves known today are developed in sands dispersed at low sea level makes it difficult to assess what might be the pattern of sand dispersal and the form of sand sheets without sea-level change. Lowering of sea level may well have been a most important process in development of ancient sheet sands.

A few areas of the world have shelves where a 'steady' current (one with longer period variability than the usual tidal and wave motions) drives sediment transport. One is the shelf of eastern South Africa whose edge lies under the Agulhas current. Here outer-shelf relict carbonate sediments display large sand waves and sand ribbons over a gravel pavement (Flemming 1978, 1980). Another such region is the northern Bering Sea where a steady northerly coastal current has constructed larger sand waves with superimposed small sand waves and linguoid ripples (Field *et al.* 1981). Thus many of the features of tidally dominated shelves are to be found under more steady marine flows.

Wave-dominated sands on shelves

Bedforms

The mechanics of sand transport examined above show that transport rates on wave-dominated shelves are relatively slow. A slight tendency to offshore transport in the midshelf region is shown by Vincent's (1984) data for the U.S.A. Atlantic shelf and the theory of Smith & Long (1976) for the Washington shelf. However the offshore spread of sand as bedload from the coastal zone is very slight, as Vincent's data suggest onshore flux at the shallowest site.

The best-documented sandy shelf with wave dominance is the Middle Atlantic Bight of the eastern U.S.A. shelf, extensively described and analysed by Swift and his colleagues and summarized in Swift (1976a, b). There is scarcely any fluvial sand input to this shelf at present and it is an environment of preservation and modification of bedforms and deposits created during the Holocene transgression. Uchupi's (1968) bathymetry of the area reveals many oblique ridges which Duane *et al.* (1972) show are detached oblique shoreface ridges. These ridges develop under the opposing forces of a wind-driven coastal current and wave surge. The convergent sediment transport yielding a ridge is analogous to that forming tidal sand banks. Swift *et al.* (1972) also recognize 'shoal retreat massifs', bodies of sand originally forming shoals which were abandoned during rising sea level. The coastal site of sand accumulation remains active, so as sea-level rises a long pile of sand extending from the outer shelf to the present coastal shoal is formed. Examples are estuary entrance shoals, e.g. Chesapeake Bay (Ludwick 1974) and shoals off the Capes, e.g. Cape Hatteras (Hunt *et al.* 1977).

The surface of this shelf also has small dune-type bedforms, mainly in water depths < 50 m, created under storm wind-driven currents (Swift *et al.* 1979). These in turn are generally mantled with ripples. The dune fields tend to be short-lived, being formed in winter and slowly flattened in summer. Swift *et al.* (1983) argue that these dune fields, when modified by wave-driven sediment transport, yield the structure known as hummocky cross-stratification. With dune heights of up to 0.3 m, at least the upper 0.3 m of the sand sheet is physically reworked. In several places shallower than 30 m, sand waves 2 m high occur, also from wind-driven currents.

In zones of intense wave action, Clifton *et al.* (1971) have shown that the seabed displays asymmetric lunate dunes and upper flow regime plane beds. This is generally in water shallower than 5 m and in the breaker zone. Some of the structures described by Clifton *et al.* (1971) also resemble the hummocky cross-stratification ascribed by Dott & Bourgeois (1982) to an oscillatory flow regime transitional to sheet flow on a plane bed. As the bed on which HCS is formed is probably irregularly undulating it has not appeared distinctive to marine geologists and has not been specially recorded. Where the shoreface is affected by relatively low wave action, a shallow planar zone is succeeded offshore by rippled sand, then increasingly muddy lamination and bioturbation in the Georgia coastal example of Howard & Reineck (1972) and the Oregon shelf of Kulm *et al.* (1975).

Storm sand dispersal and facies

The possibility of interpreting the stratigraphic record as dominantly the product of extreme events (Ager 1973) has given added importance to the deposits of extreme storms. In areas which are entirely sandy, storms may generate large scours and leave lag gravels lining them. Nearshore they may generate structures of the type described by Clifton *et al.* (1971). The material eroded nearshore is carried seaward and deposited as storm sand layers (Hayes 1967; Gadow & Reineck 1969; Reineck & Singh 1972; Nelson 1982; Aigner & Reineck 1982; Figueiredo *et al.* 1982). These layers are graded at least to the extent that they usually overlie a scoured surface with a lag and comprise sand fining to silt at the top of the layer. Those on the Texas shelf are up to 9 cm thick (Hayes 1967), the Yukon layers are 20 cm nearshore, thinning to 1 cm at 75 km distance (Nelson 1982), and the German Bight ones are thinner at 2 cm decreasing to < 1 cm (Gadow & Reineck 1969). Nelson (1982) recognizes a sequence of structures, analogous to the Bouma sequence for turbidites with which they may be confused. It comprises basal parallel laminated, middle cross and convolute laminated and upper parallel laminated sands to silts, capped by mud.

A conflict of interpretation has arisen over the mechanism by which these beds were emplaced. Hayes (1967) suggested that, following the 2.5 m storm surge caused by Hurricane Carla in 1961, catastrophic outflow from Laguna Madre generated a turbidity current that became fast enough on the steeper shoreface to flow 30 km on a slope of less than 1 in 1000 out to between 15 and 36 m water depth. Apart from the intrinsic unlikelihood of such a current achieving 'ignition' conditions for a turbidity current (Parker 1982), Morton (1981) has pointed out several inconsistencies in the story, notably that water levels in Laguna Madre were actually below normal and that nearshore currents were in the wrong direction for the transport path envisaged by Hayes (1967). As wind stress causes set-up of the sea surface towards the coast, the increased head of water there causes a compensating return flow along the bottom. This flow will be quasi-geostrophic and nearer to shore-parallel than shore-normal. Throughout the surge there is a strong flow with some seaward component. When the surge relaxes an additional strong current pulse may result if water was piled-up in an embayment (Nelson 1982). The flow speeds nearshore under hurricane winds can be in excess of 1 m s^{-1} (Morton 1981). Clearly a flow of 1 m s^{-1} in shallow water acting in one direction for a day with superimposed waves can move a lot of sand (up to 10^3 tonnes m^{-1} day^{-1}), and, for fine sand in suspension, move it tens of kilometres. Deposition of the sand will follow a period of storm wave scour of the bed and will wane as the set-up of the surface dissipates. The notion of shelf sand turbidites vigorously espoused by Hamblin & Walker (1979), popularized by Walker (1979) and adopted by Dott & Bourgeois (1982), is dynamically unreasonable and unnecessary to explain the modern shelf data.

Hayes' turbidite interpretation was applied by Howard & Reineck (1981) to sand units averaging 45 cm thick in vibrocores from the Ventura shelf, California. They thicken towards the probable source, Santa Clara River, reaching 1.9 m. The required flow direction on the shelf cannot be easily reconciled with a gravity-driven flow, and the thickness is extreme. The Santa Clara is well known as a river with rare, catastrophic floods: in 1969 it carried 50 M tonnes, 70% of which arrived in two days (Drake *et al.* 1972). This gave an initial deposit 5–15 cm thick on the mid-shelf. The dispersal of material from the mouth of the Santa Clara could have been by several storms to produce the layers observed by Howard & Reineck (1981). A single event of catastrophic dispersal is unnecessary. However storm-driven transport is thought essential as fair weather processes move mainly fine suspended sediment (Drake *et al.* 1972). We should also remember that there are explanations for graded bedding other than deposition from a waning flow. Various possibilities including grading resulting from bedform migration and from sediment liquefaction are discussed by Figueiredo *et al.* (1982).

On the areas where material is actively supplied, even where wave action is relatively intense such as the Ventura shelf, the shoreface sands pass into shelf muds at about 20 m water depth. The same is true off the Niger Delta where coastal sands give place to prodelta mud/sand lamination by at most 15 m depth and to uniform muds by 50 m (Allen 1970b). In regions of active supply (perhaps equivalent to depositional regression) and wave dominance, a nearshore sand zone gives place seawards to shelf muds with storm sand layers and then to muds on the slope or in an adjacent basin. In such regions there is no wide shelf such as that of the eastern U.S.A.

Conclusion

Research on shelf clastic sedimentation is very active. Both geologists and hydrodynamicists are contributing to the picture so that we not only have an idea of the distribution of materials and their history, but where and how fast they are moving. A start has been made on the problem of storm sediment transport, though measurements during the greatest storms are elusive! In principle, if one knows rates of supply, transport and deposition, one should be able to model change in morphology, as has been done for some aspects of coastal change (Komar 1977). However shelf morphology is so very complicated that for some years considerable extrapolations made by geologists will have to serve as our best estimate of future shelf behaviour.

Curray (1965) and Swift (1970) have made such extrapolations (Fig. 3). The present tendency on shelves is to form the beginnings of a mud blanket. On wave-dominated shelves the inner- or mid-shelf mud belts are just the beginning of mud cover that should extend out over the shelf edge. They are the basal regressive cover to the transgressive sands and relict morphologies of earlier periods of the Holocene. In cases where there is active supply and coastal progradation, such as off the Mississippi and Niger deltas, the shelf is not wide and the shelf-break not so marked as in non-depositional areas. A prograding shelf would not be as morphologically distinct as most present-day shelves whose shape bears a strong imprint of

sea level. The mud cover would grade into a slope where mass flows of mud become more common (Prior & Coleman 1980).

Under such a regime the sands would tend to: (a) be confined to the nearshore zone, down to perhaps 40 m water depth on high-energy shelves, and (b) escape from the shelf in channels feeding canyons. Yet Swift (1970) suggests a stage beyond this, the climax shelf, where sands diffuse across the shelf, probably in storms, together with suspended mud to yield a muddy sand. In this deposit one would perhaps expect a greater frequency of storm sand layers if bioturbation permits.

Strong tidal currents generally prevent deposition of mud, so such shelves should be sandy. However we have little idea of how such a shelf would evolve given a large input of sand. There is exchange of sand between nearshore parabolic banks, but offshore linear banks, abandoned by rising sea level, are thought to form closed sand circulations in many cases and thus not be supplied from the coastal zone. A prograding situation might just have a coastal zone of parabolic banks or, at lower speeds, a sand wave field succeeded offshore by shelf muds with storm sand layers. Whether in storm or tidal regions these storm sands, if infrequently emplaced, are likely to be considerably changed by burrowing, and may be altogether reduced to a homogeneous muddy sand.

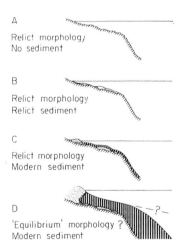 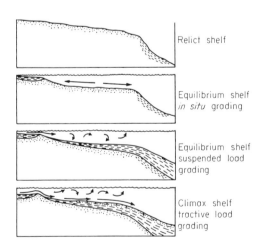

Fig. 3. Shelf evolution hypothesized by Curray (1965) (left), and Swift (1970) (right). The vertical shading in Curray's diagram denotes a predominantly muddy deposit. Although Swift's diagram speaks of 'tractive load', his data show maximum sand sizes about 250 μm and a mode at ~125 μm suggesting most of the material is transportable in suspension during storms.

References

AGER, D. V. 1973. *The Nature of the Stratigraphical Record*, 114 pp. Macmillan, London.

AIGNER, T. & REINECK, H. E. 1982. Proximality trends in modern storm sands from the Helgoland Bight (North Sea) and their implications for basin analysis. *Senckenberg. Mar.* **14**, 183–215.

ALLEN, G. P. 1972. *Étude des processus sédimentaires dans l'estuaire de la Gironde*. The`se Fac. Sci. Bordeaux. 310 pp.

ALLEN, J. R. L. 1963. Asymmetrical ripple marks and the origin of water laid cosets of cross-strata. *L'pool Manch. geol. J.* **3**, 187–236.

—— 1970a. *Physical Processes of Sedimentation*. Allen & Unwin, London. 248 pp.

—— 1970b. Sediments of the modern Niger delta: a summary and review. *In*: MORGAN, J. P. (ed.) *Deltaic Sedimentation, Modern and Ancient. Spec. Publs. Soc. econ. Paleont. Mineral. Tulsa*, **15**, 138–51.

—— 1980a. Sand waves: a model of origin and internal structure. *Sedim. Geol.* **26**, 281–328.

—— 1980b. Large transverse bedforms and the character of boundary-layers in shallow-water environments. *Sedimentology*, **27**, 317–23.

ALLERSMA, E. 1971. Mud on the oceanic shelf off Guiana. *In*: *Symp. Investigations and Resources of the Caribbean Sea and Adjacent Regions*, 193–203. UNESCO, Paris.

ANDEL, T. H. van 1960. Sources and dispersion of Holocene sediments, northern Gulf of Mexico. *In*: SHEPARD, F. P., PHLEGER, F. B. & VAN ANDEL, T. H. (eds) *Recent Sediments, Northwest Gulf of Mexico*, 44–55. American Association of Petroleum Geologists, Tulsa.

—— & POSTMA, H. 1954. *Recent Sediments of the Gulf of Paria*. Reports of Orinoco shelf expedition, vol. I. *Verh. K. ned. Akad. Wet.* **20**, 245 pp.

BAGNOLD, R. A. 1963. Beach and nearshore processes. *In*: HILL, M. N. (ed.) *The Sea* **3**. 507–28. Wiley, New York.

BELDERSON, R. H., JOHNSON, M. A. & KENYON, N. H. 1982. Bedforms. *In*: STRIDE, A. H. (ed.) *Offshore Tidal Sands*, 27–57. Chapman & Hall, London.

—— & STRIDE, A. H. 1966. Tidal current fashioning of a basal bed. *Mar. Geol.* **4**, 237–57.

BERTHOIS, L. 1946. Récherches sur les sédiments du plateau continental Atlantique. *Annls Inst. océanogr., Paris*, **23**, 1–63.

BIJKER, E. W. 1967. Some considerations about scales for coastal models with moveable beds. *Delft Hydraul. Lab. Rep. No. 50*, 142 pp.

BIRCH, G. F. 1977. Surficial sediments on the continental margin off the west coast of South Africa. *Mar. Geol.* **23**, 305–37.

BLACK, M. 1933. The precipitation of calcium carbonate on the Great Bahama Bank. *Geol. Mag.* **32**, 455–66.

BOOTHROYD, J. C. & HUBBARD, D. K. 1975. Genesis of bedforms in mesotidal estuaries. *In*: CRONIN, L. E. (ed.) *Estuarine Research, 2, Geology and Engineering*, 217–34. Academic Press, London.

BOTHNER, M. J., SPIKER, E. C., JOHNSON, P. P.,

RENDIGS, R. R. & ACRUSCAVAGE, P. J. 1981. Geochemical evidence of modern sediment accumulation on the continental shelf off southern New England. *J. sedim. Petrol.* **51**, 281–91.

BRIDGES, P. H. 1982. Ancient offshore tidal deposits. *In*: STRIDE, A. H. (ed.) *Offshore Tidal Sands*, 172–92. Chapman & Hall, London.

CASTON, V. N. D. 1972. Linear sand banks in the southern North Sea. *Sedimentology*, **18**, 63–78.

CHESTERMAN, W. D., CLYNICK, P. R. & STRIDE, A. H. 1958. An acoustic aid to sea bed survey. *Acustica*, **8**, 285–90.

CLIFTON, H. E., HUNTER, R. E. & PHILLIPS, R. L. 1971. Depositional structures and processes in the non-barred high energy nearshore. *J. sedim. Petrol.* **41**, 651–70.

COLEMAN, J. M. & WRIGHT, L. D. 1975. Modern river deltas: variability of processes and sand bodies. *In*: BROUSSARD, M. L. (ed.) *Deltas, Models for Exploration*. 99–149. Houston Geological Society.

CURRAY, J. R. 1960. Sediments and history of Holocene transgression, continental shelf, northwest Gulf of Mexico. *In*: SHEPARD, F. P., PHLEGER, F. B. & VAN ANDEL, Tj. H. (eds) *Recent Sediments, Northwest Gulf of Mexico*, 221–66. American Association of Petroleum Geologists, Tulsa.

—— 1965. Late Quaternary history, continental shelves of the United States. *In*: WRIGHT, H. E. & FREY, D. G. (eds) *The Quaternary of the United States*, 723–35. Princeton University Press.

—— 1969. History of continental shelves. *In*: STANLEY, D. J. (ed.) *The New Concepts of Continental Margin Sedimentation*. JC-VI-1–18. American Geological Institute, Washington D.C.

DALRYMPLE, R. W., KNIGHT, R. J. & LAMBIASE, J. J. 1978. Bedforms and their hydraulic stability relationships in a tidal environment, Bay of Fundy, Canada. *Nature*, **275**, 100–4.

DAVIES, J. L. 1980. *Geographical Variation in Coastal Development*. 2nd ed. Oliver & Boyd, Edinburgh. 212 pp.

DAVIES, P. J. 1979. Marine geology of the continental shelf off southeast Australia. *Bull. Bur. Miner. Resour. Geol. Geophys. Aust.* **195**.

DIETZ, R. S. 1952. Geomorphic evolution of continental terrace (continental shelf and slope). *Bull. Am. Ass. Petrol. Geol.* **36**, 1802–19.

—— & MENARD, H. W. 1951. Origin of abrupt change in slope at continental shelf margin. *Bull. Am. Ass. Petrol. Geol.* **35**, 1994–2016.

DORJES, J. & HERTWECK, G. 1975. Recent biocoenoses and ichnocoenoses in shallow-water marine environments. *In*: FREY, R. W. (ed.) *The Study of Trace Fossils*, 459–91. Springer-Verlag, New York.

——, GADOW, S., REINECK, H,-E. & SINGH, I. B. 1970. Sedimentologie und Makrobenthos der Nordergrunde und der Aussenjade (Nordsee). *Senckenberg. Mar.* **2**, 31–59.

DOTT, R. H. & BOURGEOIS, J. 1982. Hummocky

stratification: significance of its variable bedding sequences. *Bull. geol. Soc. Am.* **93**, 663–80.

DRAKE, D. E., KOLPACK, R. L. & FISCHER, P. J. 1972. Sediment transport on the Santa Barbara-Oxnard shelf, Santa Barbara Channel, California. *In:* SWIFT, D. J. P., DUANE, D. B. & PILKEY, O. H. (eds) *Shelf Sediment Transport: Process and Pattern*, 307–31. Dowden, Hutchinson & Ross, Stroudsburg.

DUANE, D. B., FIELD, M. E., MEISBURGER, E. P., SWIFT, D. J. P. & WILLIAMS, S. J. 1972. Linear shoals on the Atlantic inner continental shelf, Florida to Long Island. *In:* SWIFT, D. J. P., DUANE, D. B. & PILKEY, O. H. (eds) *Shelf Sediment Transport: Process and Pattern*, 447–99. Dowden, Hutchinson & Ross, Stroudsburg.

DYER, K. R. 1971. The distribution and movement of sediment in the Solent, southern England. *Mar. Geol.* **11**, 175–87.

EDELMAN, C. H. 1939. Petrological relationships of the sediments of the southern North Sea. *In:* TRASK, P. D. (ed.) *Recent Marine Sediments.* 343–7. American Association of Petroleum Geologists, Tulsa.

EISMA, D. & MAREL, H. W. van der 1971. Marine muds along the Guyana coast and their origin from the Amazon Basin. *Contr. Miner. Petrol.*, **31**, 321–4.

EMERY, K. O. 1968. Relict sediments on continental shelves of the world. *Bull. Am. Ass. Petrol. Geol.* **52**, 445–64.

ENGELUND, F. & HANSEN, E. 1967. *A monograph on sediment transport in alluvial streams.* Technisk Vorlag, Copenhagen. 62 pp.

FIELD, M. E., NELSON, C. H., CACCIONE, D. A. & DRAKE, D. E. 1981. Sand waves on an epicontinental shelf: northern Bering Sea. *Mar. Geol.* **42**, 233–58.

FIGUEIREDO, A. G., SANDERS, J. E. & SWIFT, D. J. P. 1982. Stormgraded layers on inner continental shelves: examples from southern Brazil and the Atlantic coast of the central United States. *Sedim. Geol.* **31**, 171–90.

FLEMMING, B. W. 1978. Underwater sand dunes along the south east African continental margin—observations and implications. *Mar. Geol.* **26**, 177–98.

—— 1980. Sand transport and bedform patterns on the continental shelf between Durban and Port Elizabeth (southeast African continental margin). *Sedim. Geol.* **26**, 179–205.

FLEMMING, N. C. & STRIDE, A. H. 1967. Basal sand and gravel patches with separate indications of tidal current and storm-wave paths, near Plymouth. *J. mar. biol. Ass. U. K.* **47**, 433–44.

GADD, P. E., LAVELLE, J. W. & SWIFT, D. J. P. 1978. Estimates of sand transport on the New York shelf using near-bottom current meter observations. *J. sedim. Petrol.* **48**, 239–52.

GADOW, S. & REINECK, H.-E. 1969. Ablandiger Sand-transport bei Sturmfluten. *Senckenberg. Mar.* **1**, 63–78.

GINSBURG, R. N. 1956. Environmental relationships of grain size and constituent particles in some south Florida carbonate sediments. *Bull. Am. Ass. Petrol. Geol.* **40**, 2384–427.

GIRESSE, P. & KOUYOUMONTZAKIS, G. 1973. Cartographie sédimentologique des plateaux continentals sud du Gabon, du Congo, du Cabinda et du Zaire. *Cah. ORSTOM, Ser. Geol.* **5**, 235–57.

GOLDBERG, E. D. & KOIDE, K. 1962. Geochronological studies of deep-sea sediments by the ionium-thorium method. *Geochim. cosmochim. Acta*, **26**, 417–50.

GRANT, W. D. & MADSEN, O. S. 1979. Combined wave and current interactions with a rough bottom. *J. geophys. Res.* **84**, 1797–808.

GUINASSO, N. L. & SCHINK, D. R. 1975. Quantitative estimates of biological mixing rates in abyssal sediments. *J. geophys. Res.* **80**, 3032–43.

HAMBLIN, A. P. & WALKER, R. G. 1979. Storm-dominated shelf deposits: the Fernie-Kootenay (Jurassic) transition, southern Rocky Mountains. *Can. J. Earth Sci.* **16**, 1673–90.

HAMILTON, D., SOMMERVILLE, J. H. & STANFORD, P. N. 1980. Bottom currents and shelf sediments, southwest of Britain. *Sedim. Geol.* **26**, 115–38.

HAYES, M. O. 1967. Hurricanes as geological agents, south Texas coast. *Bull. Am. Ass. Petrol. Geol.* **51**, 937–42.

HEATHERSHAW, A. D. 1981. Comparisons of measured and predicted sediment transport rates in tidal currents. *Mar. Geol.* **42**, 75–104.

HOUBOLT, J. J. H. C. 1968. Recent sediments in the southern bight of the North Sea. *Geologie Mijnb.* **47**, 245–273.

HOWARD, J. D. & REINECK, H.-E. 1972. Georgia coastal region, Sapelo Island, U.S.A.: sedimentology and biology, IV. Physical and biogenic sedimentary structures of nearshore shelf. *Senckenberg. Mar.* **4**, 81–123.

—— & —— 1981. Depositional facies of high-energy beach-offshore sequence: comparison with low-energy sequence. *Bull. Am. Ass. Petrol. Geol.* **65**, 807–30.

HOWARTH, M. J. 1982. Tidal currents of the continental shelf. *In:* STRIDE, A. H. (ed.) *Offshore Tidal Sands*, 10–26. Chapman & Hall, London.

HUNT, R. E., SWIFT, D. J. P. & PALMER, H. 1977. Constructional shelf topography, Diamond Shoals, North Carolina. *Bull. geol. Soc. Am.* **88**, 299–311.

HUTHNANCE, J. M. 1981. Waves and currents near the continental shelf edge. *Prog. Oceanog.* **10**, 193–226.

—— 1982. On one mechanism forming linear sand banks. *Estuarine coast. Mar. Sci.* **14**, 79–99.

ILLING, L. V. 1954. Bahamian calcareous sands. *Bull. Am. Ass. Petrol. Geol.* **38**, 1–95.

JOHNSON, M. A., KENYON, N. H., BELDERSON, R. H. & STRIDE, A. H. 1982. Sand transport. *In:* STRIDE, A. H. (ed.) *Offshore Tidal Sands*, 58–94. Chapman & Hall, London.

KENYON, N. H., BELDERSON, R. H., STRIDE, A. H. & JOHNSON, M. A. 1981. Offshore tidal sand banks as indicators of net sand transport and as potential deposits. *In*: NIO, S.-D., SHÜTTENHELM, R. T. E. & VAN WEERING, Tj. C. E. (eds) *Holocene Marine Sedimentation in the North Sea Basin. Spec. Publs Int. Ass. Sediment.* **5**, 257–68. Blackwell Scientific Publications, Oxford.

KOLDEWIJN, B. W. 1958. *Sediments of the Paria-Trinidad Shelf. Rep. Orinoco Shelf Exped. 3.* Den Haag, Mouton & Co. 109 pp.

KOMAR, P. D. 1977. Modelling of sand transport on beaches and the resulting shoreline evolution. *In*: GOLDBERG, E. D., MCCAVE, I. N., O'BRIEN, J. & STEELE, J. H. (eds) *The Sea*, **6**, 499– 513. Wiley (Interscience), New York.

KRONE, R. B. 1962. *Flume studies of the transport of sediment in esturial shoaling processes.* Hydraulic Engineering and Sanitary Engineering Research Laboratory, University of California, Berkeley. 110 pp.

—— 1963. *A study of rheologic properties of estuarial sediments. Sanit. engrg Res. Lab. Rept. Univ. California, Berkeley, 63–8,* 91 pp.

KUEHL, S., NITTROUER, C. A. & DeMASTER, D. J. 1982. Modern sediment accumulation and strata formation on the Amazon continental shelf. *Mar. Geol.* **49**, 279–300.

KULM, L. D., ROUSH, R. C., HARLETT, J. C., NEUDECK, R. H., CHAMBERS, D. M. & RUNGE, E. J. 1975. Oregon continental shelf sedimentation: inter-relationships of facies distribution and sedimentary processes. *J. Geol.* **83**, 145–75.

LEES, A. 1975. Possible influence of salinity and temperature on modern shelf carbonate sedimentation. *Mar. Geol.* **19**, 159–98.

—— & BULLER, A. T. 1972. Modern temperate-water and warm-water shelf carbonate sediments contrasted. *Mar. Geol.* **13**, M67–73.

LEOPOLD, L. B., WOLMAN, M. G. & MILLER, J. P. 1964. *Fluvial Processes in Geomorphology.* Freeman, San Francisco. 522 pp.

LÜDERS, K. 1939. Sediments of the North Sea. *In*: TRASK P. D. (ed.) *Recent Marine Sediments.* 322–42. American Association of Petroleum Geologists, Tulsa.

LUDWICK, J. C. 1974. Tidal currents and zig-zag sand shoals in a wide estuary entrance. *Bull. geol. Soc. Am.* **85**, 717–26.

—— & WELLS, J. T. 1974. Particle size distribution and small-scale bedforms on sand waves, Chesapeake Bay entrance. *Techn. Rep. Old Dom. Univ. Inst. Oceanogr. 12*, 112 pp.

McCAVE, I. N. 1971a. Wave effectiveness at the sea bed and its relationship to bedforms and deposition of mud. *J. sedim. Petrol.* **41**, 89–96.

—— 1971b. Sand waves in the North Sea off the coast of Holland. *Mar. Geol.* **10**, 199–215.

—— 1972. Transport and escape of fine-grained sediment from shelf areas. *In*: SWIFT, D. J. P., DUANE, D. B. & PILKEY, O. H. (eds) *Shelf Sediment Transport*, 225–48. Dowden, Hutchinson & Ross, Stroudsburg.

—— 1979a. Suspended material over the central Oregon continental shelf in May 1974, I: Concentration of organic and inorganic components. *J. sedim. Petrol.* **49**, 1181–94.

—— 1979b. Tidal currents at the North Hinder Lightship, Southern North Sea: flow directions and turbulence in relation to maintenance of sand banks. *Mar. Geol.* **31**, 101–14.

—— 1984. Erosion, transport and deposition of fine-grained marine sediments. *In*: STOW, D. A. V. & PIPER, D. J. W. (eds) *Fine Grained Sediments: Deep-Water Processes and Facies. Spec. Publs geol. Soc. Lond.* Blackwell Scientific Publications, Oxford. In press.

—— & LANGHORNE, D. N. 1982. Sand waves and sediment transport around the end of a tidal sand bank. *Sedimentology*, **29**, 95–110.

McLEAN, S. R. 1981. The role of non-uniform roughness in the formation of sand ribbons. *Mar. Geol.* **42**, 49–74.

MIDDLETON, G. V. & SOUTHARD, J. B. 1978. *Mechanics of Sediment Movement. Short Course Soc. econ. Paleont. Miner.* **3**, 246 pp.

MIGNIOT, C. 1968. Étude des propriétés physiques de différents sédiments très fins et de leur comportement sous des actions hydrodynamiques. *Houille blanche*, **7**, 591–620.

MOORE, D. G. & SCRUTON, P. C. 1957. Minor internal structures of some recent unconsolidated sediments. *Bull. Am. Ass. Petrol. Geol.* **41**, 2723–51.

MORGAN, J. P., LOPIK, J. R. van & NICHOLS, L. G. 1953. Occurrence and development of mud flats along the western Louisiana coast. *Tech. Rep. Cstl Stud. Inst. La St. Univ.* **2**, 34 pp.

MORTON, R. A. 1981. Formation of storm deposits by wind-forced currents in the Gulf of Mexico and the North Sea. *In*: NIO, S.-D., SHÜTTENHELM, R. T. E. & VAN WEERING, Tj. C. E. (eds) *Holocene Marine Sedimentation in the North Sea Basin. Spec. Publs int. Ass. Sediment.* **5**, 385-96. Blackwell Scientific Publications, Oxford.

NELSON, C. H. 1982. Modern shallow-water graded sand layers from storm surges, Bering shelf: a mimic of Bouma sequences and turbidite systems. *J. sedim. Petrol.* **52**, 537–45.

NEWELL, N. D. & RIGBY, J. K. 1957. Geological studies on the Great Bahama Bank. *In*: LE BLANC, R. J. & BREEDING, J. G. (eds) *Regional aspects of Carbonate Deposition. Spec. Publs Soc. econ. Paleont. Miner. Tulsa*, **5**, 15–72.

——, PURDY, E. G. & IMBRIE, J. 1960. Bahamian oolitic sand. *J. Geol.* **68**, 481–97.

NITTROUER, C. A., STERNBERG, R. W., CARPENTER, R. & BENNETT, J. T. 1979. The use of Pb-210 geochronology as a sedimentological tool: application to the Washington continental shelf. *Mar. Geol.* **31**, 297–316.

—— & —— 1981. The formation of sedimentary strata in an allochthonous shelf environment: the Washington continental shelf. *Mar. Geol.* **42**, 201–32.

NOTA, D. J. G. 1958. Sediments of the Western Guiana Shelf. *Meded. Landbouwhogesch. Wageningen*, **58**, 1–98.

OFF, T. 1963. Rhythmic linear sand bodies caused by tidal currents. *Bull. Am. Ass. Petrol. Geol.* **47**, 324–41.

PAK, H., BEARDSLEY, G. F. (Jr) & PARK, P. K. 1970. The Columbia River as a source of marine light-scattering particles. *J. geophys. Res.* **75**, 4570–8.

PARKER, G. 1982. Conditions for the ignition of catastrophically erosive turbidity currents. *Mar. Geol.* **46**, 307–27.

PINGREE, R. D. 1978. The formation of the Shambles and other banks by tidal stirring of the seas. *J. mar. biol. Ass. U. K.* **58**, 211–26.

—— & GRIFFITHS, D. K. 1979. Sand transport paths around the British Isles resulting from M_2 and M_4 tidal interactions. *J. mar. biol. Ass. U. K.* **59**, 497–513.

—— & MADDOCK, L. 1979. The tidal physics of headland flow and offshore tidal bank formation. *Mar. Geol.* **23**, 269–89.

POTTER, P. E. & PETTIJOHN, F. J. 1963. *Palaeocurrents and Basin Analysis*, Springer, Heidelberg 296 pp.

PRIOR, D. B. & COLEMAN, J. M. 1980. Sonograph mosaics of submarine slope instabilities, Mississippi River Delta. *Mar. Geol.* **36**, 227–39.

PURDY, E. G. 1963. Recent calcium carbonate facies of the Great Bahama Bank. 2. Sedimentary facies. *J. Geol.* **71**, 472–97.

REINECK, H.-E. 1963a. Sedimentgefüge im Bereich der südlichen Nordsee. *Abh. senckenb. naturforsch. Ges.* **505**, 138 pp.

—— 1963b. Der Kastengreifer. *Natur Mus., Frankf.* **83**, 102–8.

——, DORJES, J., GADOW, S. & HERTWECK, G. 1968. Sedimentologie, Faunenzonierung und Faziesabfolge vor der Ostküste der inneren Deutschen Bucht. *Senckenberg. Leth.* **49**, 261–309.

——, GUTMANN, W. F. & HERTWECK, G. 1967. Das Schlickgebiet südlich Helgoland als Beispiel rezenter Schelf-ablagerungen. *Senckenberg. Leth.* **48**, 219–75.

—— & SINGH, I. B. 1971. Der Golf von Gaeta/Tyrrhenisches Meer. 3. Die Gefüge von Vorstrand-und Schelfsedimenten. *Senckenberg. Mar.* **3**. 185–201.

—— & —— 1972. Genesis of laminated sand and graded rhythmites in storm-sand layers of shelf mud. *Sedimentology*, **18**, 123–8.

—— & —— 1980. *Depositional Sedimentary Environments*, 2nd ed. Springer-Verlag, Berlin. 549 pp.

REN MEIE, ZHANG RENSHUN & YANG JUHAI, 1984. Sedimentation of tidal mud flat in China—a case study of sedimentation on a prograding mud flat in Wanggang area, Jiangsu Province. *In: Proc. Internat. Symp. Sed. Continental Shelf with special reference to the East China Sea*, Beijing.

ROBINSON, A. H. W. 1960. Ebb-flood channel systems in sandy bays and estuaries. *Geography*, **45**, 183–99.

—— 1966. Residual currents in relation to shoreline evolution of the East Anglian coast. *Mar. Geol.* **4**, 57–84.

ROY, P. S. & THOM, B. G. 1981. Late Quaternary marine deposition in New South Wales and southern Queensland—an evolutionary model. *J. geol. Soc. Aust.* **28**, 471——89.

RUBIN, D. M. & McCULLOCH, D. S. 1980. Single and superimposed bedforms: a synthesis of San Francisco Bay and flume observations. *Sedim. Geol.* **26**, 207–31.

SHEPARD, F. P. 1963. *Submarine Geology*. 2nd ed. Harper & Row, New York. 557 pp.

——, PHLEGER, F. B. & ANDEL, Tj. H. van (eds) 1960. *Recent sediments, northwest Gulf of Mexico*. American Association of Petroleum Geologists, Tulsa. 393 pp.

SMITH, J. D. 1977. Modelling of sediment transport on continental shelves. *In:* GOLDBERG, E. D., McCAVE, I. N., O'BRIEN, J. J. & STEELE, J. H. (eds) *The Sea*, **6**, 539–77. Wiley (Interscience), New York.

—— & LONG, C. E. 1976. The effect of turning in the bottom boundary layer on continental shelf sediment transport. *Mem. Soc. R. Sci. Liège, 6e serie* **10**, 369–96.

—— & McLEAN, S. R. 1977a. Spatially averaged flow over a wavy surface. *J. geophys. Res.* **82**, 1735–46.

—— & —— 1977b. Boundary layer adjustments to bottom topography and suspended sediment. *In:* NIHOUL, J. C. J. (ed.) *Bottom Turbulence*. 123–50. Elsevier, Amsterdam.

SOUTHARD, J. B. & STANLEY, D. J. 1976. Shelf-break processes and sedimentation. *In:* STANLEY, D. J. & SWIFT, D. J. P. (eds) *Marine Sediment Transport and Environmental Management*. 351–77. Wiley (Interscience), New York.

STANLEY, D. J. & WEAR, C. M. 1978. The "mud-line": an erosion—deposition boundary on the upper continental slope. *Mar. Geol.* **28**, M19–29.

—— ADDY, S. K. & BEHRENS, E. W. 1983. The mudline: variability of its position relative to shelf break. *In:* STANLEY, D. J. & MOORE, (eds) *The Shelf Break: Critical Interface on Continental Margins. Spec. Publs Soc. econ. Paleont. Miner., Tulsa*, **33**, 279–98.

STRIDE, A. H. 1963. Current-swept sea floors near the southern half of Great Britain. *Q. Jl geol. Soc. Lond.* **119**, 175–99.

—— (ed.) 1982. *Offshore Tidal Sands*. Chapman & Hall, London. 222 pp.

——, BELDERSON, R. H. & KENYON, N. H. 1972. Longitudinal furrows and depositional sand bodies of the English Channel. *Mém. Bur. Rech. Géol. Min.* **79**, 233–40.

——, ——, & JOHNSON, M. A. 1982. Offshore tidal deposits: sand sheet and sand bank facies. *In:* STRIDE, A. H. (ed.) *Offshore Tidal Sands*, 95–125. Chapman & Hall, London.

SWIFT, D. J. P. 1970. Quaternary shelves and the return to grade. *Mar. Geol.* **8**, 5–30.

—— 1975. Tidal sand ridges and shoal retreat massifs. *Mar. Geol.* **18**, 105–34.

—— 1976a. Coastal sedimentation. *In:* STANLEY, D. J. & SWIFT, D. J. P. (eds) *Marine Sediment Transport and Environmental Management*. 255–310. Wiley, New York.

—— 1976b. Continental shelf sedimentation. *In:* STANLEY, D. J. & SWIFT, D. J. P. (eds) *Marine*

Sediment Transport and Environmental Management, 311–50. Wiley, New York.

——, FREELAND, G. L. & YOUNG, R. A. 1979. Time and space distribution of megaripples and associated bedforms, Middle Atlantic Bight, North American Atlantic shelf. *Sedimentology*, **26**, 389–406.

——, FIGUEIREDO, A. G., FREELAND, G. L. & OERTEL, G. F. 1983. Hummocky cross-stratification and megaripples: a geological double standard? *J. sedim. Petrol.* **53**, 1295–317.

——, KOFOED, J. W., SAULSBURY, F. P. & SEARS, P. 1972. Holocene evolution of the shelf surface, central and southern Atlantic coast of North America. *In*: SWIFT, D. J. P., DUANE, D. B. & PILKEY, O. H. (eds) *Shelf Sediment Transport: Process and Pattern*. 499–574. Dowden, Hutchinson & Ross, Stroudsburg.

——, STANLEY, D. J. & CURRAY, J. R. 1971. Relict sediments on continental shelves, a reconsideration. *J. Geol.* **79**, 322–46.

TERWINDT, J. H. J. 1971. Sand waves in the southern bight of the North Sea. *Mar. Geol.* **10**, 51–67.

THORN, M. F. C. 1975. Deep tidal flow over a fine sand bed. *Proc. 16th Congress, int. Ass. Hydraulic Research* **1**, 217–23.

—— & PARSONS, J. G. 1980. Erosion of cohesive sediments in estuaries: an engineering guide. *In*: *Third International Symposium on Dredging Technology*, 349–58. British Hydraulics Research Association, Cranfield.

UCHUPI, E. 1968. Atlantic continental shelf and slope of the United States—physiography. *Prof. Pap. U. S. geol. Surv. 529-C*, 30 pp.

VEEN, J. van 1936. *Onderzoekingen in de Hoofden in verband met de gesteldheid der Nederlandse kust*. Algemene Landsdrukkerij, The Hague. 252 pp.

—— 1950. Eb- en vloedschaar-systemen in de Nederlandse getijwateren. *Tijdschr. K. ned. aardrijksk. Genoot.* **66**, 303–25.

VINCENT, C. E., YOUNG, R. A. & SWIFT, D. J. P. 1981a. Bedload transport under waves and currents. *Mar. Geol. 39*, 71–80.

——, SWIFT, D. J. P. & HILLARD, B. 1981b. Sediment transport in the New York Bight, North American Atlantic shelf. *Mar. Geol. 42*, 369–398.

——, YOUNG, R. A. & SWIFT, D. J. P. 1982. On the the relationship between bedload and suspended sand transport on the inner shelf, Long Island, New York. *J. geophys. Res.* **87**, 4163–70.

—— 1984. A climatological approach to sediment transport on the continental shelf. *In*: *Proc. int. Symp. Sed. Continental Shelf with Special Reference to the East China Sea*, Beijing.

WALKER, R. G. 1979. Shallow marine sands. *In*: WALKER, R. G. (ed.) *Facies Models. Geol. Ass. Can.* pp. 75–89, Geoscience Canada, Reprint Series 1.

WANG BAOYONG 1983. Hydrodynamic specificity of nearshore sediments in the East China Sea (in Chinese). *Mar. Geol. Quat. Geol.* (Beijing) **3**, 37–46.

WANG, Y. 1983. The mudflat system of China. *Can. J. Fish. Aquat. Sci.* **40** (Suppl. 1), 160–71.

WELLS, J. T. & COLEMAN, J. M. 1981. Physical processes and fine-grained sediment dynamics, coast of Surinam, South America. *J. sedim. Petrol.* **51**, 1053–68.

—— & LUDWICK, J. C. 1974. Application of multiple comparisons to grain size on sand waves. *J. sedim. Petrol.* **44**, 1029–36.

I. N. McCAVE, School of Environmental Sciences, University of East Anglia, Norwich NR4 7TJ.

Deep-sea clastics: where are we and where are we going?

D. A. V. Stow

SUMMARY: The transition from our belief in a deep calm ocean to a recognition that deep-sea clastics other than pelagic clays exist in the oceans, spanned nearly a century. In the last three decades enormous strides have been made in understanding these sediments and their deposition. There is a continuum of processes that transfer material from shallow to deep water and rework sediments within the deep sea. These include: (1) resedimentation processes, ranging from giant rockfalls and slumps to low-density turbidity currents; (2) normal bottom currents; and (3) pelagic settling through the water column. More than fifty distinct facies have been described from the deep sea and these can be interpreted in terms of depositional process via ten standard facies models for resedimented, normal bottom current and pelagic sediments. Environmental models can be constructed for: (1) normal, faulted, carbonate and ridge-flank slope-aprons; (2) radial, elongate and fan-delta submarine fans; and (3) under- and oversupplied basin-plains. These show the generalized horizontal and vertical distribution of facies and the chief morphological elements in each of the three major deep-sea settings. Sedimentary, tectonic and sea-level changes are the main groups of factors that control deep-sea sedimentation within these separate environments. Part of the interest in deep-sea clastics stems from their demonstrable economic importance for the generation and entrapment of hydrocarbons. Many areas of deep-sea sedimentology remain to be investigated and earlier models to be refined; these advances will depend significantly on improvements in our methodology.

The systematic study of deep-sea sediments (Fig. 1) began with the voyage of the HMS *Challenger* (1872–1876) which established the general morphology of the oceans and the types of sediments they contained. Following this pioneering expedition, the cornerstone of deep-sea sedimentology was, for a long time, the paper on 'Deep-sea deposits' by Murray & Renard (1891). However, the paradigm put forward by these authors was that only pelagic clays and biogenic oozes were found in the deep sea and that all coarser-grained clastics were restricted to shallow water or continental environments.

Such belief held sway amongst many geologists for several decades while several different lines of research were conspiring to undermine its dominance. In particular, as more and more bottom samples were collected on early European and American oceanographic expeditions in the first half of the Twentieth Century (Böggild 1916; Andrée 1920; Stocks 1933; Shepard 1932, 1948; Bramlette & Bradley 1940; Arrhenius 1950) it was realized that sediments do not become uniformly finer-grained seaward across the continental shelves, and that the ocean floors have irregularities as great as any other part of the globe.

Although the existence of density undercurrents in lakes and reservoirs had been known for some time (Forrel 1885; Grover & Howard 1938), it was Daly (1936) who suggested that density currents, formed by waves stirring up sediments on the continental shelf during periods of lowered sea level, may have excavated submarine canyons as they flowed downslope. Johnson (1938) coined the term turbidity current for this type of flow. A series of flume experiments on both dilute and high-density flows by Kuenen (1937, 1950), combined with Migliorini's observations of graded sand beds in the Italian Apenines paved the way for their classic paper 'Turbidity currents as a cause of graded bedding' (Kuenen & Migliorini 1950).

This revolution in clastic sedimentology, as the turbidity current paradigm has been called (Kuhn 1970; Walker 1973), stimulated an intense period of systematic field, laboratory and oceanographic studies that has continued to the present day, and that has shown the deep sea to be anything but calm! Some of the key advances are listed below: there was early confirmation of graded sands in the deep sea (Ericson *et al.* 1951) and of the occurrence and nature of major turbidity currents (Heezen & Ewing 1952); palaeo-turbidity-current directions were measured to document the pattern of basin fill in ancient sequences (Pettijohn 1957); the understanding and interpretation of geosynclinal sediments was much improved (Kay 1951; Drake *et al.* 1959) and, following the plate tectonic revolution in geology as a whole, in the early 1960s, patterns of deep-sea sedimentation were better related to global tectonics (Mitchell & Reading 1969); the classical sequence of structures in turbidites was developed by Bouma (1962); the physics of

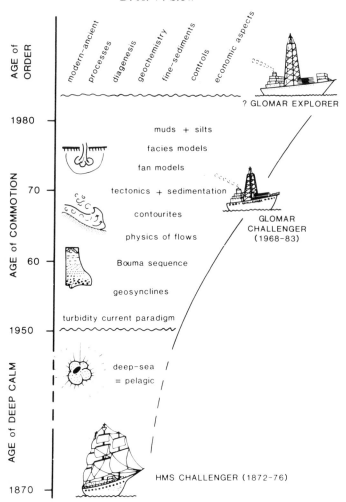

FIG. 1 Historical development of the main concepts in the study of deep-sea clastic sediments.

turbidity current flow has been much better appreciated following the experimental and theoretical work of Harms & Fahnestock (1965), Middleton (1966, 1967) and Komar (1969, 1970); deep-ocean currents were put forward as an important alternative to turbidity currents in the mid-1960s (Heezen et al. 1966; Hollister & Heezen 1972), and the characteristics of contourites more firmly established by Stow & Lovell (1979); submarine fans were characterized in detail from the present-day oceans (Normark 1970) and from ancient sequences (Mutti & Ricci Lucchi 1972); resedimented conglomerates were recognized as different from sandy turbidites but equally important (Walker 1975); similarly, the characteristics of silt and mud turbidites have only recently been detailed (Piper 1978; Stow & Shanmugam 1980; Stow et al. 1984).

The Deep-Sea Drilling Project, initiated in 1968, has seen nearly one hundred different voyages by the drillship Glomar Challenger throughout the world's oceans. These, together with many other marine and land expeditions, have clearly contributed enormously, not only to the study of deep-sea clastics but also to the related studies of pelagic and authigenic sediments, ocean history, tectonics and ocean-margin development.

Processes

For clastic sedimentary particles to accumulate in the deep sea they must undergo erosion, from land or from the seafloor, transportation and deposition. Biogenic material may be similarly eroded, but much is synthesized directly in the oceans and simply transported to and deposited

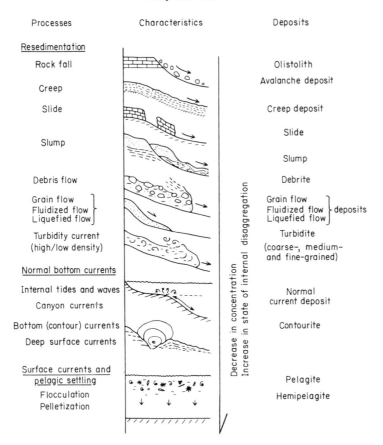

Processes	Characteristics	Deposits

Fig. 2. Schematic representation of the main transport and depositional processes in the deep sea, illustrating the concept of a process continuum.

on the seafloor. Authigenic minerals grow *in situ* at or near the sediment–water interface, although they too may be subsequently reworked. There are three main groups of processes that operate on both terrigenous and biogenic material in the deep sea (Fig. 2): resedimentation processes, movement by normal bottom currents and transport by surface currents/pelagic settling.

(1) Resedimentation

Resedimentation processes (synonymous with mass gravity transport) are all those processes that move sediment downslope over the seafloor from shallower to deeper water and that are driven by gravitational forces (Fig. 2) (Middleton & Hampton 1976; Saxov & Nieuwenhuis 1982; Hein 1982; Hill *et al.* 1984; Gorsline 1984).

Rock falls are sudden, rapid freefall events that occur only on steep slopes of faulted or carbonate margins or in the heads of submarine canyons. They are initiated by undercutting or erosion, earthquakes and other shock events. Single displaced clasts or olistoliths (up to tens of metres in size) and avalanche deposits bounce, roll and slide downslope for several tens or hundreds of metres before coming to rest.

Sediment creep is a process of slow strain due to constant load-induced stress. It is a barely perceptible but probably widespread phenomenon on many slopes, both gentle and steep, depending on the physical properties of the sediment. It appears to propagate along an internal décollement zone, and may be a precursor to slide or slump failure.

Slides and slumps involve complete sediment failure, downslope displacement and remoulding. They are widespread on slopes of all gradients over about 0.5° and range in size from less than 1 m³ to over 100 km³. They comprise mobile shear zones along which the sediment mass moves. Coherent slide blocks may be less easily recognized than contorted slump units in ancient sediments. They are triggered by various shock events or may develop progressively from sediment creep.

Debris flows are slurry-like flows of sand to boulder-size clasts supported by their own buoyancy in a muddy matrix. They can be large or small (several to tens of metres thick), occur commonly on slopes greater than 1–2°, and advance slowly downslope either continuously or intermittently for distances up to tens of kilometres.

Grain flows are characterized by grain to grain collisions and a resulting dispersive-pressure support mechanism. They require slopes in excess of about 18° and so are probably very localized and relatively small events on steep slopes such as in the heads of submarine canyons. They are not a process of long-distance transport in the deep sea.

Liquefied or fluidized flows are related processes in which the grains settle rapidly through upward-moving pore fluids. Liquefied flows may move rapidly down slopes in excess of 2–3° for a short distance only, before 'freezing' and deposition occurs. These flows rarely occur alone as a separate process in the deep sea, but are common during the final stages of deposition from high concentration turbidity currents.

Turbidity currents are very widespread throughout the deep sea over variable slopes and are of various types. High-density (50–250 g l^{-1}) and low-density (0.025–2.5 g l^{-1}) currents have been identified, as well as storm-generated flows, turbid-layer flows and mid-water density flows. Flow thickness may be from less than 10 to over 500 m, and flow velocities from 10–50 cm s^{-1} for low-density currents to over 25 m s^{-1} (70 km h^{-1}) for high-density currents. In all cases the flow is sustained and sediment supported by a process of turbulent auto-suspension. Turbidity currents can develop either from slumps and debris flows by mixing with seawater, or more directly from sediment suspensions caused by storm stirring, rivers in flood and melting glaciers.

These various flows are simply end-members of a process continuum in which all inter-gradations exist. A single resedimentation event may be initiated by seismic tremors or sediment overloading on a slope causing slumping which, by mixing with seawater, evolves through a debris flow into a turbidity current. A rock fall may trigger a sand avalanche or grain flow that deposits rapidly through a phase of liquefied or fluidized flow. Storm stirring at the shelf break

or across the shelf may generate a low-density turbidity current, and it appears that major river floods or glacial melting may initiate or augment this same process.

That a process continuum exists is emphasized by the progressive decrease in flow concentration and increase in the state of internal disaggregation from rock falls (solid) through to low-density turbidity currents. The maximum distance of transport is a few tens of metres for rock falls and probably similar for grain flows and liquefied flows, a few kilometres for slides and slumps, 50–100 km for debris flows and several 1000 km for turbidity currents.

(2) Normal bottom currents

The second group of processes (Fig. 2) are all those deep currents that actively erode, transport and deposit sediment on the seafloor but that are not driven by gravity, and may therefore flow alongslope and upslope as well as downslope (Hollister & Heezen 1972; Heezen 1977; Shepard 1973; Shepard et al. 1979; Stow & Lovell 1979; Stow 1982).

Internal tides and waves are widespread in the upper few hundreds of metres, at the thermocline and at other density discontinuities. Their effects on the bottom sediments are particularly apparent at the shelf break, in submarine canyons or in deep narrow passages.

Canyon currents have now been recorded from very many submarine canyons throughout the oceans even at depths in excess of 4000 m. They commonly flow alternately up and down the canyon, with a tidal periodicity in the deeper parts but with a higher frequency of flow reversal in the head region. Other flow periods and directions have also been recorded probably related to internal waves, surface currents, storm surges or cold-water cascading currents. Typical velocities range up to 30 cm s^{-1}.

Bottom (contour) currents are formed by cooling and sinking of surface water at high latitudes and the deep, slow, thermohaline circulation of these polar water masses throughout the world's oceans. Highly saline and warm water also flows out of the Mediterranean as an intermediate level contour current. Current intensification is caused by flow restriction through narrow passages and on the western margins of basins by the Coriolis force. These stronger currents commonly attain velocities of 10–20 cm s^{-1} (rarely over 100 cm s^{-1}), but are highly variable in direction, velocity and periodicity, and advance with a slow eddy-like progression. They may be associated with well-developed nepheloid layers.

Surface currents driven by the winds may also impinge on the seafloor at very great depths (several kilometres), such as the deep Gulf-Stream gyres of the North Atlantic or the deep Kuroshio current off Japan.

It is worth emphasizing again that, for these flow types as for the resedimentation processes, there are no rigorous boundaries between flows. Thus, internal tides and waves can cause or contribute to canyon currents, and these latter grade progressively into swifter flowing, more concentrated low-density turbidity currents. Deep, surface currents and thermohaline bottom currents are practically indistinguishable at the seafloor and, where associated with intensified nepheloid layers, grade into the lowest-density large-scale turbidity currents.

Mostly, however, the concentration (0.025–2.5 mg l^{-1}) and velocity (average 0.05–0.2 m s^{-1}) of the normal bottom currents are significantly lower than those of the various resedimentation flows. Their range of thickness (up to 2000 m) and distance of transport (up to thousands of kilometres) is variable. Such currents are clearly capable of profoundly affecting the bottom sediments and morphology, and so it is not surprising that we are discovering more and more facies and features in the deep sea that display 'fluvial-like' characteristics.

(3) Surface currents and pelagic settling

The third important depositional process in the deep sea, slow pelagic settling through the water column (Hsü & Jenkyns 1974; Jenkyns 1978; Gorsline 1984), can be considered one extreme end-member of the process continuum (Fig. 2). It is less important for clastic than for biogenic sediments as the materials involved are largely the tests of calcareous and siliceous planktonic organisms and their associated organic matter. Terrigenous elements (clays, very fine-grained quartz, volcanic dust, etc.) are transported to the open ocean in variable amounts by surface currents, winds and floating ice.

Vertical settling of the finest particles is extremely slow (10^{-4}–10^{-6} m s^{-1}), although much of the material settles more quickly (10^{-2}–10^{-3} m s^{-1}) as flocs and faecal pellets (McCave 1984). As it settles, the material is subject to dissolution of calcareous and siliceous tests, oxidation of organic matter and lateral transport by bottom and turbidity currents.

Facies

The most important features used to define different deep-sea facies are: grain size and other textural attributes, sand/mud ratio, bed thickness and geometry, internal organization of beds, biogenic and sedimentary structures, fabric composition and biota. Ideally, each facies so defined should be a unique type that forms under certain conditions of sedimentation, reflecting a particular process or environment.

However, with at least ten distinct depositional processes, a large range of environments, and sediments ranging from huge boulders to the finest clays, there are clearly a very large number of possible facies in the deep sea. More than fifty facies have been identified for clastic sediments alone (Mutti & Ricci Lucchi 1972, 1975; Walker & Mutti 1973; Carter 1975) (Fig. 3), although this degree of subdivision is clearly unnecessary for most purposes other than very detailed sedimentological interpretation.

A rather more simple grouping into seven major facies classes, based largely on grain size differences, is also shown in Fig. 3. We can identify, in both modern and ancient sediments, the following classes:

A gravels and pebbly sands (conglomerates and pebbly sandstones),
B sands (sandstones),
C sand-mud units (sandstone-mudstone units),
D silts and silt-mud units (siltstones and siltstone-mudstone units),
E muds (mudstones),
F chaotic mixed-grade deposits (rockfalls, slumps, debrites),
G oozes and arls (limestones, marlstones, cherts, etc.). ('Arl' is a term introduced by Dean *et al.* (1984) for muddy oozes and biogenic muds.)

These are modified and updated from those originally proposed by Mutti & Ricci Lucchi (1972, 1975).

It is sometimes convenient to subdivide the first five of these classes (A–E) further into disorganized and organized facies groups. The disorganized groups contain such facies as thick-bedded, massive, structureless gravels, sands and muds; irregular, thin-bedded gravel lag or coarse sand layers; and bioturbated, massive or irregularly layered, silty muds. The organized facies groups contain regularly laminated, cross-laminated, rippled and graded layers of variable bed thickness and grain size.

Facies class F is mainly disorganized and can be subdivided into three groups: exotic clasts, ranging from giant rock-fall boulders to small glacial dropstones; contorted and disturbed

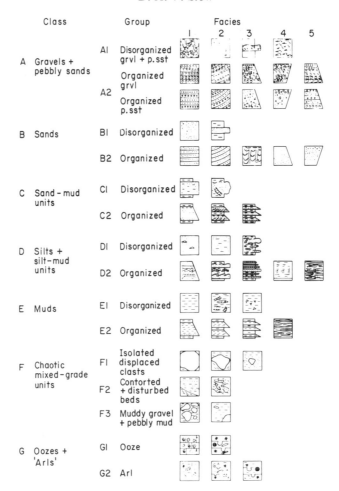

FIG. 3. The main classes and groups of sediment facies recognized in the deep sea (modified after Mutti & Ricci Lucchi 1972; Rupke 1978). Over fifty distinct facies have been identified and these are illustrated schematically. See text for further discussion.

slumps and slide masses; and pebbly muds or muddy gravels. Finally, facies class G comprises the purely biogenic (pelagic) sediments, the calcareous and siliceous oozes, and the mixed muddy oozes and biogenic muds that are so common in the deep sea. Dean *et al.* (1984) have called this latter group of sediments 'arls', adding 'sarl' (siliceous biogenic mud) and 'smarl' (calcareous and siliceous biogenic mud) to the commonly used and convenient term, marl.

Facies models

Most of the large number of separate facies shown in Fig. 3 can now be interpreted in terms of depositional process by reference to one of the facies models for resedimented, normal current deposited and pelagic sediments outlined in Figs 4 and 5. These facies models show schematically the idealized sequences and sedimentary characteristics of sediments deposited by single events or particular processes. Actual examples of some of these facies from both the recent and ancient record are shown in Figs 6 and 7. Facies models are not shown for the isolated displaced clasts, sediment creep deposits, gravel and coarse sand lags, dune-bedded sands and black shale facies, as more work is required on each of these groups before an adequate synthesis can be attempted.

Slumps (facies group F2) can involve any lithology and be very thick (tens to hundreds of metres) or very thin (a few centimetres). Laminae and beds are rolled, contorted or rotated, but sedimentary structures will often

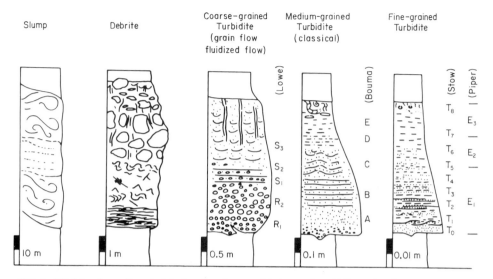

FIG. 4. Models of resedimented facies, for slumps, debrites and turbidites, showing the idealized structural sequences for debrites and for coarse-, medium- and fine-grained turbidites. The scale bars give only an indication of typical unit thicknesses, which may vary widely in practice. Grain size increases to the right for each column.

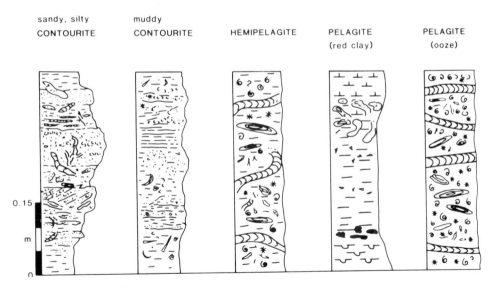

FIG. 5. Normal sedimentation facies models for contourites, hemipelagites and pelagites. Grain size increases to the right for each column.

show the way-up of the beds and the facing direction will determine the sense of the slope. No standard vertical sequence has been identified (Moore 1961; Morgenstern 1967; Lewis 1971; Saxov & Nieuwenhuis 1982).

Debrites (facies group F3) also involve mixed lithologies, with hard pebbles and boulders or soft mudstone clasts set in a muddy matrix, and

vary in thickness up to several tens of m. They may be disorganized or minimally organized with a basal zone of lensoid (?shear) lamination, a middle zone of high-angle faults and slump folds capped by convolute lamination, and an upper clast-rich zone that can show slight positive grading, water-escape pipe and dish structures and some horizontal alignment of

74   *D. A. V. Stow*

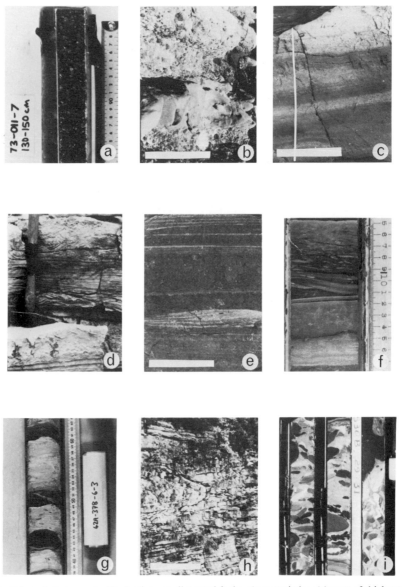

FIG. 6. Selected photographs illustrating resedimented facies characteristics: (a) part of thick coarse-grained, graded, gravel turbidite (?Holocene) from Laurentian Fan channel, western North Atlantic (Facies A3.3); (b) interbedded disorganised conglomerates and pebbly sandstones from the Upper Cretaceous of Pigeon Point, California (Facies A1/A2.1 and 2) (scale bar = 1 m); (c) parallel-laminated, thick-bedded, coarse-grained sandstone turbidite, with deep mud-scour at top surface, from the Ordovician Halifax Formation, Nova Scotia (Facies B2.1) (scale bar = 1 m); (d) typical medium-bedded, Bouma BCDE turbidite from the Tertiary of the central Apennines, Italy (Facies C2.2); (e) thin-bedded, fine-grained silt-mud turbidites from the Ordovician, Halifax Formation, Nova Scotia (Facies D2.2 and 3) (scale bar = 3 cm); (f) thin-bedded, fine-grained silt-mud turbidites from DSDP Site 530A, late Cretaceous, SE Angola Basin (Facies D2.2 and 3); (g) thin-bedded, graded mud turbidites (dark) within a pelagic ooze sequence from DSDP Site 378, Pliocene, Aegean Sea (Facies E2.3); (h) slumped beds from the Cretaceous–Tertiary Scaglia Rossa Formation, Umbro–Marchean Apennines, Italy (Facies F2.2) (Scale bar = 1 m); (i) detail of thick-bedded marl-ooze debrites from DSDP Site 530B, Plio–Quaternary, SE Angola Basin (Facies F3.1) (core width = 7 cm). Individual graded units (turbidites) shown by arrows where possible.

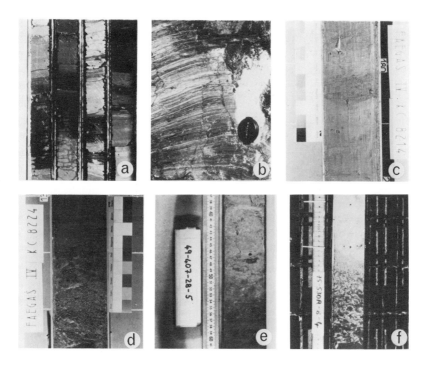

FIG. 7. Selected photographs illustrating normal sediment facies characteristics: (a) interbedded black organic-rich fissile mudstones (Facies E2.4), and green calcareous hemipelagic mudstones (marls) (Facies G2.1) from DSDP Site 530A mid-Cretaceous, SE Angola Basin (core width = 7 cm); (b) finely laminated diatomaceous pelagic ooze from the Miocene Monterey Formation California (Facies G1.2); (c) homogeneous muddy contourite (Facies E1.2) from Faro Drift, Gulf of Cadiz. (d) fine sandy contourite (lower) (Facies C1.2) overlain by mottled silt and mud contourites (Facies D1.3) and muddy contourite (upper) (Facies E1.2) from Faro Drift, Gulf of Cadiz (core width = 10 cm); (e) homogeneous muddy contourite (Facies E1.2) and irregularly laminated silty contourites (Facies D1.3) from DSDP Site 407, NE Atlantic; (f) interbedded, bioturbated, pelagic calcareous ooze (light) (Facies G1.1) and siliceous ooze (dark) Facies G1.2) DSDP Site 530A, Pliocene, SE Angola Basin.

clasts. (Abbate *et al.* 1970; Middleton & Hampton 1976; Naylor 1980; Thornton 1984).

The *coarse-grained* turbidite model is a composite including many of the facies in facies classes A and B which are each represented by one of the divisions of the idealized sequence ($R_{1, 2}, S_{1, 2, 3}$, Lowe 1982). Deposition can be by grain flow, fluidized- or liquified-flow mechanisms often following long-distance transport by turbidity currents. The lower part of the sequence can comprise either gravel or sand or any gradation of pebbly sand between the two. Thus there is a sharp, scoured base, a negatively graded lower division, an intermediate massive, stratified, graded-stratified division and an upper division with dish and pipe structures. The top is commonly sharp and flat (Walker 1975, 1978; Carter 1975; Middleton & Hampton 1976; Lowe 1979, 1982; Hein 1982).

The *medium-grained turbidite* model is the classical Bouma (1962) sequence and represents most of the facies class C and parts of B and D (there being some overlap between the three turbidite models). The five divisions are well-known: massive to graded sand (A), parallel-laminated sand (B), cross-laminated to convolute sand (C), parallel-laminated fine sand and silt (D) and massive to bioturbated mud (E) (Bouma 1962; Kuenen 1964; Walker & Mutti 1973; Hesse 1975).

The *fine-grained turbidite* model, representing much of facies classes D and E, was developed to facilitate description and interpretation of the resedimented muds and silts not adequately covered by Bouma's E division. Thus, Piper (1978) recognized a graded silt-laminated mud division (E_1), a graded mud (E_2) and an ungraded mud (E_3). Stow & Shanmugam (1980) recognized an idealized vertical sequence of silt laminae structures passing up from the graded laminated unit including fading ripples (T_0), to mud with convolute silt laminae (T_1), low-amplitude ripples (T_2), parallel distinct (T_3), parallel indistinct (T_4) and wispy laminae (T_5). These are overlain, as in the Piper sequence, by graded mud (T_6), ungraded mud (T_7) and a thin micro-bioturbated zone (T_8) (see also, Moore 1969; Rupke & Stanley 1974; Mutti 1977; Nelson *et al*. 1978; Kelts & Arthur 1981; Stow *et al*. 1984).

Each of these idealized turbidite sequences can be interpreted hydrodynamically as resulting from a single resedimentation event that deposited progressively finer grades of sediment and gave rise to different sedimentary structures as the flow velocity and carrying power decreased. A complete sequence is very rarely deposited and partial sequences are the rule (top-absent, base-absent, mid-absent, etc.). These partial sequences give rise to the many possible facies shown in Fig. 3; for example, deposition of top-absent classical turbidites (Bouma divisions A, AB, ABC) produces massive sands (facies B1.1), parallel-laminated sands (facies B2.1) or thick-bedded turbidites (facies C2.1), whereas, base-absent fine-grained turbidites (Piper divisions E_2E_3, Stow divisions T_{678}) give massive and graded mud turbidites (facies E1.1, E2.1, E2.2 and E2.3); and so on.

Different facies can thus be related to different parts of the idealized sequences and hence to a particular type of flow and to a stage in the evolution of that flow. This information then leads to an interpretation of the bathymetric, environmental or other factors controlling the location and occurrence of different flow types.

Two *contourite* models are shown in Fig. 5, one for sandy contourites and one for muddy contourites (Stow & Lovell 1979; Stow 1982). These represent the sediments *deposited* from bottom currents in the large elongate contourite drifts of the ocean basins, and also interbedded with other facies on continental rises. *Muddy contourites* appear homogeneous and thoroughly bioturbated, although rare parallel or wavy lamination, pockets of coarser material, irregular sharp contacts between silts and mud, and certain textural characteristics are evidence that deposition has been current-influenced. *Sandy contourites* (also coarse silt grade) occur in thin to medium irregular layers or beds with sharp or diffuse contacts, rare internal lamination, and very common bioturbation and burrows. Compositionally, both types generally contain planktonic and benthonic biogenic material mixed with terrigenous material, often broken or iron-stained, although other composi-

tional types are also known. A complete gradation exists between the two facies, and they commonly occur together in an irregular vertical 'sequence' that shows negative grading from a muddy through mottled silt and mud to a sandy facies and then a positive grading back to a muddy facies (Faugères *et al*. 1984). Parts of this 'sequence' occur as the individual facies E1.2, D1.3 and C1.2 (Fig. 3).

Reworking and winnowing by bottom currents results in *coarse sand* and *gravel-lag contourite* facies, shown as facies B1.2 and A1.3 in Fig. 3.

Hemipelagites are compositionally very similar to muddy contourites and also appear homogeneous, massive and thoroughly bioturbated. However, they do not show any evidence of current-control during deposition, probably have a somewhat different ichnofacies and show no vertical 'sequence' of facies or textures (Hesse 1975; Cook & Enos 1977; Hill 1981). These are the 'arls' of facies group G2.

Two contrasting *pelagite* models are shown in Fig. 5, one for oozes (facies group G1) comprising more than 70% biogenic material, and one for red clays (facies E1.3) that commonly have less than 10% biogenics. Neither of these are considered in detail here (but see Jenkyns 1978, Hoffert 1980, Thiede *et al*. 1981 and Leggett this volume).

Environments

Distinctive vertical sequences and horizontal distributions of facies and the occurrence of particular depositional processes are characteristic of specific environments in the deep sea. Amidst the great variability of the marine realm we can identify three fundamentally different environments, each with their own mixture of sedimentary, tectonic and morphological features. These are slope aprons, submarine fans and basin plains. Slope aprons probably account for the greatest volume of sediments, basin plains are areally most extensive, whereas submarine fans, having early attracted the geologists' attention, have been the most thoroughly studied to date. Nevertheless, it is now possible to begin to construct generalized environmental models for each of these different regions (e.g. Pickering 1983).

(1) Slope aprons

Slope aprons make up the region between the shelf and the basin floor, surrounding both small shelf basins and the large ocean basins where they are taken here to include the con-

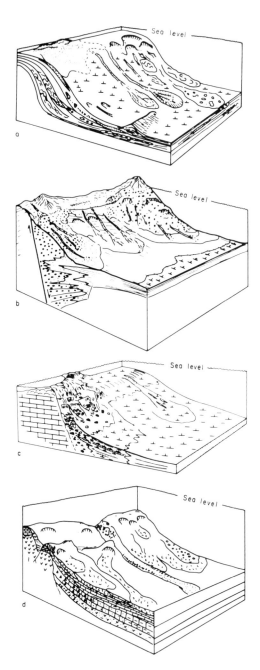

FIG. 8. Sedimentary environment models for submarine slope aprons showing schematic distribution of facies and morphological elements. No fixed scale applies: slope widths vary between 1 and 500 km; slope gradients 1–7°. (a) normal (clastic), (b) faulted, (c) carbonate, and (d) ridge-flank.

tinental rise. Broad areas of slope also occur on the flanks of mid-ocean ridges and surrounding isolated seamounts or plateaus.

They vary in width from less than 1 km to over 200 km and commonly have gentle gradients from 2 to 7°, rarely exceeding 10°. They may be erosional or depositional, smooth or rugged, and comprise a complete range of clastic and biogenic facies. The main morphological elements include a relatively abrupt shelf-break, slump and slide scars, irregular slump and debris flow masses, small straight or slightly sinuous channels and gulleys, more complex dendritic canyons, isolated lobes, mounds and drifts, and broad areas of smooth or current-moulded surface.

Although at least ten different types of slope apron can be distinguished on the basis of the morpho-tectonic setting (Emery 1977; Bouma *et al.* 1978; McIlreath & James 1978; Doyle & Pilkey 1979), there are only four major types that are significantly different in terms of their sedimentological characteristics. These are: normal (clastic), faulted, carbonate and ridge-flank slope aprons (Fig. 8).

Normal (clastic) slope aprons (Fig. 8a) range from mainly constructional, with a relatively smooth convex-concave profile built upwards and outwards by slope progradation, to mainly destructional, with erosion (slumping, sliding, etc.) on the face of the slope causing a steeper and more irregular profile to be developed. In the former, lower-energy case the slope surface tends to be smooth or current-moulded, whereas in the latter, higher-energy case the slope is often gullied and slump-scarred, with sediment lobes, debris-flow masses and slump blocks at the foot of the slope. Large canyons and channels may cut across the slope at intervals, and elongate contourite drifts be constructed near the base-of-slope.

The distribution of facies is highly irregular and dominated by finer-grained sediments (silts, muds, oozes and arls). Coarser-grained sediments (sands and gravels) occur above the mud-line, as sand-spillover sheets, in the axes of channels and in base-of-slope lobes. Slides, slumps and debrites are common throughout.

Many of the continental slopes and rises throughout the world fit into this category, as well as the slope-aprons around marginal seas and shelf basins. A particularly well-studied modern example is the Nova Scotian slope-rise system off eastern Canada (Stanley *et al.* 1972; Piper 1975; King & Young 1977; Stow 1978; Hill 1981). Ancient equivalents include the Lower Palaeozoic Meguma Group onshore Nova Scotia (Schenk *et al.* 1980; Stow *et al.* 1984), and the Carboniferous–Permian Sweetwater slope in the subsurface of Texas where oil and gas are found in small channel and lobe sandstone reservoirs.

Faulted slope aprons (Fig. 8b) are those that form across an active synsedimentary fault margin. They commonly have relatively steep portions alternating with flatter perched basins forming a stepped profile. There is an abrupt change of profile at the base-of-slope to a flat basin floor and little development of a lower slope or rise. A thick wedge of sediment accumulates in a narrow trough at the foot of the slope as a result of continued down-faulting.

Steeper portions of the slope may be bare of sediment or have a relatively thin veneer, with frequent slump-scars, slump masses and short-lived shallow channels. A roughly slope-parallel arrangement of coarse to fine grained facies may be developed, often built up from numerous small fans or lobes that overlap along the length of the slope. However, much lateral facies variability occurs as a result of the non-uniform, periodic nature of fault activity.

Faulted slope-aprons develop mainly along active compressional and strike-slip margins, as well as young rifting margins such as parts of the Red Sea rift system. Modern examples include the Gulf of Guinea slope (Delteil *et al.* 1974) and the western margin of the Tyrrhenian Sea (Wezel *et al.* 1981). Several ancient examples are known from the later Jurassic early rifting phase around the North Atlantic margins; these include the Wollaston Foreland Group of eastern Greenland (Surlyk 1978), and the Brae oilfield and related subsurface systems of the northern North Sea (Stow *et al.* 1982).

Carbonate slope aprons (Fig. 8c) have either a steep off-reef profile, often with stepped portions of submarine cliffs, or a gentle convex-concave off-platform profile. These are the by-pass and depositional margins identified by McIlreath & James (1978). Sediments are thin or absent on the steeper parts which are fringed by a periplatform, calcirudite talus wedge that grades rapidly downslope into calcarenites, calcilutites and pelagic-hemipelagic limestones. Channels and canyons locally funnel coarser sediments into deeper water. The gentler slopes have a more irregular facies distribution of finer-grained calciturbidites, bottom current and pelagic deposits. Slumps, debris flow masses, channels and lobes further complicate the arrangement of facies.

Such slope aprons are found surrounding coral reefs and carbonate platforms throughout the lower latitude regions of the world. The best studied modern examples in terms of sediments are those around the Bahamas (Mullins & Neumann 1979), and the Belize barrier and atoll reefs (James & Ginsburg 1979). Many ancient carbonate slope-apron systems have been described, two well-known examples being the Cambro-Ordovician Hales Limestone Formation of central Nevada (Cook & Taylor 1977), and the Cow Head Breccia of the same age in western Newfoundland with spectacular giant limestone clasts in slope-apron megabreccias and debrites (Hubert *et al.* 1977).

Ridge-flank slope aprons are the very low-gradient, highly dissected slopes of oceanic ridges, linear island chains, oceanic plateaus and isolated seamounts. They have a distinctive irregular to concave profile with ocean-crust basement highs, perched basins and transverse fracture zone valleys. Sediment cover is usually thin and irregular, comprising both biogenic (calcareous or siliceous) and volcanigenic material in slump, debrite, turbidite and pelagite facies.

Modern examples are summarized by Kelts & Arthur (1981) from many DSDP sites. More detailed studies are given for the Walvis Ridge slope-apron in the South Atlantic (Stow 1984), and the Mid-Atlantic Ridge flank in the vicinity of Gibbs Fracture Zone (Faugères *et al.* 1982). Ancient examples are best known from the sediments associated with ophiolites, although these are not often well preserved (Nisbet & Price 1974).

(2) Submarine fans

Submarine fans are distinctive constructional features at the foot of slopes, both in small shelf or marginal basins and in the large ocean basins. Unlike slope aprons, which are continuous parallel to the margin, fans are isolated bodies that develop seaward of a major sediment source (river, delta, glacier, etc.) or main supply route (canyon, gulley, trough, etc.).

They are also very variable in size, from a radius of little more than 1 km to a length of more than 2000 km, and have gradients similar to those of slopes, decreasing from the upper to lower fan region. They comprise one or more feeder channels or canyons, tributary and distributary channels, abandoned half-filled channels, slump and slide scars and blocks, debris flow masses, broad channel levees, lobes built up at the ends of channels and distributaries, and relatively smooth or current-moulded interchannel and inter-lobe areas.

A number of different fan models have been developed over the past 15 years (Normark 1970, 1978; Nelson & Nilsen 1974; Mutti & Ricci

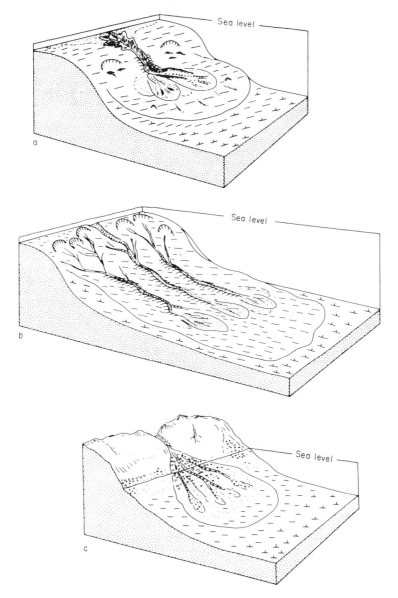

Fɪɢ. 9. Sedimentary environment models for submarine fans showing schematic distribution of facies and morphological elements. No fixed scale applies: fan radius normally <150 km for A, <1500 km for B and <15 km for C; steepest parts of upper fans rarely exceed 10°, lower fans <0.5°. (A) Radial, (B) elongate, and (C) fan-delta.

Lucchi 1972, 1975; Walker 1978; Stow 1981; Howell & Normark 1982) from studies of both modern and ancient systems. However, there appear to be two principal end-member types developed in deeper water (Fig. 9a, b), with all possible gradations between the two (Stow *et al*. 1984), and a third shallow-water type (Fig. 9c). All these types may be substantially modified by morphological confinement and synsedimentary

tectonics (Scott & Tillman 1981).

Radial fans (also called 'sandy', 'low-efficiency', 'canyon-fed', 'small', 'restricted-basin') have a true fan-like shape arranged concentrically about a small apex, a single feeder channel or canyon, a concave-convex-concave longitudinal profile, and a hummocky mid-fan region characterized by distributary channels and lobes. There is both an elongate and concentric distribution of coarse-to fine-grained facies. Slumps, slides and debrites occur in the lower slope, upper fan and channel margin areas. Coarse-grained turbidites are mainly confined to channels and lobes, whereas fine-grained turbidites occur throughout together with hemipelagic and pelagic facies.

Examples of modern radial fans include many of the smaller ones from the west coast of North America such as La Jolla, Navy, Redondo, Coronado, San Lucas and Nitinat (Normark 1970, 1978). The Carboniferous Pesaguero Fan of northern Spain (Rupke 1977) and some of the Upper Miocene Stevens Sandstone fans that provide hydrocarbon reservoirs in subsurface California, (Scott & Tillman 1981) may be ancient equivalents.

Elongate fans (also called 'muddy', 'high-efficiency', 'delta-fed', 'large', 'open-basin') are longitudinally extended perpendicular to the margin, commonly with a broad head region, two or more main feeder channels, a complex tributary-distributary system, an irregular-concave-smooth profile, and large terminal lobes constructed at the ends of channels on to the lower fan region. The sediment distribution is more elongate than concentric, and there is usually a high proportion of mud to sand. The various resedimented and pelagic facies are otherwise similar to those of radial fans.

Modern examples include the giant (3000 km long) Bengal fan (Curray & Moore 1971) and the Indus fan (Jipa & Kidd 1974) amongst others. In reality, there is a complete gradation between the different fan types. Stow *et al.* (1984) plot a number of modern fans on a triangular diagram with the apices as radial fans, elongate fans and normal slope-apron systems. Ancient examples probably include the Eocene Butano Sandstone system of California (Nelson & Nilsen 1974), and the Precambrian Kongsfjord Formation of northern Norway (Pickering 1982).

Fan deltas (also called 'short-headed delta-front fans') develop as the subaqueous part of alluvial fans that prograde from highlands directly into a standing body of water (lake or sea). They are mostly relatively small, pear-shaped in outline and with an ephemeral system of shallow braided channels radiating down-slope from the fan head. Sediments are mainly coarse-grained in the upper regions and in the channels, with muddier sediments on the levees, interchannel and more distal areas.

Modern fan deltas have been reviewed by Westcott & Ethridge (1980), well-studied examples being the Yallahs fan-delta off south-east Jamaica, and a number of fan-deltas along the south-east coast of Alaska (Boothroyd &

Nummedal, 1978). Ancient fan-deltas are also well-known from rocks of all ages around the world. Westcott & Ethridge (1983) review these briefly, and document in more detail an example from the Wagwater Trough of east central Jamaica.

(3) Basin plains

Basin plains are the flattest and deepest of our three deep-sea environments, and may vary widely in their areal extent up to the size of the major ocean basins, and in their depth to the deep floors of submarine trenches. They generally have a very gentle relief and merge gradually or more abruptly with the surrounding slopes and isolated seamounts or other basement highs.

Their main morphological elements include the extreme distal portions of submarine fans, channels and lobes, isolated intra-basinal channels, ridges and drifts, structurally-controlled grabens, morphologically restricted passages, and very large areas of smooth or current-modified seafloor.

Several different basin classifications have been proposed using the criteria of composition, depth, restrictedness, fill geometry or morpho-tectonic setting (Dott & Shaver 1974; Hesse & Butt 1976; Gorsline 1978; Ballance & Reading 1980; Pilkey *et al.* 1980). From the point of view of their sedimentary characteristics, the most important controls are the basin size, the sediment supply and source area, and tectonic activity. On this basis, we recognize two end member basin plain types, those that are undersupplied and those that are oversupplied, and a complete gradation between the two.

Undersupplied basin plains (Fig. 10a) are commonly large, open basins far from land in a low-relief relatively stable tectonic setting. They have a low sediment supply to area ratio, and fill very slowly with simple sediment drape or current drift forms. The facies are dominantly fine-grained, terrigenous and biogenic, and overall thicknesses rarely exceed about 1 km.

Most of the major abyssal plains of the ocean basins are of this kind, including the Sohm and Hatteras abyssal plains in the North Atlantic (Horn *et al.* 1971, 1972; Pilkey *et al.* 1980). Although these Atlantic plains receive an enormous volume of terrigenous material derived from very large drainage basins, their very great size maintains their undersupplied character.

Oversupplied basin plains (Fig. 10b), by contrast, are mostly rather small, often partly confined and located in tectonically active areas.

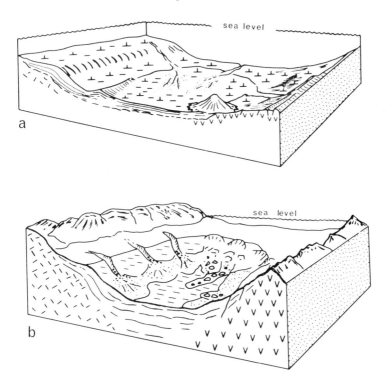

FIG. 10. Sedimentary environment models for submarine basin plains showing schematic distribution of facies and morphological elements. No fixed scale applies: basin plain areas vary from <200 km² to >100,000 km², gradients flat to very low. (A) Undersupplied, (B) oversupplied.

One or more of the basin margins may be a zone of active synsedimentary faulting, and sediment supply is large compared with the basin floor area. The basins fill up rapidly with progradational, mounded and onlap fill geometries and an overlapping, somewhat chaotic distribution of coarse and fine-grained facies is developed.

Modern basin plains of this type occur in regions of active compressive or strike-slip tectonics, and are particularly well known from the Californian borderland (Gorsline *et al.* 1984) and the New Zealand continental margin (Spörli 1980). The classical ancient examples are the Oligo–Miocene periadriatic foredeep basin plains of Italy (Ricci Lucchi 1975, 1978).

(4) Morphological elements and vertical sequences

The first steps in interpreting the palaeoenvironment from an ancient rock series are to identify the facies correctly and then to assess, as far as exposure and structural complexity will allow, the vertical arrangement and horizontal distribution of those facies. We are just begin-

ning to recognize particular vertical sequences and horizontal associations that we believe are characteristic of specific morphological elements in the deep sea (e.g. Rupke 1978; Walker 1978).

Some of these sequences are shown schematically in Fig. 11. Several types of canyon or channel-fill are recognized: a blocky, massive coarse-grained fill of canyon or proximal channel; more regular fining-upward sequences of mid-slope or mid-fan channels; packets of sands deposited in distributary-type channels; and a blocky to fining-upward mud-dominated channel-fill. Regular coarsening-upward sequences appear typical of more proximal sandy (suprafan) lobes and probably also of proximal muddy lobes, whereas distal (terminal) sandy and muddy lobes are commonly more symmetrical. Other mounds on the seafloor include contourite drifts with an irregular variation of more or less sandy and silty hemipelagic-like muds, and slump or debris flow masses with a chaotic assemblage of slumps and debrites. Irregular sequences also characterize levees, interchannel, smooth slope and several basinal environments. The main differences

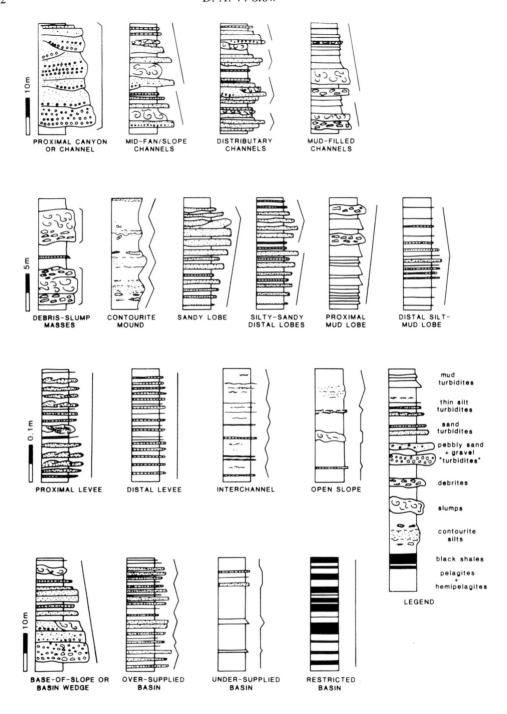

FIG. 11. Typical vertical sequences of turbidites and associated sediments from various morphological elements in the different deep-sea environments. Fining-upward, coarsening-upward, blocky, symmetrical and irregular sequence types are indicated by the lines to the right of lithological columns.

between these settings are the dominant facies types: sandy, silty or muddy turbidites, hemipelagites and pelagites, or black shales. Tectonically controlled fining-upward sequences are common on faulted slope-aprons or in over-supplied basins.

These are generalized sequences associated with specific elements in the deep sea. They do show variations of scale, sedimentary materials and regularity and it is not always easy to make a definitive interpretation. It is still more difficult to reconstruct the larger palaeoenvironment as this requires the vertical or lateral association of several of the required elements. It must also be remembered that isolated channel, lobe or other sequences may develop in any of the major environments.

Controls and rates

A whole new dimension to our understanding of the deep sea is added when we consider the factors that control the development of different processes, facies or environments (Stow *et al.* 1982, 1984; Howell & Normark 1982), the rates at which these controls operate (Howell & Von Huene 1980; Blatt *et al.* 1980; Stow *et al.* 1984), and also the material budgets involved (Gorsline 1978, 1984). Three primary controls, as well as a number of secondary controls, on deep-sea sedimentation can be identified: (1) sediment

type and supply, (2) tectonic setting and activity, and (3) sea level. These controls are not, of course, entirely independent; for example, tectonic factors influence sediment supply or local sea-level changes, and so on (Fig. 12).

(1) Sedimentary controls

One of the key sedimentary variables is, clearly, the type of sediment available for deposition or redeposition. The most important types are clastics, including gravel, sand and mud grade terrigenous materials, and biogenics, including skeletal or reefal platform debris, calcareous and siliceous oozes. Locally, evaporites, volcaniclastics, organic-carbon-rich and other sediments are also significant. The volume and rate at which sediments are made available for deposition is another important variable. For example, major rivers can provide a large and rapid supply, glaciers a somewhat lesser supply, and pelagic settling a relatively low supply of sediment. Finally, the number and position of input points exerts a considerable influence over the morphology developed.

The secondary factors that influence sediment type and supply are also illustrated schematically in Fig. 12. Rates of accumulation and denudation, together with the periodicity or frequency of significant sedimentary aspects are shown in Fig. 13.

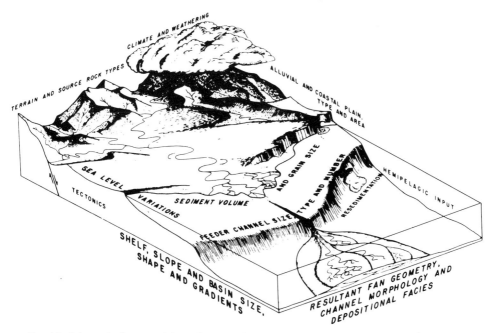

FIG. 12. Schematic diagram of the major controls on sedimentation in the deep sea (from Stow *et al.*, 1982, 1984).

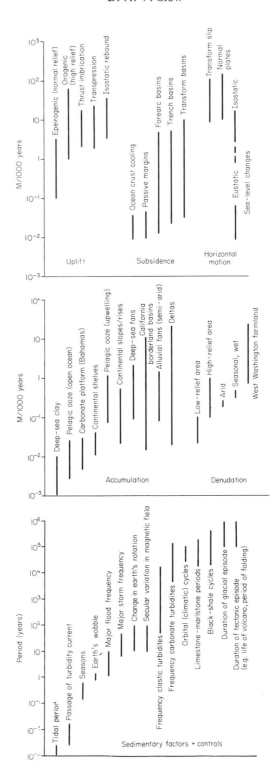

Fig. 13. Rates (in m 1000 yr^{-1}) of tectonic and sedimentary processes and periods (in years) of sedimentary factors that influence sedimentation in the deep sea.

(2) Tectonics

The broad tectonic settings in which the various deep-water systems can develop include: mature passive margins (American or African types); immature or rifting passive margins; convergent margins with arc or trench systems; transform margins; marginal seas and back-arc basins; intraoceanic basins on the flanks of ridges and seamounts; and intracratonic basins on continental shelves and within continents.

These tectonic settings exert a first-order control on the types of system developed by affecting the regional stress regime, rates of uplift and denudation, drainage patterns, coastal plain and shelf widths, slope gradients, gross sediment budgets, the morphology of receiving basins, and local sea-level changes. The style and frequency of seismic activity and faulting both in the original and transitional source areas are also of primary significance. This degree of tectonic activity varies temporally and spatially within and between the main tectonic settings, but is most pronounced in convergent, transform and young passive margins.

Secondary factors involved in tectonic activity include the rates of horizontal and vertical motion, the maturity of the margin, and the relationship of a particular setting to neighbouring plates. The rates of motion can be critical. If deposition rates are slower than tectonic rates, growth will be controlled by tectonism rather than by sedimentary factors such as fluctuating gradients and migrating channels, distributaries, and terminal lobes. Some of these rates are indicated in Fig. 13.

(3) Sea level

Fluctuation in sea level not only affects the nearshore realm of sedimentation, but also profoundly influences deep-sea depositional and resedimentation patterns. Shoreline sources such as rivers or littoral drift cells either may have direct access to basin slopes during periods of low sea level or indirect access through paralic and continental shelf environments during periods of high sea level.

Sea level changes may be global (eustatic) or regional in nature. Eustatic fluctuations occur as a result of a change in the total volume of ocean basins or a change in the volume of seawater. The volume of ocean basins is affected by four secondary factors (Pitman 1979): sediment input, continental collision and subduction, growth of seamount chains, swelling and shrinking of mid-ocean ridge sytems.

Mostly, the rates of sea-level change effected (Fig. 13) appear too small in comparison with tectonic rates of uplift and subsidence to have any appreciable effect on active margins. However, prolonged eustatic changes, due to changes in ocean-ridge volume for example, acting in consort with passive margin subsidence will have a significant effect on sea level along the more stable coastlines. And also, the locking-up of very large amounts of water in expanded polar ice sheets during glacial periods and its release during warm climatic epochs, can cause enormous sea-level fluctuations of up to 10 m 1000 yr^{-1}.

Regional fluctuations in sea level as a result of local tectonic and isostatic factors can be very much greater than most eustatic changes (Fig. 13), as evidenced by the rates of tectonic processes discussed in the preceding section. They are not always readily distinguished from and may completely mask the eustatic change. However, various attempts have been made to chronicle the eustatic as distinct from local sea level fluctuations through time, and the currently popular world wide sea-level curve is that of Vail & Mitchum (1979). Clearly, during periods of very high stands the deep-sea sedimentation regime is likely to contrast markedly with that during very low stands.

Economic aspects

The demonstrable importance of deep-sea sediments as a source of great economic potential has certainly provided a new impetus to their study, as well as a new focus for much research. The materials they contain that have so far attracted the most attention are: oil and gas, ferromanganese nodules, metalliferous sediments and ores, and the uranium series elements of black shales. Most of these are in the pelagic and associated facies discussed by Leggett (this volume); I will only briefly consider the potential hydrocarbon resources in this paper.

There have been numerous important hydrocarbon discoveries in deep-sea sediments of Devonian to Tertiary age (Table 1) in the U.S.A. (Walker 1978) in the North Sea (Woodland 1975; Illing & Hobson 1981) and off the Brazilian coast (Tofelli & Barros 1979). There are also vast new areas of petroleum potential beneath present-day continental margins, where canyon, fan or other reservoirs may become economically viable in the near future (Yarborough *et al.* 1977; Mattick *et al.* 1978; Wilde *et al.* 1978).

These reservoirs are mostly found in the

TABLE 1. *Hydrocarbon reservoirs in submarine fans and associated sequences*

Author		Area	Other details	Reference
Barbat	1958	Los Angeles Basin, California		In: L. G. Weeks, (ed.) Habitat of Oil
Sullwold	1960	Los Angeles Basin, California		AAPGB, 44, 433–57.
Yerkes et al.	1965	Los Angeles Basin, California	Tarzana Fan, late Miocene	USGS Prof. Pap., 420A,
Mayuga	1970	Los Angeles Basin, California	Wilmington Oilfield	AAPG Mem. 14, 158–84
Gardet	1971	Los Angeles Basin, California		AAPG Mem. 15, 298–308
Weser	1978	Los Angeles and Ventura Basins		AAPG Stud. Geol. 7, 227–42
Nagle & Parker	1971	Ventura Basin, California		AAPG Mem. 15, 254–97
Hsü	1977	Ventura Basin, California	Ventura Field, lower Pliocene turbidites	AAPGB 61, 137–68
Sullwold	1961	Great Valley, California	Stevens Sandstone, Bakersfield arch, Miocene	In: Peterson & Osmond (eds) Geometry of Sandstone Bodies
Martin	1963	Great Valley, California	Rosedale Channel, late Miocene	AAPGB, 47, 441–56
Seiden	1975	Great Valley, California	Asphalto, San Joaquin Valley, Miocene	Pac. Sect. Am. Ass. Petrol. Geol. 40th AGM
Sanem & Soddard	1965	Great Valley, California	Stevens Sand, San Joaquin Valley, Miocene	AAPGB, 49, 1089
Callaway	1968	Great Valley, California	W. San Joaquin Valley, Miocene	AAPG–SEPM Pac. Sect. Guidebk, 43rd AGM
Bazeley	1972	Great Valley, California	San Emido Nose Field, Miocene	AAPG Mem. 16, 297–316
Maher et al.	1975	Great Valley, California	Naval Petroleum Reserve, Miocene	USGS Prof. Pap. 912, 109 pp.
MacPherson	1978	Great Valley, California	San Joaquin Valley, Bakersfield Arch, Miocene fans	AAPGB, 62, 2243–74
Edmonson	1965	Sacramento Valley, California		San Joaquin geol. Soc. select. Pap. 3, 36–61
Dickas & Payne	1967	Sacramento Valley, California	Upper Palaeocene Channel	AAPGB 51, 873–82
Weagant	1972	Sacramento Valley, California	Grimes Gas Field	AAGP Mem. 16, 428–39
Paine	1966	Louisiana and Gulf of Mexico		Trans. Gulf-Cst Ass. geol. Socs 16, 261–74
Sabate	1968	Louisiana and Gulf and Mexico	Pleistocene oil and gas fields	Trans. Gulf-Cst Ass. geol. Socs 18, 373–86
Benson	1971	Louisiana and Gulf and Mexico	Oligocene	Trans. Gulf-Cst Ass. geol. Socs 21, 1–14
Hoyt	1959	Texas		Trans. Gulf-Cst Ass. geol. Socs 9, 41–50

Author	Year	Location	Field/Formation	Reference
Galloway & Brown	1973	Texas	Carboniferous	*AAPGB* **57**, 1185–218
Bloomer	1977	Texas		*AAPGB* **61**, 344–59
Dixon	1972	Pennsylvania	Bradford Oilfield	*Mem. Am. Ass. Petrol. Geol. Eastern Section Field Trip Gdbk*
Vedros & Visher	1978	Arkona Basin, Oklahoma	Red Oak Sandstone, Carboniferous	*In:* Stanley & Kelling (eds) *Sedimentation in Canyons, Fans and Trench*
Jacka *et al.*	1968	Delaware	Bell Canyon Formation	*Pac. Sect. Gdbk Am. Ass. Petrol. Geol. Soc. econ. Paleont. Mineral., Tulsa, Permian Basin Section Guide I*
Thomas *et al.*	1974	North Sea	Forties Field, Palaeocene	*AAPGB* **58**, 396–406
Parker	1975	North Sea	Tertiary sands	*In:* A. W. Woodland (ed.) 447–52
Fowler	1975	North Sea	Montrose Field, Palaeocene	*Petroleum and the Continental Shelf of N. W. Europe*, 467–76
Walmsley	1975	North Sea	Forties Field, Palaeocene	*Shelf of N. W. Europe*, 467–52
De'ath & Schuyleman	1981	North Sea	Magnus Field, Jurassic	*In:* Illing & Hobson (eds) *Petroleum geology of the Continental Shelf of N. W. Europe*
Carman & Young	1981	North Sea	Forties Field, Palaeocene	*In:* Illing & Hobson (eds) *Petroleum geology of the Continental Shelf of N. W. Europe*
Heritier *et al.*	1981	North Sea	Frigg Field, Eocene	*In:* Illing & Hobson (eds) *Petroleum geology of the Continental Shelf of N. W. Europe*
Kessler *et al.*	1980	North Sea	Cod Field, Palaeocene	NPF, Geilo Volume
Skjold	1980	North Sea	Balder Field, Palaeocene	NPF, Geilo Volume
Stow *et al.*	1980	North Sea	Brae Field, Jurassic	*Mem. Am. Ass. Petrol. Geol. mtg*, Houston, 1980
Stow *et al.*	1982	North Sea	Brae Field, Jurassic	*J. Petrol. Geol.* **5**, 129–68; **6**, 103–4
Tofelli & Barros	1979	Brazil continental maryin	Various fields	*1st Conyr. Brasil. Petrol.* **1:11**, 51–61
Ojeda	1982	Brazil continental maryin	Various fields	*AAPGB* **66**, 732–49
Casnedi	1983	Italy, Periadriatic basins	Pliocene Cellino Formation	*AAPGB* **67**, 359–70

coarser-grained facies (sands and gravels) of canyons, channels, lobes, slope wedges and over-supplied basins. Such reservoir facies, however, are commonly associated with muddy deposits that may be sufficiently enriched in marine or terrigenous-derived organic matter to provide an adequate source of oil or gas after burial and thermal maturation. Furthermore, the particular organization of potential reservoir and source facies in certain deep-sea settings (fans for example) provides the possibility of suitable stratigraphic or structural-stratigraphic traps forming with time.

Future directions

If the next three decades are as productive as the past three, then we will cover a very large amount of ground in deep-sea studies. But just what ground will that be, where will it lead and how will we carry out the research? These questions are addressed briefly in this concluding section.

(1) Areas of study

Although we may have now reached at least some concensus in the five areas discussed in this review paper, there remains much room for improvement to, and refinement of, the working models outlined here. In particular, the following developments may be anticipated: development of rigorous physical models of processes and of the critical conditions for erosion, transport and deposition; emphasis on the effects of normal currents in the deep sea, particularly on the growth of contourite drifts and fluvial-like morphologies; work on fine-grained facies, dune-bedded sands and black shales; elaboration of the slope and basin environmental models, work on the shelf-slope interface, and further emphasis on characterizing individual morphological elements; further exploration of sedimentary controls, rates and budgets and a more critical understanding of their effects; and determination of the relationships between source-rock organofacies and lithofacies, the effects of diagenesis on reservoir rocks, and the geometry of reservoir bodies.

Other quite different avenues of study will doubtless also be pursued, particularly in fields where we are almost but not quite at the synthesis and model-construction stage. These include: the petrography and geochemistry of both coarse- and fine-grained deep-sea facies, and the development of post-depositional diagenetic models; the influence of organisms on sediments in terms of primary productivity, organic matter content and particle coatings, erosion and suspension, and bioturbation; and the long-awaited marriage between studies of ancient and modern sediments.

(2) Methods of research

In many ways, even with the advent of deep-ocean drilling technology and the voyages of the *Glomar Challenger*, our studies of deep-sea sediments have remained relatively simple. I suspect that the next three decades will bring immense sophistication to both the methodology and technology we use. Again, I am guessing as well as being selective when suggesting that the advances listed below will prove the most constructive: technological developments in, for example, marine-based observation of smaller and smaller sea-bottom features approaching the scale of observations on land, and laboratory-based studies at the ultramicroscopic scale; the use of actualistic experiments on the seafloor (or in lakes, perhaps), as well as of actual observations on the water column and recently deposited sediments; specific programmes designed to test sedimentological models using closely spaced drill sites or sea to land drill transects, and to examine in detail particular areas of the seafloor using a multidisciplinary approach; the computer handling of increasingly large data-sets as well as the systemization of data produced in different countries and different laboratories; a communications breakthrough that overcomes the many barriers between laboratories, countries and languages, and that renders accessible large volumes of information that are otherwise comparatively lost.

The deep-sea sedimentological community is already grappling with these areas and methods of study. It will perhaps not be too long now before we can claim to have entered the 'Age of Order'.

ACKNOWLEDGMENTS: My thanks to the very many people who have contributed much through their helpful discussions to the synthesis of these ideas and, in particular, to Drs Kevin Pickering and David Piper. Special thanks are also due to Patricia Stewart and Marcia Wright for typing the manuscript and Diana Baty for drafting the figures. Peter Crimes, Ron Pickerill and Pat Brenchley reviewed an earlier version of the manuscript.

References

ABBATE, E., BORTOLOTTI, V. & PASSERINI, P. 1970. Olistostromes and olistoliths. *Sediment. Geol.* **4**, 521–57.

ANDREE, K. 1920. *Geologie des Meeresbodens.* v. 2 Bortraeger (Publ.), Leipzig, 689 pp.

ARRHENIUS, G. 1950. The Swedish Deep-Sea Expedition. The geological material and its treatment with special regard to the eastern Pacific. *Geol. Fǒr. Stockh. Fǒrh.* **72**, 185–91.

BALLANCE, P. F. & READING, H. G. (eds) 1980. *Sedimentation in Oblique-slip Mobile Zones. Spec. Publs int. Ass. Sediment.* **4**, 524 pp. Blackwell Scientific Publications, Oxford.

BLATT, H., MIDDLETON, G. & MURRAY, R. 1980. *Origin of Sedimentary Rocks.* 2nd ed. Prentice-Hall, New Jersey. 782 pp.

BÖGGILD, O. B. 1916. Meeresgrundproben der *Siboga* expedition. *Sigboda-Expedite Monographie*, **65**, 1–50.

BOOTHROYD, J. C. & NUMMEDAL, D. 1978. Proglacial braided outwash: a model for humid alluvial fan deposits. *Mem. Can. Soc. Petrol. Geol.* **5**, 641–68.

BOUMA, A. H. 1962. *Sedimentology of some Flysch Deposits.* Elsevier, Amsterdam. 168 pp.

——, MOORE, G. T. & COLEMAN, J. M. (eds) 1978. Framework, facies and oil-trapping characteristics of the upper continental margin. *Am. Ass. Petrol. Geol. Studies Geol. No.7.*

BRAMLETTE, M. N. & BRADLEY, W. H. 1940. Lithology and geologic interpretations. Pt 1, *In:* *Geology and Biology of North Atlantic Deep-Sea Cores. Prof. Pap. U. S. geol. Surv.* **196**, 1–24.

CARTER, R. M. 1975. A discussion and classification of subaqueous mass-transport with particular application to grain-flow, slurry-flow and fluxoturbidites. *Earth Sci. Rev.* **11**, 145–77.

COOK, H. E. & ENOS, P. (eds) 1977. *Deep-water Carbonate Environments. Spec. Publs Soc. econ. Paleont. Miner. Tulsa,* **25**.

—— & TAYLOR, M. E. 1977. Comparison of continental slope & shelf environments in the Upper Cambrian and lowest Ordovician of Nevada. *In:* COOK, H. E. & ENOS, P. (eds) *Deep-water Carbonate Environments. Spec. Publs. Soc. econ. Paleont. Miner., Tulsa,* **25**, 51–82.

CURRAY, J. R. & MOORE, D. G. 1971. Growth of the Bengal Deep Sea Fan. *Bull. geol. Soc. Am.* **82**, 563–72.

DALY, R. A. 1936. Origin of submarine 'canyons'. *Am. J. Sci.* **31**, 401–20

DEAN, W. E., STOW, D. A. V., BARRON, E. & SCHALLREUTER, R. 1984. A revised sediment classification for siliceous-biogenic, calcareous and non-biogenic components. *Init. Rep. Deep Sea Drill. Proj.* 75, U.S. Government Printing Office, Washington DC.

DELTEIL, J. R., VALERY, P., MONTADERT, L., FONDEUR, C., PATRIAT, P. & MASCLE, J. 1974. Continental margin in the northern part of the Gulf of Guinea. *In:* BURK, C. A. & DRAKE, C. L. (eds) *The Geology of Continental Margins,*

297–311. Springer-Verlag, New York.

DOTT, R. H. & SHAVER, R. H. (eds) 1974. *Modern and Ancient Geosynclinal Sedimentation. Spec. Publs. Soc. econ. Paleont. Mineral., Tulsa,* **19**.

DOYLE, L. J. & PILKEY, O. H. (eds) 1979. *Geology of Continental Slopes. Spec. Publs Soc. econ. Paleont. Mineral., Tulsa,* **27**.

DRAKE, C. L., EWING, M. & SUTTON, G. H. 1959. Continental margins and geosynclines: the east coast of North America, north of Cape Hatteras. *In:* AHRENS L. H. *et al.* (eds) *Physics and Chemistry of the Earth* 3, 110–98. Pergamon Press, Oxford.

EMERY, K. O. 1977. Stratigraphy and structure of pull-apart margins. *Am. Ass. Petrol. Geol. Continuing Education Course Notes,* **5**, 131–20.

ERICKSON, D. B., EWING, M. & HEEZEN, B. C. 1951. Deep-sea sands and submarine canyons. *Bull. geol. Soc. Am.* **62**, 961–5.

FAUGÈRES, J. C., GAYET, J., GONTHIER, E., POUTIERS, J. & NYANG, I. 1982. La Dorsale Médio-Atlantique entre 43° et 56°N: faciès et dynamique sédimentaire dans plusieurs types d'environnements au Quaternaire Récent *Bull. Inst. géol. Bassin d'Aqu. Bordeaux,* **31**, 195–216.

——, STOW, D. A. V. & GONTHIER, E. 1984. Contourite drift moulded by deep Mediterranean outflow. *Geology,* **12**, 296–300.

FORREL, F. A. 1885. Les ravin sous-lacustres des fleuves glaciaires. *C. R. hebd. Seanc. Acad. Sci., Paris,* **101**, 725–8.

GORSLINE, D. S. 1978. Anatomy of margin basins. *J. sedim. Petrol.* **48**, 1055–68.

—— 1984. Some thoughts on problems of fine-grained sediment production and transport. *In:* STOW, D. A. V. & PIPER, D. J. W. (eds) *Fine-Grained Sediments: Deep-Water Processes and Facies. Spec. Publ. geol. Soc. Lond.* Blackwell Scientific Publications, Oxford. In press.

—— *et al.* 1984. Studies of fine-grained sediments in the Californian continental borderland. *In:* STOW, D. A. V. & PIPER, D. J. W. (eds) *Fine-Grained Sediments: Deep-Water Processes and Facies. Spec. Publ. geol. Soc. Lond.* Blackwell Scientific Publications, Oxford. In press.

GOT, H. *et al.* 1981. Sedimentation on the Ionian active margin (Hellenic Arc)—provenance of sediments and mechanisms of deposition. *Sediment. Geol.* **28**, 243–72.

GROVER, N. C. & HOWARD, C. S. 1938. The passage of turbid water through Lake Mead. *Trans. Am. Soc. civ. Engrs* **103**, 720–90.

HARMS, J. C. & FAHNESTOCK, R. K. 1965. Stratification, bedforms and flow phenomena (with an example from the Rio Grande). *In:* MIDDLETON, G. V. (ed.) *Primary Sedimentary Structures and Their Hydrodynamic Interpretation. Spec. Publs Soc. econ. Paleont. Miner. Tulsa,* **12**, 84–115.

HEEZEN, B. C. (ed.) 1977. Influence of abyssal circulation on sedimentary accumulations in space and time. *Mar. Geol.* **23** (special issue).

—— & EWING, M. 1952. Turbidity currents and submarine slumps, and the 1929 Grand Banks earthquake. *Am. J. Sci.* **250**, 849–73.

——, HOLLISTER, C. D. & RUDDIMAN, W. F. 1966. Shaping of the continental rise by deep geostrophic contour currents. *Science*, **152**, 502–8.

HEIN, F. J. 1982. Depositional mechanisms of deep-sea coarse clastic sediments, Cap Enragé Formation, Quebec. *Can. J. Earth Sci.* **19**, 267–87.

HESSE, R. 1975. Turbiditic and non-turbiditic mudstone of Cretaceous flysch sections of the East Alps and other basins. *Sedimentology*, **22**, 387–416.

—— & BUTT, A. 1976. Paleobathymetry of Cretaceous turbidite basins of the East Alps relative to the calcite compensation level. *J. Geol.* **34**, 505–33.

HILL, P. R. 1981. *Detailed morphology and late Quaternary sedimentation on the Nova Scotian slope south of Halifax.* Unpublished Ph.D. Thesis, Dalhousie University, Halifax, Canada. 331 pp.

——, MORAN, K. M. & BLASCO, S. M. 1984. Creep deformation of slope sediments in the Canadian Beaufort Sea. *In:* BENNETT, R. H., BEA, R. G., & HOOPER, J. (eds) *Proc. SEPM-NORDA Research Con. on Seafloor Stability. Geomarine Lett. Spec. Publ.*, in press.

HOFFERT, M. 1980. *Les "argiles rouges des grands fonds" dans le Pacifique centre-est: authigenèse, transport, diagenèse.* Thesis, Université Louis Pasteur de Strasbourg, Memoire No. 61, 231 pp.

HOLLISTER, C. D. & HEEZEN, B. C. 1972. Geologic effects of ocean bottom currents: western North Atlantic. *In:* GORDON, A. L., (ed.) *Studies of Physical Oceanography,* Vol. 2, *Gordon & Breach, London.* 37–66.

HORN, D. R., EWING, M., HORN, B. M. & DELACH, M. N. 1971. Turbidites of the Hatteras & Sohm Abyssal Plains, Western North Atlantic. *Mar. Geol.* **11**, 287–323.

——, —— & EWING, M. 1972. Graded bed sequences emplaced by turbidity currents north of 20°N in the Pacific, Atlantic & Mediterranean. *Sedimentology*, **18**, 247–75.

HOWELL, D. G. & HUENE, R. von 1980. *Tectonics and sediment along active continental margins.* Short Course Soc. econ. Paleont. Mineral. San Francisco.

—— & NORMARK, W. R. 1982. Sedimentology of submarine fans. *Mem. Am. Ass. Petrol. Geol.* **31**, 365–404.

HSÜ, K. J. & JENKYNS, H. C. (eds) 1974. *Pelagic Sediments: on Land and under the Sea. Spec. Publs int. Ass. Sediment.* **1**, Blackwell Scientific Publications, Oxford. 447 pp.

HUBERT, J. K., SUCHECKI, R. K. & CALLAHAN, R. K. M. 1977. The Cowhead Breccia: sedimentology of the Cambro-Ordovician continental margin, Newfoundland. *In:* COOK, H. E., & ENOS, P. (eds) *Spec. Publs Soc. econ. Paleont. Mineral., Tulsa.* **25**, 125–54.

ILLING, L. V. & HOBSON, G. D. 1981. *Petroleum geology of the continental shelf of Northwest Europe.* Heyden, London.

JAMES, N. P. & GINSBURG, R. N. 1979. *The Seaward Margin of Belize Barrier and Atoll Reefs. Spec. Publs int. Ass. Sediment.* **3**. Blackwell Scientific Publications, Oxford. 193 pp.

JENKYNS, H. C. 1978. Pelagic environments. *In:* READING, H. G. (ed.) *Sedimentary Environments and Facies,* 314–71. Blackwell Scientific Publications, Oxford.

JIPA, D. & KIDD, R. B. 1974. Sedimentation of coarser grained interbeds in the Arabian Sea and sedimentation processes pf the Indus Cone. *In: Init. Rep. Deep Sea drill. Proj.* **23**. 471–92. U. S. Government Printing Office, Washington D. C.

JOHNSON, D. 1938. The origin of submarine canyons. *J. Geomorph.* **1**, 230–43.

KAY, M. 1951. *North American Geosynclines. Mem. geol. Soc. Am.* **48**, 143 pp.

KELTS, K. & ARTHUR, M. A. 1981. Turbidites after ten years of deep-sea drilling—wringing out the mop? *In:* WARME, J. E., DOUGLAS, R. G. & WINTERER, E. L. (eds) *The Deep Sea Drilling Project: A Decade of Progress. Spec. Publs Soc. econ. Paleont. Mineral., Tulsa,* **32**, 91–127.

KING, L. H. & YOUNG, I. F. 1977. Paleocontinental slopes of East Coast Geosyncline (Canadian Atlantic Margin) *Can. J. Earth Sci.* **14**, 2553–64.

KOMAR, P. D. 1969. The channelised flow of turbidity currents with application to Monterey Deep-Sea Fan Channel. *J. geophys. Res.* **74**, 4544–58.

—— 1970. The competence of turbidity current flow. *Bull. geol. soc. Am.* **81**, 1556–62.

KUENEN, Ph. H. 1937. Experiments in connection with Daly's hypothesis on the formation of submarine canyons. *Leidse geol. Med.* **8**, 327–35.

—— 1950. *Marine Geology.* Wiley, New York. 568 pp.

—— 1964. Deep-sea sands and ancient turbidites. *In:* BOUMA, A. H. & BROUWER, A. (eds) *Turbidites—Developments in Sedimentology,* **3**, 3–33.

—— & MIGLIORINI, C. I. 1950. Turbidity currents as a cause of graded bedding. *J. Geol.* **58**, 91–127.

KUHN, T. S. 1970. *The Structure of Scientific Revolutions.* University of Chicago Press. 210 pp.

LEGGETT, J. K. (ed.) 1982. *Trench-Forearc Geology: Sedimentation and Tectonics on Modern and Ancient Active Plate Margins. Spec. Publ. geol. Soc. Lond.* **10**. Blackwell Scientific Publications, Oxford. 600 pp.

LEWIS, K. B. 1971. Slumping on a continental slope inclined at 1°–4°. *Sedimentology*, **16**, 97–100.

LOWE, D. R. 1979. Sediment gravity flows: their classification and some problems of application to natural flows and deposits. *In:* DOYLE, L. J. & PILKEY, O. H. (eds) *Geology of Continental Slopes. Spec. Publs Soc. econ. Paleont. Mineral., Tulsa,* **27**, 75–82.

—— 1982. Sediment gravity flows: II. Depositional models with special reference to the deposits of high-density turbidity currents. *J. sedim. Petrol.* **52**, 279–97.

MATTICK, R. E. GIRARD, O. W., SCHOLLE, P. A. & GROW, J. A. 1978. Petroleum potential of U. S. Atlantic Slope, Rise and Abyssal Plain. *Bull. Am. Ass. Petrol. Geol.* **62**, 592–608.

McCAVE, I. N. 1984. Erosion, transport and deposition of fine-grained marine sediment. *In*: STOW, D. A. V. & PIPER, D. J. W. (eds) *Fine-Grained Sediments: Deep-Water Processes and Facies. Spec. Publ. geol. Soc. Lond.* Blackwell Scientific Publications, Oxford. In press.

McILREATH, I. A. & JAMES, N. P. 1978. Facies models 13. Carbonate slopes. *Geosci. Can.* **5**, 189–99.

MIDDLETON, G. V. 1966. Experiments on density & turbidity currents II. Uniform flow of density currents. *Can. J. Earth Sci.* **3**, 627–37.

—— 1967. Experiments on density and turbidity currents III. Deposition of sediment. *Can. J. Earth Sci.* **4**, 475–505.

—— & HAMPTON, M. 1976. Subaqueous sediment transport and deposition by sediment gravity flows. *In*: STANLEY, D. J. & SWIFT, D. J. P. (eds) *Marine Sediment Transport and Environmental Management*, 197–218. Wiley (Interscience), New York.

MITCHELL, A. H. G. & READING, H. G. 1969. Continental margins, geosynclines and ocean-floor spreading. *J. Geol.* **77**, 629–46.

MOORE, D. G. 1961. Submarine slumps. *J. sedim. Petrol.* **31**, 343–57.

—— 1969. Reflection profiling studies of the Californian continental borderland: structure and Quaternary turbidite basins. *Spec. Pap. geol. Soc. Am.* **107**, 1–142.

MORGENSTERN, N. R. 1967. Submarine slumping and the initiation of turbidity currents. *In*: RICHARDS, A. F. (ed.) *Marine Geotechnique*, 189–220. University of Ilinois Press, Urbana.

MULLINS, H. T. & NEUMANN, A. C. 1979. Deep carbonate bank margin structure and sedimentation in the northern Bahamas. *In*: DOYLE, L. J. & PILKEY, O. H. (eds) *Geology of Continental Slopes. Spec. Publs Soc. econ. Paleont. Mineral., Tulsa*, **27**, 165–92.

MURRAY, J. & RENARD, A. F. 1891. Report on deep-sea deposits based on specimens collected during the voyage of HMS *Challenger* in the years 1873–1876. *In*: *Challenger Reports*. HMSO., Edinburgh. 525 pp.

MUTTI, E. 1977. Distinctive thin-bedded turbidite facies and related depositional environments in the Eocene Hécho Group (South-central Pyrenees, Spain). *Sedimentology*, **24**, 107–31.

—— & RICCI LUCCHI, F. 1972. Le torbiditi del'Appenino settentrionale: introduzione all' analisis di facies. *Memorie Soc. géol. Ital.* **11**, 161–99.

—— & —— 1975. Turbidites of the northern Apennines: Introduction to facies analysis. *Am. geol. Inst.* Reporting Series, **3**, 21–36.

NAYLOR, M. A. 1980. Origin of inverse grading in muddy debris flow deposits—a review. *J. sedim. Petrol.* **50**, 1111–6.

NELSON, C. H. & NILSEN, T. H. 1974. Depositional trends of modern and ancient deep-sea fans. *In*: DOTT, R. H. & SHAVER, R. H. (eds) *Modern and Ancient Geosynclinal Sedimentation. Spec. Publs Soc. econ. Paleont. Mineral., Tulsa*, **19**, 69–91.

——, NORMARK, W. R., BOUMA, A. H. & CARLSON, P. R. 1978. Thin-bedded turbidites in modern submarine canyons and fans. *In*: STANLEY, D. J. & KELLING, G. (eds) *Sedimentation in Submarine Canyons, Fans and Trenches*, 177–89. Dowden, Hutchinson & Ross, Stroudsburg.

NISBET, E. G. & PRICE, I. 1974. Siliceous turbidites: bedded cherts as redeposited, ocean ridge-derived sediments. *In*: HSÜ, K. J. & JENKYNS, H. C. (eds) *Pelagic Sediments: on Land and under the Sea. Spec. Publs int. Ass. Sediment.*, **1**, 351–66. Blackwell Scientific Publications, Oxford.

NORMARK, W. R. 1970. Growth patterns of deep-sea fans. *Bull. Am. Ass. Petrol. Geol.* **54**, 2170–95.

—— 1978. Fan valleys, channels and depositional lobes on modern submarine fans: characters for recognition of sandy turbidite environments. *Bull. Am. Ass. Petrol. Geol.* **62**, 912–31.

PETTIJOHN, F. J. 1957. *Sedimentary Rocks*, 2nd ed., Harper, New York. 718 pp.

PICKERING, K. T. 1982. The Kongsfjord Formation—a late Precambrian submarine fan in NE Finnmark, N Norway. *Norg. geol. Unders.* **367**, 77–104.

—— 1982. The shape of deep-water siliciclastic systems—a discussion. *Geomar. Lett.* **2**, 41–46.

PILKEY, O. H., LOCKER, S. D. & CLEARY, W. J. 1980. Comparison of sand-layer geometry on flat floors of 10 modern depositional basins. *Bull. Am. Ass. Petrol. Geol.* **64**, 841–56.

PIPER, D. J. W. 1975. Late Quaternary deep water sedimentation off Nova Scotia and the western Grand Banks. *In*: YORATH, C. J. PARKER, E. R. & GLASS, D. J. (eds) *Canada's Continental Margins. Can. Soc. Petrol. Geol.* **4**, 195–204.

—— 1978. Turbidites, muds and silts on deep-sea fans and abyssal plains. *In*: STANLEY, D. J. & KELLING, G. (eds) *Sedimentation in Submarine Canyons, Fans and Trenches*, 163–76. Dowden, Hutchinson & Ross, Stroudsburg.

PITMAN, W. C. 1979. The effect of eustatic sea level changes on stratigraphic sequences at Atlantic margins. *In*: WATKINS, J. S., MONTADERT, L. & WOOD DICKERSON, P. (eds) *Geological and Geophysical Investigations of Continental Margins. Mem. Am. Ass. Petrol. Geol.* **29**, 453–60.

RICCI LUCCHI, F. 1975. Miocene paleogeography and basin analysis in the Periadriatic Apennines. Reprinted from *Geology of Italy* (ed. SQUYRES COY) Petrol. Explor. Soc. Libya. 111 pp.

—— 1978. Turbidite dispersal in a Miocene deep-sea plain: the Marnoso-Arenacea of the Northern Apennines. *Geol. Mijnb.* **57**, 559–76.

RUPKE, N. A. 1977. Growth of an ancient deep-sea fan. *J. Geol.* **85**, 725–44.

—— 1978. Deep clastic seas. *In*: READING, H. G. (ed.) *Sedimentary Environments and Facies*, 372–95. Blackwell Scientific Publications, Oxford.

—— & STANLEY, D. J. 1974. Distinctive properties of

turbiditic and hemipelagic mud layers in the Algero-Balearic Basin, western Mediterranean Sea. *Smithson. Contr. Earth Sci.* **13**, 40 pp.

SAXOV, S. & NIEUWENHUIS, J. K. (eds) 1982. *Marine Slides and other Mass Movements*, 353 pp. Plenum Press, New York.

SCHENK, P. E., LANE, T. E. & JENSEN, L. R. 1980. Palaeozoic history of Nova Scotia—a time trip to Africa (or South Africa?). *Geol. Ass. Can. Field Trip Guidebk* **20**, 82 pp.

SCOTT, R. M. & TILLMAN, R. W. 1981. Stevens Sandstone (Miocene), San Joaquin Basin, California. *In*: SIEMERS, C. T., TILLMAN, R. W. & WILLIAMSON, C. R. (eds) *Deep-Water Clastic Sediments: a Core Workshop. Soc. econ. Paleont. Mineral. Core Workshop No. 2*, San Francisco, 116–248.

SHEPARD, F. P. 1932. Sediments of the continental shelves. *Bull. geol. Soc. Am.* **43**, 1017–40.

—— 1948. *Submarine Geology*. Harper & Row, New York. 348 pp.

—— 1973. *Submarine Geology*. 3rd ed. Harper & Row, New York. 517 pp.

——, MARSHALL, N. F., McLOUGHLIN, P. A. & SULLIVAN, G. G. 1979. *Currents in Submarine Canyons and other Seavalleys. Stud. Geol. Am. Ass. Petrol. Geol.* **8**, 179 pp.

SPÖRLI, K. B. 1980. New Zealand and oblique-slip margins: tectonic development up to and during the Cainozoic. *In*: BALLANCE, P. F. & READING, H. G. (eds) *Sedimentation in Oblique–Slip Mobile Zones. Spec. Publs int. Ass. Sediment.* **4**, 147–70. Blackwell Scientific Publications, Oxford.

STANLEY, D. J., SWIFT, D. J. P., SILVERBERG, N., JAMES, N. P. & SUTTON, R. G. 1972. Late Quaternary progradation and sand "slipover" on the outer continental margin off Nova Scotia, southeast Canada. *Smithson. Contr. Earth Sci.* **8**, 88 pp.

STOCKS, T. 1933. *Die Echoloprofile Wissenschaftliche Ergebrisse der Deutschen Atlantischen Expedition auf den "Meteor" 1925–1927*. Vol. 2. Walter de Gruyter, Berlin.

STOW, D. A. V. 1978. Regional review of the Nova Scotian outer margin geology. *Marit. Sed.* **14**, 17–32.

——1981. Laurentian Fan: morphology, sediments, processes and growth pattern. *Bull. Am. Ass. Petrol. Geol.* **65**, 375–93.

—— 1982. Bottom currents and contourites in the North Atlantic. *Bull. Inst. géol. Bassin d'Aqu.* **32**, 151–66.

—— 1984. Cretaceous to Recent submarine fans in the SE Angolan Basin. *In*: HAY, W. W. & SIBUET, J. C. *et al.* (eds) *Init. Rep. Deep Sea drill. Proj.* **75**.

—— & LOVELL, J. P. B. 1979. Contourites: their recognition in modern and ancient sediments. *Earth Sci. Rev.* **14**, 251–91.

—— & SHANMUGAM, G. 1980. Sequence of structures in fine-grained turbidites: comparison of recent deep-sea and ancient flysch sediments. *Sediment. Geol.* **25**, 23–42.

——, BISHOP, C. D. & MILLS, S. J. 1982. Sedimentology of the Brae Oil Field, North Sea: fan models and controls. *J. Petrol. Geol.* **5**, 129–48.

——, HOWELL, D. G. & NELSON, C. H. 1984. Sedimentary, tectonic and sea-level controls on submarine fans and debris apron turbidite systems. *Spec. Publ. Geomar. Lett.*, in press.

——, ALAM, M. & PIPER, D. J. W. 1984. Sedimentology of the Halifax Formation, Nova Scotia: Lower Palaeozoic fine-grained turbidites. *In*: STOW, D. A. V. & PIPER, D. J. W. (eds) *Fine-Grained Sediments: Deep-Water Processes and Facies. Spec. Publ. geol. Soc. Lond.* Blackwell Scientific Publications, Oxford.

SURLYK, F. 1978. Submarine fan sedimentation along fault scarps on tilted fault blocks (Jurassic/Cretaceous boundary, East Greenland). *Bull. Grøn. Geol. Unders.* **128**, 108 pp.

THIEDE, J., STRAND, J.-E. & AGDESTEIN, T. 1981. The distribution of major pelagic sediment components in the Mesozoic and Cenozoic North Atlantic Ocean. *In*: WARME, J. E., DOUGLAS, R. G. & WINTERER, E. L. (eds) *The Deep Sea Drilling Project: A Decade of Progress. Spec. Publs Soc. econ. Paleont. Miner., Tulsa*, **32**, 67–90.

THORNTON, S. E. 1984. Basin model for hemipelagic sedimentation in a tectonically active continental margin: Santa Barbara Basin, California continental borderland. *In*: STOW, D. A. V. & PIPER, D. J. W. *Fine-Grained Sediments: Deep-Water Processes and Facies. Spec. Publ. geol. Soc. Lond.* Blackwell Scientific Publications, Oxford. In press.

TOFELLI, L. C. & BARROS, M. C. 1979. Esforcos e resultados de pesquisas de trapas estratigraphicas mus bracias sedimentaires brasileirus. *1st Congresso Brasileiro de Petroleo*, **1** (11), 51–62, Instituto Brasileiro de Petroleo.

VAIL, A. R. & MITCHUM, R. M. 1979. Global cycles of relative changes of sea level from seismic stratigraphy. *In*: PAYTON, C. E. (ed.) *Seismic Stratigraphy—applications to hydrocarbon exploration. Mem. Am. Ass. Petrol. Geol.* **29**, 469–72.

WALKER, R. G. 1973. Mopping up the turbidite mess. *In*: GINSBURG, R. N. (ed.) *Evolving Concepts in Sedimentology*, 1–37. John Hopkins Press, Baltimore.

—— 1975. Upper Cretaceous resedimented conglomerates at Wheeler Gorge, California: Description and Field Guide. *J. sedim. Petrol.* **45**, 105–12.

—— 1978. Deep water sandstone facies and ancient submarine fans: Models for exploration for stratigraphic traps. *Bull. Am. Ass. Petrol. Geol.* **62**, 932–66.

—— & MUTTI, E. 1973. Turbidite facies and facies associations. *In*: MIDDLETON, G. V. & BOUMA, A. H. (eds) *Turbidites and Deep Water Sedimentation. Soc. econ. Paleont. Mineral. Pacific. Sec. Short Course Notes*, Anaheim, 119–58.

WESTCOTT, W. A. & ETHRIDGE, F. G. 1980. Fan-delta

sedimentology and tectonic setting—Yallahs Fan Delta, southeast Jamaica. *Bull. Am. Ass. Petrol. Geol.* **64**, 374–99.

—— & —— 1983. Eocene fan delta/submarine fan deposition in the Wagwater Trough, east-central Jamaica. *Sedimentology*, **30**, 235–48.

WEZEL, F. C., SAVELLI, D., BELLAGAMBA, M., TRAMONTANA, M. & BARTOLE, R. 1981. Plio-Quaternary depositional style of sedimentary basins along insular Tyrrhenian margins. *In*: WEZEL, F. C. (ed.) *Sedimentary Basins of Mediterranean Margins*, 239–69. CNR Italian Project of Oceanography.

WILDE, P., NORMARK, W. R. & CHASE, T. E. 1978. Channel sands and petroleum potential of Monterey deep-sea fan, California. *Bull. Am. Ass. Petrol. Geol.* **62**, 967–83.

WOODLAND, A. W. (ed.) 1975. *Petroleum and the Continental shelf of Northwest Europe.* Applied Science, Barking, 501 pp.

YARBOROUGH, H., EMERY, K. O., DICKINSON, W. R., SEELY, D. R., DOW, W. G., CURRAY, J. R. & VAIL, P. R. 1977. *Geology of Continental Margins. Am. Ass. Petrol. Geol., Cont. Educ. Course Note Ser.* 5.

DORRIK A. V. STOW, Grant Institute of Geology, West Mains Road, University of Edinburgh, Edinburgh EH9 3JW. Present address: Geology Department, University Park, Nottingham.

Deep-sea pelagic sediments and palaeo-oceanography: a review of recent progress

J. K. Leggett

SUMMARY: Understanding of present and past patterns of sedimentation, circulation, fertility, biogeography and chemistry in the oceans has advanced greatly in the last decade. In achieving this a diverse range of geological and geophysical techniques has been drawn on. Egress zones of active hydrothermal systems venting sulphide-laden hot water at spreading centres, long predicted on theoretical and other grounds, were discovered using deep-towed geophysical packages and have been fruitfully studied using submersibles. The chemistry of the venting solutions, and the apparent scale of the hydrothermal processes, have engendered a rethink of the oceanic fluxes of many elements. Continuing deep-sea drilling increasingly refines reconstructions of continental-drift events and the consequent changes in circulation patterns. The Deep-Sea Drilling Project planning has become more problem-specific, and on 'palaeo-environment' legs, sediment successions have been cored more continuously than was earlier the case. As a result, palaeo-oceanographers have been provided with a flood of data germane to the reconstruction of past behaviour of the CCD, to appreciation of past states of oxygenation of the oceans and of carbon preservation patterns (the 'black-shale problem'), to documentation and interpretation of hiatus patterns, to calculation of global sediment budgets through time, and to calibration of the overall cooling of the Cenozoic hydrosphere. Isotopic determinations and organic geochemistry have become key tools in these studies. Global sea-level stands, clearly important as a regulator of conditions in the pelagic realm (though to an arguable extent) have been documented using seismic stratigraphy.

Attempts at synthesis of the complex spectrum of palaeo-oceanographic changes evident through time in the pelagic realm stress the rhythmicity with which the oceans have oscillated between broadly uniformitarian and non-uniformitarian conditions. Syntheses build towards identification of driving forces and responses.

In the past decade our knowledge of deep-sea pelagic sediments has expanded considerably, and given birth to a new discipline in geology: palaeo-oceanography. This has been due chiefly to the successful progress of the Deep-Sea Drilling Project (DSDP) (e.g. Arthur 1979, Berger 1979). Recently, understanding of deep-sea sedimentary processes has been augmented by exciting discoveries made using submersibles.

Jenkyns (1978) has provided a thorough review of pelagic sedimentary environments, processes, and palaeo-oceanography, covering both the modern oceans and the outcrop record of ancient oceans. Summary papers on specific topics are available in volumes edited by Hsü & Jenkyns (1974), Talwani et al. (1979) and Warme et al., (1981)., Kennett (1982) has produced a comprehensive textbook, and a recent review of the carbonate pelagic environment by Scholle et al. (1983) is superbly illustrated with colour plates. The first part of this paper is a summary, highlighting advances made since the publication of Jenkyn's review, of some of the most interesting current fields of investigation. Where possible the more important techniques and principles are introduced, but the interested student should refer to the literature cited above for critical details: an essay of this length and attempted scope is of necessity eclectic. In particular, treatment of palaeo-oceanographic events is largely confined to the Cretaceous–Recent interval, and the emphasis on deep-sea metallogenesis is a result of the economic onus of this volume. References are highly selective, and in each section attention is drawn to relevant review papers.

Part one of the paper builds towards a catalogue of the more important *events* in the Cretaceous–Recent oceanic domain. These are shown in simplified form in Table 1. At the same time, the intention is to give an appreciation of how complex the *controls* on pelagic sedimentation actually are. Part two of the paper summarizes recent developments in an attempt to sift a measure of order from the innumerable possible interactions between the controlling parameters: geographic, hydrographic, climatic, chemical and biological. This field of endeavour has been termed *systemic stratigraphy* (Berger & Vincent 1981). Despite the difficulty of recognizing obvious one-to-one 'cause-and-effect' relationships, a degree of order is indeed apparent in the oceanic realm, in that the sediments record a vacillation of conditions in the oceans between states which might, as a starting point at least, be regarded as

broadly uniformitarian and broadly non-
uniformitarian. These secular variations
(rhythms) occur on several time scales ranging
from several tens of thousands of years to
several hundreds of millions of years. Their
elaboration has been the recent objective of the
more adventurous geological oceanographers
and palaeo-oceanographers, and the final
section of this paper summarizes ideas on
rhythmicity in the oceanic realm.

Deep-sea hydrothermal deposits

Background

Hydrothermal activity along constructive
plate boundaries was first proposed on
theoretical grounds more than 15 years ago.
Since then, a veritable library of evidence has
built up confirming such activity along
spreading centres, large and small, around the
globe. Metal-rich sediments precipitated from
the hydrothermal solutions were first recovered
in the 1960s: from hot, saline pools in the axial
trough of the Red Sea rift (Miller *et al.* 1966) and
from the crest of the East Pacific Rise (Boström
& Peterson 1966). Numerous subsequent
discoveries were made on the East Pacific Rise,
on the Mid-Atlantic Ridge, on the Mid-Indian
Ridge and on small oceanic spreading centres
such as in the Gulf of California; these have
been augmented by recovery of metal-rich, basal
pelagic sections in numerous open-ocean DSDP
sites (summary in Cronan 1980).

Such direct evidence is supplemented by geophysical observations, such as heat-flow values
which suggest that convective flow is occurring
at and near spreading centres, decreases in upper
crustal seismic velocities which probably indicate high crustal porosity associated with intense
fissuring near ridge crests, and swarms of micro-
earthquakes which may reflect passage of fluids
through the crust (e.g. McDonald & Luyendyk
1981). Petrological/geochemical observations
provide further evidence: widespread hydro-
thermally altered basalts have been recovered in
dredged and drilled samples, low ^{18}O contents
in oceanic intrusive rocks testify to high-
temperature fluid–rock interactions (Muehlenbachs & Clayton 1976), and concentrations of
^3He, indicative of mantle degassing, have been
located in plumes in the water column above
spreading centres (e.g. Lupton *et al.*, 1977).
Discoveries of supra-ophiolitic sediments of
similar chemistry to those recovered in the
extant oceans indicate that such hydrothermal
processes operated in ancient spreading centres
(e.g. Fleet & Robertson 1980 and references
therein).

However, until relatively recently no sites of
active hydrothermal venting had been discovered. In 1976 the first such discovery was
made on the Galapagos Rift (Corliss *et al.* 1979).
In 1979 further discoveries were made on the
East Pacific Rise at 21°N (Francheteau *et al.*
1979, McDonald & Luyendyk 1981). More recently, active vents have been discovered in the
Gulf of California (Lonsdale *et al.* 1980).

Additionally, zones of hydrothermal egress
through a pelagic sediment blanket—i.e. on
oceanic crust away from the crest of a spreading
centre—have been located on the flank of the
Galapagos Rise (Honnorez *et al.* 1981 and
references therein).

Vents on the Galapagos Rift and East Pacific Rise

Their presence predicted by bottom-water
temperature anomalies and measurements of
heat flow, and their location tied down using a
deep-tow geophysical instrument package, the
Galapagos warm-water springs were discovered
on a 2.5 km deep portion of the rift where the
spreading rate is 3.5 cm year^{-1} using the
American submersible *Alvin* (summary in
Corliss *et al.* 1979). Four vent areas along a 3 km
stretch of the axial rift range from 30 to 100 m
across, and emit water varying in temperature
from 7 to 17°C, laden with a shimmering, thin
milky precipitate, at rates around 2–10 l s^{-1} and
in plumes extending to more than 180 m above
the bottom. From these emissions Fe and Mn
oxy-hydroxides and nontronite (an Fe-rich
smectite) precipitate, and a thin layer of Mn
oxide builds up on the surrounding rocks. Most
surprising of all, the active vents are oases
around which members of a previously unknown community of animals cluster. This community comprises large clams, mussels, limpets,
tube worms (Pogonopherans) and ophidioid
fish. At the base of the food chain which
sustains the community are chemosynthetic
bacteria, richly concentrated in the warm waters
of the springs. It is, therefore, the only known
community which exists without recourse to
photosynthetic sources of energy.

Using the French submersible *Cyana* in
1978, the joint French–American–Mexican
(CYAMEX) expedition to the East Pacific Rise
at 21°N found no active vents, despite locating a
Galapagos-type animal community. However,
they did find 'inactive' mounds of weathered
Zn, Fe and Cu sulphides in a small graben
600–700 m west of the rise axis (Francheteau *et
al.* 1979). In 1979 an American cruise (RISE) to
the same general area used a deep-towed camera
and thermometer sled (ANGUS) to locate a

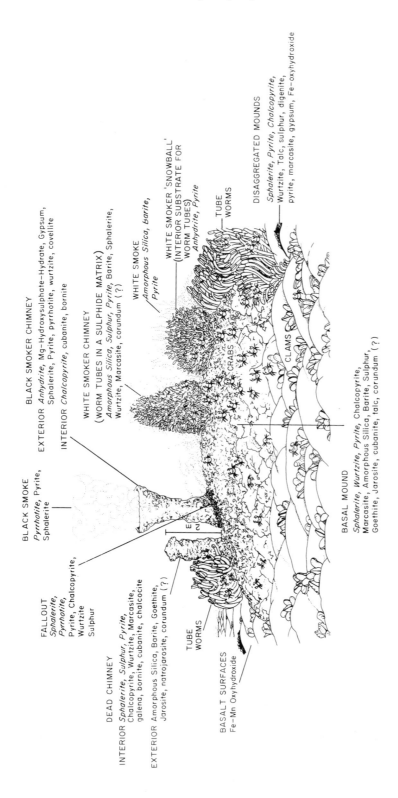

Fig. 1. Idealized scene in an East Pacific Rise hydrothermal field (after McDonald & Luyendyk 1981 and Haymon & Kastner 1981). Major mineralogical components in inclined script.

string of twenty-five active vents. These are situated at 2.6 km depth along a 6.2 km long segment of ridge axis in a narrow 100–200 m wide band within the zone of most recent magma extrusion; the spreading rate is 6 cm year^{-1}, intermediate by Pacific standards but higher than the Galapagos, some 3000 km to the southeast. *Alvin* investigated eight of the vent sites. The fortunate divers observed not just the vent communities, but hot springs much more impressive than those previously seen on the Galapagos Rise. At 21°N chimneys up to 10 m high and 40 cm wide, made of solid sulphide, blast out billowing black clouds of sulphide-laden water at temperatures of up to about 350°C, far in excess of any previously recorded deep-ocean water temperatures. These ex-halates, which during the first attempts at temperature measurement melted the housing of *Alvin*'s thermometer probe, are not actually boiling, since the pressure at the depth of the vents is roughly 275 times atmospheric pressure. The chimneys sit on mounds of worm tube-encrusted sulphide aggregates rising up to 2 m above the basaltic substrate, and are of several kinds (Fig. 1). 'Live' chimneys (those emitting hot water) are classed as 'smokers' or 'non-smokers', depending on whether or not sulphides are suspended in the venting water. Smokers are 'black' or 'white', depending on the colour of the precipitates. Black smokers are the hotter, vent more rapidly (at rates of several metres per second) and are given a wide berth by the mound animals. White smokers are cooler (<20–*c*. 300°C), have slower flow rates, and organisms very similar to those of the Galapagos vents cluster round them. 'Snowball' vents are a type of white smoker encrusted by a spherical mass of white polychaete worm tubes inhabited by living worms.

Chemistry and mineralogy

Magmatism at the spreading centre is the motor which drives the hydrothermal system, and the high fracture permeability of the oceanic crust around the axial rift is the plumbing. The heated seawater reacts with the basalt through which it circulates. Much of the seawater sulphate is precipitated out or reduced to sulphide, and Mg hydroxyl ions are removed and sequestered in hydrothermal clays. In this way the seawater is converted to a reducing acidic solution capable of leaching Ca, Si, Mn, Fe, Li and other cations from the basalt. These include the elements which have given rise to interest in commercial quarters, such as Zn and Cu. The initial chemistry of a hydrothermal solution therefore depends on the water/rock ratio (i.e.

the amount of seawater which has flushed through, and reacted with, a given amount of hot basalt).

Variations in the chemistry of vented sol-utions are explained by variable mixing of hot sulphur-rich, oxygen-poor, low-pH hydro-thermal fluids, and ambient cold, relatively sulphur-poor and oxygen-rich, high-pH bottom water ('ground' water). The 21°N East Pacific Rise high-temperature solutions do not appear to have mixed significantly with downwards-percolating groundwater, and are thus seen as the hydrothermal end-member, in particular because they are devoid of magnesium (Edmond *et al.* 1982). Variations in the chemistry of venting solutions between adjacent hydro-thermal fields in the Galapagos dive area indicate variable mixing between ground water and a primary hydrothermal fluid, the com-position and temperature of which had been predicted as very similar to the 21°N exhalates before these were actually discovered (Edmond *et al.* 1979).

The mineralogy of the East Pacific Rise vent components is summarized in Fig. 1. Pyrite and sphalerite precipitate from the hot, sulphide-laden solutions as coatings on active chimneys, along with anhydrite, which precipitates from heated ambient seawater. Intimately intergrown with this anhydrite is a phase identified in seawater-heating experiments, but never pre-viously described in nature: Mg-hydroxysulphate-hydrate (Haymon & Kastner 1981). In addition to pyrite and sphalerite, the black smoke con-tains metastable pyrrhotite, which is rapidly recrystallized to pyrite or marcasite. The cooler emissions of the white smokers contain par-ticulate amorphous silica, barite and pyrite. Venting solutions disseminate manganese (which is more soluble), and possibly some silica, over broad areas on the East Pacific Rise: the Mn is ultimately oxidized and precipitates as Mn-oxyhydroxide and Si as nontronite.

In 'dead' chimneys anhydrite dissolves and the sulphate assemblage is dominated by barite and alteration products such as jarosite and natrojarosite. The basal mounds largely comprise fallen and weathered chimney debris. The lifespan of individual chimneys appears to be rather short (perhaps 10–20 years) and the structures are highly unstable: weathering may in time transform a 21°N-type sulphide assemblage into one resembling sulphide ore bodies in Tethyan ophiolites (Strens & Cann 1982).

Animal communities

At the base of the food chain in the newly discovered ecosystem are chemosynthetic bacteria which concentrate in abundance in the warm vent emissions. Here they oxidize H_2S to form elemental sulphur and various sulphates, harnessing the energy so liberated to synthesize CO_2 into organic matter. Feeding on the bacteria are a variety of molluscs. In the Galapagos dive area, where the communities were first observed, clams cover large patches of the ocean floor around the vents, commonly filling depressions between pillows. They resemble vesicomyids, which have a considerable zoogeographic and bathymetric range, but are larger than normal (15–30 cm in length). Mussels up to 15 cm are particularly common in certain of the Galapagos vent fields, and have an anatomy distinctly different from other known mytilids, both shallow- and deep-water forms. Limpets of an entirely new gastropod family attach to rock surfaces not occupied by the mussels.

The most striking animals in underwater photos of the vent areas are chitinous tube worms belonging to the class Vestimentifera of the phylum Pogonophora. The tubes range up to *c.* 3 m in length, and the worms have an extendable red plume of fused tentacles, the tissues of which contain haemoglobin. They attach to the rocks directly in the flow of warm waters from the vents. Sea anemones, serpulid worms, and galatheid and brachyuran crabs are also present. Ophidioid fish swim in the rising plumes of warm water. Such was the density of organisms in some vent areas that Corliss and his colleagues were moved to name one Galapagos vent field the Garden of Eden.

In the 21°N vent areas each colony occupies an area roughly 30 m wide and 100 m long. With the exception of brown mussels, the communities are essentially the same as those in the Galapagos Rise (McDonald & Luyendyk 1981).

Hydrothermal mounds

In the early 1970s deep-tow investigations of the Galapagos Spreading Centre at 86°W had located numerous rows of mounds, subparallel to the spreading axis, above faults in the ocean crust and within a band of high heat flow some 18–30 km south of the axis. These mounds have subsequently proved to be the result of the same process of convective flow of seawater through the ocean crust as that already described: the difference from the ridge-crest activity is that the crust is from 500,000 to 900,000 years old in the region of the mounds and the hydrothermal fluids escape into a sediment cover 25–50 m thick. Mounds visited by *Alvin* varied in height

from <1 m to >20 m. Most were roughly circular, but some formed ridges up to 200 m long. Temperatures were high inside the mound sediment (12–20°C), and though no temperature anomalies occur in the water column above the mounds, a flow of warm water (4–10°C) could be initiated by punching holes through the thin crusts of black manganese oxides which mantle the mounds (Corliss *et al.* 1979).

Subsequently, the *Glomar Challenger* visited the Galapagos mounds twice to investigate their internal structure. The drilling showed that mounds up to 25 m high can form in less than a few 100,000 years by interaction of upward-percolating hydrothermal solutions and the pelagic sediments. The mounds typically comprise three units. At the top is a unit of siliceous foraminiferal nannofossil ooze, usually capped by an oxidized yellow or orange surface layer (Unit A, 0.3–3 m). This in general overlies, with a sharp contact, a layer of interbedded pelagic and hydrothermal sediment (green nontronitic clay), usually capped by Mn-oxide crusts, (Unit B, 13–28 m), which in turn grades progressively downwards into a foraminiferal nannofossil ooze (Unit C, 7–13 m), underlain by basement (Honnorez *et al.* 1981).

The hydrothermal layers of unit B lack carbonate and siliceous organisms, and are confined to the vicinity ($\leqslant 100$ m) of mounds. The biogenic siliceous component of the pelagic sediments decreases more rapidly with depth at mound sites than off-mound sites. These and other observations indicate that the ambient pelagic sediments have been dissolved and replaced by interaction with hydrothermal solutions which percolate slowly upward through them, so forming the mounds. Interestingly, the uppermost few metres of the basement rocks below the mounds are unaltered, and for this reason Honnorez *et al.* (1981) believe that the hydrothermal solutions must tap much deeper ($\geqslant 1$ km) rocks at higher temperatures.

Recent discoveries elsewhere

The submersible programme has recently been extended to the Gulf of California, where a youthful extension of the East Pacific Rise rifted Baja California from mainland Mexico in Pliocene times. The axial rift is a series of *en echelon* short spreading centres linked by lengthy transforms. The Guaymas Basin, in one portion of the rift, was drilled on DSDP Leg 64. There, active volcanism occurs in an area of rapid sedimentation. Cores at Site 477 penetrated a 30 m sill intruded into diatomaceous turbidites. Below the sill anhydrite and pyrite have grown in the sediment; oxygen-isotope studies indicate temperatures of formation equivalent to the greenschist facies, suggesting the presence of a magma chamber not far below (Kastner 1982). On a recent diving programme an active hydrothermal vent was discovered on the site of the DSDP hole, complete with chimney. (M. Kastner, pers. comm. 1983). Elsewhere along the rift, vent fields akin to those at 21°N have been discovered, along with the oases of life (including new species) and new pagoda-like chimneys (Lonsdale *et al.* 1980).

Widespread hydrothermal activity may not be limited to actively spreading mid-ocean ridges. Active hydrothermal springs are known from submarine volcanoes (see Cronan 1980), and a hydrothermal manganese deposit was dredged in 1981 from the flanks of the Tonga–Kermadec Ridge, an active arc in the SW Pacific (Cronan *et al.* 1982). So far, island arcs and other submarine volcanic settings have been little explored compared with mid-ocean ridges.

Geochemical implications

The volume of water which passes through the intrusion zone in the ridge axes has been estimated by Edmond *et al.* (1982) based on the deficit of measured conductive heat loss from young oceanic crust against that predicted from thermal models of the plates, assuming that the 'missing' heat was removed by convective flow at 300°C (as sampling of venting solutions and oxygen-isotope investigations of altered oceanic basalts suggest). The estimate suggests that a volume of water equivalent to the whole ocean must circulate through the high-temperature intrusion zone in the ridge axes (i.e. through the 300°C isotherm) every 8–10 m.y. Though the exact magnitude is contested in some quarters the apparent scale of the process has major implications for the chemical budget of the oceans. The former view that seawater in general maintains an equilibrium between input processes (mainly river efflux) and output processes (deposition of sediments and low-temperature chemical reactions between seawater and ocean floor) is seriously deficient for many elements.

Using Galapagos Rise data as a generalization for all ridge-crest hydrothermal systems, the flux of elements into or out of the ocean can be computed, and compared to the river input (Edmond *et al.* 1982). Notwithstanding the limitations of the assumption, it appears that ridge-crest hydrothermal activity must be the major sink for Mg, a significant sink for alkalinity and the major source for sedimentary Mn and Li; it also provides a significant proportion of Si and Ba, plus some Ca and CO_2.

Sea-level changes

That global stands of sea level have oscillated markedly through Phanerozoic time has long been recognized by geologists, and both the geological causes of these changes and the responses to them have been the subject of great debate. One approach to assessment of the relative extent of flooding of the continents by epicontinental seas at various times in the Phanerozoic has been to plot fossil shorelines on sequential palaeogeographic maps (e.g. Hallam 1977). The vast amounts of seismic reflection data now available, chiefly as a result of commercial activities in offshore sedimentary basins, has of late allowed an alternative approach. Vail *et al.* (1977), in a seminal publication on the applications of seismic stratigraphy to hydrocarbon exploration, propose that changes in coastal onlap patterns on continental margins are a reliable index of the eustatic (worldwide) component of sea-level changes through time. Their arguments for the synchrony of many sea-level changes around the world involved publication of a series of charts showing global cycles of relative changes of sea level (Fig. 2). These charts largely make use of oil company seismic data, but the less well-constrained pre-Late Triassic curves use the onland geology of North America. Though the Vail *et al.* (1977) 'curves' have not been universally accepted (e.g. Watts 1982), they have proved successful in improving stratigraphic analyses within basins, and in estimation of the geological age of strata prior to drilling (eustatic low stands and high stands of known age will produce characteristic alternating responses in sedimentary architecture which show up in the seismic records).

Of particular interest to palaeo-oceanographers is a marked hierarchy in the cycles of relative sea-level change:

FIG. 2. Global cycles of relative change of sea level (after Vail *et al.* 1977). (a) First- and second-order cycles (from which the smoothed first-order cycle curve is derived) on the same scale. (b) Late Triassic–Recent third-order cycles. Cretaceous cycles (hachured area) are second order; third-order cycles were not released for publication by Vail *et al.* (c) Cenozoic third-order cycles.

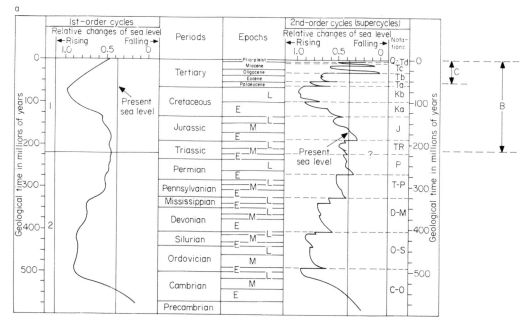

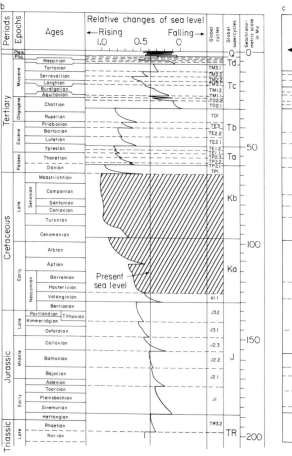

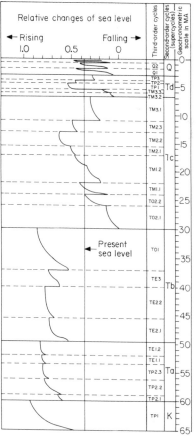

First-order cycles have durations of 200–300 Ma. There were two of them during the Phanerozoic, with high-stand peaks in the Early Palaeozoic (from Late Cambrian time onwards) and Late Cretaceous, and with low-stand peaks in the late Precambrian and Early-middle Triassic (Fig. 2a). Pitman (1978) has argued that the Mesozoic–Recent cycle matches well with overall rates of generation of new oceanic crust (sea-floor spreading rates).

Second- and third-order cycles are of 10–80 and 1–10 Ma duration, respectively. There are more than fourteen second-order and *c*. eighty third-order cycles in the Phanerozoic, not counting late-Palaeozoic cyclothems. An interesting feature of the curves is the evident secular withdrawal of the seas from the continents during Cenozoic time. Also of interest, both second- and third-order cycles are inferred to have relatively slow rise phases, and relatively fast fall phases.

Vail and his co-workers feel able to approximate the amplitudes of the relative changes. Their best estimate for the highest stand is *c*. 350 m above present sea level, at the end of Campanian time (Late Cretaceous). The most prominent lowstands were during Early Jurassic, middle Oligocene and late Miocene, being about 150, 250 and 200 m, respectively, below present sea level. The charts are scaled from +1 to −1 using the maxima (Late Cretaceous and Oligocene).

Global ridge volume changes would appear to be of sufficient magnitude and duration to account adequately for the first-order cycles and most of the second-order cycles. This is difficult to check for pre-Mesozoic curves since sea-floor spreading evidence (ocean-crust magnetic anomalies) is not available. Glaciation and deglaciation phenomena are a second popular explanation for global sea-level fluctuations, but are probably too short-lived to explain more than a few of the second-order cycles. They are, however, an attractive explanation for the third-order cycles, especially those in the late Neogene (Vail *et al.* 1977). A third alternative has been put forward recently by Schlanger et al. (1981). Highlighting recent documentation of copious off-axis volcanism during DSDP legs in the western Pacific, Schlanger *et al.* propose that mid-plate volcanism and concomitant shallowing of the sea floor could have been a major factor in causing global Cretaceous transgressions.

Whatever their causes, the global sea-level vacillations have had significant oceanographic and sedimentological effects in the oceans. Among the more prominent of these, highstands appear to correlate with increased biotic diversity, climatic amelioration, decreased mid-water oxygen concentrations, and shoaling of the CCD in open-ocean settings (e.g. Fischer & Arthur 1977). Additionally, Vail *et al.* (1980) have argued that their global lowstands coincide with widespread non-deposition (hiatuses) on continental rises and abyssal plains—though Tucholke (1981) questions the validity of this assertion. However, these are generalizations: effects of changing global sea levels obviously have superimposed on them other forcing parameters such as changing continental distributions and various internal feedback processes arising from global climate changes. The interaction of these factors is examined in the second part of this paper.

Drift history and oceanographic responses

Since the first DSDP leg in 1968 the sites drilled (more than 600) have encompassed every ocean save the Arctic. As a consequence the history of Cretaceous–Cenozoic continental drift is now much better known. In concert with palaeomagnetic determinations this knowledge has given us increasingly useful serial palinspastic reconstructions (e.g. Smith *et al.* 1981) from which to consider likely oceanographic differences between Cretaceous–Cenozoic oceans and the modern ocean. Those drift events which are believed to have most influenced the oceanographic history of the world ocean are summarized in Table 1. Reconstructions of the Palaeocene and Oligocene world ocean (Fig. 3) provide a framework for the following brief synopsis, distilled from numerous references (for summaries see Ul Haq 1981; Thiede 1981).

The Cretaceous oceans differed considerably from Cenozoic and Recent oceans. Global sea levels were higher and the disposition of continents was such that the dominant feature of the modern southern ocean, the circum-Antarctic current, was not established. High-latitude temperatures were warmer, thermal gradients were lower and the climate was generally equable. During the early Tertiary there was a somewhat abrupt transition to cooler high-latitude temperatures, steeper latitudinal thermal gradients and accentuated seasonality. Patterns of bottom-water distribution, so important in the development of the Cenozoic–Recent oceans, must have been very different in Cretaceous times. Bottom water was most likely produced by the sinking of warm, salty water formed by evaporation in the then extensive low- and mid-latitude marginal seas, rather than by sinking of cold, dense water

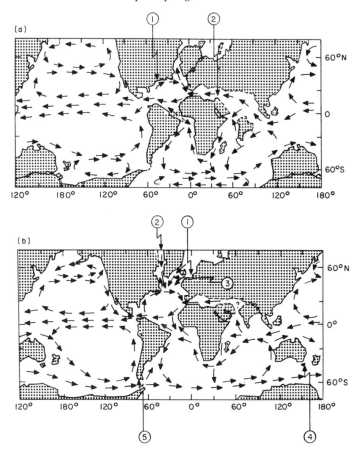

FIG. 3. Schematic reconstructions of the Palaeocene and Oligocene world oceans and patterns of surface circulation (after Ul Haq 1981, and references quoted therein). Features mentioned in text: (a) Palaeocene (1) Proto Gulf Stream; (2) Tethys current; (b) Oligocene (1) Norwegian–Greenland Sea; (2) North Labrador Passage; (3) Greenland–Iceland–Faroe ridge; (4) South Tasman Rise; (5) Drake Passage.

formed in higher-latitude marginal seas as at present (Arthur & Natland 1979; Brass *et al.* 1982).

Drift events during the prolonged fragmentation of Gondwanaland also ensured that the Cretaceous oceans were much less well mixed than were more recent oceans. Submarine edifices divided the early South Atlantic into discrete basins as Africa and South America moved apart. North of the Walvis–Sao Paolo Ridge, evaporites deposited on new oceanic crust reflect a denser, more saline water mass than that which existed south of the ridge in the Cape-Argentine Basin, itself partially restricted behind the Falkland Plateau. Sporadic spillover of dense waters from these basins, and others like them, is an attractive explanation for episodic stagnation in the Cretaceous Atlantic (Thierstein & Berger 1979; Arthur & Natland

1979; Berger & Thierstein 1979). Foundering of the barriers as the South Atlantic widened led to progressively more complete mixing of the water masses.

During Palaeocene times circulation patterns established during the Cretaceous largely maintained themselves (Fig. 3a). In the absence of a circum-Antarctic current system, clockwise subpolar gyres are inferred in the S Atlantic and Pacific oceans. The circum-global, tropical, Tethys Current was still the significant feature it had been in Jurassic-Cretaceous times, dispersing the characteristic Tethyan marine biotas widely. In the N Pacific two gyres have been inferred on the basis of experimental studies (Ul Haq 1981, and references therein).

By late Oligocene times (Fig. 3b) there had been many changes, and the global surface circulation pattern had essentially evolved its

TABLE 1. *Summary chart of major events and trends in the geology, geography, chemistry and biology of the Cretaceous–Cenozoic world ocean*

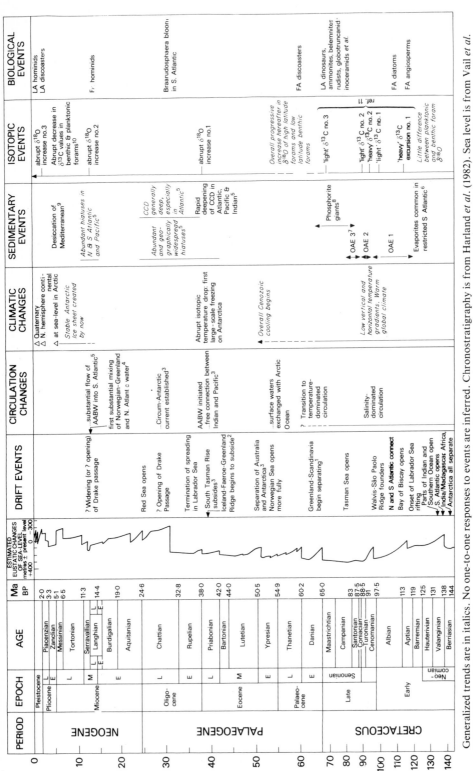

Generalized trends are in italics. No one-to-one responses to events are inferred. Chronostratigraphy is from Harland *et al.* (1982). Sea level is from Vail *et al.* (1977) (Nb. *eustatic* element shown: involves superimposition of continental-margin subsidence on relative sea-level-change curve). For timing of events project across to age column opposite first word of annotation. Annotations are from Harland *et al.* (1982) or general accounts cited in text, except where referenced. For similar tabulations involving more detailed chemical and biological data see Arthur (1979) and Arthur & Jenkyns (1981).

Abbreviations: △ Glaciation, FA first appearance, LA last appearance, AABW Antarctic Bottom Water, NADW North Atlantic Deep Water, OAE Oceanic anoxic event.

References: [1]Talwani & Eldholm (1977), [2]Talwani & Udintsev (1976), [3]Kennett (1977), [4]Eldholm & Thiede (1980), [5]Ramsay (1977), [6]Arthur & Natland (1979), [7]Jenkyns (1980), [8]Arthur & Jenkyns (1981), [9]Hsü *et al.* (1973), [10]Savin *et al.* (1975).

present-day features. In the N Atlantic, Greenland and Scandinavia had begun separating in the early Palaeocene, and by early Eocene times the Norwegian–Greenland Sea may have been exchanging surface waters with the Arctic Ocean. The North Labrador Passage, which began to open in the Palaeocene, was the route for transfer of relatively warm Atlantic water into the Arctic by middle Eocene times. The Greenland–Iceland–Faroe Ridge, a volcanic high across oceanic crust between Scotland and Greenland was for a long time a barrier to deep-water exchange in the developing Atlantic. It began sinking in late Eocene times, but there was no significant mixing of Norwegian–Greenland Sea and N Atlantic waters, and hence establishment of the modern Atlantic deep-water (NADW) flow, until mid Miocene times (Thiede 1981).

In the tropics, the Tethyan current was maintained until late Eocene times, finally to be interrupted by a complicated series of collisions of the Arabian and Indian plates, plus associated microcontinents and arcs, against Eurasia.

Formation of new ocean floor between Australia and Antarctica began in early Eocene times, and subsidence of the South Tasman Rise in late Eocene times finally allowed free connection between the Indian and Pacific Oceans. At this stage Antarctica was sufficiently isolated to allow the first large-scale freezing at sea level, and the cold, dense, oxygenated northward-flowing, Antarctic bottom water (AABW) was initiated near the Eocene/Oligocene boundary. The circum-Antarctic Current was not fully established until the Drake Passage opened between South America and Antarctica, an event which probably took place sometime in Oligocene times (it is difficult to date exactly because magnetic anomalies are poorly defined) (Kennett 1977).

Clearly plate tectonic processes, in the form of 'drifting and spreading events', have played a great if not overriding part in influencing the development of ocean-water circulation, both deep and shallow, and of Cenozoic climatic patterns. Continental dispersal since Cretaceous times has produced the present, essentially meridional, continental configuration, disrupted the circum-global tropical surface flow of Cretaceous–Palaeogene times, and given rise to the Neogene–Recent circum-polar, southern hemisphere surface flow. It has also contributed to the build-up of Oligocene–Recent ice. Secular withdrawal of seas from the Cretaceous shelf areas, itself believed to be a first-order response to decrease in volume of the global spreading centres (e.g. Hallam 1977) reduced the area of the net-evaporation zone in low- and mid-latitudes, increasing contrasts in global thermal gradients. Acting in concert with the circulation changes arising from drift events, this finally led to the creation of a stable Antarctic ice sheet in Miocene times (Shackleton & Kennett 1975).

Carbonate stratigraphy

Carbonate-facies patterns in the deep sea

That carbonate is dissolved in the deeper reaches of the ocean basins was known from the earliest dredge hauls of the *Challenger* expedition in the last century. The level of the facies boundary between carbonates and deeper-water siliceous oozes and red clays became known as the calcite compensation depth (CCD). The position of the CCD is a function of supply rate and dissolution rate of calcite (largely coming from calcareous micro-organisms); these in turn depend on the fertility of upper waters and on the state of saturation of deep waters (principally their alkalinity and CO_2 content). As such, the past position of the CCD is likely to be much affected by changes in circulation patterns, temperature, and salinity of ocean water such as those arising from the changing continental dispositions and global sea-level fluctuations outlined in the previous sections. An early discovery of the DSDP was that this is very much the case. Tracking the variations in magnitude and configuration of the CCD, both within and between the major ocean basins, has become a critical topic in palaeo-oceanographic investigations.

Figure 4 is a schematic representation of the principal controls in distribution of carbonate facies in an ocean basin, based on the Pacific. Below areas of open-ocean high productivity the CCD is depressed, as is the case below the Pacific equatorial divergence. Polewards, the CCD rises due to the presence of corrosive bottom waters. It also rises against the continental margins, where phytoplankton productivity is high and the high CO_2 levels in the pore waters of the resulting organic-rich sediments are not favourable to carbonate preservation. Diluting of carbonate by terrigenous sediment may also in part explain the rise. The CCD averages 4500 m depth in modern open-ocean environments, being deepest (down to 5500 m) in the well-flushed, CO_2-poor N Atlantic and shallowest in the poorly mixed, CO_2-rich N Pacific (rising to above 4000 m).

Complexities in dissolution processes arise from the variable nature of calcareous pelagic

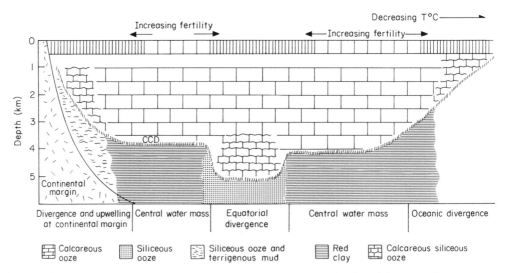

FIG. 4. Schematic diagram (based on Pacific Ocean) illustrating distribution of deep-sea sedimentary facies and characteristics of the CCD relative to major patterns of oceanic circulation and near-surface fertility (after Ramsay 1977).

organisms. The aragonitic pteropods dissolve at relatively shallow depths: <300 m in the Pacific and 2000–2500 m in the Atlantic. The low-Mg calcite organisms, foraminifera and coccoliths, dissolve selectively depending on the robustness of shells. In addition to the CCD, a second even more complex level of dissolution is recognizable in most places: the lysocline. This is defined as the level of maximum change in the composition of foraminiferal assemblages through dissolution (Berger & Winterer 1974). Recent work has shown the need to view oceanic sediment sections in terms of a hierarchy of dissolution levels: the aragonite lysocline, the aragonite compensation depth, the calcite lysocline and the calcite compensation depth (e.g. Winterer & Bosellini 1981). Most is known about the CCD, however, since being a carbonate–non-carbonate-facies boundary it is easiest to track in sediment sections.

Reconstructing the past behaviour of the CCD

Establishing the past position of the CCD depends on the uniform rate at which a portion of ocean crust subsides down the flanks of a mid-ocean ridge as sea-floor spreading progresses. Once suitable corrections are made for isostatic adjustment of the basement to the sediment load, this 'age–depth constancy' curve provides a template for reading CCD palaeodepths for carbonate–non-carbonate boundaries in sediment sections. Using well-spaced sites, a palaeodepth-palaeolatitude matrix allows recon-

struction of the past relief of the CCD for any time interval for which there is sufficient DSDP drill-hole data. For historically important reviews see Berger & Winterer (1974) and van Andel (1975).

Ramsay (1977) has provided a synthesis of the vast amount of data bearing on reconstruction of the past behaviour of the CCD which arose from the first half of the DSDP programme. Similar syntheses (e.g. van Andel et al., 1977) differ in detail but not in the main conclusions. The CCD has varied both between the ocean basins and within them. Figure 5(i) shows the variation within the Pacific. Though the distinctive shape of the CCD curve—i.e. depressed beneath the equatorial zone and shallowing towards higher latitudes, being deeper at mid-latitudes in the South Pacific— has been maintained since the late Eocene, there are several striking fluctuations. Most notable of these is the drop at all latitudes between late Eocene and late Oligocene. Subsequent to the Oligocene the CCD has continued to be depressed north of 10°N, but shallowed south of 10°S in the early Miocene (note also the smoothing of the high latitude portion of the curve at that time), before continuing to be depressed. The equatorial belt of carbonate deposition was at its widest in the late Oligocene, when the CCD was everywhere low. After the early Miocene, the bilateral symmetry of the belt was distorted, probably as a result of variations in the topography of the East Pacific Rise.

In Figure 5(ii) the CCD variations in the Pacific

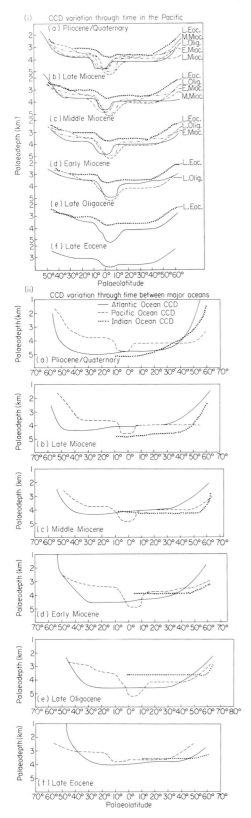

FIG. 5. Cenozoic CCD variations (i) within the Pacific, and (ii) comparison between Atlantic, Indian and Pacific oceans (after Ramsay 1977).

are compared with those in the Atlantic and Indian Oceans. There are a number of trends which are detectable in all oceans and which therefore reflect events of global importance. The deepening of the CCD between late Eocene and Oligocene is the most prominent common characteristic of all oceans, though it is less pronounced in the Indian Ocean than in the others. Subsequent to the Oligocene the pattern of fluctuation in CCD depth in the Atlantic has been broadly similar to that in the Pacific, and the post-late Miocene deepening at low latitudes is common to all the oceans. The post-Oligocene downward migration of the CCD in the N Pacific is also a feature in the N Atlantic and Indian oceans.

However, the shapes of the curves and the average CCD levels are clearly different in each of the oceans. Most interestingly, the relatively shallow Pacific CCD and deep Atlantic CCD have existed since the late Eocene, a pattern which is especially marked in the northern hemisphere. At that time, therefore, the Atlantic must already have had its present more 'lagoonal' characteristics relative to the Pacific (i.e. an outflow of oxygen-rich deep water, and low rates of calcite dissolution). According to Ramsay (1977) this is probably best explained by the Paleocene–early Eocene opening of the Norwegian–Greenland Sea, and advection of cold high-latitude water into the Atlantic. The end-Eocene drop in the CCD in all the oceans is best explained by the development of sea-ice in the Oligocene and the introduction of highly oxidized 'young' bottom-waters, which are not as corrosive to the tests of foraminifera and coccoliths as 'old' bottom-waters rich in CO_2 (e.g. Kennett 1977).

Such explanations, in addition to finding support from refined appreciation of drift history and from oxygen-isotope palaeo-temperature studies (see later section), also draw on the distribution of hiatuses in pelagic sections. The inception of a deep-water circulation system should in general be reflected by an increase in hiatuses in sediment sections deposited along its path, and variations in the vigour of flow should similarly be indexed. Though complexities arise because carbonate dissolution gradients have increased during Neogene times (Heath *et al.* 1977), and sedimentation rates influence hiatus development (Moore & Heath 1977) hiatuses are in general geographically widespread in the late Eocene and Oligocene: a pattern which may reflect encroachment of northward-flowing AABW in the Atlantic and Pacific oceans (Ramsay 1977). Similarly, southward-flowing N Atlantic deep water (NADW) may account for middle Miocene hiatuses in the Atlantic.

Interpretation of CCD fluctuations in terms of influx of corrosive bottom-water currents is attractive, but the overall parallelism of the Atlantic and Pacific CCD may indicate the operation of other, more 'global' effects (Berger 1979). These cannot easily be accounted for. The oft-quoted relationship of the CCD to global sea-level stands holds that during transgressions carbonate is locked up on shelves, causing an overall global CCD rise; and that during regressions carbonate is transferred from the shelves of the seas, more being available for pelagic micro-organisms so that the CCD falls (e.g. Berger & Winterer 1974). However, sedimentation rate patterns do not seem to support this simple picture (Davies *et al.* 1977). Global sea-level changes may act in concert with circulation changes (themselves a product of varying drift configurations and climatic changes) to provide the principal control on CCD behaviour. However, the position of the present CCD has not yet been adequately modelled from first principles, and interpretations of past positions must in consequence be undertaken with considerable circumspection (Berger 1979).

Carbon in deep-sea sediments

Carbon-rich sediments in general require an interplay of high supply rate and/or low content of dissolved oxygen in the ocean-water column in order to accumulate. They are therefore not common on the floor of the modern well-aerated oceans, being found only on certain outer continental shelf and upper-slope areas under zones of higher-than-usual phytoplankton productivity—e.g. off Angola, western India, and Peru. However, 'black shales' (a blanket term for any marly or siliceous sediment with an organic content of more than 1%) have been retrieved in DSDP cores from a surprising number of sites. In the Atlantic, Pacific, Indian and Antarctic oceans they represent a wide variety of palaeo-bathymetric settings including the deep ocean floor (central N Atlantic) and isolated oceanic plateaus (mid-Pacific). Their origin poses an interesting non-uniformitarian problem. With minor exceptions port-Palaeozoie

black shales (e.g. early Miocene black shales deposited on the African continental margin off Morocco, DSDP site 397) are of Cretaceous age, and most cluster in three time intervals: late Barremian to Albian, around the Cenomanian–Turonian boundary and (to a lesser extent) in the Coniacian to Santonian. Interestingly, black shales in ancient ocean floor, continental margin, and continental shelf successions preserved on land for the most part fall in the same age range (summary in Jenkyns 1980). Evidently organic carbon was much more liable to be preserved in the world ocean at certain times than at others. Schlanger & Jenkyns (1976) have termed these times 'oceanic anoxic events'.

Petrographic and geochemical studies of the organic matter (OM) in the Cretaceous black shales have proved instructive. Kerogen (the insoluble fraction of OM) usually amounts to *c.* 95–99% of the total OM in Cretaceous black shales, and can be studied directly in whole-rock samples by pyrolysis, which distinguishes the type of OM. Optical examination and vitrinite reflectance provide information on the stage of thermal maturation, and elemental analysis and infrared spectrometry yield the chemical composition of isolated kerogen. Kerogen is classified by plotting atomic ratios of H/C against O/C (Tissot *et al.* 1974). Organic matter in Cretaceous black shales falls in three categories, defined by decreasing H/C and increasing O/C ratios. Marine or planktonic OM (type 2) comprises mainly an amorphous organic ('sapropelic') material, is H-rich and O-poor and has a high potential for oil generation. Terrestrial OM (type 3) is moderately degraded (less H, more O), and has an appreciable component of recognizable plant debris. Some kerogens comprise recycled OM, which has been deeply oxidized in subaerial conditions ('residual' OM). Type-1 kerogens are those in oil shales and other H-rich deposits, so far not found in DSDP holes (summary in Tissot *et al.* 1979).

Cretaceous black shales recovered during the DSDP fall in four regional groups: central N Atlantic, northern N Atlantic, S Atlantic–Indian, and Pacific (summary in Weissert 1981). The most extensive investigations of OM type have been carried out on Atlantic black shales. In the central N Atlantic, Barremian to Albian black shales in deep basinal environments average 1–4% organic carbon, with local values of up to 30%. In eastern sites (*c.* 3 km palaeo-depth) the OM is predominantly of marine origin, and in western sites (4–5 km depth range) the OM is predominantly of terrigenous origin. According to some works however, common fine lamination and general absence of bioturbation indicate that bottom waters were frequently anoxic during deposition, and the degree of preservation of OM appears to have been dependent on the intensity of anoxia (summary in Weissert 1981). Palynological studies led Habib (1982) to conclude that most of the OM came from deltaic systems and that rate of supply of OM and rate of supply of host sediments, if sufficiently high, could lead to the formation of black-shale facies regardless of the degree of oxicity of the bottom waters.

In the northern N Atlantic black shales were deposited on the European passive margin in water depths of a few hundred metres to a few kilometres at various times. In Barremian–Albian shales organic carbon does not exceed 2%, and is entirely of terrestrial origin; much of it seems to have been introduced by turbidity currents, and hence the black shales do not necessarily indicate bottom-water anoxia. Late Albian to Cenomanian black shales, in which organic carbon content locally rises to 9%, have a component of marine OM.

Isolated basins formed during the irregular fragmentation of Gondwanaland were the sites of common black-shale deposition in the S Atlantic and Indian oceans during the Aptian–Albian interval, during most of which time there was probably no marine connection between the S and N Atlantic. Reconstructions of palaeodepth suggest that the sites drilled varied from a few hundred metres to a few kilometres. Organic matter (OM) is of mixed terrigenous and marine origin (summary in Weissert 1981).

In the Pacific no carbonaceous sediments have been recovered from deep basinal settings (excepting an as-yet-unpublished instance from the Mariana Basin drilling of Leg 89), but black shales drilled on rises (Shatsky Rise, Manahiki Plateau, Hess Rise) at first glance would seem to indicate anoxia at depths of 2–3 km in the open ocean during the same broad time intervals that black shale deposition was common in the Atlantic (Schlanger & Jenkyns 1976). The OM in these deposits is mostly derived from aquatic marine organisms, but according to Thiede *et al.* (1982) open-ocean anoxia may not be necessary to explain them. Thiede *et al.* invoke passage of the Mid-Pacific mountains under the equatorial divergence with its high surface-water productivity, acting in concert with certain structural and hydrographic factors coinciding with a time of relatively sluggish bottom-water circulation, to explain the organic-rich beds.

In view of the composite provenance of their OM, what is the best explanation for the origin of the Cretaceous black shales? The clearly demarked time intervals during which the majority of the black shales were deposited (oceanic anoxic events or OAEs) invite some kind of 'global' explanation. Schlanger & Jenkyns (1976) pointed out that the OAEs correspond to eustatic (global) elevations of sea level. Increased phytoplankton productivity in surface waters would have been the result of the increased area of nutrient-rich shelf seas. Oxidative consumption of the increasing amounts of OM sinking from the upper mixed layer would cause expansion of the oxygen-minimum layer (Fig. 6a), in both vertical and lateral senses. In time, this layer could grow to encompass oceanic rises and even basin floors. Wherever it intersected the sediment–water interface, the preservation of C-rich sediments would be favoured. Schlanger & Jenkyns (1976) further suggested that the terrestrial component of the mixed marine/terrestrial OM in many Cretaceous black-shale sequences derives from the incorporation of land plant material which had colonized lowland areas prior to the transgressions.

The Schlanger–Jenkyns model is appealing because it relates two essentially global phenomena (transgressions and widespread stagnation in oceans) without being 'universalist'. For example, the intensity of oxygen-minimum layer expansion would doubtless have varied with the degree of productivity increase, both between oceans and within oceans, allowing for considerable variation in preservation of carbonaceous sediment. One problem is that it is not immediately clear why only *some* of the numerous Cretaceous eustatic transgressions (Fig. 3) should have caused OAEs.

Various workers have considered alternative spatially restricted models, which may have acted independently or in concert with the global sea-level model. Barred basins of the Black Sea-type stagnate by simply being unable to overturn water below the upper mixed layer (Degens & Ross 1974), or possibly by entrainment of H_2S from mid-water anoxic layers by sinking dense plumes (Southam *et al.* 1982); in either case dissolved oxygen is rapidly used up as OM is produced and sinks through the water column (Fig. 6b). Although that model is probably inappropriate to most of the Atlantic black shales because of the continuously changing configuration of the ocean basins during sea-floor spreading, isolation of water masses may have been important in other ways. During isolation, water masses in adjacent basins can evolve to contain waters of differing salinity and density; release of such waters can cause salinity

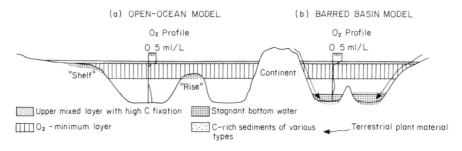

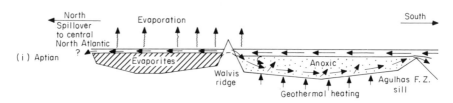

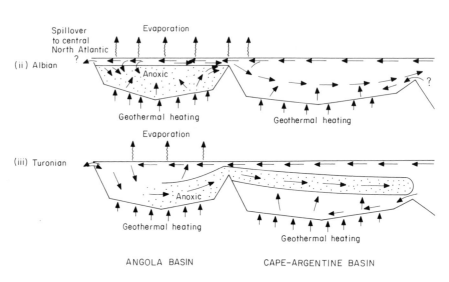

FIG. 6. Models for the origin of Cretaceous black shales. (a) and (b) after Schlanger & Jenkyns 1976, (c) after Natland 1978. For further explanation see text.

stratification and stagnation once barriers founder as the continents drift (Thierstein & Berger 1978). Natland (1978) and Arthur & Natland (1979) have applied this model to the prime example, the S Atlantic (Fig. 6c). There, the intra-oceanic Walvis-São Paolo Ridge isolated the Angola–Brazil Basin until late Albian time. Evaporites accumulated until mid-Aptian time. During this process, dense, hyper-saline water flowed into the basin, developing a stratified water mass so that carbonaceous mudstones were able to accumulate in anoxic deep-water areas. To the south, the Cape-Argentine Basin was stagnant during most of Aptian time, being partially restricted behind the Falkland Plateau. Sporadic overflow of dense brines from the Angola–Brazil Basin may have contributed to the stagnation (Fig. 6c). During Albian times, as the S Atlantic widened and a sill present along the Agulhas Fracture Zone was deepened, the Cape-Argentine Basin became oxygenated. Intermittent salinity stratification and anoxia continued in the more restricted, though no longer isolated, Angola–Brazil Basin from late Albian to Coniacian time (Fig. 6c).

Models of a similar nature have been applied to stagnation events elsewhere. Ryan & Cita (1977), among many others, have explained the brief periods of stagnation in the Quaternary Mediterranean (indicated by thin sapropel layers in pelagic sections) by salinity stratification below a brackish-water lid derived from river input during interglacials. Additionally, models involving water masses of differing density provide one possible explanation for

centimetre- to millimetre-scale carbonaceous–non carbonaceous cyclicity, a common feature of black-shale deposits. These may arise from a pulsing supply of saline bottom water from the 'reservoir' basin: in the case of the late Cretaceous black shales, as a result of spillover of saline water from the Angola Basin (Thierstein & Berger 1978).

Isotope stratigraphy

Oxygen

Overall progressive cooling of surface temperatures on Earth during Cenozoic times, with short periods of amelioration during the Eocene and Miocene, has long been appreciated

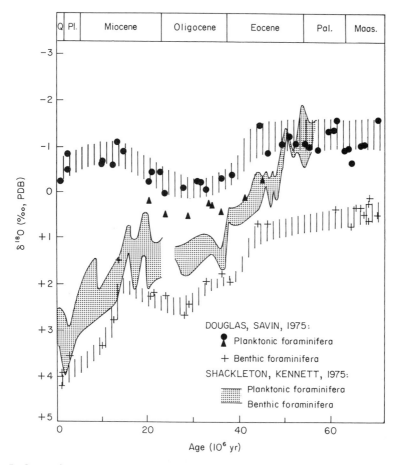

Fig. 7. Oxygen-isotope stratigraphy of calcareous foraminifera in Pacific sediments (reflecting decreasing palaeotemperature of ocean water with increasing $\delta^{18}O$). The Douglas & Savin (1975) data are from low latitudes (DSDP Leg 32) and Shackleton & Kennett (1975) data are from high southern latitudes (DSDP Leg 29). Planktonic foraminifera give surface palaeotemperature, benthic foraminifera give bottom-water palaeotemperature. Note narrow range of values in the early Eocene, and increasing separation since. For further explanation see text.

by stratigraphers. One of the most exciting developments in palaeo-oceanography has been the use of oxygen isotopes to document this process in the oceans. The $\delta^{18}O$ of calcareous shells gives an indication of the temperature of the water in which they were secreted (see summary in Kennett 1982). Thus, calcareous shells of benthic foraminifera may be used as an index of bottom-water temperatures and shells of planktonic foraminifera as an index of surface-water temperatures.

Figure 7, a classic diagram which has been reproduced in legions of review articles, summarizes some of the key early work and illustrates several of the major lessons of oxygen-isotope studies. The Douglas & Savin (1975) data are from low latitudes and the Shackleton & Kennett (1975) data are from high southern latitudes. First, the familiar Cenozoic cooling trend is clearly a high-latitude pheno- menon. Low-latitude planktonic foraminifera show that, with the exception of the Oligocene, surface waters in the tropics must have remained about as warm as they were in the latest Cretaceous. The Cenozoic cooling trend is recorded by benthic foraminifera at low latitudes and by both benthic and pelagic foraminifera at high latitudes. Superimposed on it are several prominent steps. Of these, the abrupt $\delta^{18}O$ drop at about the Eocene– Oligocene boundary is perhaps the most striking. This is probably best explained by the onset of Antarctic bottom-water formation at tempera- tures close to zero. However, there is no common agreement on how the temperature drop took place, or why it happened so abruptly (within 100,000 years according to Kennett & Shackleton 1976).

The mid-Miocene data show a drop towards heavier $\delta^{18}O$ values in high-latitude foraminifera and in low-latitude benthic foraminifera. This is held to indicate both a temperature drop at the bottom-water source, and a glacial effect (locking up of ^{16}O in ice) due to rapid ice expansion on Antarctica (Shackleton & Kennett 1975). Interestingly, low-latitude planktonic foraminifera indicate a concurrent warming trend in the tropics. Again, palaeo-oceano- graphers find difficulty in arriving at a consensus over the possible explanations for this pattern. The prime candidates (one of which or all-in-combination may have been responsible) are a change in the entire system of circulation (the opening of the Drake Passage possibly occurred at this time), high-latitude albedo increase due to ice build-up, or low-latitude albedo decrease due to Miocene transgression. The absence of reliable data on the exact age of

opening of the all-important Drake Passage (between S America and Antarctica), which allowed the development of the circum- Antarctic current, is a continuing source of frustration in this question.

The final major event in the oxygen-isotope stratigraphic record is a sudden change to heavier values in late Pliocene times. This is widely interpreted as an indication of the onset of continental glaciation in the northern hemisphere. No radical change in geological setting is apparent at this time: the Pliocene world of 3 Ma was much as it is today. Rather the glaciation was probably the result of a cumulative series of positive feedback effects arising from general climatic deterioration. Mountain building and increased volcanic activity during Pliocene times may have been contributory factors (summary in Berger *et al.* 1981).

Oxygen-isotope studies can also be brought fruitfully to bear on stratigraphy of an order of magnitude more refined resolution. Emiliani (1978) has related a 'sawtooth' pattern of $\delta^{18}O$ values in a Pleistocene core from the Pacific to water temperature changes arising from glacial– interglacial cycles. With a period of average *c.* 100,000 years, the cycles in the core have a characteristic shape involving rather gradual $\delta^{18}O$ increase (reflecting glacial build-up) and relatively rapid $\delta^{18}O$ drop (reflecting more abrupt deglaciation). Currently, numerous studies of similar resolution are being under- taken on Neogene cores. These have been greatly facilitated by the introduction (since DSDP Leg 64 in 1978) of the hydraulic piston corer to the inventory of equipment aboard *Glomar Challenger*. Unlike sediments recovered by the normal rotary coring technique, uncon- solidated and semi-consolidated material can be retrieved undisturbed with this device.

Carbon

Carbon isotopes provide a tool for the palaeo- oceanographer equally as exciting as oxygen isotopes, though more difficult to interpret. Isotopic ratios of dissolved carbon in the oceans are shifted by two processes: extraction of ^{13}C into various carbon reservoirs and additional supply of ^{12}C into the ocean from such reservoirs. The most likely reservoirs available for rapid exchange are terrestrial plant matter, soil carbon derived from land plants, marine organic sediments, atmospheric CO_2, and carbonate sediments. The complicated 'balance books' of these reservoirs, and effects of trans- gression and regression on exchange between them, are summarized by Scholle & Arthur (1980).

Carbon-isotope stratigraphy, like oxygen-isotope stratigraphy, reveals both cycles of variation (see next section) and isotopic 'events'. Of these, the terminal Miocene 'event' is one of the most prominent. Both planktonic and benthic foraminifera from the Indian, Pacific and S Atlantic oceans record an abrupt shift to more negative $\delta^{13}C$ values *c.* 6 Ma BP (summary in Berger *et al.* 1981). This shift, which occurred over a *c.* 200,000 year period, marks the onset of the famous Mediterranean desiccation event. This, the so-called 'Messinian salinity crisis' is probably the most spectacular DSDP discovery. The record of sedimentation from its deep basins suggests that the whole Mediterranean dried up at the end of the Miocene, creating huge waterfalls at the Gibraltar straits and allowing deposition of vast amounts of evaporites (e.g. Hsü *et al.* 1973). Global regression during the Messinian may have introduced large amounts of isotopically light carbon from eroded soil and land plant material into the ocean as a major contributory factor to the negative $\delta^{13}C$ values of ocean waters at the end Miocene C isotope event (Berger *et al.* 1981).

Scholle & Arthur (1980) describe similar $\delta^{13}C$ excursions from widely distributed sections of Cretaceous pelagic limestone in the circum-Atlantic–western Tethyan region. Recognizing at least seven major events, all correlatable over wide areas and, interestingly, occurring at stage boundaries (Table 1), they stress the association of the two pronounced 'heavy' excursions (near the Aptian–Albian and Cenomanian–Turonian boundaries) with 'oceanic anoxic events'.

Rhythmicity in the oceanic realm

From the foregoing sections it will be clear that in terms of its geography, circulation, sedimentation, chemistry and biology the present world ocean represents but one phase in the evolution of a highly complex and continuously evolving realm. 'Non-uniformitarian' overprints of varying degree have been a persistent feature of the evolution of our planet, and nowhere is the record of their effects on the atmosphere, biosphere and hydrosphere better preserved than in the condensed sedimentary records of the open-ocean basins.

The flood of data from DSDP-related research over the last decade has led to a widespread realization that broadly uniformitarian and non-uniformitarian conditions have alternated with each other through Phanerozoic time, on several time scales. Palaeo-oceanographers have attempted to articulate the parameters involved in these rhythmic oscillations. A seminal treatment of this kind was published by Fischer & Arthur in 1977, who recognized two alternating modes with a *c.* 32 Ma variation period over the last 200 Ma. Their 'polytaxic' phases are, in general, times of warmer seas, gentler latitudinal and vertical temperature gradients, overall dampening of the rate of oceanic circulation as well as the vigour of deep currents, expansion of the oxygen minimum zone, elevation of the CCD, and highly diverse and complex animal communities. Polytaxic 'peaks' occur in mid-Triassic, Liassic, Late Jurassic, Aptian–Albian, Senonian, Eocene and mid-Miocene times. 'Oligotaxic' phases are those times when the ancient world ocean resembled the modern one—i.e. when seas in general were cooler, temperature gradients were more extreme, the ocean waters were well ventilated with only local regions of significant oxygen depletion (below areas of locally high primary productivity associated with upwelling), when the CCD was lower, and when in the biosphere extinction rates of nekto-planktonic organisms were relatively high, global diversity was low and communities reduced in complexity. Fischer & Arthur (1977) recognize oligotaxic peaks in late Triassic, mid-Jurassic, Neocomian, Cenomanian–Turonian, Palaeocene, Oligocene, and Pliocene–Quaternary times.

As the names given to the modes imply, manifestation of the rhythmicity is most marked in the oceanic biosphere (Fischer 1981). The most impressive rhythmicity is in planktonic foraminiferal diversity, though Fischer & Arthur (1977) also present apparently substantiating data from phytoplankton, ammonites, and chordates. Interestingly, superpredators such as galeoid sharks are at their most diverse in the record approximately at diversity peaks and 'disaster forms' (species which proliferated in adverse conditions) are a feature of oligotaxic peaks. Of the latter, the coccolith *Braarudosphaera* is perhaps the most spectacular: it is a genus currently limited to obscure bays, living in conditions unsuitable to pelagic organisms, in Maine, Panama and Japan. It is a rare component of Upper Jurassic to Recent open-ocean chalks and oozes, but during crises becomes an appreciable component of the sediment. The Oligocene *Braarudosphaera* chalk of the S Atlantic is perhaps the best-known example.

The geochemical and geological data on which Fischer & Arthur (1977) draw most heavily are summarized in Fig. 8. (For similar, more recent compilations of data see: Arthur 1979; Scholle & Arthur 1980; Arthur & Jenkyns 1981; Scholle *et al.* 1983.) Oxygen-isotope measurements of palaeotemperature for the Mesozoic use data from belemnites from NW Europe. These vary between 'cold-water' and 'warm-water' states fairly convincingly. For Cenozoic information Fischer & Arthur (1977) make use of the foraminiferal data alluded to previously. The rapid Oligocene cooling corresponds to an oligotaxic peak, though it is part of a clear overall cooling trend affecting the whole of the Cenozoic.

Skeletal carbon-isotope data interestingly follow the polytaxic–oligotaxic rhythms fairly closely, though there are exceptions and the exact causes of the variations are not simple (Fischer & Arthur 1977; Scholle & Arthur 1980). Photosynthetic fixation selectively favours ^{12}C over ^{13}C, and OM in live organisms and in sediments is therefore enriched in the light isotope. Hence, as a generalization, $\delta^{13}C$ values in ocean waters might be expected to be heavier in polytaxic times, when primary productivity is high and when appreciable amounts of $\delta^{12}C$-enriched OM are being sequestered unoxidized in sediments because of well-developed oxygen minima. (Foraminiferal tests are, of course, secreted using dissolved carbonate in the ocean water, and thus, since organic deposition of carbonates only slightly favours the heavier isotope, they record the isotopic signature of the ocean-water carbon reservoir.) Other processes may complicate this issue, however. In particular, supply of carbon isotopes to the ocean may vary significantly, both in quantity and proportion of isotopes, and temporal variations in the rates of carbonate deposition in shelf areas are also likely to change the isotope ratios (see discussions in Fischer & Arthur 1977; Scholle & Arthur 1980; Berger *et al.* 1981).

The other data illustrated in Fig. 8 also appear to fit with the rhythms outlined by Fischer & Arthur (1977), and there is no doubt that the concept of rhythmically alternating variations in mode in the pelagic realm is in general terms a valid one. However, numerous enigmas remain, not least of which are the driving mechanisms for the rhythmicity, and the explanation for its apparent uniformity during Cenozoic time. Fischer & Arthur (1977) observe that their variations in mode show no relation to plate tectonic dispersal patterns or magnetic reversal history. There is, they assert, some resemblance to eustatic oscillations and, of course, to terrestrial climates (though to what extent these may themselves be a product of eustacy remains an unanswered question). Another problem is that one of the 'oceanic anoxic events', at the Cenomanian–Turonian boundary, corresponds to an oligotaxic, not a polytaxic, phase. Additionally, there is no obvious reason why hiatus development should follow the rhythms in a simple way, and recent syntheses indicate that they do not (Heath *et al.* 1977). Initiation of density-driven flows of high-latitude water are a prime cause of hiatuses in open-ocean pelagic settings, and these are caused by circulation changes in response to continental-drift events, as outlined previously (e.g. Ramsay 1977).

Rhythmic variations of broadly similar character occur in the Early Palaeozoic marine realm (though their documentation lacks isotopic data). They are not of uniform period, varying between *c.* 5–45 Ma. (Leggett *et al.* 1981). Wilde & Berry (1982) have argued for progressive ventilation of the oceans through the Phanerozoic as a result of major glaciations. This could provide one reason for the absence of a clear pre-Mesozoic cyclicity similar to that defined by Fischer & Arthur (1977) for Mesozoic–Cenozoic times.

Recently, Fischer (1980, 1981) has emphasized the importance of two other notable sets of cycles. Major cycles marked principally by oscillations of global climate and consequent biotic diversity, which he terms 'greenhouse' and 'icehouse' states, have a period of around 300 Ma. The Phanerozoic pattern of these compares closely to the first-order relative sea-level cyclicity of Vail *et al.* (1977). According to Fischer these cycles may—because of a close correspondence between volcanism and fluctuations in level—be attributed to changes in the pattern and vigour of mantle convection.

The importance of a further family of cycles, in the 10,000—100,000 year range, is increasingly being recognized. These, following suggestions made long ago by Milankovitch, may be related to the Earth's orbital perturbations (Hays *et al.* 1976; Fischer 1981). They are manifested most clearly by millimetre- to centimetre-scale bedding rhythms, commonly reflecting pulses in plankton productivity and oscillations in sea-floor dissolution. Stable isotope studies indicate rather regular temperature fluctuations (Hays *et al.* 1976; Imbrie & Imbrie 1979).

Though most clearly displayed in Quaternary sediments, influence of the so-called Milankovitch rhythms is increasingly being invoked in Cenozoic and Mesozoic sections. For example, de Boer (1982) describes detailed oxygen-isotope studies of cyclic carbonate-rich and marly beds from the Middle Cretaceous pelagic sequence of the Apennines which suggest regular changes in temperature; these he believes are caused by

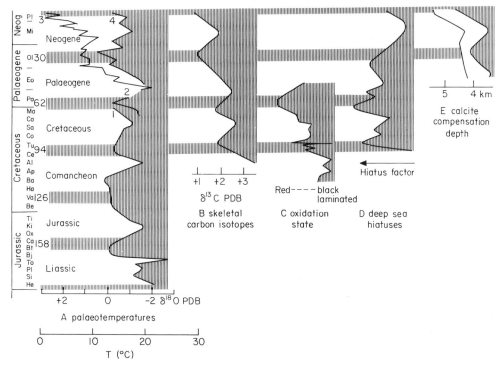

FIG. 8. Synoptic diagram showing chemical and geological parameters used to define 32 Ma cyclicity in Jurassic–Recent marine realm (from Fischer & Arthur 1977 and source papers quoted therein). Oligotaxic modes highlighted. (A) Palaeotemperatures, using oxygen isotopes in the following fossil groups: (1) belemnites from NW Europe, (2) globigerinids from the S Atlantic, (3) globigerinids from the southern Pacific (showing progressive cooling of high-latitude waters), (4) globigerinids from the tropical Pacific. (B) Synthesis of carbon isotopes from both planktonic and benthic foraminifera in pelagic carbonates. (C) Relative oxidation state of sediments in a pelagic sequence, the Scisti a Fucoidi and Scaglia formations of the Gubbio area, Italy considered representative of fluctuations widespread in the upper pelagic realm of the world oceans. (D) Semi-quantitative representation of deep-sea distribution in DSDP cores. (E) CCD variations generalized from DSDP data. The two lines bracket the range in values from different oceans.

fluctuations of the velocity of ocean-water circulation which resulted from shifts of the caloric equator due to astronomical influences.

Whatever the problems involved, the concept of rhythmic variations of a variety of scales will act as a framework on which constantly arriving new data will be hung, and as a sounding-board as palaeo-oceanographers pursue the outstanding questions in their science: those involving the root causes of major palaeo-oceanographic changes.

Systemic stratigraphy

It will be clear from the foregoing sections that perhaps the most important task facing synthesizers of data arising from deep-sea drilling is to define the driving forces of the changes in oceanographic and climatic conditions. To do this, the independent environmental variables must be identified and

correlated with the dependent ones.

Global sea level is clearly important and the links between sea-level stands and changes in climate, ocean-water temperature, carbonate sedimentation, productivity and carbon-isotope fraction have been discussed already. However, it is equally clear that to view all palaeo-oceanographic and climatic changes as a one-to-one product of rise and fall of sea level (consequent on vacillations in the rate of production of new ocean crust and thereby on perturbations of mantle convection rates, and thus in the final analysis on little-understood variations in heat production in the Earth's interior) is too great an over-simplification (e.g. Berger 1979).

In recent papers, palaeo-oceanographers at Scripps Institute of Oceanography have initiated an approach which draws analogues from systems analysis, viewing the whole climate-producing ocean-atmosphere-biosphere (the

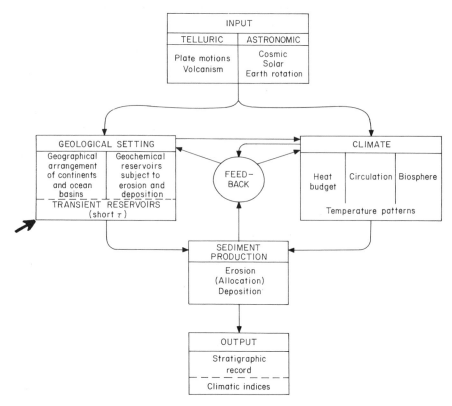

FIG. 9. The 'exogenic system' (the climate-producing, ocean-atmosphere–biosphere–lithosphere system) depicted as a machine with internal feedback control, receiving variable input from external sources and with feedback from climatic variations. Instability is introduced by isolating temporary storage bins (arrow) whose contents become available all of a sudden, mainly through endogenic changes in geography (e.g. rapid regression, or opening at a crucial strait). For further explanation see text (From Berger *et al.* 1981).

'reactive' part thereof) – lithosphere system (otherwise simply known as the 'exogenic' system) as a cybernetically* controlled machine* (Berger *et al.* 1981). This has been dubbed 'systemic stratigraphy' (Berger & Vincent 1981). Figure 9 summarizes the elements of this approach. There are two 'operator' subsystems: 'geological setting' and 'climate'. These correspond respectively to the 'plumbing' and the 'valves and parts' in that climate (being highly variable) controls the short-term performance of the machine, while the limits are set by 'geological setting'. The third subsystem is 'exogenic production', which writes the geological record ('output').

The machine varies in performance in three ways. First, input (Fig. 9) is variable. In this essay, I have considered the important telluric aspects of this variable input, but not the

astronomic aspects. The latter will be a new and recurring theme of the palaeo-oceanographic palaeoclimatic literature over the next few years, but has received little attention to date (see next section). Second, feedback processes are important. They come from the 'exogenic production' subsystem to both 'geological setting' and 'climate'—e.g. as changes in geochemical reservoir sizes or changes in organic preservation patterns. 'Climate' is highly responsive to feedback and can produce its own feedback effects—e.g. through albedo changes arising from ice build-up. However, the third method of varying performance is perhaps the most important. It involves the sudden introduction of 'transient reservoirs' into the geological setting sub-unit. This idea is an extension of Berger & Thierstein's postulation that the rapid merging of previously isolated

*Cybernetics: science of communication, in both animals and machines.

water reservoirs can explain many palaeo-oceanographic events in the Phanerozoic (Thierstein & Berger 1978, Berger & Thierstein 1979). This 'build-up and collapse' hypothesis holds that geographic changes are responsible both for initial isolation of the temporary 'storage bins', and for the eventual release of their contents (though global regression or the opening of a crucial strait). Of the candidates for this process, perhaps the proposal of an end-Cretaceous 'Arctic spillover' would be the most impressive if proved valid: it holds that the release of a brackish-water reservoir in the Arctic was the prime cause of the famous biotic crises (Gartner & Keany 1978), but is of course only one of a legion of explanations for that event. An increasingly popular explanation, stressing the presence in pelagic sediments at the Cretaceous–Tertiary boundary of iridium and other noble metals (elements richly concentrated in chondritic and iron meteorites but virtually absent in rocks of the Earths crust), invokes the impact of an asteroid or comet (Alvarez *et al.* 1980: Hsü 1981).

The 'build-up and collapse' concept has been extended to carbon reservoirs. Berger *et al.* (1981) suggest that erosion of carbon-rich sediments of shallow marine or coastal origin during regression might conceivably have delivered enough CO_2 to acidify the ocean and thus lead to increased pCO_2 in the atmosphere. This would cause warming, and oppose the effects of albedo increase from exposing more land as a result of regression, adding more complexity and difficult-to-quantify feedback links to the system. Similarly, carbon locked up in carbonaceous sediments deposited in a widely stratified ocean (a likely result of release of waters of anomalous density) would mean that carbon could enter the oceans easily, but not leave them (as gas). Atmospheric CO_2 content would drop, leading to another possible feedback complication: atmospheric cooling, which could then increase convective mixing of the ocean, releasing CO_2, and raising temperatures once more.

The possible permutations and combinations are quite clearly appreciable. The systemic stratigraphy approach is a useful means of organizing knowledge as more data becomes available with which to assess which parameters are the 'chickens' and which are the 'eggs', and which of the chickens are the larger.

Discussion: a forward look

The aspects of deep-sea pelagic sediments and palaeo-oceanography selected for particular attention in this summary include those which are likely to be the most exciting growth areas over the next few years.

Clearly the recent discoveries of hot springs and sulphides in the Pacific are only the first page of a book which largely remains to be written. The 'pay-off' of dive time *v.* discoveries has been surprisingly high in the submersible programmes undertaken to date. This should ensure that funding will be available for future diving on other parts of the Pacific spreading centres and on ridges in other ocean basins, including the back-arc basins. Though eventual exploitation of ocean-floor mineral deposits may be in the minds of some, the more immediate scientific benefits include refined appreciation of chemical fluxes in the oceans and the provision of analogues for past hydrothermal systems. Additionally, students of the oases of life associated with the hydrothermal systems were quick to realize the possible implications for modelling environments germane to the origin and early history of life on earth (e.g. Corliss *et al.* 1981), and this promises to be a continuing avenue of fruitful research.

Though sadly it appears inevitable at the time of writing that DSDP activities will soon enter a period of abeyance, there are clearly many problems which merit further investigation using current *Glomar Challenger*-type drilling. First, our understanding of the palaeogeographic evolution of the planet is still patchy in some areas. More drilling is needed to tie down the evolution of straits which have been critical to the exchange of water masses, and the Drake Passage is the prime candidate for this. The Arctic, whose sedimentary record would prove particularly informative in this respect, is unfortunately likely to remain undrilled for some time to come.

Second, additional documentation of global sea-level changes is required, and further drill sites on passive continental margins have much potential if carefully selected. In particular, carbonate platforms such as the Blake Plateau preserve a splendid and readily datable series of lowstand-related unconformities.

Almost any drill site on the continental rises and in open-ocean settings provides additional useful information with which to further assess the exact effects of global sea-level stands, provided a sufficiently continuous sediment section is recovered. Principal questions here are the further interpretation of patterns of hiatus development and construction of meaningful global sediment budgets. An obvious area in which more data is needed is the sedimentary record of abyssal circulation. Considering their likely importance as stratigraphic records of variations in circulation patterns, the contourite drifts (such as the Blake Outer Ridge) have been relatively little studied (see Stow, this volume).

The 'black-shale problem' remains a source of interest and controversy to many. One problem is the non-uniformitarian nature of the conditions under which widespread black shales were deposited. However, this may not be so insuperable a barrier as might at first be imagined. Much of the present Pacific would be anaerobic down to a depth of 2000 m if there were only 2 ml l^{-1} less oxygen in deep water below the thermocline; this reduction in oxygen content could be derived simply from warming the source area of deep waters according to Berger (1979). Thus, oxygenation of the past oceans might have been a matter of delicate balance, but with rather spectacular effects arising from disturbance of the balance. Data compiled from middle Cretaceous organic-rich sediments in the S Atlantic by Thiede & van Andel (1977) suggest that carbon preservation fluctuated considerably at all depths, suggesting a fluctuation in oxygen content around a low, critical level. Berger (1979) has argued from this that perhaps neither a single expanded oxygen minimum model nor a restricted basin model is applicable; rather, that the mid-Cretaceous S Atlantic developed a non-uniformitarian 'multilayer' profile of oxygen minima. An additional problem, of course, is that geochemical studies show that not all

organic-rich sediments record serious depletion of oxygen in bottom waters. Resolution of the models and refinement of the instructive information available from the organic geochemistry of black shales can only be facilitated by further drilling.

Perhaps the single most exciting tool for future research might prove to be the hydraulic piston corer. Sediments recovered with this device have the potential for isotope stratigraphic studies of the Neogene which rival those initially undertaken in the Pleistocene in terms of stratigraphic resolution. This will become important as palaeo-oceanographers increasingly consider the significance of the 20,000–100,000 year rhythms (Milankovitch rhythms) evident on a fine scale in many pelagic rocks. Assessment of the influence on processes in the pelagic realm of astronomical phenomena and of the possible effect of excursions of the Earth's geomagnetic field, are an additional goal of the more synthesis-orientated palaeo-oceanographers in the coming years.

ACKNOWLEDGMENTS: There are many geologists much more qualified than I to attempt a review such as this. The first draft was enormously improved by the comments of four of them: Mike Arthur, Hugh Jenkyns, Tony Moorby and Alastair Robertson. However, they bear no responsibility for the material included or left out of the final version.

References

ALVAREZ, L. W., ALVAREZ, W., ASARO, F. & MICHEL, H. V. 1980. Extraterrestrial cause for the Cretaceous-Tertiary extinction. *Science*, **208**, 1095–108.

ANDEL, T. H. VAN 1975. Mesozoic/Cenozoic calcite compensation depth and the global distribution of calcareous sediments. *Earth planet. Sci. Lett.* **26**, 187–95.

——, THIEDE, J., SCLATER, J. G. & HAY, W. W. 1977. Depositional history of the South Atlantic Ocean during the last 125 million years. *J. Geol.* **85**, 651–98.

ARTHUR, M. A. 1979. Paleoceanographic events—recognition, resolution and reconsideration. *Rev. Geophys. space Phys.* **17**, 1474–94.

—— & JENKYNS, H. C. 1981. Phosphorites and paleoceanography. *Oceanologica Acta, Proc. 26th Int. Geol. Congress, Geology of Oceans Symposium, Paris*, 83–96.

—— & NATLAND, J. H. 1979. Carbonaceous sediments in the North and South Atlantic: the role of stable stratification of Early Cretaceous basins *In*: TALWANI, M., HAY, W. & RYAN, W.B.F.

(eds) *Deep Drilling Results in the Atlantic Ocean: Continental Margins and Paleo-environment. Maurice Ewing Series 3*, 375–401. American Geophysical Union, Washington D. C.

BERGER, W. H. 1979. Impact of deep-sea drilling on paleoceanography. *In*: TALWANI, M., HAY, W. & RYAN, W. B. F. (eds) *Deep Drilling Results in the Atlantic Ocean: Continental Margins and Paleo-environment. Maurice Ewing Series 3*, 297–314. American Geophysical Union, Washington D. C.

—— & THIERSTEIN, H. R. 1979. On Phanerozoic mass extinctions. *Naturwiss.* **66**, 46–7.

—— & VINCENT, E. 1981. Chemostratigraphy and biostratigraphic correlation: exercises in systemic stratigraphy. *Oceanologica Acta, Proc. 26th Int. Geol. Congress, Geology of Oceans Symposium, Paris*, 115–27.

——, —— & THIERSTEIN H. R. 1981. The deep-sea record: major steps in Cenozoic ocean evolution. *In*: WARME, J. E., DOUGLAS, R. G. & WINTERER, E. L. (eds) *The Deep Sea Drilling Project: A Decade of Progress. Soc. econ. Pal. Min. Spec.*

Publ. **32**, 489–504.

—— & WINTERER, E. L. 1974. Plate stratigraphy and the fluctuating carbonate line. *In*: HSÜ, K. J. & JENKYNS, H. C. (eds) *Pelagic Sediments on Land and Under the Sea. Spec. Publ. Int. Ass. Sedimentol.* **1**, 1–48.

DE BOER, P. L. 1982. Some remarks about the stable isotope composition of cyclic pelagic sediments from the Cretaceous in the Apennines (Italy). *In*: SCHLANGER, S. O. & CITA, M. B. (eds) *Nature and Origin of Cretaceous Carbon-rich Facies.* Academic Press, London. 229 pp.

BÖSTROM, K. & PETERSON, M. N. A. 1966. Precipitates from hydrothermal exhalations on the East Pacific Rise. *Econ. Geol.* **61**, 1258–65.

BRASS, G. W., SOUTHAM, J. R. & PETERSON, W. H. 1982. Warm saline bottom waters in the ancient ocean. *Nature, Lond.* **296**, 620–3.

CLIMAP PROJECT MEMBERS. 1976. The surface of the Ice Age Earth. *Science*, **191**, 1131–7.

CORLISS, J. B., DYMOND, J., GORDON, L. I. *et al.* 1979. Submarine thermal springs on the Galapagos Rift. *Science*, **203**, 1073–82.

——, BAROSS, J. A. & HOFFMAN, S. E. 1981. An hypothesis concerning the relationship between submarine hot springs and the origin of life on Earth. *Oceanologica Acta, Proc. 26th Int. Geol. Congress, Geology of Oceans Symposium, Paris*, 59–69.

CRONAN, D. S. 1980. *Underwater Minerals*, Academic Press, London. 364 pp.

——, GLASBY, G. P., MOORBY, S. A., THOMPSON, J., KNEDLER, K. E. & MCDOUGALL, J. C. 1982. A submarine hydrothermal manganese deposit from the south-west Pacific island arc. *Nature, Lond.* **298**, 456–8.

DAVIES, T. A., HAY, W. W., SOUTHAM, J. R. & WORSLEY, T. R. 1977. Estimates of Cenozoic oceanic sedimentation rates, *Science*, **197**, 53–5.

DEGENS, E. T. & ROSS, D. A. 1974. *The Black Sea—Geology, Chemistry, and Biology.* Am. Ass. Petrol. Geol. Mem. **20**.

DOUGLAS, R. G. & SAVIN, S.M. 1975. Oxygen and carbon isotope analyses of Tertiary and Cretaceous microfossils from Shatsky Rise and other sites in the North Pacific Ocean. *Init. Rep. Deep Sea drill. Proj.* **32**, 509–20.

EDMOND, J. M., MEASURES, C., MCDUFF, R. E., CHAN, L. H., COLLIER, R., GRANT, B., GORDON, L. I. & CORLISS, J. B. 1979. Ridge crest hydrothermal activity and the balances of the major and minor elements in the ocean: the Galapagos data. *Earth planet. Sci. Lett.* **46**, 1–18.

——, VON DAMM, K. L., MCDUFF, R. E. & MEASURES, C. I. 1982. Chemistry of hot springs on the East Pacific Rise and their effluent dispersal. *Nature, Lond.* **297**, 187–91.

ELDHOLM, O. & THIEDE, J. 1980. Cenozoic continental separation between Europe and Greenland. *Palaeogeogr. Palaeoclimatol. Palaeoecol.* **30**, 243–59.

EMILIANI, C. 1978. The cause of the ice ages. *Science*, **123**, 1061–6.

FISCHER, A. G. 1980. Gilbert-bedding rhythms and geochronology. *In*: YOCHELSON, E. L. (ed.) *The Scientific Ideas of G. K. Gilbert. Geol. Soc. Am. Spec. Publ.* **183**, 93–104.

—— 1981. Climatic oscillations in the biosphere. *In*: NITECKI, M. (ed.) *Biotic Crises in Ecological and Evolutionary Time*, p. 103.

—— & ARTHUR, M. A. 1977. Secular variations in the pelagic realm. *In*: COOK, H. E. & ENOS, P. (eds) *Deep Water Carbonate Environments. Soc. econ. Pal. Min. Spec. Publ.* **25**, 19–50.

FLEET, A. J. & ROBERTSON, A. H. F. 1980. Ocean ridge metalliferous and pelagic sediments of the Semail Nappe, Oman. *J. geol. Soc. London*, **137**, 403–22.

FRANCHETEAU, J., NEEDHAM, H. D., CHOUKROUNE, P. *et al.* 1979. Massive deep-sea sulphide ore deposits discovered on the East Pacific Rise, *Nature, Lond.* **277**, 523–8.

GARTNER, S. & KEANY, J. 1978. The terminal Cretaceous event: a geologic problem with an oceanographic solution. *Geology*, **6**, 708–12.

HABIB, D. 1982. Sedimentary supply origin of Cretaceous black shales. *In*: SCHLANGER, S. O. & CITA, M. B. (eds) *Nature and Origin of Cretaceous Carbon-rich Facies*, 113–28. Academic Press.

HALLAM, A. 1977. Secular changes in marine inundation of USSR and North America through the Phanerozoic. *Nature, Lond.* **269**, 769–72.

HARLAND, W. B., COX, A. V., LLEWELLYN, P. G. PICKTON, C. A. G., SMITH, A. G. & WALTERS, R. 1982. *A Geologic Time Scale*, Cambridge University Press, Cambridge. 131 pp.

HAYMON, R. M. & KASTNER, M. 1981. Hot spring deposits on the East Pacific Rise at 21°N: preliminary description of mineralogy and genesis. *Earth planet. Sci. Lett.* **53**, 363–81.

HAYS, J. D., IMBRIE, J. & SHACKLETON, N. J. 1976. Variations in the earth's orbit: pacemaker of the ice ages. *Science*, **194**, 1121–32.

HEATH, G. R., MOORE, T. C. & ANDEL, T. H. VAN 1977. Carbonate accumulation and dissolution in the equatorial Pacific during the past 45 million years. *In*: ANDERSEN, N. R. & MALAHOFF, A. (eds) *The Fate of Fossil Fuel CO_2 in the Oceans*, 627–90. Plenum Press, New York.

HONNOREZ, J., VON HERZEN, R. P., BARRETT, T. J. *et al.* 1981. Hydrothermal mounds and young ocean crust of the Galapagos: preliminary deep sea drilling results, Leg 70. *Bull. geol. Soc. Am.* **92**, 457–72.

HSÜ, K. J. 1981. Origin of geochemical anomalies at Cretaceous–Tertiary boundary. Asteroid or cometary impact? *Oceanologica Acta, Proc. 26th Int. Geol. Congress, Geology of Oceans Symposium, Paris*, 129–33.

—— & JENKYNS, H. C. (eds) 1974. *Pelagic sediments on land and under the sea. Spec. Publ. Int. Ass. Sedimentol.* **1**, 447 pp.

——, CITA, M. B. & RYAN, W. B. F. 1973. The origin of the Mediterranean evaporites. *Init. Rep. Deep Sea drill. Proj.* **13**, 1203–31.

IMBRIE, J. & IMBRIE, K. P. 1979. *Ice Ages—Solving the Mystery*, 224 pp. Enslow, Short Hills, New Jersey.

JENKYNS, H. C. 1978. Pelagic Environments. *In*:

READING, H. G. (ed.) *Sedimentary Environments and Facies*, pp. 314–71. Blackwell Scientific Publications, Oxford.

—— 1980. Cretaceous anoxic events: from continents to oceans. *J. geol. Soc. London*, **137**, 171–88.

KASTNER, M. 1982. Evidence for two distinct hydrothermal systems in the Guaymas Basin. *Init. Rep. Deep Sea drill. Proj.* **64**, 1143–57.

KENNETT, J. P. 1977. Cenozoic evolution of Antarctic glaciation, the circum-Antarctic Current, and their impact on global paleoceanography. *J. Geophys. Res.* **82**, 3843–60.

—— 1982. *Marine Geology*. Prentice Hall, Englewood Cliffs, New Jersey. 812 pp.

—— & SHACKLETON, N. J. 1976. Oxygen isotopic evidence for development of the psychosphere 38 m. yr ago. *Nature, Lond.* **260**, 513–5.

LEGGETT, J. K., McKERROW, W. S., COCKS, L. R. M. & RICKARDS, R. B. 1981. Periodicity in the early Palaeozoic marine realm. *J. geol. Soc. London*, **138**, 167–76.

LONSDALE, P. F., BISCHOFF, J. L., BURNS, V. M., KASTNER, M. & SWEENEY, R. E. 1980. A high-temperature hydrothermal deposit on the seabed at a Gulf of California spreading center. *Earth planet Sci. Lett.* **49**, 8–20.

LUPTON, J. E., WEISS, R. F. & CRAIG, H. C. 1977. Mantle helium in hydrothermal plumes in the Galapagos Rift. *Nature, Lond.* **266**, 603.

McDONALD, K. C. & LUYENDYK, B. P. 1981. The crest of the East Pacific Rise. *Scient. Am.* **244**, 100–16.

MILLER, A. R., DENSMORE, C. D., DEGENS, E. T., HATHAWAY, J. C., MANHEIM, F. T., McFARLIN, P. F., POCKLINGTON, R. & JOKELA, A. 1966. Hot brines and recent iron deposits of the Red Sea. *Geochim. Cosmochim. Acta*, **30**, 341.

MOORE, T. C. & HEATH, G. R. 1977. Survival of deep-sea sedimentary sections. *Earth planet. Sci. Lett.* **37**, 71–80.

MUEHLENBACHS, K. & CLAYTON, R. N. 1976. Oxygen isotope composition of the oceanic crust and its bearing on seawater. *J. Geophys. Res.* **81**, 4365–9.

NATLAND, J. H. 1978. Composition, provenance and diagenesis of Cretaceous clastic sediments drilled on the Atlantic continental rise off southern Africa. *Init. Rep. Deep Sea drill. Proj.* **40**, 1025–62.

PITMAN, W. C. 1978. Relationship between eustacy and stratigraphic sequence of passive margins. *Bull. geol. Soc. Am.* **89**, 1389–403.

RAMSAY, A. T. S. 1977. Sedimentological clues to palaeo-oceanography. *In*: RAMSAY, A. T. S. (ed.) *Oceanic Micropalaeontology*, pp. 1371–1453. Academic Press, London.

RYAN, W. B. F., & CITA, M. B. 1977. Ignorance concerning episodes of oceanwide stagnation. *Mar. Geol.* **23**, 197–215.

SCHLANGER, S. O., & JENKYNS, H. C. 1976. Cretaceous oceanic anoxic events: causes and consequences. *Geol. Mijnbouw*, **55**, 179–84.

——, —— & PREMOLI-SILVA, I. 1981. Volcanism and vertical tectonics in the Pacific basin related to global Cretaceous transgressions. *Earth planet. Sci. Lett.* **52**, 435–49.

SCHOLLE, P. A. & ARTHUR, M. A. 1980. Carbon isotope fluctuations in Cretaceous pelagic limestones: potential stratigraphic and petroleum exploration tool. *Bull. Am. Ass. Petrol. Geol.* **64**, 67–87.

——, —— & EKDALE, A. A. 1983. Pelagic Environments. *In*: SCHOLLE, P. A., BEBOUT, D. G. & MOORE, C. H. (eds) *Carbonate Depositional Environments*. Am. Assoc. Petrol. Geol. Mem. **33**, 620–91.

SHACKLETON, N. J. & KENNETT, J. P. 1975. Paleotemperature history of the Cenozoic and initiation of Antarctic glaciation: oxygen and carbon isotope analyses in DSDP Sites 277, 279 and 281. *Init. Rep. Deep Sea drill. Proj.* **29**, 743–55.

SMITH, A. G., HURLEY, A. M. & BRIDEN, J. C. 1981. *Phanerozoic Palaeocontinental Maps*, Cambridge University Press, Cambridge. 102 pp.

SOUTHAM, J. R., PETERSON, W. H. & BRASS, G. W. 1982. Dynamics of anoxia. *Palaeogeog. Palaeoclim. Palaeoecol.* **40** 183–98.

STRENS, M. R. & CANN, J. R. 1982. A model of hydrothermal circulation in fault zones at Mid Ocean Ridges. *Geophys. J. R. astr. Soc.* **71**, 225–40.

TALWANI, M. & ELDHOLM, O. 1977. Evolution of the Norwegian–Greenland Sea. *Bull. geol. Soc. Am.* **88**, 969–99.

——, HAY, W. W. & RYAN, W. B. F. (eds) 1979. Deep sea drilling results in the Atlantic Ocean: continental margins and paleoenvironments. *Maurice Ewing Series 3*, 437. Am. Geophys. Union, Washington D. C.

—— & UDINSTEV, G. 1976. Tectonic synthesis. *Init. Rep. Deep Sea drill. Proj.* **38**, 1213–40.

THIEDE, J. 1981. Palaeo-oceanography, margin stratigraphy and palaeophysiography of the Tertiary North Atlantic and Norwegian–Greenland Seas. *In*: KENT, P. *et al.* (eds) *The Evolution of Passive Continental Margins*, pp. 177–85. Royal Society of London, London.

—— & ANDEL, T. H. VAN 1977. The paleoenvironment of anaerobic sediments in the Late Mesozoic South Atlantic Ocean. *Earth planet. Sci. Lett.* **33**, 301–9.

——, DEAN, W. E. & CLAYPOOL, G. E. 1982. Oxygen-deficient depositional paleoenvironments in the Mid-Cretaceous tropical and subtropical central Pacific Ocean. *In*: SCHLANGER, S. O. & CITA, M. B. *Nature and Origin of Cretaceous Carbon-rich facies*, pp. 79–100. Academic Press, London.

THIERSTEIN, H. R. & BERGER, W. H. 1978. Injection events in earth history. *Nature, Lond.* **276**, 461–4.

TISSOT, B., DURAND, B., ESPITALIÉ, J. & COMBAZ, A. 1974. Influence of nature and diagenesis of organic matter in formation of petroleum. *Bull. Am. Ass. Petrol. Geol.* **58**, 499–506.

——, DEROO, G. & HERBIN, J. 1979. Organic matter in Cretaceous sediments of the North Atlantic: contribution to sedimentology and paleogeography. *In*: TALWANI, M., HAY, W. W. & RYAN, W. B. F. (eds) *Deep Sea Drilling Results in the*

Atlantic Ocean: Continental Margins and Paleo-environment. Maurice Ewing Series 3, 362–74. American Geophysical Union, Washington D. C.

TUCHOLKE, B. E. 1981. Geological significance of seismic reflectors in the deep western North Atlantic Basin. *In:* WARME, J. E., DOUGLAS, R. G. & WINTERER, E. L. (eds) *The Deep Sea Drilling Project—A Decade of Progress. Soc. econ. Pal. Min. Spec. Publ.* **32,** 23–37.

UL HAQ, B. 1981. Paleogene paleoceanography: Early Cenozoic oceans revisited. *Oceanologica Acta, Proc. Int. Geol. Congress, Geology of Oceans Symposium, Paris,* 71–82.

VAIL, P. R., MITCHUM, R. M. & THOMPSON, S. III. 1977. Seismic stratigraphy and global changes of sea level. IV. Global cycles of relative changes of sea level. *In:* PAYTON, C. (ed.) *Seismic Stratigraphy—Applications to Hydrocarbon Exploration. Am. Ass. Petrol. Geol. Mem.* **26,** 83–97.

——, —— SHIPLEY, T. H., & BUFFLER, R. T. 1980. Unconformities of the North Atlantic. *Phil. Trans. R. Soc.* **294,** 137–55.

WARME, J. E., DOUGLAS, R. G. & WINTERER, E. L. (eds) 1981. *The Deep Sea Drilling Project: A Decade of Progress. Soc. econ. Pal. Min. Spec. Publ.* **32.**

WATTS, A. B. 1982. Tectonic subsidence, flexure and global changes of sea level. *Nature, Lond.* **297,** 469–74.

WEISSERT, H. 1981. The environment of deposition of black shales in the Early Cretaceous: an ongoing controversy. *In:* WARME, J. E., DOUGLAS, R. G. & WINTERER, E. L. (eds) *The Deep Sea Drilling Project: A Decade of Progress. Soc. econ. Pal. Min. Spec. Publ.* **32,** 547–60.

WILDE, P. & BERRY, W. B. N. 1982. Progressive ventilation of the oceans—potential for return to anoxic conditions in the post-Paleozoic. *In:* SCHLANGER, S. O. & CITA, M. B. (eds) *Nature and Origin of Cretaceous Carbon-rich Facies,* pp. 209–224. Academic Press, London.

WINTERER, E. L. & BOSELLINI, A. 1981. Subsidence and sedimentation on Jurassic passive continental margin, southern Alps, Italy. *Bull. Am. Ass. Petrol. Geol.* **65,** 394–421.

J. K. LEGGETT, Department of Geology, Imperial College of Science and Technology, London SW7 2BP.

Facies analysis of volcaniclastic sediments: a review

R. J. Suthren

SUMMARY: A number of problems are encountered in the study and interpretation of sequences of volcaniclastics (sediments or sedimentary rocks composed predominantly of volcanic particles). These problems include the lack of modern analogues, the complex interaction of volcanic and sedimentary processes, and extensive diagenetic alteration. Volcaniclastic sediments are produced and emplaced by *autoclastic* (mechanical breakage during magma movement), *hydroclastic* (water/magma interaction), *pyroclastic* (magmatic explosion) and *epiclastic* (sedimentary erosion, transport and deposition) processes, or by any combination of these. The systematic study of lithology, texture and primary structure sequences, and their lateral and vertical variations, is leading to the erection of facies models for recent volcaniclastics. These models are likely to have important applications in the interpretation of ancient sequences of volcaniclastic sedimentary rocks, as well as practical applications in mineral and petroleum exploration and volcano prediction.

This paper is intended as a personal view of the state of the art in the field of volcaniclastic rocks, rather than a comprehensive review. Some of the problems inherent in the study of these rocks are outlined, and followed by reviews of the characteristics by which the most important volcaniclastic facies may identified, and of important recent developments in the study of each facies. It is not the intention of this paper to consider in depth the physical processes of volcaniclastic sedimentation, but rather to outline the resulting sequences and their features, and to show how they may be recognized in the geological record. In conclusion, a number of important recent contributions to the facies analysis of volcaniclastics are introduced. A good review is given by Lajoie (1979), and the present account concentrates on developments since that time. The author's research has been mainly concerned with interpretation of ancient subaerial sequences, and the reader may therefore find the article biased in this direction.

Volcaniclastic as defined by Fisher (1961) refers to all clastic sediments and rocks, regardless of depositional process, whose particles are predominantly of volcanic origin. The term 'sediments' is here interpreted in the broadest possible sense to include all fragmentary volcanics, ranging from those emplaced by magmatic processes through to the products of 'normal' sedimentary processes.

Rigorous facies analysis of volcaniclastic sequences is in its infancy compared with other branches of sedimentology. Until very recently, little was known of lateral and vertical facies variations: e.g. none of the currently available volcanology texts describes rock sequences in any detail.

However, the last 5 years have seen the beginning of the development of facies models comparable with those for non-volcanic sediments, though not yet so proliferous nor so sophisticated. Thus, a sedimentological approach to volcaniclastic successions is beginning to pay off. Most recently proposed facies models come from North and Central America and New Zealand, but from rocks of widely differing ages. Examples which may be cited include a model for submarine volcanogenic sedimentation in the Lesser Antilles arc (e.g. Sigurdsson *et al.* 1980) and a model for subaerial forearc sedimentation in Guatemala (e.g. Kuenzi *et al.* 1979; Vessell & Davies 1981). Remarkably, considerable progress has recently been made in the much older volcaniclastic sequences of Archaean greenstone belts in the Canadian Shield (e.g. Tassé *et al.* 1978).

The problems of studying volcaniclastics

Methods

One serious problem is that the methods used in studying modern volcaniclastics are often not applicable to ancient sequences (by 'ancient' the present author normally means pre-Tertiary). A good example is the use of grain-size distributions, obtained by sieving unconsolidated material, to determine source locations and to distinguish emplacement mechanisms (e.g. Walker 1971). Whilst the distribution of large clasts has been used with considerable success in the study of ancient sequences (e.g. Tassé *et al.* 1978), granulometric analyses of the finer fractions in thin section are of limited use for two main reasons. Firstly, the extremely irregular grain shapes and wide range of densities of volcanic fragments—e.g. from low-density pumice to pyroxene crystals—make analysis difficult. Correlation between areas of volcanic grains in thin section and weights of sieve-size fractions is practically impossible. Secondly, alteration all too often modifies the original grain-size distribution, particularly in producing secondary fine-grained matrix.

In the past, then, there has been a tendency for workers on recent volcaniclastics to use methods and examine aspects which are not applicable to ancient sequences. The interpretation of ancient deposits relies mainly on the recognition and critical analysis of characteristics and relationships observed in the field, particularly vertical and lateral facies changes, structures, and overall geometry. Fortunately, a number of workers on recent sequences are now taking this approach, and this is leading to the erection of facies models applicable to the interpretation of ancient examples.

Observation

Further problems are posed in the direct study of processes, particularly those involved in explosive eruptions. Approach to a working volcano-sedimentary system is rather more difficult and hazardous than, say, observation of an area of active intertidal sedimentation. Some volcanologists have taken considerable risks to make observations (e.g. Perret 1935). Danger is not the only problem: frequently, the obscuring effect of clouds of steam and ash hinders useful observations. The problems are even more acute in the study of subaqueous eruptions.

Eruption frequency

Another problem is the infrequency of major eruptions. For example, no scientist has yet observed an ignimbrite-forming eruption, and the largest closely observed eruption yielded less than 5 km³ of material, which is very small in comparison with some eruptions represented in the geological record. For example, single basaltic lava flows with volumes greater than 600 km³ are known (Jepsen *et al.* 1980), and individual pyroclastic flows have produced flow units with volumes up to 3000 km³ (Smith & Roobol 1982). Compared with these examples, the 18 May 1980 eruption of Mount St Helens (Lipman & Mullineaux 1982), which produced less than 1 km³ of ejecta, pales into insignificance. Thus, the understanding of large, infrequent, violently explosive eruptions must

depend heavily on the study of their products (Walker 1981b).

Processes

Interpretation is often more difficult in the case of volcaniclastic sediments than non-volcanic sediments, because of the greater number of possible variables involved in their formation. On top of the whole spectrum of sedimentary, climatic and tectonic variables, these sediments are subject to a variety of volcanic processes. Frequently, there is close interaction between volcanic and sedimentary processes. For example, fluvial deposition in active volcanic areas is often directly related to major eruptions, which provide abundant sediment, and often generate their own intense rainstorms. This leads to greatly increased run-off and sediment transport.

Alteration

Alteration is a particular problem in the interpretation of ancient sequences. Diagenesis tends to be particularly rapid in the case of volcaniclastic sediments, because of their chemically unstable mineral assemblages. If the warm climate and high rainfall of some modern volcanic areas are added to this, rates of chemical weathering become very high. Diagenesis not only destroys textures, sometimes rendering grains indistinguishable from matrix, but may completely obliterate sedimentary structures too. Some of the most important early diagenetic reactions include: glass → clay, glass → zeolite, and plagioclase → clay + calcite.

These hydration reactions result in an increase in the amount of fine-grained matrix present (Surdam & Boles 1979), so that in most ancient volcaniclastics, primary and secondary matrix cannot reliably be distinguished. To give an example of the rates at which these reactions proceed, modern volcaniclastic debris-flow deposits in Guatemala, which on deposition contained less than 1% clay, have 20% clay matrix within 3–5 years of deposition (Vessell & Davies 1981). In younger rocks, the secondary matrix consists largely of zeolites and mont-morillonite, whilst in older rocks the assemblage is altered to chlorite, sericite and quartz.

Besides near-surface chemical weathering, burial diagenesis and sea-floor alteration may occur. Volcaniclastic sediments are frequently deposited in zones of high heat flow at plate boundary settings. Intense alteration commonly occurs where convecting hydrothermal systems form above subvolcanic intrusions. Such areas

TABLE 1. *Grain-size classification of volcanic fragments and rocks (based on Fisher 1961; Lajoie 1979)*

Grain size (mm)	ϕ	Pyroclasts	Pyroclastic rock	Non-genetic rock term	Wentworth Scale	
256	-8	Blocks	Pyroclastic breccia	Volcanic breccia	Boulders	
128	-7	and bombs	and			
64	-6		agglomerate		Cobbles	
32	-5			and		Gravel
16	-4	Lapilli	Lapillistone	Volcanic	Pebbles	
8	-3			conglomerate		
4	-2					
2	-1				Granules	
1	0	Coarse	Coarse	Volcanic	VC	
0.5	1				C	
0.25	2				M	Sand
0.125	3	Ash	Tuff	Sandstone	F	
0.063	4				VF	
0.031	5	Fine	Fine	Volcanic		
0.016	6				Silt	
0.008	7	Ash	Tuff	Siltstone		
0.004	8			Volcanic claystone	Clay	

are also prone to low-grade, wet regional metamorphism. It may be difficult or impossible to distinguish the products of the various types of alteration outlined.

The composition and flow of ground water are probably more important than depth of burial and temperature in controlling the distribution of diagenetic mineral phases (Surdam & Boles 1979). Permeability and porosity are greatly reduced or eliminated during early diagenesis when hydration reactions are dominant, resulting in clay rims around grains, and pore-filling by zeolites, phyllosilicates and calcite (Galloway 1979).

Terminology

Because of the relative youth of volcaniclastic sedimentology, there is still no generally agreed system for the description and classification of volcaniclastic sediments and rocks. Fisher's (1961) grain-size description scheme (Table 1) is,

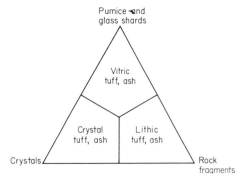

FIG. 1. Classification of pyroclastics based on the nature of their components. After Pettijohn *et al.* (1973).

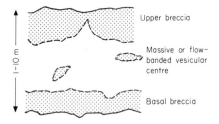

FIG. 2. A typical sequence through an autobrecciated lava flow of basic or intermediate composition. Model applies to both aa and block lava. After Krauskopf (1948) and MacDonald (1972).

however, in general use. It has the advantage of combining a set of terms for purely pyroclastic materials with a parallel, non-genetic scheme. In addition to these terms, the present author advocates the use of standard Wentworth Scale terms (e.g. coarse sand) to enable a more accurate definition of grain size.

A lithological classification should be based not only on grain size, but also on particle composition in terms of vitric, crystal and lithic components (Fig. 1). Fuller discussions on volcaniclastic petrography and classification are provided by Pettijohn *et al.* (1973, Chapter 7), Schmincke (1974), Pettijohn (1975, Chapter 9), Wright *et al.* (1980, 1981b) and Schmid (1981).

Besides the descriptive terminology outlined above, a series of genetic terms is used to describe volcaniclastic rocks and processes. The scheme adopted here is derived from the summaries of Fisher (1961, 1984), and Lajoie (1979), and four groups of volcaniclastics are defined: *autoclastic*, *hydroclastic*, *pyroclastic* and *epiclastic*. The processes leading to their formation are outlined below:-

Autoclastic. Mechanical breakage and related processes during lava movement.

Hydroclastic. Quenching of magma in contact with water, ice or wet sediment, generally resulting in copious generation of steam. This may involve processes ranging from non-explosive granulation by thermal shock, to violently explosive phreatic eruptions. *Hydroclastic* is synonymous with *hyaloclastic* as used by some authors: the former term is preferable, as it stresses the vital role of water.

Pyroclastic. Explosive disruption of magma resulting from release of magmatic gases, and with a wide range of eruptive styles from gentle to extremely violent.

Epiclastic. Weathering and sedimentary

reworking of older volcanic rocks.

There are two 'grey areas' within this classification. Firstly, the products of reworking of unconsolidated pyroclastic debris are considered by some workers to be pyroclastic ('secondary pyroclastics'), and by others as epiclastics. Secondly, the presence of water is important in explosive volcanism, and it may have three main effects (Walker, 1982a):

1 Hydrostatic pressure may inhibit or prevent explosive eruptions in subaqueous environments—the depth at which this occurs is the *pressure compensation level* (PCL) (Fisher, 1984). The PCL varies according to the composition and volatile content of the magma, but may lie in the range 500–1000 m. Above the PCL, abundant hydroclastic and pyroclastic materials are formed, whilst below the PCL, lava flows are dominant.

2 Water/magma interactions may be the sole cause of *phreatic* explosions.

3 When water has access to a vent in which magmatic explosions are taking place, the explosions may be enhanced by water/magma interaction, resulting in *phreatomagmatic* explosions, which thus involve both pyroclastic and hydroclastic processes.

The rigorous application of the genetic classification scheme given above may only be possible for Quaternary sequences (Wright *et al.* 1980, 1981b)

Autoclastic deposits

Autobrecciation is a phenomenon common to lavas of all compositions. As the top of a subaerial lava flow cools, it may form a brittle crust, which fractures as the hotter lava beneath continues to move. This results in the formation of a layer of monolithologic blocks on the flow top. In some basaltic lavas, the blocks are highly vesicular and spinose (aa). In other basaltic

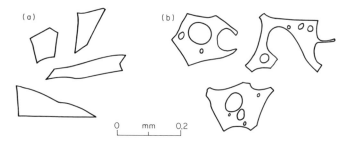

FIG. 3. Typical grain shapes for hydroclastic deposits. (a) hyaloclastites; (b) hyalotuffs. After Honnorez & Kirst (1975).

lavas, and those of intermediate and acid composition, the blocks are less vesicular, and have more regular, rectilinear shapes (*block lava*). The structure and mode of formation of aa and block lava are very similar, except that most block lavas flow more slowly, due to their higher viscosities (MacDonald 1972).

Movement of the flow top results in abrasion between the blocks, causing rounding, and the formation of sand- to dust-sized material. Most of the fragmentation and attrition occurs in narrow zones at lava channel margins (Krauskopf 1948). As the flow advances, the upper parts, moving at higher velocities, cause the flow front to oversteepen. Material avalanches down the flow front, both from the blocky top and from the massive, viscous central portion of the flow. The latter then advances over the resulting 'carpet' of autoclastic material, and the sequence shown in Fig. 2 results. The lower and upper breccias may have a matrix of fragmental lava produced by attrition. Alternatively, the blocks may be set in a lava matrix, resulting from intrusion of lava into the breccia, or from the sinking of blocks into the viscous centre of the flow. In some cases, hot blocks weld together. In intermediate and acid lavas, the centre of the flow is frequently flow-banded and flow-folded, and may contain isolated breccia pockets. The relative thickness of the breccias to that of the massive centre is highly variable: in its distal parts, the entire lava flow may be brecciated. On steep slopes, lava-flow units are often much thinner, and may move partly by sliding, resulting in isolated lenses of massive lava surrounded by breccia (Krauskopf 1948) or in hot lava avalanche breccias (Francis *et al*. 1974).

The processes outlined above may result in complex interlayering of massive lava and several types of breccia (Krauskopf 1948). Autoclastic deposits may be recognized by their association with massive lava, and their monolithologic nature.

Hydroclastic deposits

Hydroclastic processes involve magma/water interactions which range from relatively gentle to violently explosive. Honnorez & Kirst (1975) suggested useful terms for the products of the end members of this range of processes. *Hyaloclastites* (*sensu stricto*) are formed by non-explosive granulation of volcanic glass when quenched by water or wet sediment. *Hyalotuffs* are produced by explosive magma/water interaction—i.e. phreatic and phreatomagmatic explosions. Some workers have used *hyaloclastite* in a looser sense to include all hydroclastic deposits, but it is suggested that the more restricted use should be adopted. There are distinct differences in grain shape between hyaloclastites and hyalotuffs (Fig. 3). Hyaloclastites are generally less vesicular, and most grains are bounded by roughly planar fractures, with relatively few corners. The more vesicular hyalotuff grains have boundaries dominated by concave and convex surfaces, many of them bubble walls. The resulting ragged shapes have many more inflections than hyaloclastite grains. Honnorez & Kirst (1975) give quantitative methods, based on these properties, for distinguishing hyaloclastites and hyalotuffs.

Pillow breccias and hyaloclastites

There is a common subaqueous association of pillow lavas and various breccia and hyaloclastite facies derived from them, particularly in sequences of basaltic composition. There are some similarities between autoclastic and hydroclastic breccias—e.g. most are monolithologic, and the two are often intimately associated (Lajoie, 1979). Hydroclastic breccias may be distinguished by glassy clasts with chilled margins, and the presence of pillow fragments. A sequence common in pillow-lava piles is that shown in Fig. 4a, where there is a gradual upward increase in the degree of fragmentation, from pillow lava, through isolated-pillow

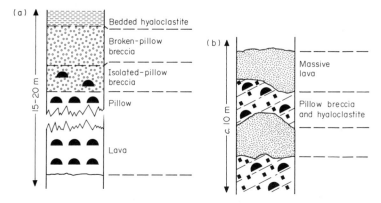

FIG. 4. Sequences in pillow lava/hyaloclastite associations of basaltic composition: (a) lavas erupted under water (after Carlisle, 1963); (b) lavas flowed from land into intertidal zone, with resulting alternation of massive lava (low tide) and pillow breccia/hyaloclastite foresets (intermediate and high tides). Based on Jones & Nelson (1970), Furnes & Fridleifsson (1974) and Furnes & Sturt (1975).

breccia, into broken-pillow breccia, and finally bedded hyaloclastite. This crude fining-upward sequence, which is often incomplete, is interpreted as the result of subaqueous extrusion of basaltic lava: the upper parts of the flow, in contact with water, were subject to more extensive *in situ* quenching and granulation than the lower parts. (Information from Carlisle 1963, and Lajoie 1979.)

A similar facies association results from the subaqueous effusion of acid lavas, except that pillows are generally absent. The massive acid lava flow or dome is mantled by a carapace of angular blocks of autoclastic and hydroclastic origin (Pichler 1965; Lajoie 1979). Breccias may also form by slope failure in subaqueous lavas of all compositions, resulting in talus deposits, debris flows and turbidites (Bevins & Roach 1979; Furnes & Fridleifsson, 1979). The finer fractions of hydroclastic material are frequently reworked by current and wave action, and are represented by bedded hyaloclastite at the top of the sequence.

Distinctive facies associations are produced when lava flows into water: the examples studied are mainly basaltic. The subaerial portion of the lava is massive, and passes downwards into pillow lava, pillow breccia and hyaloclastite produced by fragmentation of the lava on contact with water (Fig. 4b). The downward passage is gradational, and the degree of fragmentation decreases downwards, resulting in a crude coarsening-upward sequence. The pillow lava, breccia and hyaloclastite are commonly organized into steeply dipping (*c.* 30°) foresets (Fig. 4b), representing the progradation of a lava delta (Jones & Nelson 1970; Allen *et al.* 1982). Individual prograding sequences may be

up to 100 m thick. Changes in water level, or vertical tectonic movements, give rise to complex alternations of facies (Jones & Nelson 1970). In some recent examples, it has been possible to relate facies associations to lavas flowing into intertidal areas. The change in water level through the tidal cycle causes fluctuations in the level of the contact between the overlying subaerial lava and the underlying pillow lava and hydroclastic material. The flow rate of the lava is also a very important factor in determining the complex geometry of these sequences (Furnes & Fridleifsson 1974; Furnes & Sturt 1976).

Hyalotuffs

Violent hydroclastic explosions cause fragmentation and pulverization of lava (Tazieff 1973), resulting in a poorly sorted hyalotuff rich in fines. This poor sorting, and the absence of grains which were liquid or plastic on eruption— e.g. lava droplets, spatter, bombs—enable the distinction between hyalotuffs and subaerial basaltic tuffs (Walker & Croasdale 1972). *Surtseyan* explosions (Walker & Croasdale 1972), where water has access to the vent, build up low cones (*ash rings* or *tuff rings*) in which the hyalotuffs form thin, well-defined, laterally continuous beds. There is a change from steeply in-dipping beds near the vent to gently out-dipping beds on the flanks of the cone. Commonly, a steep unconformity separates in-dip and out-dip beds, due to repeated collapse of the inner walls of the vent during eruption.

Where lava flows into water, rootless cones (*littoral cones* of Fisher 1968) may be rapidly built up by phreatic eruptions.

Magma/wet sediment interaction

Interactions between magmas of all compositions and wet, unconsolidated or partially lithified sediment produce hydroclastic lithologies resembling both hyaloclastites and hyalotuffs. The most common occurrence of such material is along the margins of igneous bodies—e.g. the contacts of a sill or dyke, or the base of a lava flow or ignimbrite (Schmincke, 1967). A gradational passage (downwards, upwards or laterally) from massive igneous rock into pillowed magma and finally into a mixed magma/sediment breccia (*pépérite*) is frequently observed. This transition is analogous to that in subaqueously erupted pillow basalts (Fig. 4a). Spectacular illustrations of such relationships are provided by Snyder & Fraser (1963). In some cases, pépérite may form the entire thickness of an intrusion. Pépérites are typically unsorted, and contain angular fragments, pillows and irregular tongues of magmatic material in a disrupted sediment matrix. The magma bodies commonly have chilled margins, with prismatic joints normal to the cooling surfaces, and sediment-filled fractures (Schmincke, 1967).

Kokelaar (1982) proposed that fluidization of wet sediment is an important process during the emplacement of igneous bodies, and may explain the removal of large volumes of sediment with little disturbance of the remaining host. Kokelaar's model accounts simply for complex contact relationships and space problems. Fluidization of wet sediment can only occur at shallow depths, and sections are documented where fluidized sediment has been erupted at the sediment–water interface as debris flows and turbidity currents.

Pyroclastic deposits

Pyroclastic deposits, the products of magmatic explosions, have been extensively studied in both modern and ancient examples. It is recognized that there are three main groups of pyroclastic rocks, produced by pyroclastic fall, pyroclastic flow and pyroclastic surge, although there may be overlap between these processes (Walker 1981b). Pyroclastic flow and pyroclastic surge, for instance, are simply the end members of a range of gas-rich sediment gravity flows, which vary from concentrated laminar and plug flows through to dilute turbulent currents. Thus it can often be difficult to distinguish the deposits of the different pyroclastic processes. The end members, however, do have distinctive features: one is the relationship to the underlying topography

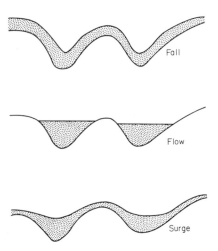

FIG. 5. Relationships of the three main groups of pyroclastic deposits to the underlying topography. After Wright *et al.* (1980). Not to scale.

(Fig. 5). The overall geometry of a pyroclastic deposit is one of the most important features used to interpret its origin, but unfortunately is not always determinable in ancient sequences.

No one characteristic is diagnostic of a particular type of pyroclastic deposit, and the most accurate interpretation will be reached by consideration of a range of properties (Walker 1981b). A good summary of the general characteristics of fall, flow and surge deposits is given by Schmincke (1974), whilst a useful compilation of recent work on pyroclastics is that of Self & Sparks (1981). Important contributions to the understanding of pyroclastic deposits are resulting from the 1980 and later eruptions of Mount St Helens, and many of these are contained in the volume edited by Lipman & Mullineaux (1982).

Pyroclastic fall deposits

Pyroclastic fall encompasses the fallout products of subaerial eruptions, emplaced into either a subaerial or subaqueous environment. It also includes the deposits of subaqueous magmatic and phreatomagmatic explosions. Initially, pyroclasts are ejected into air or water as an *eruption column*, and in most cases are then transported laterally away from the vent as an *ash plume*, by processes including lateral expansion of the column, laterally directed explosions, winds and currents (Walker 1981b). True pyroclastic fall deposits, produced solely by vertical settling from suspension in air or water, are probably very rare. In nearly all cases, the characteristics of the final deposit are partially due to the interfering effects of wind, rain, water currents and waves.

Subaerial pyroclastic fall (air-fall)

The distribution and geometry of a fall deposit is closely related to the height of the eruption column: in general, the higher the eruption column, the more widespread the deposit. Thus, low eruption columns tend to deposit pyroclasts near the vent as steep-sided cones. High (several tens of kilometres) columns produce laterally extensive sheetlike *plinian fall* deposits (Walker 1973), whose distribution is strongly controlled by high velocity winds in the upper atmosphere. Such winds are very effective in transporting fine ash and dust over long distances. It is an interesting paradox that the products of the most powerful plinian eruptions are often thin sheets which appear rather insignificant in the geological record (Walker 1981a).

Since the distal limits and volume of a pyroclastic fall deposit are commonly difficult to determine, Walker (1973) expresses the dispersal of the deposit as the area (D) enclosed by the $0.01\ T_{max}$ isopach (where T_{max} = maximum thickness of the deposit). Pyroclastic fall deposits, as redefined by Walker (1973) occupy different fields on a plot of D against the degree of fragmentation (F) (Fig. 6). For example, *hawaiian* fall deposits have low D and low F, whilst *plinian* deposits are characterized by high D and low to moderate F. D ranges from $<10\ km^2$ for strongly cone-building deposits to

$>1000\ km^2$ for strongly sheet-forming deposits. In both cases, the thickness of the fall deposit decreases systematically away from the vent, unless modified by other factors.

Most pyroclastic fall sediments show good to moderate sorting of grain size, and systematic changes away from the vent are noticed: there is a decrease in grain size, an improvement in grain-size sorting, and a variation in compositional sorting (*fractionation* of Walker 1981a). Since ash particles have a wide range of densities, sorting in terms of fall velocity (σ_v) is often better than the size sorting (σ_ϕ) of the same sample. Ideally, there is an exponential decrease in grain size away from the vent, but complicating factors often result in less regular grain-size distributions. These factors operate particularly on the fine-ash fraction.

Fine ash and dust are produced in great volumes by violently explosive eruptions—including pyroclastic flow, plinian, vulcanian and phreatomagmatic eruptions (Walker 1973, 1981c). The products of all these eruptions will have a high F. Perhaps the greatest volumes of dust are generated by large ignimbrite eruptions, when fines are produced in the eruption column, and within moving ash flows (Walker 1981a). Vitric dust is elutriated from the ash-flow up into the overriding ash cloud, from which it may eventually be deposited as air-fall—*co-ignimbrite ash* of Sparks & Walker (1977).

Several processes may lead to premature deposition of fine ash from eruption columns and ash plumes. Fine ash may fall near the vent, either as a result of weak explosions, or if

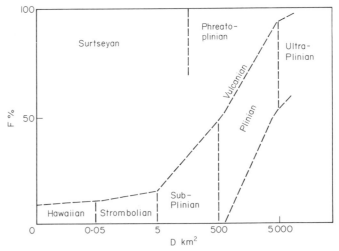

FIG. 6. Discrimination of pyroclastic fall types on a plot of dispersion against degree of fragmentation. D is the area enclosed by the $0.01\ T_{max}$ isopach, where T_{max} = maximum thickness of deposit. F is the percentage of the deposit finer than 1 mm (0ϕ) at the point where the axis of dispersal crosses the $0.01\ T_{max}$ isopach. After Walker (1973) and Wright *et al.* (1980).

deposited on the upwind side of the vent. The fine-ash fraction may be washed out of the eruption cloud by *rain flushing*. Rain-flushed ash falls as mud, shows poor sorting and fractionation, and frequently contains accretionary lapilli (spheroidal particles composed of concentric fine-ash layers accreted round a nucleus) and vesicles. The ash thickness may be controlled by the location of the rainfall, rather than the position of the volcano (Walker 1981a). Thus, thickness distributions should be used with caution when attempting to locate the source areas for fine tuffs. In major phreatic and phreatomagmatic eruptions, the eruption column may contain considerable amounts of water and/or steam: this leads to flushing, and to the nucleation of fine ash to form accretionary lapilli or more irregular aggregates (Walker 1981a). The aggregates and lapilli are often only loosely held together, by moisture, electrostatic charges or mechanical interlocking (Carey & Sigurdsson 1982), consequently they are likely to disaggregate during flight or on landing, and there may be no trace of them in the resulting deposit. This can give a misleading impression of the original grain size and sorting, and an apparently abnormal thickness distribution of the pyroclastic fall deposit. Formation and disintegration of ash aggregates has been well documented from recent eruptions of Soufrière, St Vincent (Brazier *et al.* 1982) and Mount St Helens (Sorem 1982; Carey & Sigurdsson 1982). The mechanism outlined may account for the apparent poor sorting of some ancient fine-grained fall units.

Air-fall deposits are commonly well bedded, and bed thickness may decrease away from the vent. In general, bedding develops as a result of discontinuity in the explosive activity. Vertical facies variations depend on varying eruption intensity, and thus are often non-systematic. Major plinian air-fall deposits are thought to be produced by continuous gas-blast eruptions, and show little or no bedding (Lajoie 1979). They are characterized by a high content of coarse, dense pumice, usually of rhyolitic or dacitic composition, and wide dispersal (high *D*) (Walker 1982b). The maximum size of pumice and lithic clasts decreases systematically away from the vent, as do the median diameter and crystal content, and this information may assist in the location of the source vent (Walker 1981a). All these features suggest powerful, open-vent eruptions which produce eruption columns several tens of kilometres high. In some cases, there is evidence of coarsening upwards, and many deposits show an overall compositional zoning, from more acidic at the base to more

basic at the top—this represents the emptying of a zoned magma chamber (Walker 1982a, b). Thickness is not a reliable indicator of vent position, as the maximum thickness of a plinian deposit may be developed several tens of kilometres downwind from the vent (Walker 1981a).

Phreatoplinian fall deposits, produced by violent phreatomagmatic eruptions involving silicic magma, share some features in common with plinian deposits (Self & Sparks 1978). The former have a similar widepread dispersal from high eruption columns, resulting in extensive ash sheets (high *D*), but their overall grain size is much finer (high *F*), and sorting poorer, so that they are fine-grained even near source. They may contain accretionary lapilli (Self & Sparks 1978). Similar deposits may result when large pyroclastic flows, on entering the sea, initiate major phreatoplinian eruptions (Walker 1979).

Vulcanian fall deposits result from brief, cannon-like explosions which produce small-volume, thin but widely dispersed fine-grained ash deposits (high *F*, moderate to high *D*) from high (5–20 km) eruption columns (Wright *et al.* 1980; Walker 1982a). Such deposits are commonest on composite andesitic and basaltic volcanoes. Much of the tephra is non-juvenile, and sorting is poor near the vent, where boulder-size blocks and scoria may accumulate. It is uncertain whether ground water plays an important role in these explosive eruptions (Self *et al.* 1979), but much of the fine ash is produced by 'milling' in the vent (Walker 1982a).

It has recently been recognized (Sparks & Wright 1979; Suthren & Furnes 1980; Wright 1980; Wolff & Wright 1981) that welding occurs in pyroclastic fall deposits as well as in ash-flow tuffs. Welding is common in the proximal portions of Plinian fall deposits (Walker 1982a), and is most intense where the deposit is thickest. Welding is favoured by high accumulation rates, low viscosity and high-temperature magma (Sparks & Wright 1979). These conditions are most easily satisfied by peralkaline magmas with a low volatile content, although welded calc-alkaline air-fall tuffs are known. Welded air-fall tuffs may be found either within thick sequences of predominantly non-welded air-fall tuff, or interbedded with massive ash-flow tuffs. Welding may also be induced in a previously deposited ash by an external heat source, such as an overlying lava flow or massive ash-flow tuff (Schmincke 1967; Suthren & Furnes 1980). Although welded air-fall tuffs may superficially appear similar to welded ash-flow tuffs, a number of features enable the two to be distinguished. Welded air-fall tuffs commonly grade laterally and vertically into typical non-welded air-fall tuffs. They show internal bedding, and mantle pre-existing topography, and there are abrupt changes in grain size, clast type and degree of welding from bed to bed. Some beds are graded, and sorting is often better than for pyroclastic flow deposits (Sparks & Wright 1979; Suthren & Furnes 1980).

To conclude this section on subaerial pyroclastic fall deposits, it is pertinent to summarize the features which enable their recognition in the geological record. These are outlined by Walker (1981b). Pyroclastic falls mantle all but the steepest topography, with uniform thickness over depressions and highs (Fig. 5). The bedding is generally parallel-sided and laterally continuous, and both normal and reverse grading may be present. Thickness and grain size may decrease systematically away from the vent, and most fall units show good to moderate sorting. In their proximal parts, block impact structures ('bomb sags') may be present, and the tuffs may be welded. Difficulties could be encountered in distinguishing air-fall tuffs from (i) parallel-bedded pyroclastic surge deposits, and (ii) parallel-bedded reworked volcaniclastic sediments, although the latter are likely to contain less juvenile vitric material (e.g. pumice), and may show rounding of the grains.

Subaqueous pyroclastic fall

Subaqueous ash-fall deposits are much less well documented than their subaerial equivalents, although recent work on deep-sea ash layers has helped to remedy this deficiency. Ash falling out through a water column tends to become well bedded, and often finely laminated (Lajoie 1979). Subaqueous ash-fall deposits may originate either from subaerial or submarine explosive eruptions; the two main sources of fine-grained submarine fall layers are major plinian eruptions, and the ash-cloud overriding subaqueous pyroclastic flows and sediment gravity flows (Fisher 1984). The distribution of these deposits is influenced by wind and current action, particularly in the case of fine ash and pumice. The dispersal and delayed sinking of pumice may result in beds with pumice-rich tops, or scattered pumice within marine sediments (Huang 1980). Submarine ash layers are often very thin, but may extend for hundreds or even thousands of kilometres from source. In some cases, they may be directly correlated with major subaerial eruptions (e.g. Carey & Sigurdsson 1978; Ninkovich *et al.* 1978). Correlation of ash layers in deep-sea cores is effected by detailed petrographical and geochemical (especially electron microprobe) analysis of the component glass shards and

crystals. Frequently, submarine tephra occurs intermixed with the 'background' sediment: the formation and preservation of discrete ash layers are favoured by high pelagic sedimentation rates and the absence of burrowing organisms (Huang 1980).

Pyroclastic flow deposits

Pyroclastic flow (or *ash flow*) deposits include the products of a variety of gravity-driven mass flows originating directly from explosive eruptions. The main components of these flows are volcanic gas and primary magmatic particles, ranging in size from dust to boulders, and predominantly of acidic composition, although flows of all compositions are recorded. Such deposits are emplaced either into subaerial or subaqueous environments. The deposit of a single pyroclastic flow is a *flow unit*. A commonly used term for such deposits is *ignimbrite*: an ignimbrite may include one, several or many flow units. Important early contributions to the understanding of pyroclastic flow deposits were made by Smith (1960) and Ross & Smith (1961). In recent years, the literature on these deposits, and on models for their eruption and emplacement, has become very extensive. Most of this work has concentrated on subaerial pyroclastic flows, and their subaqueous counterparts are much less well known.

Pyroclastic flows form part of a continuum, in which other end members are water-laden debris flows, cold rock avalanches, and the much more dilute pyroclastic surges. All gradations between these types of flows occur. Pyroclastic flows themselves include a range of mechanisms, from turbulent to viscous suspensions, but it may not always be possible to distinguish the deposits of these different types of flows (Lajoie 1979). High particle concentrations are thought to be the main cause of poor sorting of pyroclastic flow deposits, rather than turbulence. Flow units which show compositional zoning from base to top must have been non-turbulent (e.g. Wright & Walker 1981), although some high-velocity pyroclastic flows may be turbulent (Walker *et al.* 1980b).

Subaerial pyroclastic flows

Rhyolitic pyroclastic flow deposits are the most voluminous, and form a major part of subaerial acidic volcanoes, particularly in destructive continental-margin settings. Individual pyroclastic flow units may exceed 1000 km^3 in volume.

A review of emplacement mechanisms is presented by Sheridan (1979). There is now a consensus that most pyroclastic flows form by the collapse of eruption columns, and travel as hot, concentrated laminar flows or plug flows, with high particle : gas ratios, in which large clasts are supported in a fluidized gas/ash matrix of finite yield strength (Walker 1981b, 1982a). A sheared basal layer is usually present. Large volume flows probably form by continuous collapse of a high-intensity eruption column, whilst smaller volume flows are generated by intermittent collapse of columns from discrete explosions (Wright *et al.* 1980, 1981b).

Most pyroclastic flows observed by geologists have been small volume *nuées ardentes*, and it has been tempting to use these as analogues for pyroclastic flows several orders of magnitude larger. Many nuées ardentes result from the explosive or gravitational collapse of high viscosity lava flows or domes on steep-sided volcanoes, forming thick, unsorted deposits with blocks up to several tens of metres across. Grain flow may be an important mechanism in their deposition. In view of these differences in volume, grain size and eruptive and depositional mechanisms, nuées ardentes deposits may not be an appropriate analogue for the much larger ignimbrites which are more commonly encountered in the geological record (information from Wright *et al.* 1980, 1981b; Smith & Roobol 1982; Walker 1982a).

The distribution and lateral and vertical facies changes in ignimbrites are controlled by differences in velocity and mobility (dependent on eruption intensity and column height: Wright & Walker 1977), and in the degree of inflation of the flow, which will tend to become deflated at its distal end due to the convective loss of gas and fine ash (Sheridan 1979). Poorly expanded, low-velocity flows from low eruption columns are strongly confined by topography. Their deposits have a high aspect ratio, up to 1:300 (Wright 1981), and are often *compound*, consisting of many separate flow units (Wright 1981; Wilson & Walker 1981). Deposition from a low eruption column favours heat retention and welding (Sparks & Wilson 1976).

High-velocity or expanded pyroclastic flows are capable of surmounting passes up to 1500 m high (Walker *et al.* 1980b). Their deposits often consist of a single widespread flow unit with a low aspect ratio (may be as low as 1:100,000), which partially mantles the underlying topography, and shows crude layering. Such flow units may originally have been deposited over the whole landscape, followed by incomplete 'draining' into topographic lows (Walker *et al.* 1980a; Walker, 1981a; Wilson & Walker, 1982b). These thin, widespread ignimbrites result from collapse of high eruption columns, and are often underlain by plinian fall deposits. Much heat is lost to the atmosphere, and so these deposits are commonly non-welded (Sparks & Wilson 1976).

Many pyroclastic flows separate into a dense basal avalanche or *underflow* and an overriding dilute, turbulent *ash cloud*, which expands and rises by convective buoyancy (Sheridan 1979).

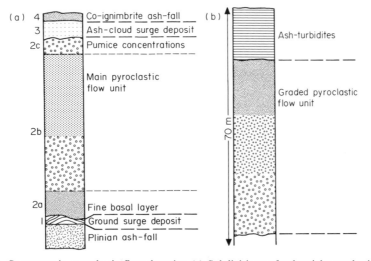

FIG. 7. Sequences in pyroclastic flow deposits: (a) Subdivisions of subaerial pyroclastic flow and associated deposits. Scale and relative thicknesses of subdivisions are highly variable. Based on Sparks *et al.* (1973), with minor modifications after Fisher (1979). (b) Sequence in a subaqueous pyroclastic flow unit and associated deposits. After Fiske & Matsuda (1964).

This is just one type of *flow transformation* (Fisher, 1984). The basal avalanche is the true pyroclastic flow, moving by laminar or plug flow (Smith & Roobol 1982). The ash cloud often separates completely from the avalanche —it is much less confined by topography, and may cover a much larger area. Up to half the erupted mass may be contained in the ash cloud (Walker 1979). Thus, in some cases, the pyroclastic flow deposit proper is overlain by or interbedded with its ash cloud deposits, whilst in other areas the two are not found together. Transport and deposition mechanisms of the ash cloud are varied: fines may settle out as a *co-ignimbrite ash fall* (Sparks & Walker 1977), or the cloud may travel as a pyroclastic flow or pyroclastic surge (*ash-cloud surge* of Fisher 1979; Fisher *et al.* 1980). High density, poorly expanded pyroclastic flows may not separate into an underflow and an ash cloud (Buck 1981).

Now that an outline of ignimbrite emplacement mechanisms has been given, it may be useful to describe the characteristics which may enable their recognition in the geological record: these are summarized by Walker (1981b). Because they are deposited by gravity-driven flows, many ignimbrites are ponded in depressions, and have a planar top (Fig. 5). Thus, there is an irregular thickness variation away from the vent. Sorting and internal bedding are generally absent, and there may be evidence for high temperatures, such as welding, vapour-phase crystallization, carbonization of wood, or directional magnetic properties. A valuable contribution was made by Sparks *et al.* (1973) who suggested an idealized vertical sequence through a single pyroclastic flow unit (Fig. 7a). In many cases this sequence is incompletely developed. Below many ignimbrites formed by column collapse is a plinian fall unit. Above this there may be a thin plane- or cross-bedded layer (layer 1) which represents a dilute pyroclastic surge (*ground surge*—see next section) which preceded the main flow. Lithic/crystal-rich *ground breccia* may also occur at this level, due to settling out of coarse pyroclasts from the flow (Druitt & Sparks 1982). The *basal layer* (2a) is finer grained than the rest of layer 2, and is present even at steep or vertical contacts. It forms by shearing at the flow margins, and frequently shows reverse grading. Layer 2b is the main body of the pyroclastic flow unit, transported by concentrated laminar or plug flow above the sheared basal layer. It is characteristically unsorted and matrix-supported, and the clasts rarely exceed a few tens of centimetres across. The base of 2b is commonly rich in lithic fragments, which sank through the flow during transport. These show a decrease in size and abundance both upwards and away from source (Walker 1979). In contrast, pumice fragments often show reverse grading, being segregated towards the top of the flow unit by their buoyancy (layer 2c). If the overriding ash cloud separates from the pyroclastic flow, there may be a thin ash cloud surge layer (3), consisting of lenses of laminated or cross-bedded ash (Fisher 1979)—this layer is often absent near source. At the top of the sequence, the elutriated fines may settle out to form a thin, widespread co-ignimbrite ash-fall, with low preservation potential. In areas distant from source, this may be the only representative of a major ignimbrite eruption.

The model of Sparks *et al.* (1973), outlined above, has proved to be generally applicable, although the relative thicknesses of the layers show considerable lateral variation. More recent work has shown the presence of distinctive facies which occur as proximal, distal and lateral equivalents to the sequence of Sparks *et al.* (1973).

The loss of fine ash from a pyroclastic flow results in concentration of denser components, particularly crystals and lithic fragments, within the flow (Walker 1972). In extreme cases, the loss of fines may result in clast-supported flow deposits, and the crystal : vitric ratio may increase 10-fold (Walker 1972, 1979). Much of the fine fraction may be generated not at the vent, but by fragmentation and attrition within the flow (Walker 1979). This is supported by the common occurrence of rounded pumice fragments in ignimbrites (Druitt & Sparks 1982). The loss of fines is one way in which pyroclastic flows become fractionated, particularly in their distal reaches.

Fractionation may also occur close to source by the settling out of large clasts which the pyroclastic flow is not sufficiently competent to carry. Thus, coarse deposits up to 25 m thick accumulate at or near the site of column collapse: these are termed *co-ignimbrite lag fall* (Wright & Walker 1977). They are clast-supported, poorly sorted and show normal coarse-tail grading and rapid lateral facies variations. Like the corresponding pyroclastic flow units, into which they pass laterally or vertically, they have a thin, reverse-graded basal layer, and may show a compositional zoning corresponding to the reversed stratigraphy of the magma chamber. They are thought to be the thick, proximal equivalents of the concentrations of smaller lithic blocks which occur at the base of layer 2b of a typical pyroclastic flow unit (Druitt & Sparks 1982).

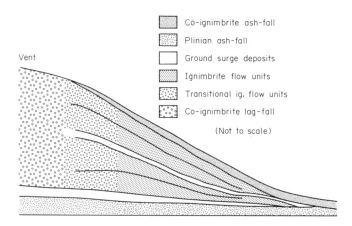

Vent

☐ Co-ignimbrite ash-fall
☐ Plinian ash-fall
☐ Ground surge deposits
☐ Ignimbrite flow units
☐ Transitional ig. flow units
☐ Co-ignimbrite lag-fall

(Not to scale)

FIG. 8. A facies model for ignimbrites, showing vertical and lateral (proximal to distal) facies changes. Based on Bandelier Tuff, New Mexico (Wright *et al.* 1981a). Vertical scale greatly exaggerated.

Many subaerial ignimbrites also display welding—the cohesion of hot, viscous glassy fragments. This may vary in intensity from incipient, to a *eutaxitic* or *parataxitic* texture in which glass and pumice grains are highly flattened, or in extreme cases to a homogeneous glass. Pyroclastic flow deposits may cool either as single flow units (*simple cooling units*), or as several flow units emplaced in close succession (*compound cooling unit*) (Smith 1960). Welded cooling units generally have a thin non-welded base and a thicker non-welded top. The degree of welding is controlled chiefly by emplacement temperature and the thickness of the cooling unit (Smith 1960).

Ancient examples of bedded ash-flow tuffs, both welded and non-welded, are known (e.g. Sparks 1976; Suthren & Furnes 1980). They usually occur as single beds or groups of several beds within sequences of massive ash-flow tuffs. These beds may be the lateral or distal equivalents of thick flow units, or represent the proximal 'smear' left by a thick flow on a slope too steep for it to come to rest.

The recognition of a variety of facies in sub-aerial ignimbrites has led to further development of the facies model of Sparks *et al.* (1973). In the model of Wright *et al.* (1981a) (Fig. 8), co-ignimbrite lag-fall is the dominant near-vent facies, and passes gradationally away from the vent into ignimbrite flow units. These become thinner and fewer in number in the distal region. The ignimbrite sequence may be underlain by plinian fall and ground surge deposits.

Walker (1981a) and Wilson & Walker (1982) believe that pyroclastic flows, like turbidity currents, have a head, a body and a tail. These authors suggest that the head of the flow, fluidized by ingestion of air, deposits layer 1. The bulk of the deposit—layer 2—is deposited by the body and tail of the flow. Ingestion of air at the sides of the flow may result in deposits which show the combined features of layers 1 and 2.

Subaqueous pyroclastic flows

Pyroclastic flows in the subaqueous environment originate either from subaqueous explosive eruptions, or when subaerial pyroclastic flows enter a body of water (Fisher, 1984). The distribution of their deposits is controlled by subaqueous topography (Huang 1980). Subaqueous pyroclastic flows, in which the particles are dispersed in gas, often pass gradationally into various types of subaqueous sediment gravity flow in which water is the dispersal medium: these include debris flows, grain flows and ash turbidites. The deposits of subaqueous pyroclastic flows and other types of mass flows (dealt with in a later section) may be indistinguishable (Lajoie 1979; Huang 1980). However, the products of some subaqueous pyroclastic flows show evidence of high temperatures. Heat conservation and strong welding are not confined to subaerial ignimbrites, but also occur in their subaqueous counterparts (Francis & Howells 1973; Lajoie 1979). Unlike the former, some of the latter are welded right down to their lower contacts, which in some cases are highly irregular and transgressive in relation to the underlying sediments. This is a result of either large scale loading and foundering (Francis & Howells 1973) or of wet sediment fluidization at the base of the flow (Kokelaar 1982).

In both subaqueous and subaerial pyroclastic

flow deposits, the bulk of each flow unit comprises massive, unsorted, matrix-supported, pumice-rich tuff. However, many subaqueous pyroclastic flow units pass upwards and distally into bedded tuffs deposited by thin sediment gravity flows, fallout from suspension, and reworking (Fiske 1963; Fiske & Matsuda 1964; Francis & Howells 1973; Lajoie 1979; Busby-Spera 1982). Some of these bedded deposits originate from the dilute, turbulent ash/water cloud which separates from the dense pyroclastic underflow (Fisher, 1984). In other cases, mixing of the dense pyroclastic flow head with water may produce more dilute sediment gravity flows (cf. Hampton 1972).

The characteristics of the flow units are very variable, depending on flow conditions, including velocity, density and viscosity, and on distance from source (Lajoie 1979). Some show normal grading by size or density both upwards and away from source. There is frequently an increase in bed thickness downflow—often high velocity and low viscosity near the vent result in little deposition in this area. Sequences of primary sedimentary structures may show a systematic variation from proximal to distal regions (Tassé et al. 1978; Lajoie 1979).

Pyroclastic surge deposits

Pyroclastic surges are dilute particulate flows, in which the particles are kept in suspension in turbulent gas (Walker 1981b), and are known only from the subaerial environment. There is thought to be a complete gradation from high concentration pyroclastic flows to low concentration pyroclastic surges (Wright et al. 1980). Pyroclastic surges can be classified into two groups: (i) hot, dry surges from pyroclastic column collapse, and (ii) cool, wet surges from phreatic or phreatomagmatic column collapse. Sparks & Walker (1973) realized that many pyroclastic flows separate into a dense avalanche and a relatively low-density hot, dry surge. Such surges may either precede the pyroclastic flow, as a *ground surge*, or may separate from the overlying ash cloud, to produce an *ash-cloud surge*. Thus a ground surge deposit will underlie the main pyroclastic flow unit, whilst an ash-cloud surge deposit will lie above it. Ground surge may also originate from explosions which produce neither a vertical eruption column nor a pyroclastic flow (Fisher 1979; Fisher et al. 1980). Cool, wet surges were first recognized during the 1965 eruption of Taal, in the Philippines (Moore 1967) and are often referred to as *base surges*.

The deposits of pyroclastic surges are relatively thin and fine grained, and they have a low preservation potential unless buried quickly—e.g. below an associated pyroclastic flow unit (Sparks & Walker 1973; Fisher 1979). They have a number of distinctive characteristics, summarized by Walker (1981b). Surges are not entirely confined to topographic lows, and their deposits tend to drape the topography, although they are thickest in depressions and thinnest over highs (Fig. 5). There may, however, be a general decrease in thickness and grain size away from source. The base of a surge deposit is commonly erosional, and U-shaped channels are frequently seen—these may be cut by lobes in the head of the surge cloud (Fisher 1977).

In their other features, the products of cool, wet surges may differ from those of hot, dry surges. Base surge deposits are well bedded, and often show the spectacular development of unidirectional sedimentary structures of both lower and upper flow regimes (Crowe & Fisher 1974), including plane beds, sometimes with reverse grading, low-angle cross-bedding, climbing dunes, pinch and swell, antidunes, and chute and pool structures. All except the plane beds are grouped together as *sandwave beds* (Sheridan & Updike, 1975). Excellent examples of these structures are illustrated by Schmincke et al. (1973). Due to deceleration of the surge away from source, there may be a down-surge decrease in the wavelength of the bedforms (Moore 1967). The sequences of sedimentary structures also change in a down-surge direction. Near source, massive beds and sandwave beds are found, accompanied at intermediate distances by plane beds. In the distal parts of the surge deposit sandwave beds are often absent. These lateral changes are attributed to a change from dominantly turbulent and fluidized flow in proximal areas, to dominantly laminar grain flow, inertial flow and turbulent flow in distal areas, as the surge deflates and decelerates (Sheridan & Updike 1975; Wohletz & Sheridan 1979).

Base surge deposits also show considerable grain-size variations from one bed to the next, although individual beds tend to be well sorted (Wright et al. 1980). The water and steam content results in cohesive fine ash which commonly forms accretionary lapilli or vesicles, or plasters the up-vent side of obstacles in the path of the surge (Walker 1981b).

By contrast, the deposits of hot, dry surges often have poorly developed bedding. Sorting is moderate to good, and they are depleted in fines. There may be evidence for high temperature, such as the charring of wood and, in rare instances, welding (Walker 1981b). Ash-cloud surge deposits often consist of a breccia containing clasts up to 70 cm across, sometimes with a fine-grained basal layer, overlain by fine tuff which may be cross-bedded. The breccia is regarded as the deposit of a secondary dense underflow, produced by the settling of coarse, dense material to the base of the moving ash cloud surge (Fisher *et al.* 1980; Fisher & Heiken 1982). Such deposits may be recognized by their stratigraphic position above, or as discontinuous lenses within, a pyroclastic flow unit, and may be overlain by a thin co-ignimbrite ash-fall. They tend to become more continuous and thicker away from source, and often show segregation into crystal-rich and crystal-poor laminations (Fisher 1979; Fisher *et al.* 1980).

Although many of the features described would seem to make identification of ancient surge deposits relatively easy, very few are recorded in the literature. In large part, this is probably due to their volumetric insignificance as compared with other types of pyroclastic sediments, and their low preservation potential. Some of the characteristics outlined are common to other types of deposit, and this may hinder identification of surge deposits. For example, many pyroclastic fall deposits show plane-bedding and good sorting, whilst cross-bedding and related structures are very common in volcaniclastic sediments reworked by traction currents.

Reworked volcaniclastic deposits

The term *reworked* is used in a broad sense here, to include sediments derived almost directly from pyroclastic eruptions on the one hand, through to epiclastic sediments formed by the weathering and erosion of consolidated volcanic rocks on the other.

Once volcanic activity has ceased, most subaerial volcanic piles are partially or completely destroyed by denudation, and may only be recognized in the marine sedimentary sequences derived from them, and perhaps as eroded remnants of sub-volcanic intrusions (Bailes 1980; Francis 1983). Francis (1983) suggests that the central and proximal parts of a volcano are most prone to destruction, and consequently the facies deposited in these areas have a low preservation potential. The distal portions are most likely to be preserved if the reworked volcaniclastic sediments accumulate in a subsiding basin adjacent to the volcanic complex. Large, composite volcanoes seldom exceed 3 km in height, whilst erosion rates may exceed 1 km Ma^{-1} in volcanic regions of high relief. The rate of erosion of unconsolidated material may be several orders of magnitude greater than that for lithified volcanic products.

Sediment gravity flow deposits

Sediment gravity flows (excluding pyroclastic flows) include a wide range of gravity-driven mass flows which consist of a mixture of volcanic particles and water. Many fall into the 'grey area' between pyroclastic and epiclastic processes: e.g. there is often direct *flow transformation* from gas-supported pyroclastic flow to water-supported mass flow. In other cases, mass flows remobilize piles of loose pyroclastic sediment, whilst yet others transport weathered and eroded epiclastic sediment (Fisher, 1984). The range of processes includes subaerial and submarine landslides and slumps, concentrated debris flows (mudflows), grain flows, and more dilute, turbulent sediment/water mixtures, including turbidity currents. The deposits of this spectrum of processes may be difficult to distinguish from each other. In areas of active explosive volcanism, sediment supply rates are high, and slopes often steep, so that sediment gravity flows are a common phenomenon in both subaerial and subaqueous environments.

Subaerial sediment gravity flows

The most common type of sediment gravity flow in the subaerial volcanic environment is the volcanic mudflow or *lahar*. Many lahars result from the passage of a hot pyroclastic flow into water, such as a river or lake, so that valley-confined pyroclastic flows may pass distally into hot lahars, as seen in the 1980–1982 eruptions of Mount St Helens (Harrison & Fritz 1982; Lipman & Mullineaux 1982). Primary volcanic triggering of hot lahars may also occur when eruptions take place through a crater lake, or beneath or on to ice or snow (Crandell 1971). Cold lahars are produced when rain falls on to slopes of unconsolidated pyroclastic debris: instability and failure may result from high pore-water pressures, or may be triggered by seismic or explosive activity. The common near-vent hydrothermal alteration of volcanic rocks to form clay minerals assists in this process (Crandell, 1971).

Plug flow is the dominant mechanism in subaerial mudflows. The finite yield strength of the clay/water matrix allows large clasts to be transported, but buoyancy is also important in supporting these clasts, and increasing the competence of the flow to carry large particles (Hampton 1979). The diameter of the largest clasts (MPS, or maximum particle size) is typically high in relation to bed thickness (BTh). Gloppen & Steel (1981) quote BTh:MPS ratios of 2:1 to 4:1 for non-volcanic subaerial debris flows. The deposits of such flows are generally unsorted, and either clast- or matrix-supported, although the clasts may be well rounded, and show normal coarse-tail grading. Shearing at the base of the flow often results in a fine-grained basal layer, which may be reverse-graded (Schmincke 1974).

Subaerial mudflows grade, with increasing water content, through torrential mudfloods and streamfloods into high energy fluvial processes. The sediment flows produced by explosive eruptions on Mount St Helens in 1982 (Harrison & Fritz 1982) may be transitional between mass flow and streamflow. Their deposits, unlike those of mudflows, contain little mud, are clast-supported in part, and show some internal stratification. However, the accumulation of large boulders at the top of the flow units indicates a concentrated flow with matrix strength. Each flow unit shows a 3-fold division, into a lower massive to graded layer with large clasts, a finer, crudely bedded central layer, and an upper layer of matrix-supported boulders and logs. Similar deposits are described by Vessell & Davies (1981).

Subaqueous sediment gravity flows

A comprehensive review of subaqueous volcaniclastic sediment gravity flows and their deposits is provided by Fisher (1984). They range from concentrated debris flows through to dilute turbidity currents, and Fisher has made the important observation that flows often change their character during transport. This is reflected by lateral changes in grain size, sorting and primary structure sequences of their deposits (Tassé *et al*. 1978). Such *flow transformation* (Fisher, 1984) may occur when there is a change from turbulent to laminar flow or vice versa, when a mass flow is segregated by gravity into a concentrated laminar underflow and overriding dilute turbulent cloud, or by mixing with water at the flow margins. As examples, Fisher (1984) suggests that turbidity currents may be generated at the fronts of slumps or debris flows (cf. Hampton 1972), and that hot pyroclastic flows may pass laterally into lahars or turbidity currents. It will often be difficult to distinguish sediment gravity flow deposits generated directly by eruptions, from those which consist of reworked pyroclastics (e.g. Wright & Mutti 1981) unless a passage into primary pyroclastic flow deposits can be traced laterally.

The characteristics and sequences of a variety of non-volcanic subaqueous mass-flow deposits are well known (e.g. Middleton & Hampton 1973). Their volcaniclastic counterparts are likely to be very similar, and are not described in detail here. It should be mentioned, however, that subaqueous debris flows are less viscous and less competent than their subaerial counterparts, and have a higher percentage of fine-grained matrix. BTh:MPS ratios vary from 3:1 to 10:1, and the clasts may be imbricated (Gloppen & Steel 1981).

Descriptions of actual examples of subaqueous mass flow deposits are rare, but Mitchell (1970) describes and interprets a variety of submarine volcanic sediment gravity flows from an island-arc setting. Volcaniclastic turbidites are particularly common in such settings, and may be deposited as submarine fans, supplied from point sources such as submarine channel mouths (Bailes 1980). Alternatively, there may be a linear sediment input from the arc into the basin along the entire basin margin, in which case non-fan turbidites develop (Storey & MacDonald 1984).

Shallow water and subaerial reworking

Most of the processes involved in shallow water and subaerial reworking of volcaniclastic sediments are similar or identical to those operating in non-volcanic environments, and are not considered further. There are, however, some general points worth making. Volcaniclastic sediments may have rather different hydrodynamic properties from siliciclastic sediments, because of the wide range of size, shape, density and composition of their particles.

Erosive regimes often predominate in subaerial volcanic areas, especially where rainfall and relief are high, leading to the erosion of steep-sided rills, gullies and canyons (or *barrancos*) into loose or partially lithified pyroclastics. Such features are common in ancient deposits. Rainfall may be concentrated as intense storms, some of which are generated directly by eruptions: the rapid upward transport of moist air by an eruption column leads to precipitation, which is enhanced by the abundance of fine ash particles on which raindrops or hailstones will nucleate preferentially. Such storms lead to short periods of high run-off, erosion and sediment transport, and deposition by mass flows and fluvial torrents on extensive outwash plains (*volcanic aprons*), in which alluvial fans and cones pass distally into flood plains of low to high sinuosity rivers.

The volcaniclastic sediment supplied at the proximal end of a fluvial system is both texturally and mineralogically immature, and will undergo considerable modification during transport. Volcanic cobbles and boulders may become rounded over distances of as little as 3 km, partly by abrasion by smaller particles when the larger clasts are stationary (Pearce 1971). Downstream changes in mineralogy result from the selective destruction of chemically unstable and physically weak igneous minerals, and by dilution by sediment of non-volcanic provenance (Blatt 1978).

Case studies

In this section, recent research in two active volcanic areas is reviewed. Both are from destructive plate margin settings, and in both the sediment is supplied largely from pyroclastic eruptions. There is close interaction between volcanic and sedimentary processes, and volcanic activity is a vital control on sedimentation. Both studies have resulted in detailed facies models, which may be applicable to the interpretation of ancient volcaniclastic sequences.

Lesser Antilles arc, fore-arc and back-arc

The Lesser Antilles form an island arc, in a region where the oceanic lithosphere of the western Atlantic is subducting westwards beneath the Caribbean plate. The account which follows is compiled from the work of Carey & Sigurdsson (1978, 1980, 1984), Sigurdsson *et al.* (1980) and Sigurdsson & Carey (1981). The predominantly calc-alkaline subaerial explosive activity provides a major sediment source, largely from major pyroclastic-flow eruptions. Some 527 km³ sediment has been produced over the last 100,000 years, and 1,000,000 m³ from the 1979 eruption of Soufrière, St Vincent, alone.

Most of this sediment is rapidly eroded and transported, either eastwards into the Atlantic (30%), or westwards into the back-arc area of the Grenada Basin (70%), where a sediment pile 7 km thick has accumulated over 47 Ma. Submarine volcaniclastic sediment occurs both as discrete layers, and dispersed through the 'background' pelagic and hemipelagic sediments.

Cored sequences show a markedly asymmetrical distribution of volcaniclastic facies. Whilst almost all of the ash in the Grenada Basin was deposited from sediment gravity flows, most of that in the western Atlantic settled out as pyroclastic falls (Fig. 9). The deposits in the Grenada Basin include debris flow, grain flow and turbidite units, some of which accumulated as small submarine fans. Most units are only a few centimetres thick. It is believed that most of the sediment gravity flows were produced by flow transformation of subaerial pyroclastic flows which entered the sea, and in some cases, the cored ash layers can be matched with their proximal equivalents on land. Some eruptions, such as that of Soufrière, St Vincent in 1902 (Carey & Sigurdsson 1978) contributed ash both westward as flows, and eastward as pyroclastic fall. This particular unit, transported by wind and ocean currents, can be traced 1000 km eastward from its source.

Carey and Sigurdsson and their co-workers conclude that the asymmetrical facies distribution is controlled by eruptive style, prevailing winds and submarine topography (Fig. 9). Any ash reaching the upper atmosphere via a high eruption column will be carried east by the prevailing anti-trade winds, and eventually deposited as ash fall on the Atlantic floor, after

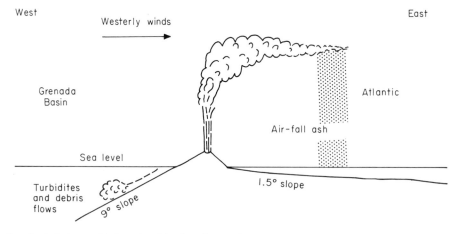

FIG. 9. A facies model for volcaniclastic sedimentation in and around the Lesser Antilles Arc. After Sigurdsson *et al.* (1980).

further redistribution by ocean waves and currents. Sediment gravity flows are favoured by relatively steep slopes, such as that descending westwards into the Grenada Basin. The gentle eastward slope to the Atlantic may not be sufficient to sustain such flows, so that ash-flows reaching the sea on this side of the arc may be largely reworked by shallow water processes.

Central American subaerial arc and fore-arc

The area described is the volcanic mountain chain and subaerial fore-arc on the Pacific side of Central America, where the oceanic Nazca plate is subducted northeastwards beneath the continental margin. The importance of explosive volcanism as a sediment supplier in this region was recognized in the classic study of Paricutin volcano, Mexico, by Segerstrom (1950), who gives a comprehensive account of the erosion, transport and deposition of ash in the subaerial environment. He observed rapid rill and channel cutting and transportation by intermittent streams and sheetfloods active during wet-season storms. As each storm flood waned, de-position took place in alluvial fans, fluvial channels and flood plains, in crater lakes, and in valley lakes dammed by alluvial fans or lava flows. Mass movement, including landslides and mudflows, was also important, and during the dry season, aeolian transport of fine ash was observed.

Further east, in Guatemala, similar processes and their resultant depositional facies have recently been observed at first hand. This account is compiled from the publications of Vessell et al. (1977), Davies et al. (1978, 1979a, b) and Vessell & Davies (1981), based on eruptions of Fuego volcano in the early 1970s, and work by Kuenzi et al. (1979), concerned with the longer-term fate of sediments derived from the 1902 eruption of Santa Maria. At a distance of 30–100 km from the ocean, a chain of active andesitic volcanoes rises to 4000 m. The fore-arc is 140–250 km wide, including a coastal plain 20–40 km wide, and is split up by faults normal to the subduction system into small fore-arc basins, each some 150 km long and 50–100 km wide. Subsidence and sedimen-tation rates vary from one basin to another, but up to 15,000 m of marine and continental sediments have accumulated in them.

The wet tropical climate results in rainfall of up to 500 cm year^{-1} on the volcanoes, and 150–250 cm year^{-1} on the coastal plain. This is concentrated into short periods during the May–October wet season, and much falls as torrential tropical rainstorms. Individual storms may last 2–3 days, and produce more than 100 cm rain high on the volcano. The major sediment source is from pyroclastic flow and fall eruptions which rapidly produce large volumes of material of basic to intermediate composi-tion. These loose unvegetated sediment piles are easily eroded and transported as bedload by the extremely high run-off from rainstorms. Within minutes, the discharge of small rivers may increase by several orders of magnitude, resulting in torrential streamfloods and debris flows capable of carrying huge volumes of very coarse debris (Table 2). Such flows are the dominant agents of transport and deposition on the volcano flanks.

Mineralogy and texture of the sediments change markedly during transport downslope. Because of the high transport rates, prede-positional chemical weathering is unimportant, but many of the glassy lithic clasts are broken down by shattering during transport. Grain size decreases exponentially downstream, due to the

TABLE 2. *Data illustrating the changes observed in a river carrying flood flows produced by torrential rainstorms. Drainage area 420 km^2 (after Vessell* et al. *1977)*

Width	Normal: 10 m	Flood: 320 m
Depth	Normal: 0.3 m	Flood: 3.5 m
Discharge	Normal: 5.5 m^3 s^{-1}	Flood: 2200 m^3 s^{-1}
Velocity		Flood: up to 7.6 m s^{-1}
Max. Clast size		Flood: 3 m
Transport rate		Flood: up to 100 t s^{-1}
Total sediment discharge for one storm		390,000 t
Sediment thickness produced by 2 h flood event		1 m

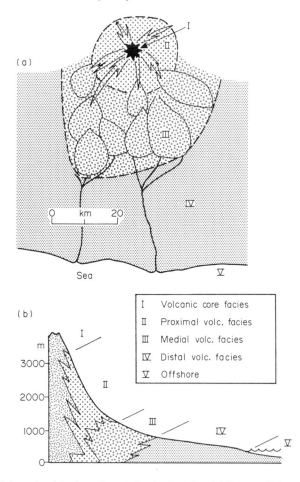

FIG. 10. A model for volcaniclastic sedimentation in the subaerial forearc of Guatemala. (a) Facies map; (b) Cross-section. Vertical scale greatly exaggerated, horizontal scale as for map. I. The volcanic core facies is mainly basaltic andesite lava. II. The proximal volcanic facies consists of unwelded valley-fill ignimbrites. III. The medial volcanic facies is dominated by alluvial fan deposits. IV. The distal volcanic facies consists of braided to sinuous fluvial deposits, passing seawards into deltaic and shoreline sediments. Note the prograding relationships between facies. After Vessell & Davies (1981).

decreasing competence of the flood flows, and boulder roundness improves.

Sedimentary processes also change downslope as the gradient decreases, and this is reflected in the depositional facies. Vessell & Davies (1981) define four facies on the basis of lithology, grain size, sedimentary structures and textures, and depositional processes and environments (Fig. 10). These are: (i) a *volcanic core facies*, (ii) a *proximal volcanic facies*, (iii) a *medial volcanic facies*, and (iv) a *distal volcanic facies*. The facies grade laterally into each other.

The *volcanic core facies* comprises the volcanic cone itself, and it composed largely of intermediate to basic lava flows. The *proximal volcanic facies*, extending up to 15 km from the cone, is built up of non-welded, unsorted breccias, which are the products of valley-confined pyroclastic flows. Scree and thin air-fall ashes also occur.

Apart from thin air-fall ashes, the medial and distal facies consist almost entirely of reworked sediments derived from the unconsolidated pyroclastic flow deposits. The deposition of the *medial volcanic facies* takes place chiefly in alluvial fans fed from the valleys followed by the pyroclastic flows. Debris flows and torrential high concentration streamflows deposit poorly sorted, unbedded coarse breccias, which extend up to 20 km from the parent pyroclastic flow deposit. More dilute streamflows deposit poorly sorted, parallel-bedded coarse sands and conglomerates, including coarse boulder lags, in fan channels during floods. These sediments are increasingly important towards the distal ends of the fans.

The alluvial fans pass gradually downslope into the braided and meandering river systems which dominate the *distal volcanic facies* of the coastal plain. Again, most of the sedimentation occurs during floods, particularly when large volumes of sediment are available. At such times, high-gradient, low-sinuosity braided streams dominate. Massive or plane-bedded coarse sands are deposited in upper flow regime conditions. High sediment supply rates from major eruptions lead to rapid sediment accumulation: in one example, the river bed was raised 10–15 m within 3 years after an eruption. This rapid aggradation frequently leads to the damming of tributary valleys, in which lakes may form. The diversion of fluvial channels into the lakes results in the progradation of simple deltas of Gilbert type over lacustrine muds. At the seaward end of the fluvial system, rapid deltation follows catastrophic eruptions. One coastal delta prograded 7 km in 20 years, following the 1902 eruption of Santa Maria.

Once sediment supply slows down, these high-constructive deltas are rapidly eroded and redistributed as strand-plain sand sheets. Small, wave-dominated, high-destructive arcuate deltas may form. On the fluvial plain, low sediment supply rates cause the river systems to aggrade slowly, or to become incised. They may also change from braided to meandering, and deposit cross-bedded medium to fine sands, often arranged in fining-upward sequences.

It should be clear from the preceding account that sediment supply and style of sedimentation are closely controlled by the frequency of major eruptions. This results in cyclic deposition, with short periods of high sedimentation rates separated by longer intervals when deposition proceeds much more slowly. Vessell & Davies (1981) break down this cyclic sedimentation into four phases for the Fuego area:

1 During the *inter-eruption phase*, sedimentation rates are low, incised meandering streams are common, and coastal deltas are eroded. This phase typically lasts 80–125 years.

2 The *eruptive phase* lasts less than 1 year, and is dominated by pyroclastic flow and fall.

3 The *fan-building phase* starts shortly after the eruption, and lasts for some 2 years. Intense rainstorms remobilize the loose pyroclastic sediments, and alluvial fans are rapidly built up by flood processes.

4 The *braiding phase* starts at the same time as the fan-building phase, but lasts 20–30 years. The supply of large volumes of sediment into the coastal plain fluvial systems, largely by flash floods, results in a change from incised meandering channels to rapidly aggrading braided channels. Rapid delta progradation also takes place. Phases (3) and (4) only occur in response to major eruptions which produce 60,000,000 m^3 or more of ejecta.

With time, and a continuing sediment supply, the whole sedimentary system is prograding in a seaward direction (Fig. 10b), producing a thick sequence which coarsens upward overall. The facies model produced by Vessell, Davies, Kuenzi and their co-workers accounts for the distribution of facies in time and space, and is likely to be valuable in interpreting ancient sequences of subaerially reworked volcaniclastic rocks. An obvious development of the model would be to examine the voluminous offshore sequences.

Applications

There are a number of important applications of facies analysis of volcaniclastic sediments. Thin, widespread fall layers from high eruption columns can often be recognized over very large areas (e.g. Ninkovich *et al.* 1978), and thus provide valuable stratigraphic markers. In ancient sedimentary sequences, they are often represented by bentonites. Since they are produced by essentially instantaneous events, they are excellent chronostratigraphical markers. Where several such ash beds occur, they may be useful in unravelling the time-stratigraphical relationships of sedimentary sequences in which there are lateral facies changes and diachronous units (e.g. Allen & Williams 1981). Correlation of thin tuff markers may also have economic applications, particularly where other stratigraphical information is lacking: surtseyan tuff layers have been used as markers in North Sea oil wells (e.g. Malm *et al.* 1979).

Volcaniclastic sandstones are clearly of interest as potential petroleum reservoir rocks, since they are often found as thick folded and faulted sequences in major sedimentary basins. However, because of their chemically unstable mineral assemblages, diagenesis often leads to a very considerable reduction of porosity and permeability, sometimes within as little as 2000 years of deposition (Davies *et al.* 1979a).

Volcaniclastic facies analysis has also proved very useful in locating base metal deposits, such as porphyry coppers. These are often related to hydrothermal circulation above sub-volcanic intrusions at intermediate to acidic volcanic centres. Volcaniclastic successions often show systematic changes in grain size, bed thickness, primary structure sequences and other attributes away from the source vent, and these lateral variations may therefore be used to attempt to locate the source (and hopefully an economic ore body). This has been done with considerable success in the Archaean greenstone belt sequences of Canada (Lajoie 1977), and the lateral variation in mean maximum clast size has proved a particularly useful and rapid exploration tool (Fox 1977).

Another vitally important use of volcaniclastic facies analysis is in determining the eruptive history of a volcano (e.g. Walker 1981b). Once eruption style and frequency and other aspects of the activity are understood, it may be possible to make predictions about the future behaviour of the volcano. Since many active volcanoes occur in densely populated areas, it is of considerable social importance to assess possible volcanic hazards, and to plan for mitigation of them.

Conclusions

Volcaniclastic sediments are formed and emplaced by four main groups of processes: (i) autoclastic brecciation during magma movement, (ii) magma/water interactions which produce hydroclastic material, (iii) magmatic explosions of pyroclastic material, and (iv) sedimentary reworking, either of loose pyroclastic debris, or of epiclastic sediment derived from earlier volcanic rocks. These groups of processes do not fall into neat compartments, and there are gradations and overlaps between them, due to a complex interaction of volcanic and sedimentary processes. Thus, their products may not always be easily distinguished from each other. A systematic approach to the analysis of volcaniclastic sedimentary sequences is proving successful, however, and the first facies models are appearing in the literature. Thus, although rigorous facies analysis of volcaniclastics is in its infancy when compared with siliciclastic and carbonate rocks, the child may be said to be alive and well, and growing fast.

References

ALLEN, C. C., JERCINOVIC, M. J. & ALLEN, J. S. B. 1982. Subglacial volcanism in north-central British Columbia and Iceland. *J. Geol.* **90**, 699–715.

ALLEN, J. R. L. & WILLIAMS, B. P. J. 1981. Sedimentology and stratigraphy of the Townsend Tuff Bed (Lower Old Red Sandstone) in South Wales and the Welsh Borders. *J. geol. Soc. London*, **138**, 15–29.

BAILES, A. H. 1980. Origin of early Proterozoic volcaniclastic turbidites, south margin of the Kisseynew sedimentary gneiss belt, File Lake, Manitoba. *Precambrian Res.* **12**, 197–225.

BEVINS, R. E. & ROACH, R. A. 1979. Early Ordovician volcanism in Dyfed, SW Wales. *In*: HARRIS, A. L., HOLLAND, C. H. & LEAKE, B. E. (eds) *The Caledonides of the British Isles—Reviewed. Spec. Publ. geol. Soc. London*, **8**, 603–9.

BLATT, H. 1978. Sediment dispersal from Vogelsberg basalt, Hessen, West Germany. *Geol. Rdsch.* **67**, 1009–15.

BRAZIER, S., DAVIS, A. N., SIGURDSSON, H. & SPARKS, R. S. J. 1982. Fall-out and deposition of volcanic ash during the 1979 explosive eruption of the Soufrière of St. Vincent. *J. Volcan. geoth. Res.* **14**, 335–59.

BUCK, M. D. 1981. Peralkaline ignimbrite sequences on Mayor Island, New Zealand. *In*: SELF, S. & SPARKS, R. S. J. (eds) *Tephra Studies*, pp. 337–45. D. Reidel, Dordrecht, Holland.

BUSBY-SPERA, K. 1982. Depositional features of rhyolitic and andesitic volcanic rocks of the Mineral King submarine caldera complex, southern Sierra Nevada, California. Abstract,

Volcanic Processes in Marginal basins. Geol. Soc. London. Ord. Gen. Meeting, Keele, Sept. 1982.

CAREY, S. N. & SIGURDSSON, H. 1978. Deep-sea evidence for distribution of tephra from the mixed magma eruption of the Soufrière on St Vincent, 1902: ash turbidites and air-fall. *Geology*, **6**, 271–4.

—— & —— 1980. The Roseau Ash: deep-sea tephra deposits from a major eruption on Dominica, Lesser Antilles Arc. *J. Volcan. geoth. Res.* **7**, 67–86.

—— & —— 1982. Influence of particle aggregation on deposition of distal tephra from the May 18, 1980 eruption of Mount St Helens Volcano. *J. geophys. Res.* **87**, 7061–72.

—— & —— 1984. A model of volcanogenic sedimentation in marginal basins. *In*: KOKELAAR, B. P. & HOWELLS, M. F. (eds) *Marginal Basin Geology: Volcanic and Associated Sedimentary and Tectonic Processes in Modern and Ancient Marginal Basins. Spec. Publ. Geol. Soc. London*, **16**, 37–58. Blackwell Scientific Publications, Oxford.

CARLISLE, D. 1963. Pillow breccias and their aquagene tuffs, Quadra Island, British Columbia. *J. Geol.* **71**, 48–71.

CRANDELL, D. R. 1971. Postglacial lahars from Mt Rainier, Washington. *Prof. Pap. U. S. geol. Surv.* **677**.

CROWE, B. M. & FISHER, R. V. 1973. Sedimentary structures in base surge deposits with special reference to cross-bedding, Ubehebe Craters, Death Valley, California. *Bull. geol. Soc. Am.* **84**, 663–84.

DAVIES, D. K., ALMON, W. R., BONIS, S. B. & HUNTER, B. E. 1979a. Deposition and diagenesis of Tertiary–Holocene volcaniclastics, Guatemala. *In*: SCHOLLE, P. A. & SCHLUGER, P. R. (eds) *Aspects of Diagenesis, Spec. Publ. Soc. econ. Paleont. Mineral. Tulsa*, **26**, 281–306.

——, QUEARRY, M. W. & BONIS, S. B. 1978. Glowing avalanches from the 1974 eruption of the volcano Fuego, Guatemala. *Bull. geol. Soc. Am.* **89**, 369–84.

——, VESSELL, R. K., MILES, R. C., FOLEY, M. G. & BONIS, S. B. 1979b. Fluvial transport and downstream sediment modification in an active volcanic region. *In*: MIALL, A. D. (ed.) *Fluvial Sedimentology. Mem. Can. Soc. Petrol. Geol. Alberta*, **5**, 61–84.

DRUITT, T. H. & SPARKS, R. S. J. 1982. A proximal ignimbrite breccia facies on Santorini, Greece. *J. Volcan. geoth. Res.* **13**, 147–71.

FISHER, R. V. 1961. Proposed classification of volcaniclastic sediments and rocks. *Bull. geol. Soc. Am.* **72**, 1409–14.

—— 1968. Puu Hou littoral cones, Hawaii. *Geol. Rdsch.* **57**, 837–64.

—— 1977. Erosion by volcanic base-surge density currents: U-shaped channels. *Bull. geol. Soc. Am.* **88**, 1287–97.

—— 1979. Models for pyroclastic surges and pyroclastic flows. *J. Volcan. geoth. Res.* **6**, 305–18.

—— 1984. A review of submarine volcanism, transport processes and deposits. *In*: KOKELAAR, B. P. & HOWELLS, M. F. (eds) *Marginal Basin Geology: Volcanic and Associated Sedimentary and Tectonic Processes in Modern and Ancient Marginal Basins. Spec. Publ. Geol. Soc. London*, **16**, 5–28. Blackwell Scientific Publications, Oxford.

—— & HEIKEN, G. 1982. Mt Pelée, Martinique: May 8 and 20, 1902, pyroclastic flows and surges. *J. Volcan. geoth. Res.* **13**, 339–71.

——, SMITH, A. L. & ROOBOL, M. J. 1980. Destruction of St Pierre, Martinique, by ash-cloud surges, May 8 and 20, 1902. *Geology*, **8**, 472–6.

FISKE, R. S. 1963. Subaqueous pyroclastic flows in the Ohanepecosh Formation, Washington. *Bull. geol. Soc. Am.* **74**, 391–406.

—— & MATSUDA, T. 1964. Submarine equivalents of ash flows in the Tokiwa Formation, Japan. *Am. J. Sci.* **262**, 76–106.

FOX, J. S. 1977. Rapid pyroclastic mapping in base metal exploration. *Bull. Can. Inst. Mining Metall.* 173–178.

FRANCIS, E. H. 1983. Magma and sediment. II. Problems of interpreting palaeovolcanics buried in the stratigraphic column. *J. geol. Soc. London*, **140**, 165–83.

—— & HOWELLS, M. F. 1973. Transgressive welded ash-flow tuffs among the Ordovician sediments of NE Snowdonia, North Wales. *J. geol. Soc. London*, **129**, 621–41.

FRANCIS, P. W., ROOBOL, M. J., WALKER, G. P. L., COBBOLD, P. R. & COWARD, M. 1974. The San Pedro and San Pablo volcanoes of northern Chile and their hot avalanche deposits. *Geol. Rdsch.* **63**, 357–88.

FURNES, H. & FRIDLEIFSSON, I. B. 1974. Tidal effects on the formation of pillow lava/hyaloclastite deltas. *Geology*, **2**, 381–4.

—— & —— 1979. Pillow block breccia—occurrences and mode of formation. *Neues Jahrb. Geol. Palaeontol. Monatshefte*, **3**, 147–54.

—— & STURT, B. A. 1976. Beach/shallow marine hyaloclastite deposits and their geological significance—an example from Gran Canaria. *J. Geol.* **84**, 439–53.

GALLOWAY, W. E. 1979. Diagenetic control of reservoir quality in arc-derived sandstones: implications for petroleum exploration. *In*: SCHOLLE, P. A. & SCHLUGER, P. R. (eds) *Aspects of Diagenesis. Spec. Publ. Soc. econ. Paleontol. Mineral. Tulsa*, **26**, 251–62.

GLOPPEN, T. & STEEL, R. J. 1981. The deposits, internal structure and geometry in six alluvial fan-fan delta bodies (Devonian–Norway)—a study in the significance of bedding sequence in conglomerates. *In*: ETHRIDGE, F. G. & FLORES, R. M. (eds) *Recent and Ancient Non-Marine Depositional Environments. Spec. Publ. Soc. econ. Paleont. Mineral. Tulsa*, **31**, 31–45.

HAMPTON, M. A. 1972. The role of subaqueous debris flows in generating turbidity currents. *J. sediment. Petrol.* **42**, 775–93.

—— 1979. Buoyancy in debris flows. *J. sediment. Petrol.* **49**, 753–8.

HARRISON, S. & FRITZ, W. J. 1982. Depositional features of March 1982 Mount St Helens sediment flows. *Nature, Lond.* **299**, 720–2.

HONNOREZ, J. & KIRST, P. 1975. Submarine basaltic volcanism: morphometric parameters for discriminating hyaloclastites from hyalotuffs. *Bull. Volcanol.* **39**. 441–65.

HUANG, T. C. 1980. A volcanic sedimentation model: implications of processes and responses of deep-sea ashes. *Mar. Geol.* **38**, 103–22.

JEPSEN, H. F., KALSBEEK, F. & SUTHREN, R. J. 1980. The Zig-Zag Dal Basalt Formation, North Greenland. *Rapp. Grønlands geol. Unders.* **99**, 25–32.

JONES, J. G. & NELSON, P. H. H. 1970. The flow of basalt lava from air into water—its structural expression and stratigraphic significance. *Geol. Mag.* **107**, 13–9.

KOKELAAR, B. P. 1982. Fluidization of wet sediments during the emplacement and cooling of various igneous bodies. *J. geol. Soc. London*, **139**, 21–33.

KRAUSKOPF, K. B. 1948. Lava movement at Paricutin volcano, Mexico. *Bull. geol. Soc. Am.* **59**, 1267–84.

KUENZI, W. D., HORST, O. H. & MCGEHEE, R. V. 1979. Effect of volcanic activity on fluvial-deltaic sedimentation in a modern arc-trench gap, southwestern Guatemala. *Bull. geol. Soc. Am.* **90**, 827–38.

LAJOIE, J. 1977. Sedimentology: a tool for mapping 'mill-rock'. *Geosci. Can.* **4**, 119–22.

Facies analysis of volcaniclastic sediments 145

—— 1979. Facies models. XVII. Volcaniclastic rocks. *In*: WALKER, R. G. (ed.) *Facies Models. Geoscience Can. Reprint Series*, **1**, 191–200.

LIPMAN, P. W. & MULLINEAUX, D. R. (eds) 1982. The 1980 eruptions of Mount St. Helens, Washington. *Prof. Pap. U. S. geol. Surv.* 1250.

MACDONALD, G. A. 1972. *volcanoes*. Prentice-Hall, Englewood Cliffs, New Jersey. 510 pp.

MALM, O. A., FURNES, H. & BJØRLYKKE, K. 1979. Volcaniclastics of Middle Jurassic age in the Statfjord oil-field of the North Sea. *Neues Jahrb. Geol. Palaeontol. Monatshefte*, **10**, 607–18.

MIDDLETON, G. V. & HAMPTON, M. A. 1973. Sediment gravity flows: mechanics of flow and deposition. *In*: MIDDLETON, G. V. & BOUMA, A. H. (eds) *Turbidites and Deep-Water Sedimentation*, Soc. econ. Paleont. Mineral. Tulsa, Pacific Section Short Course, 1–38.

MITCHELL, A. H. G. 1970. Facies of an Early Miocene volcanic arc, Malekula Island, New Hebrides. *Sedimentology*, **14**, 201–43.

MOORE, J. G. 1967. Base surge in recent volcanic eruptions. *Bull. Volcanol.* **30**, 337–63.

NINKOVICH, D., SPARKS, R. S. J. & LEDBETTER, M. T. 1978. The exceptional magnitude and intensity of the Toba eruption, Sumatra: an example of the use of deep-sea tephra layers as a geological tool. *Bull. Volcanol.* **41**, 286–98.

PEARCE, T. H. 1971. Short distance fluvial rounding of volcanic detritus. *J. sediment. Petrol.* **41**, 1069–72.

PERRET, F. A. 1935. The eruption of Mt Pelée, 1929–1932. *Publ. Carnegie Inst. Washington*, **458**,

PETTIJOHN, F. J. 1975. *Sedimentary Rocks*, 3rd ed. Harper and Row, New York. 628 pp.

——, POTTER, P. E. & SIEVER, R. 1973. *Sand and Sandstone*. Springer-Verlag, Heidelberg. 618 pp.

PICHLER, H. 1965. Acid hyaloclastites. *Bull. Volcanol.* **28**, 293–311.

ROSS, C. S. & SMITH, R. L. 1961. Ash-flow tuffs: their origin, geologic relations and identification. *Prof. Pap. U. S. geol. Surv.* **366**.

SCHMID, R. 1981. Descriptive nomenclature and classification of pyroclastic deposits and fragments: recommendations of the IUGS Subcommission on the Systematics of Igneous Rocks. *Geology*, **9**, 41–3.

SCHMINCKE, H.-U. 1967. Fused tuff and pépérites in south-central Washington. *Bull. geol. Soc. Am.* **78**, 319–30.

—— 1974. Pyroclastic rocks. *In*: FÜCHTBAUER, H. *Sediments and Sedimentary Rocks*, **1**, 160–89. E. Schweizerbart'sche Verlagsbuchhandlung, Stuttgart.

——, FISHER, R. V. & WATERS, A. C. 1973. Antidune and chute and pool structures in the base surge deposits of the Laacher See area, Germany. *Sedimentology*, **20**, 553–74.

SEGERSTROM, K. 1950. Erosion studies at Paricutin, State of Michoacan, Mexico. *Bull. U. S. geol. Surv.* **965-A**, 1–164.

SELF, S. & SPARKS, R. S. J. 1978. Characteristics of widespread pyroclastic deposits formed by the interaction of silicic magma and water. *Bull. Volcanol.* **41**, 196–212.

—— & —— 1981. *Tephra Studies*. D. Reidel, Dordrecht, Holland. 481 pp.

——, WILSON, L. & NAIRN, I. A. 1979. Vulcanian eruption mechanisms. *Nature, Lond.* **277**, 440–3.

SHERIDAN, M. F. 1979. Emplacement of pyroclastic flows: a review. *In*: CHAPIN, C. E. & ELSTON, W. E. (eds) *Ash-Flow Tuffs. Spec. Pap. geol. Soc. Am.* **180**, 125–36.

—— & UPDIKE, R. G. 1975. Sugarloaf Mountain tephra—a Pleistocene rhyolitic deposit of base-surge origin in northern Arizona. *Bull. geol. Soc. Am.* **86**, 571–81.

SIGURDSSON, H. & CAREY, S. N. 1981. Marine tephrochronology and Quaternary explosive activity in the Lesser Antilles Arc. *In*: SELF, S. & SPARKS, R. S. J. (eds) *Tephra Studies*, pp. 255–80. D. Reidel, Dordrecht, Holland.

——, SPARKS, R. S. J., CAREY, S. N. & HUANG, T. C. 1980. Volcanogenic sedimentation in the Lesser Antilles Arc. *J. Geol.* **88**, 523–40.

SMITH, A. L. & ROOBOL, M. J. 1982. Andesitic pyroclastic rocks. *In*: THORPE, R. S. (ed.) *Andesites: Orogenic Andesites and Related Rocks*, pp. 415–33. Wiley, Chichester.

SMITH, R. L. 1960. Ash flows. *Bull. geol. Soc. Am.* **71**, 795–842.

SNYDER, G. L. & FRASER, G. D. 1963. Pillowed lavas. I. Intrusive layered lava pods and pillowed lavas, Unalaska Island, Alaska. *Prof. Pap. U. S. geol. Surv.* **454-B**, 1–23.

SOREM, R. K. 1982. Volcanic ash clusters: tephra rafts and scavengers. *J. Volcan. geoth. Res.* **13**, 63–71.

SPARKS, R. S. J. 1976. Grain size variations in ignimbrites and implications for the transport of pyroclastic flows. *Sedimentology*, **23**, 147–188.

——, SELF, S. & WALKER, G. P. L. 1973. Products of ignimbrite eruptions. *Geology*, **1**, 115–8.

—— & WALKER, G. P. L. 1973. The ground surge deposit: a third type of pyroclastic rock. *Nature (Phys. Sci.) Lond.* **241**, 62–4.

—— & —— 1977. The significance of vitric-enriched air-fall ashes associated with crystal-enriched ignimbrites. *J. Volcan. geoth. Res.* **2**, 329–41.

—— & WILSON, L. 1976. A model for the formation of ignimbrite by gravitational column collapse. *J. geol. Soc. London*, **132**, 441–51.

—— & WRIGHT, J. V. 1979. Welded air-fall tuffs. *In*: CHAPIN, C. E. & ELSTON, W. E. (eds) *Ash-Flow Tuffs. Spec. Pap. geol. Soc. Am.* **180**, 155–66.

STOREY, B. & MACDONALD, D. I. M. 1984. Processes of formation and filling of a Mesozoic back-arc basin on the island of South Georgia. *In*: KOKELAAR, B. P. & HOWELLS, F. (eds) *Marginal Basin Geology: Volcanic and Associated Sedimentary and Tectonic Processes in Modern and Ancient Marginal Basins. Spec. Publ. Geol. Soc. London*, **16**, 207–218. Blackwell Scientific Publications, Oxford.

SURDAM, R. C. & BOLES, J. R. 1979. Diagenesis of volcanic sandstones. *In*: SCHOLLE, P. A. & SCHLUGER, P. R. (eds) *Aspects of Diagenesis*.

Spec. Publ. Soc. econ. Paleont. Mineral. Tulsa, **26**, 227–42.

SUTHREN, R. J. & FURNES, H. 1980. Origin of some bedded welded tuffs. *Bull. Volcanol.* **43**, 61–71.

TASSÉ, N. LAJOIE, J. & DIMROTH, E. 1978. The anatomy and interpretation of an Archean volcaniclastic sequence, Noranda Region, Quebec. *Can. J. Earth Sci.* **15**, 874–88.

TAZIEFF, H. 1973. About deep-sea volcanism. *Rend. Soc. Ital. mineralog. petrol.* **29**, 427–36.

VESSELL, R. K. & DAVIES, D. K. 1981. Nonmarine sedimentation in an active fore-arc basin. *In*: ETHRIDGE, F. G. & FLORES, R. M. (eds) *Nonmarine Depositional Environments, Spec. Publ. Soc. econ. Paleont. Mineral. Tulsa*, **31**, 31–45.

——, ——, FOLEY, M. G. & BONIS, S. B. 1977. Sedimentology and hydrology of flood flows on the active volcano Fuego, Guatemala. *Abstr. with Programs geol. Soc. Am.* **9**, 1210–1.

WALKER, G. P. L. 1971. Grain size characteristics of pyroclastic deposits. *J. Geol.* **79**, 696–714.

—— 1972. Crystal concentration in ignimbrites. *Contrib. Mineral. Petrol.* **36**, 135–46.

—— 1973. Explosive volcanic eruptions—a new classification scheme. *Geol. Rdsch.* **62**, 431–46.

—— 1979. A volcanic ash generated by explosions where ignimbrite entered the sea. *Nature, Lond.* **281**, 642–6.

—— 1981a. New Zealand case histories of pyroclastic studies. *In*: SELF, S. & SPARKS, R. S. J. (eds) *Tephra Studies*, pp. 317–30. D. Reidel, Dordrecht, Holland.

—— 1981b. Volcanological applications of pyroclastic studies. *In*: SELF, S. & SPARKS, R. S. J. (eds) *Tephra Studies*, pp. 391–403. D. Reidel, Dordrecht, Holland.

—— 1981c. Generation and dispersal of fine ash and dust by volcanic eruptions. *J. Volcan. geoth. Res.* **11**, 81–92.

—— 1982a. Eruptions of andesitic volcanoes. *In*: THORPE, R. S. (ed.) *Andesites: Orogenic Andesites and Related Rocks*, pp. 403–13. Wiley, Chichester.

—— 1982b. Plinian eruptions and their products. *Bull. Volcanol.* **44**, 223–40.

—— & CROASDALE, R. 1972. Characteristics of some basaltic pyroclastics. *Bull. Volcanol.* **35**, 303–17.

——, HEMING, R. F. & WILSON, C. J. N. 1980a. Low-aspect ratio ignimbrites. *Nature, Lond.* **283**, 286–7.

——, WILSON, C. J. N. & FROGGATT, P. C. 1980b. Fines-depleted ignimbrite in New Zealand—the product of a turbulent pyroclastic flow. *Geology*, **8**, 245–9.

WILSON, C. J. N. & G. P. L. WALKER. 1981. Violence in pyroclastic flow eruptions. *In*: SELF, S. & SPARKS, R. S. J. (eds) *Tephra Studies*, pp. 441–8. D. Reidel, Dordrecht, Holland

—— & —— 1982. Ignimbrite depositional facies: the anatomy of a pyroclastic flow. *J. geol. Soc. London*, **139**, 581–91.

WOHLETZ, K. H. & SHERIDAN, M. F. 1979. A model of pyroclastic surge. *In*: CHAPIN, C. E. & ELSTON, W. E. (eds) *Ash-Flow Tuffs. Spec. Pap. geol. Soc. Am.* **180**, 177–94.

WOLFF, J. A. & WRIGHT, J. V. 1981. Rheomorphism of welded tuffs. *J. Volcanol. geoth. Res.* **10**, 13–34.

WRIGHT, J. V. 1980. Stratigraphy and geology of the welded air-fall tuffs of Pantelleria, Italy. *Geol. Rdsch.* **69**, 263–91.

—— 1981. The Rio Caliente Ignimbrite: analysis of a compound intraplinian ignimbrite from a major late Quaternary Mexican eruption. *Bull. Volcanol.* **44**, 189–212.

—— & MUTTI, E. 1981. The Dali Ash, Island of Rhodes, Greece: a problem in interpreting submarine volcanigenic sediments. *Bull. Volcanol.* **44**, 153–67.

——, SELF, S. & FISHER, R. V. 1981a. Towards a facies model for ignimbrite-forming eruptions. *In*: SELF, S. & SPARKS, R. S. J. (eds) *Tephra Studies*, pp. 433–9. D. Reidel, Dordrecht, Holland.

——, SMITH, A. L. & SELF, S. 1980. A working terminology of pyroclastic deposits. *J. Volcan. geoth. Res.* **8**, 315–36.

——, —— & —— 1981b. A terminology for pyroclastic deposits. *In*: SELF, S. & SPARKS, R. S. J. (eds) *Tephra Studies*, pp. 457–63. D. Reidel, Dordrecht, Holland.

—— & WALKER, G. P. L. 1977. The ignimbrite source problem: significance of a co-ignimbrite lag-fall deposit. *Geology*, **5**, 729–32.

—— & —— 1981. Eruption, transport and deposition of ignimbrites: a case study from Mexico. *J. Volcan. geoth. Res.* **9**, 111–31.

R. J. SUTHREN, Department of Geology and Physical Sciences, Oxford Polytechnic, Headington, Oxford OX3 0BP.

Shallow-marine carbonate facies and facies models

M. E. Tucker

SUMMARY: Shallow-marine carbonate sediments occur in three settings: platforms, shelves and ramps. The facies patterns and sequences in these settings are distinctive. However, one type of setting can develop into another through sedimentational or tectonic processes and, in the geologic record, intermediate cases are common. Five major depositional mechanisms affect carbonate sediments, giving predictable facies sequences: (1) tidal flat progradation, (2) shelf-marginal reef progradation, (3) vertical accretion of subtidal carbonates, (4) migration of carbonate sand bodies and (5) resedimentation processes, especially shoreface sands to deeper subtidal environments by storms and off-shelf transport by slumps, debris flows and turbidity currents.

Carbonate platforms are regionally extensive environments of shallow subtidal and intertidal sedimentation. Storms are the most important source of energy, moving sediment on to shoreline tidal flats, reworking shoreface sands and transporting them into areas of deeper water. Progradation of tidal flats, producing shallowing upward sequences is the dominant depositional process on platforms. Two basic types of tidal flat are distinguished: an active type, typical of shorelines of low sediment production rates and high meteorologic tidal range, characterized by tidal channels which rework the flats producing grainstone lenses and beds and shell lags, and prominent storm layers; and a passive type in areas of lower meteorologic tidal range and higher sediment production rates, characterized by an absence of channel deposits, much fenestral and cryptalgal peloidal micrite, few storm layers and possibly extensive mixing-zone dolomite. Fluctuations in sea-level strongly affect platform sedimentation.

Shelves are relatively narrow depositional environments, characterized by a distinct break of slope at the shelf margin. Reefs and carbonate sand bodies typify the turbulent shelf margin and give way to a shelf lagoon, bordered by tidal flats and/or a beach-barrier system along the shoreline. Marginal reef complexes show a fore-reef—reef core—back reef facies arrangement, where there were organisms capable of producing a solid framework. There have been seven such phases through the Phanerozoic. Reef mounds, equivalent to modern patch reefs, are very variable in faunal composition, size and shape. They occur at shelf margins, but also within shelf lagoons and on platforms and ramps. Four stages of development can be distinguished, from little-solid reef with much skeletal debris through to an evolved reef-lagoon-debris halo system. Shelf-marginal carbonate sand bodies consist of skeletal and oolite grainstones. Windward, leeward and tide-dominated shelf margins have different types of carbonate sand body, giving distinctive facies models.

Ramps slope gently from intertidal to basinal depths, with no major change in gradient. Nearshore, inner ramp carbonate sands of beach-barrier-tidal delta complexes and subtidal shoals give way to muddy sands and sandy muds of the outer ramp. The major depositional processes are seaward progradation of the inner sand belt and storm transport of shoreface sand out to the deep ramp.

Most shallow-marine carbonate facies are represented throughout the geologic record. However, variations do occur and these are most clearly seen in shelf-margin facies, through the evolutionary pattern of frame-building organisms causing the erratic development of barrier reef complexes. There have been significant variations in the mineralogy of carbonate skeletons, ooids and syn-sedimentary cements through time, reflecting fluctuations in seawater chemistry, but the effect of these is largely in terms of diagenesis rather than facies.

Through the study of recent sediments and their ancient counterparts, it is possible to synthesize the facies distributions into various *facies models*. These summaries of facies patterns can be extremely useful when new sequences are being examined; the facies models do have a predictive quality which can be important when particular facies-types are being sought, as in petroleum and mineral exploration. Modern marine carbonates have been studied intensively over the last two decades and, in particular, the results of researches in the Bahamas, Florida, the Arabian Gulf and Shark Bay Western Australia have contributed much to the understanding and interpretation of ancient carbonates. The data from these modern carbonate environments go a long way towards providing the basis for useful facies models. However, there are three important deficiencies of the recent sediment record which must be borne in mind when trying to produce generally applicable carbonate facies models:

(1) As a consequence of the Pleistocene glaciation and associated major fluctuations in

sea level over the last 1 Ma, in most areas where carbonates are forming today, sedimentation only began 4–5000 years ago. Thus relic topographies exert a strong control on sedimentation and in some areas (mostly low latitude, deeper-water, mid- to outer continental shelves) relic carbonate sediments abound. A steady-state situation with an equilibrium between sediments and environments has frequently not been attained.

(2) Sea level at the present time is relatively low compared with much of the geologic record. Thus there are now no extensive low-latitude shallow seas (epeiric or epicontinental seas) where carbonates are accumulating, comparable with the many instances in the past of whole cratons being covered by knee-deep marine waters.

(3) Modern carbonate sediments are almost entirely produced by biogenic processes, apart from ooids and possibly some lagoonal lime muds. The organism types contributing their skeletons to limestones have varied drastically throughout the Phanerozoic due to changing fortunes: the evolution of new groups and demise of others. The roles played by organisms have also changed through time; this is particularly important when considering reef limestones. In addition, the dominant mineralogic composition of organism skeletons and inorganic $CaCO_3$ precipitates has varied through time, in response to fluctuations in seawater–atmosphere chemistry.

As a result of the above three points, facies models for carbonates cannot be derived entirely from studies of recent carbonate sedimentary processes and products; essential information has to come from the rock record. In addition, as alluded to above, in some instances facies models have to allow for the evolutionary pattern of carbonate-secreting organisms through time.

The important papers or compilations in the field of carbonate facies and facies models are: Irwin (1965), Shaw (1964), Purser (1973a), Laporte (1974), Heckel (1972, 1974), Logan (1974), Bathurst (1975), Ginsburg (1975), Wilson (1975), Sellwood (1978), James (1979), Asquith (1979), Toomey (1981) and Flügel (1982).

Major controls on carbonate sedimentation

There are two overriding controls on carbonate sedimentation: (1) geotectonics and (2) climate.

The geotectonic context is of paramount importance. It controls one of the prime requisites for carbonate sedimentation, the lack of siliciclastic material, by determining hinterland topography and drainage. Geotectonics also determines the depositional setting, and three types are distinguished and discussed later: the platform, shelf and ramp. However, the setting can be modified considerably, once carbonate sedimentation is established. Geotectonics controls the orientation of shorelines and platform-shelf margins and with climate this determines the energy level and direction of wind-waves, storm and tidal currents, as well as the circulation pattern and location of upwelling, nutrient-rich zones. Both geotectonics and climate control the position and fluctuations of sea-level. This is of great significance to the production of carbonate sediment and the resulting facies mosaic. Rates of subsidence and uplift, which also affect sea-level transgressions and regressions, and the location of positive and negative areas, are also determined by geotectonic factors. Climate is important in terms of seawater salinity, especially where lagoons are involved. Salinity is a major factor, for many organisms cannot tolerate any deviations from stenohaline conditions.

Where optimal conditions exist for the growth of organisms with carbonate skeletons, then it appears that the carbonate production rate is fairly constant, regardless of the types of organism involved (e.g. Smith 1973; Hallock 1981). Production rates determined for benthic foraminifera, corals and coralline algae on seaward reef flats are around 1.5–4.5 kg $CaCO_3$ m^{-2} year^{-1}, equivalent to a carbonate deposition rate of 0.5–1.5 mm year^{-1}. Rates are somewhat lower in back-reef lagoons (0.1–0.5 mm year^{-1}). Production rates can be much higher on the reef front, 6 m 1000 years^{-1} has been recorded (6 mm year^{-1}). The point to note is that the carbonate production rates are determined by ocean physico-chemistry, rather than organic-biological factors.

Depositional processes and facies sequences: constant sea level

Where carbonate sedimentation takes place without any change in sea level, there are five principal depositional processes which lead to the formation of characteristic facies sequences (see Fig. 1).

(1) Tidal flat progradation results largely from deposition of shallow subtidal sediments on flat-marginal beach ridges and on the flats themselves during major storms. Trapping and some precipitation

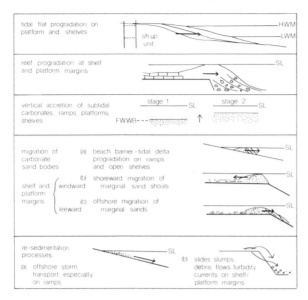

FIG. 1. The principal depositional processes of carbonate sediments. The typical settings in which these processes operate are also noted.

of sediment by algal mats on the flats are important. Some carbonate (and other minerals) can be precipitated inorganically on tidal flats in an arid climate. The net result is a shallowing-upward sequence (further discussed later) of intertidal sediments overlying subtidal sediments. In detail there are often variations in the microfacies of these shallowing-upward sequences, depending on the type of tidal flat, energy level and climate, etc.

(2) Reef progradation is important at shelf-breaks and platform margins and mostly involves seaward growth of the reef over its storm-produced talus (fore-reef slope).

(3) Vertical accretion of subtidal carbonates can take place when sediment production rates are high. Shallowing-upward sequences are produced, of deeper subtidal facies giving way to shallower subtidal facies (and of course intertidal facies could follow naturally).

(4) Migration of carbonate sand bodies is significant in relatively high-energy locations, giving beach-barrier–tidal delta complexes, especially on ramps, and sand shoals, especially at shelf-breaks and platform margins. Under constant sea-level beach-barrier–tidal delta complexes will prograde offshore if there is a good supply of sediment (i.e. high organic productivity in the shoreface zone or abundant ooid formation in the tidal deltas). With sand shoals, their shoreward

migration into the shelf-lagoon or platform interior is important in windward locations, giving rise to quiet-water, below fair-weather wave-base packstones and wackestones passing up into above wave-base storm or tide-dominated grainstones. On leeward margins, offshore, basinward transport of skeletal sands is significant and can lead to progradation of the margin itself.

(5) Offshore storm transport and deposition of shoreface carbonate sediment is very important on ramps, less so on shelves and platforms. Other resedimentation processes, slides, slumps, debris flows and turbidity currents, all of which may be storm or seismically induced, are important at shelf-breaks and platform margins.

When there are fluctuations in sea level, either through eustatic or local tectonic effects, then many more facies patterns can arise. These are discussed in succeeding sections.

Depositional settings of shallow-marine carbonates

Shallow-marine carbonates are being and have been deposited in a wide range of geotectonic settings. Three basic depositional settings of shallow-marine carbonates can be defined (see Fig. 2).

(1) The *platform*: a very extensive (10^2–10^4 km wide), quite flat cratonic area covered by a shallow (epeiric) sea. Seawards, a platform is

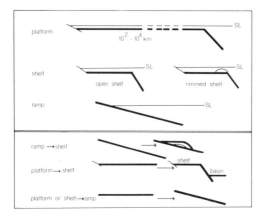

FIG. 2. The three major depositional settings of shallow-marine carbonates, platform, shelf and ramp, along with the common transitions from one to the other.

bounded by a margin which may have a gentle or steep slope. (2) The *shelf*: a far less extensive ($10-10^3$ km wide) area characterized by the presence of a distinct shelf-break, where the gradient increases dramatically into an adjacent basin. Although often relatively flat, there can be substantial gradients on the shelf itself, and many are rimmed—i.e. they have a barrier of reefs or carbonate sand shoals along the shelf-break, with a shelf lagoon behind. (3) The *ramp*: a gently sloping surface, passing seawards into deeper and deeper water.

Initially, these three depositional settings are determined by geotectonics, but once carbonate sedimentation is established, then one type of setting can be transformed into another, either through the natural processes of carbonate sedimentation itself, or through further geotectonic effects. Omitting sea-level fluctuations, there are three common patterns: (1) a ramp may develop into a shelf, especially through reef growth; (2) a platform or shelf may develop into a ramp through differential subsidence along a hinge line, and (3) a platform may develop into a shelf through contemporaneous fault movements (Fig. 2).

Carbonate facies patterns on platforms

Carbonate platforms are very extensive areas of negligible topography. Although non-existent today, shallow epeiric seas covered the cratons many times during the geologic record. Examples include the Cambrian and Ordovician of North America, the Upper Dinantian and Jurassic of parts of western Europe, and some of the Tertiary of the Middle East. Water depths on the cratons were generally less than 5–10 m, so that shallow subtidal to intertidal environments dominated. The intertidal areas would have consisted of tidal flats many kilometres to tens of kilometres wide. These would have developed extensively in the platform interiors, with supratidal flats beyond, giving way to the peneplained land surface where subaerial processes such as pedogenesis and karstification would have operated. Tidal flats would also have developed around slightly more positive areas upon the platform. Apart from local shoals of skeletal sand, the subtidal would also be a near-flat surface but probably with slightly deeper and slightly shallower areas reflecting pre-existing topography on the craton or the effects of differential subsidence.

It is generally accepted that the epeiric seas had only small tidal ranges. Tidal currents would have been insignificant on the open platform, but perhaps quite strong in any channels of the broad intertidal zone. For much of the time, platforms would have been very quiet, low energy environments, with only wind–wave activity. Fairweather wave-base would have been quite shallow, less than 5 m. The platform margins on the other hand have been sites of much turbulence for much of the time (e.g. Mazzullo & Friedman 1975). Tidal currents and waves from open ocean swell would have been very important, especially if an abrupt change in slope existed at the platform margin, causing all wave and current energy to be dissipated over a short distance. Sand bodies and reefs could well have been developed along the platform margin, as occur along many shelf-breaks (see later section) and these could further have reduced circulation on the platform itself. The dominant process affecting platform sedimentation would have been storms, their frequency, direction and magnitude controlled by climatic factors. Severe storms can raise sea level by several metres and give rise to currents reaching 1 m s^{-1}. On a craton-sized platform, storm winds blowing persistently from one quadrant will pile up water in a down-wind direction. Where normally quite shallow water exists (< 2 m), the platform floor itself could be exposed as the sea is blown off it. Strong surges would cross the platform after the storm subsided and the sea returned to its normal level.

During storms, the platform interior tidal flats would be flooded and much shallow subtidal sediment deposited upon them. In the subtidal, skeletal debris would be transported and sorted during storms and post-storm surges, and deposited to give grainstone beds. Winnowed shell lags (rudstones) would be left after the passage of storm currents and waves.

In general terms, the facies pattern of a carbonate platform would thus consist of skeletal-peloidal wackestones with lenticular grainstones in the platform interior (tidal flat deposits), skeletal-peloidal grainstones and packstones of the shallow subtidal (above fairweather wave-base) and skeletal packstones and wackestones with grainstone horizons in the deeper subtidal (below fairweather wave-base). Below storm wave-base, skeletal wackestones would dominate with thin beds of storm-derived skeletal packstone-grainstone.

Under constant sea level, apart from some aggrading of the shallow subtidal sediments through simple skeletal carbonate production, the dominant depositional process would be progradation of the tidal flats. (The movement of shallow subtidal sand shoals during storms could also be important.)

Modern platform carbonates

Although there are no modern examples of the very extensive platforms of the past, we can get an indication of what sedimentation must have been like from the studies of the interior of the Great Bahama Bank (e.g. Shinn *et al.* 1969; Hardie 1977; Gebelein *et al.* 1980). To the west of Andros Island, there occur protected tidal flats and a shallow subtidal platform. Tidal range is very low (0.46 m) and wind–wave activity is also weak since Andros Island acts as a barrier to the dominant and persistent northeasterly trade winds. Occasional winter storms from the west to north produce strong waves in spite of the shallowness of the platform (average depth 5 m). Sedimentation is largely controlled by the rare storm events.

Tidal flats are complex areas of many subenvironments: tidal channels, beach ridges (hammocks), levées, ponds, intertidal flats, areas of surficial crusts and algal marshes (often freshwater). Parts of the tidal flat are permanently subaqueous; other areas are exposed for some of the tidal cycle or for certain seasons of the year. To describe the fluctuations in water cover, an exposure index has been introduced by Ginsburg *et al.* (1977) to indicate the percentage exposure of a subenvironment over a year.

Two distinct types of tidal flat occur on the west side of Andros Island (Fig. 3): one type, which can be termed an *active tidal flat*, is dominated by tidal channels (comprising 15% of the tidal flat complex), draining ponds, intertidal flats and algal marshes; the other type, a

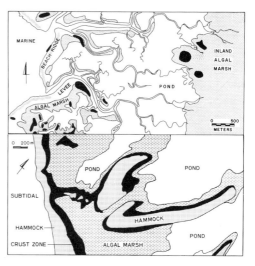

FIG. 3. The subenvironments of (a) an active and (b) a passive tidal flat from the western side of Andros Island, Bahamas. (a) After Hardie (1977), (b) after Gebelein *et al.* (1980).

passive tidal flat has virtually no tidal channels and consists of broad depressions separated by former beach ridges rising 1–2 m above normal high water mark. The depressions are variably occupied by water to form ponds with intertidal flats and algal marshes around.

At their seaward margin, both types of tidal flat have a low beach ridge which is constructed of sediment thrown up from the shallow subtidal during storms. The sediment of the present and former beach ridges is largely skeletal-peloidal grainstone-packstone with lamination. Algal marsh sediments are dominantly algal laminites with laminoid fenestrae, but skeletal-peloidal layers (up to 7 cm thick) of storm origin are intercalated. Pond sediments are chiefly lime mud, much of which is pelletized by gastropod-annelid defecation. Surrounding intertidal flats may have surficial cemented crusts, often of dolomite, and consist of peloidal lime mud with many irregular fenestrae (birdseyes).

The shallow subtidal offshore from the tidal flats is an area of peloids and skeletal grains in the shoreface zone, where affected by wind waves, and skeletal peloidal lime muds below fairweather wave base (>2–5 m). Algal micritization of grains is common and algal mats may partly cover the surface. Bioturbation is ubiquitous. Skeletal-peloidal sands do occur through sediment reworking and winnowing during storms.

The sequences generated through deposition on the two types of tidal flat will be different

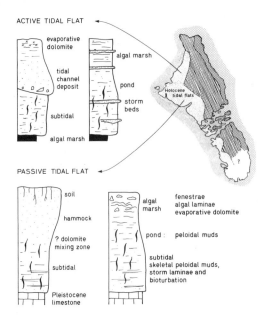

FIG. 4. The sequences of an active and passive tidal flat from the west side of Andros Island.

(Fig. 4). The migration of the tidal channels on the active type will give rise to skeletal lag deposits and skeletal-peloidal grainstones, with cross-bedding if it is not extensively bioturbated. Areas not affected by the channels will consist of pond and marsh deposits (lime muds, fenestral peloidal wackestones, algal laminites) plus prominent storm layers (thin packstones-grainstones). The passive tidal flat will consist largely of pond and algal marsh deposits, with few storm layers (since the marginal beach ridge protects the tidal flat), but with no suggestion of any sediment reworking by tidal channels. Zones of beach ridge sediments will occur within the tidal flat sequence.

Dolomite crusts may develop in both types of flat, but they are more likely to form intra-clasts and edgewise conglomerates on the active flat through tidal channel reworking. Poorly ordered dolomite of possible meteoric-marine mixing-zone origin has been reported from the shallow subsurface of the passive-type flat (Gebelein et al. 1980). The occurrence of beach ridges on the passive flat, with their freshwater lenses below, should lead to a wider development of mixing-zone dolomite there, compared to the active flat which will have mainly marine to hypersaline groundwaters.

Under a more arid climate than that of the Bahamas, evaporites would precipitate in the areas of high exposure index. Such evaporites

are accumulating along the Trucial Coast, Arabian Gulf. In upper intertidal areas, dis-coidal gypsum crystals are being precipitated and in the supratidal (sabkha) these are being replaced by anhydrite; farther landwards enterolithic anhydrite may form. In extremely arid locations, halite (and even potash salts) could precipitate in supratidal areas as crusts and beds in depressions (supratidal salinas). On a passive tidal flat, evaporites would accumulate extensively where waters in ponds and ground-waters beneath slightly higher areas were only infrequently replenished with seawater during major storms. Lenticular gypsum and nodular anhydrite could be expected to form within the sediment, and bedded evaporites could pre-cipitate within the ponds.

The reasons for the two types of tidal flat are thought to be related to sediment supply and shoreline orientation with regard to winter storms. Compared with the passive flats, the active flats are adjacent to shallow subtidal areas of lower sediment production rates, and they are oriented such that they receive the full force of storm winds and waves (i.e. the meteorologic tidal range is high). By contrast, the shoreline fronting the passive flats is more oblique to on-coming storms (the meteorologic tidal range is low) and sediment production rates appear to be higher.

Shallowing-upward cycles of carbonate platforms

Carbonate platforms respond dramatically to sea-level changes. Left to their own devices, with constant sea level and no subsidence or uplift, platforms will build up to sea level and just above, through progradation of tidal flats and vertical accretion of shallow subtidal sediments into shallower depths. The typical sequence produced through sedimentation on a platform is thus a shallowing-upward sequence of subtidal through to intertidal and supratidal deposits. A relative drop in sea level will expose the platform to supratidal-subaerial processes, namely sabkha evaporite precipitation if the climate is arid and there is still a source of seawater, or to soil formation, such as calcrete development if a semi-arid climate, or karstification if more humid. With a relative rise in sea-level, subtidal environments are widely established over a plat-form, with tidal flats at the distant shoreline.

There are numerous accounts of shallowing-upward cycles in the geologic record, demon-strating that subtidal environments were re-peatedly established by the periodic flooding of platforms through transgressive events. For examples see Coogan (1969), Wilson (1975), Ginsburg (1975) and Somerville (1979). Detailed studies show that the cycles are not all the same. Microfacies analysis reveals differences within one cycle when traced laterally across a platform, and between cycles in a vertical sequence. Frequently the cycles of one particular stage or substage of a geological period have features in common, which are different in the cycles of a succeeding stage.

As an example, the recent work of Gray (1981) can be cited. In the Llangollen area of mid-Wales, shallowing-upward cycles are developed in the Asbian and Brigantian stages of the Upper Dinantian, Lower Carboniferous (Fig. 5). Asbian sediments were deposited in the Llangollen and Oswestry embayments, separated by the Berwyn High. Brigantian sediments were deposited uniformly over the whole area. When cycles are traced towards the platform interior, gradual but distinct changes are observed in addition to a general shorewards thinning of each cycle. Away from the open platform, the transgressive phase (a below fairweather wave base, thin argillaceous packstone-wackestone facies, which forms the lower part of each cycle, gradually reduces in thickness. Sequences of more proximal areas tend to have shallow subtidal (above fairweather wave base) sediments in their lower parts. The 'regressive' phase tidal flat sediments (fenestral, cryptalgal peloidal wackestones) appear to increase in thickness as cycles are traced shorewards and may comprise the whole cycle in very proximal areas. Lateral variations are also seen in the nature of emergence horizons at the top of each cycle: in proximal areas, palaeokarstic surfaces are usually developed (possibly above a calcrete). These pass distally into 'sutured discontinuity surfaces', interpreted as the product of intertidal, rather than wholly subaerial, dissolution and erosion (cf. Read & Grover 1977).

The lateral variations in cycle form are primarily a function of the gradient of the platform. The transgressions appear to have been relatively rapid, and during the transgression, a basal bed was developed in distal to medial areas. Sedimentation after the initial transgression was determined by depth, especially relative to wave base (see Fig. 6). Differences between cycles of different stages relate to the magnitude of the transgressions. For example, compared with Asbian cycles, those of the Brigantian are dominated by thin-bedded below wave-base packstones-wackestones. This indicates that the transgressions were more widespread, resulting in a greater depth of water over the platform.

Within the Upper Dinantian sequence of mid-

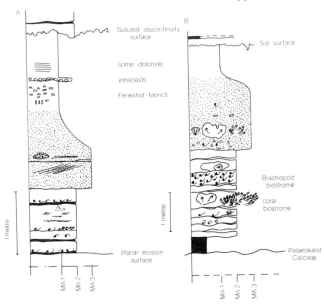

FIG. 5. Two typical shallowing-upward cycles from the Lower Asbian (Tynant) and Brigantian (Trefor) of the Llangollen area, mid-Wales. Microfacies associations: MA.1 calcisphere wackestone (tidal-flat facies), MA.2 algal packstone-wackestone (below fairweather wave-base sandy mud and muddy sand facies); MA.3 algal grainstones (shoreface, above wave-base sand facies). After Gray (1981).

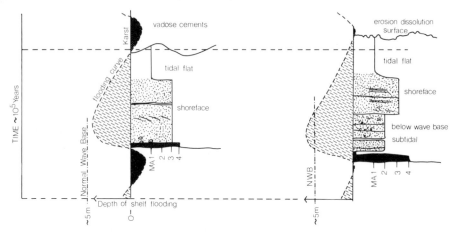

FIG. 6. Diagram interpreting the lateral variation of a cycle in terms of depth of shelf flooding. Cycle on left is more proximal (farther towards platform interior) whereas cycle on right is more distal. After Gray (1981).

Wales, and elsewhere in the U.K., it is not uncommon to find cycles 'missing' in proximal areas, or unrecognizeable in more distal areas. Problems of stratigraphic correlation can result. Sedimentologically, this can be explained with reference to Fig. 7. The absence of sediments of a particular cycle in proximal areas results from weak transgressions that did not extend landwards as far as earlier and later ones. Where this happened, the subaerial processes operated for a longer time (over several cycles) and very marked palaeokarstic horizons can develop (multiple palaeokarsts). In more distal platform areas, it could happen that the prograding tidal flats did not arrive, so that subtidal conditions were maintained throughout. Fluctuations from below to above wave base may be recognized by careful study of microfacies (e.g. Jefferson 1980), or it may be that sea-level changes were not substantial enough on the open platform to cause any major modification to the sedimentary facies.

There has been much discussion as to the underlying mechanism causing repetition of the shallowing-up cycle. Once flooded, tidal flats will prograde across a platform and the thickness of the cycle as well as the facies sequence through the cycle will depend on the depth after the initial transgression. Tidal flat progradation has been referred to as depositional regression, but regression can also occur through a eustatic sea-level fall or through slight uplift of the craton. Such an external sea-level fall can be recognized by the absence or impoverishment of peritidal facies, but with the development of emergence phenomena (palaeokarsts, calcretes, vadose diagenesis) directly upon subtidal sediments. In most of the British

Carboniferous shallowing-up cycles, and it seems in many others in the geological record, this is not seen. It appears that the shallowing is largely due to depositional processes. Carbonates can build up relatively quickly—modern depositional rates of non-reef carbonates are 0.5–1 m 10^3 years^{-1}, and rates of tidal flat progradation are rapid too—several km 10^3 years^{-1}. It is probable that much time is represented by the palaeokarstic surfaces; 50,000 years has been suggested by Walkden (1974). The length of each British Dinantian cycle has been variously estimated as 0.2–0.5 Ma. Most shallowing-up cycles appear to be the result of episodic transgressions, involving periodic sea-level rises of only a few metres. In the Upper Dinantian of the U.K. this appears to have happened some 30 or more times.

A question often asked is whether the transgressions are induced by local tectonics, i.e. episodic subsidence of the craton, or through eustatic sea-level rises (see for example Ramsbottom 1977; George 1978). The latter can be induced by fluctuations in ocean basin volume through plate tectonic processes, or in the volumes of glacial ice at the poles. However, the former process, which can give a rate of sea-level change of 1 cm 10^3 years^{-1} is three orders of magnitude slower than glacio-eustatic changes (e.g. up to 10 m 10^3 years^{-1} for the Flandrian transgression) (Donovan & Jones 1979). Modern subsidence rates can reach 2.5 m 10^3 years^{-1}. Thus, both glacio-eustacy and platform subsidence can produce rates of sea-level rise which exceed modern carbonate production rates (maximum 1 m 10^3 years^{-1}). To distinguish between a local tectonic and a eustatic control for the sea-level rises, the lateral continuity of

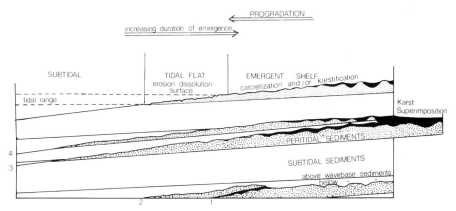

FIG. 7. Model for development of shallowing-upward cycles in a broadly transgressive sequence. The lateral variation of each cycle and the vertical variations between cycles are seen to be reflections of the degree of flooding of each transgression. After Gray (1981).

the cycles is important. If individual cycles can be correlated from one platform to another then a eustatic control is likely. However, platforms within a region may all be undergoing episodic subsidence of a similar magnitude, especially if they all occur within the same overall geotectonic framework, so that shallowing-up cycles of a similar thickness and type would occur on all platforms. These cycles would be difficult to distinguish from those of eustatic origin. The exact cause of repeated shallowing-up cycles on carbonate platforms is thus far from clear.

Although shallowing-upward cycles characterize ancient platform carbonates, there are instances where formations consist of cycles showing the opposite trend, of deepening up. Intertidal fenestral, algal laminated limestones pass up into shallow subtidal facies which may be capped by an emergence horizon, with a soil and karstic solution effects, before the overlying tidal flat unit of the next cycle. These transgressive cycles reflect times of slow to moderate sea-level rise and rapid sea-level fall. They are not common in the geologic record, but they are well known in the Alpine Triassic, where they have been called Lofer cycles (Fischer, 1964).

Carbonate shelf facies and models

A shelf is a relatively narrow (tens to hundreds of kilometres), but frequently laterally extensive, generally shallow-water depositional environment, which has a well-defined margin (from which things can fall off!), adjacent to a deeper water basin. Some shelf-margins are fault-bounded. Usually there is a rapid increase in gradient from the shelf (modern shelf gradient

125 cm km^{-1}) to the slope into the adjoining basin. Where shelves border oceans they are frequently sites of long-term downwarp so that thick sedimentary packets can develop there, if sedimentation can keep pace with the subsidence.

Where starved of siliciclastic sediment, carbonates will accumulate on the shelf and at the present time two major types of carbonate shelf are recognized: open shelves and rimmed shelves (Ginsburg & James 1974). Open shelves slope gently away from the shoreline and depths at the shelf-break are substantial (50–200 m). Carbonate sediments on the outer shelf are largely relic; sedimentation was unable to keep pace with rising sea-level during the Flandrian transgression. Rimmed shelves, which are the modern counterparts of many ancient carbonate shelves, are characterized by the development of reefs and carbonate sand bodies along the shelf margin (Fig. 8). Depths are shallow adjacent to the shelf-break, or the area may even be subaerial if islands have formed. The shelf margin is a turbulent, high energy zone where oceanic waves (swell), storm waves, and possibly tidal currents, impinge on the seafloor. Organic productivity is highest under these conditions, especially if the sea is fertile through upwelling. Much precipitation of $CaCO_3$ in the form of ooids and cements occurs along shelf margins. Behind the rim, there is usually a shelf lagoon. This will vary in its degree of protection from the marginal turbulent zone, depending on how well the reefs and sands of the margin act as a barrier. At one extreme, a true shelf-lagoon will exist, being a very quiet environment with poor circulation and perhaps hypersalinities during dry seasons. It will only be affected by major storm events. At the other, an open shelf will

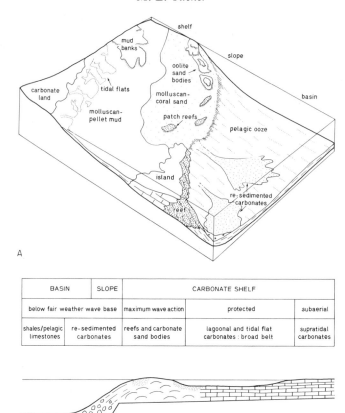

BASIN		SLOPE	CARBONATE SHELF		
below fair weather wave base		maximum wave action	protected		subaerial
shales/pelagic limestones	re-sedimented carbonates	reefs and carbonate sand bodies	lagoonal and tidal flat carbonates : broad belt		supratidal carbonates

FIG. 8. (a) Facies model of shelf carbonates. After Coogan (1969). (b) Lateral facies distribution on a carbonate shelf.

exist, subject to continuous wave and tidal motions.

The shoreline at the inner margin of the shelf may be dominated by tidal flats, especially if there is a significant tidal range (meso-macro-tidal), or by a beach-barrier-tidal delta coastline if wave energy is substantial (determined by prevailing climate and coastline orientation) and tidal range low (micro-meso-tidal). Most likely this would occur in an open shelf setting and the facies pattern would be similar to a carbonate ramp (see that section).

Modern shelf carbonates

These have been studied extensively in the Caribbean (the Bahamas, Florida, Belize, Jamaica, etc.) and off Queensland (the Great Barrier reef). The south Florida shelf can be briefly described here as a good example of a carbonate shelf (for details see Enos & Perkins 1977).

The south Florida shelf is 6–35 km wide, with a shallow shelf break at a depth of 8–18 m. The gradient on the shelf is very low, 1:1000, whereas the seaward slope initially has a gradient of 1:40 (1½°). The shelf, with a NE–SW orientation, is subject to winter storms from the NE and summer trades from the SE, so that water movement is dominantly on-shelf. The shelf-break serves to focus the wave energy so that water circulation is vigorous there. Tidal range is 50–90 cm at the shelf-break. Most carbonate sedimentation is taking place along the seaward margin of the south Florida shelf. Reefs dominated by corals and calcareous algae form a belt up to 1 km wide along the shelf margin. Not all reefs are living, but this appears to be part of a pattern, of shifting zones of active reef growth. Much skeletal carbonate sand occurs on the outer shelf and this can form distinct sand bodies with sand waves, tidal channels and spillover lobes. Much skeletal debris is concentrated behind the reefs in back-reef talus piles. Storm movement of reef-rubble may lead to the formation of low islands. Much reef debris is transported off shelf, down chutes to form wedges up to 12 m thick on the shallow slope. Shoreward of the shelf margin, there are patch reefs and skeletal sand shoals and these give way to bioturbated muddy sands and sandy muds in the quieter water, inner shelf. Although sedimentation on the inner shelf is strongly controlled by a relic topography of Pleistocene limestones (which form the Florida Keys), one feature of note is the presence of seagrass-algal stabilized mud-banks on the seaward side of the Keys and within Florida Bay (e.g. Rodriguez Bank, Turmel & Swanson 1976). Lime mud is largely derived from disintegration of calcareous green algae. Tidal flats and mangrove swamps are irregularly developed along the shoreline and around the Florida Keys.

The facies pattern seen on the south Florida shelf is thus one of shelf-marginal reefs (framestones-boundstones) and carbonate sands (oolitic and skeletal grainstones), giving way to skeletal packstones and wackestones of the protected inner shelf lagoon, with skeletal mudstones of the mud-banks, and then tidal flat facies of the shoreline. This pattern is identical to that of many ancient carbonate shelf sequences.

Where the sea-level stand is stable for a substantial period of time, three depositional processes operate: (1) either seaward progradation of tidal flats over shelf lagoon sediments, if a protected shelf *or* seaward progradation of a beach-barrier–tidal-delta complex, if an open, high wave energy, shelf; (2) shoreward progradation of shelf marginal carbonate sand bodies into the shelf lagoon and (3) seaward progradation of the marginal reefs, reef debris fans and sand bodies, constructively extending the shelf-margin.

Tidal flat progradation generates a shallowing-up unit, and this has been discussed in the carbonate platform section. Beach-barrier-tidal delta progradation is discussed in the carbonate ramp section. Processes (2) and (3) are dealt with in succeeding sections.

Shelf margin reefs and other reef facies

There is a vast literature on reefs, both modern and ancient, and much discussion has arisen as to what defines a reef. The argument arises largely because most ancient reefs cannot be compared directly with modern reefs. Organisms contributing to reef limestones have changed through time so that the character of the reef has changed too. The term 'carbonate buildup' is now widely used to denote a carbonate body of restricted extent that possessed some topographic relief. A reef is a little more specific, generally denoting a buildup where there is much skeletal material, some in growth position with perhaps some acting as a framework. To some workers a certain degree of wave resistance is necessary for the term reef to be applied. Although many different types of reef have been recognized, two broad types are barrier reefs, occurring especially at shelf (and platform) margins, and patch reefs or reef mounds occurring there too, but also in the shelf lagoon and perhaps on the platform. For a recent review see Longman (1981) and other papers in Toomey (1981), also James (1979), James & Ginsburg (1980), Laporte (1974), Frost *et al.* (1977), Stoddart (1969), Perkins (1975) and Schroeder & Zankl (1974).

The best-known modern shelf-marginal reefs are formed chiefly by scleractinian corals and coralline red algae. The corals produce an organic framework to the reef and the crustose algae cement, bind and hold the reef together. Coral growth forms and species occur in distinct zones over the reef, with strong, solid corals growing in the most turbulent zones and more leafy, delicate corals, living in slightly deeper less agitated waters. In addition to the red algae, other encrusters include bryozoans, foraminifera, serpulids and bivalves. Other organisms, especially molluscs, echinoids and fish, find homes within the crevices of the reef. The latter are common with most, especially the larger ones, being growth cavities, where the framework skeletal organisms have enclosed space. Small cavities abound and many of these are produced by boring organisms such as lithophagid bivalves, clionid sponges, polychaete worms and endolithic algae and fungi. The cavities are frequently partially or wholly filled with internal sediment. The precipitation of carbonate cements is very important in modern shelf-marginal reefs where seawater is constantly being pumped through by wave action. Acicular aragonite and bladed and micritic high-Mg calcite lithify loose carbonate sediment in reef crevices and line and fill cavities (e.g. James *et al.* 1977; James & Ginsburg 1980).

Coral and algal growth can be very rapid (6 m 10^3 years^{-1} recorded) and in areas of low turbidity, much turbulence, abundant light and fertile seas, such as occur along many tropical shelf margins, a strong, wave-resistant ridge (reef) develops. This rises above the shoreward shelf lagoon where lower sedimentation rates prevail. Under conditions of stable sea-level, the

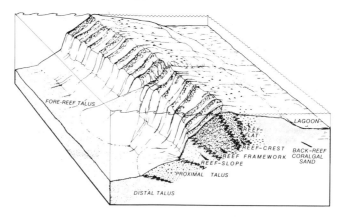

FIG. 9. The facies of a modern shelf-margin reef complex. This model can only be applied to reefs in the geologic record where a strong framework existed. From Longman (1981), with the permission of the author and Society of Economic Paleontologists and Mineralogists.

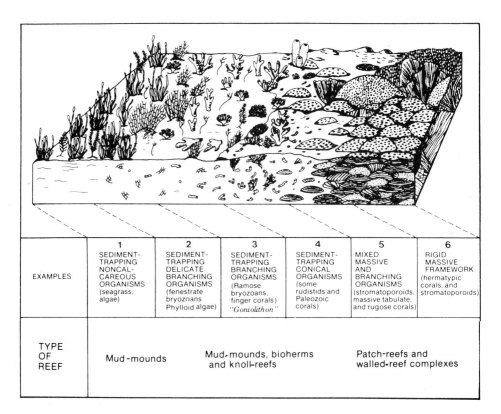

EXAMPLES	1 SEDIMENT-TRAPPING NONCAL-CAREOUS ORGANISMS (seagrass, algae)	2 SEDIMENT-TRAPPING DELICATE BRANCHING ORGANISMS (fenestrate bryozoans Phylloid algae)	3 SEDIMENT-TRAPPING BRANCHING ORGANISMS (Ramose bryozoans, finger corals) "*Goniolithon*"	4 SEDIMENT-TRAPPING CONICAL ORGANISMS (some rudistids and Paleozoic corals)	5 MIXED MASSIVE AND BRANCHING ORGANISMS (stromatoporoids, massive tabulate, and rugose corals)	6 RIGID MASSIVE FRAMEWORK (hermatypic corals, and stromatoporoids)
TYPE OF REEF	Mud-mounds	Mud-mounds, bioherms and knoll-reefs			Patch-reefs and walled-reef complexes	

FIG. 10. The roles played by organisms in constructing carbonate buildups and the types of buildup produced. From Longman (1981) with the permission of the author and the Society of Economic Paleontologists and Mineralogists.

zone of active reef growth (the reef crest) prograles seawards over an apron of debris eroded from the reef during storms (the fore-reef). A steep to overhanging wall of active coralgal growth can exist here. A relatively flat surface, around 1 m below low water mark develops behind the reef as it prograles and this reef flat gives way to further reef debris, generated by and washed over during storms, in the back reef area. Reef rubble can be piled up to form islands along the shelf margin.

Where sea level rises relatively slowly, it is possible for reef growth to keep pace and a substantial vertical thickness of reef rock can be formed. Such preferential vertical growth along the shelf margin will lead to the development of a relatively deep shelf-lagoon, with below wave-base sedimentation. A rapid sea-level rise is likely to kill off marginal reefs. Reefs cannot cope with significant falls in sea level either. Growth could only occur in the shallow subtidal on the front of the former reef. Exposure of the reef would lead to the development of emergence phenomena and the shelf itself would become subaerial if the sea-level drop was sufficient. Thick reef limestones do not form during a period of relative sea-level fall.

Studies of the Recent have given us a well-defined facies model for reefs (Fig. 9): the fore-reef zone of reef-slope proximal talus and distal talus, the reef itself (or reef core) of reef framework and reef crest, and the back-reef zone of reef flat and back-reef sand. This facies model can be applied to many ancient shelf-marginal reef complexes, but in detail there are often departures. This is especially the case where the reef core is concerned, since in many instances the organisms involved did not have the ability to produce a rigid framework and/or encrusting and binding organisms were not present. In those instances, a wave-resistant reef (i.e. a true reef) could not form.

Many organisms have contributed in a variety of ways to carbonate buildups through geologic time. The organisms have varied from non-calcareous grasses and algae through delicate colonial bryozoans, calcareous algae and corals, to robust conical rudistids and rugose corals, to massive or laminar stromatoporoids and tabulate and rugose corals. The organisms' effects have varied from simple trapping of sedimentary grains through to formation of a rigid framework (see Fig. 10).

The record of organisms capable of producing large, rigid, branching, massive or tabular skeletons has not been continuous. There have been seven major phases (Fig. 11) (James 1979; Longman 1981): (1) Middle and Upper Ordovician bryozoan-stromatoporoid-tabulate coral reefs, (2) Silurian and Devonian stromatoporoid-tabulate coral reefs, (3) Upper Permian sponge-calcareous algal reefs, (4) Upper Triassic and (5) Upper Jurassic scleractinian coral–stromatoporoid reefs, (6) Upper Cretaceous rudist bivalve reefs and (7) Tertiary–Quaternary scleractinian coral–red algal reefs. On a broad scale,

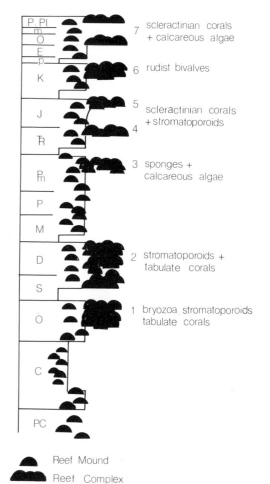

7 scleractinian corals + calcareous algae

6 rudist bivalves

5 scleractinian corals + stromatoporoids

4

3 sponges + calcareous algae

2 stromatoporoids + tabulate corals

1 bryozoa, stromatoporoids tabulate corals

Reef Mound

Reef Complex

FIG. 11. The Phanerozoic record of reef mounds (patch reefs) and reef complexes, with the dominant organisms that have contributed towards formation of the latter. After James (1979).

all these reefs can show similar facies patterns to modern coralgal reefs (phase 7), with the best development along shelf margins, a seaward fore-reef talus apron, and a shoreward back-reef skeletal sand facies passing into a shelf lagoon. In detail there will still be many differences, principally in how the organisms behaved and interacted in the reef core facies, but also in such features as the degree of syn-sedimentary cementation and the extent of bioerosion, especially by boring organisms.

There are many carbonate buildups in the geologic record which were not shelf-marginal reef complexes, but nevertheless had topographic relief and often exerted an influence on the surrounding seafloor. In shelf environments these have been referred to as bioherms, patch reefs, reef mounds, knoll reefs and banks. (Mud mounds are a further type but they most commonly developed in somewhat deeper water locations on ramps or slopes to platforms and shelves, and so are discussed under the later carbonate ramp section.)

Reef mounds are just as variable as marginal reefs. Typically though, there is little framework to a reef mound but *in situ* skeletons may be common. Much of the mound normally consists of skeletal debris, along with carbonate mud. Sediment trapping was the likely cause of many reef mounds with bryozoans, crinoids, finger corals, rudistids and some algae all able to take this role. Around the reef mounds, which are generally massive and lenticular, there usually occur well-bedded limestones of skeletal debris. Some of this would be derived from the buildup, washed off during storms, but much is generated in the slightly deeper off-mound waters, and may thus have a different faunal composition.

At the present time, patch reefs are common in shelf lagoons behind marginal reefs, as in the Caribbean for example, and they also occur on modern carbonate ramps, such as in the Arabian Gulf. Modern patch reefs usually have coral frameworks and calcareous algal crusts consolidating the structure, just like the marginal, barrier reefs. A zonation of corals also occurs with more robust and sturdy forms on the upper part of the reef and more delicate forms around the base. Many other organism types are associated and bioerosion is just as important. The patch reefs give rise to a halo of skeletal sand, which is frequently asymmetric, with the greater portion located on the leeward side.

The location of many patch reefs appears to be random, dependent on suitable substrates for coral attachment and then a period of rapid growth to establish the buildup. The same probably applies to many ancient reef mounds, that restricted areas of suitable substrate were colonized by the framework-building or sediment-trapping organisms and that once established, the mounds' growth was assured. However, it is clear that the location of some patch reefs and reef mounds was determined by slight topographic highs. These could have resulted from a more positive area of basement or a storm-deposited shoal of carbonate sand. Some modern patch reefs have formed upon slight rises in the underlying relic topography.

Patch reefs of the Arabian Gulf have developed where the seafloor is locally elevated and this is frequently due to salt diapirism (Purser 1973b). Although the depositional setting is basically a carbonate ramp, the patch reefs have

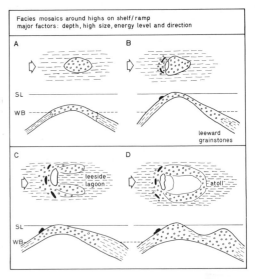

FIG. 12. Facies models showing the development of patch reefs and associated facies on a topographic high on a shelf or ramp (based on the work of Purser 1973b).

much in common with those on shelves and platforms in the Caribbean. Of interest in terms of facies models, is that a spectrum of reef development can be discerned, with four stages giving four distinct facies models (Fig. 12). The initial stage is the development of carbonate sands on the high through enhanced carbonate skeletal production on this shallow agitated area and winnowing out of lime mud. Sand distribution is asymmetric, sedimentation taking place mainly in a downwind direction. Stage 2 is the establishment of a reef on the windward side of the high, giving rise to more rapid production and accumulation of skeletal sand. In stage 3, reef growth is well advanced, producing more of a barrier to waves and currents, so that downwind tails of reef debris from the margins of the reef produce a leeside lagoon. Island formation is likely from storm piling-up of reef talus. In stage 4, the reef has grown most or all the way around the topographic high to enclose a lagoon. Lime muds may accumulate here in this atoll-like stage.

There are numerous descriptions of ancient equivalents of patch reefs and it is clear that there is much variation in shape and size, as well as internal structure. As an example, the Silurian reefs of the Great Lakes area have been imaginatively divided on the basis of shape into blue spruce, spread-eagle, mammary gland, haystack and lime kiln types by Shaver (1977). The shape is determined by sea-level fluctuations, rates of subsidence and time available for reef

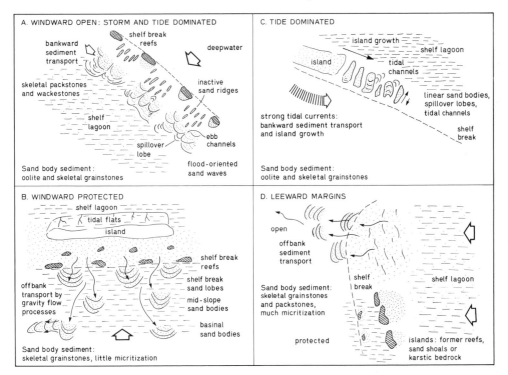

FIG. 13. Facies models of shelf-marginal carbonate sand bodies, where shelf-break orientation relative to waves, storms and tidal currents is the controlling factor (based on work of Hine *et al.* 1981).

growth. There were strong trade winds in this area during the Silurian so that reefs are frequently asymmetric and their debris is preferentially developed on one side (the lee) and atolls with lagoons formed (Lowenstam, 1950). The various facies models of Fig. 12, derived from modern patch reefs can thus be applied. Silurian platform sedimentation was relatively fast in inter-reef areas so that a substantial differential topography did not develop between the reef mounds and the surrounding seafloor. However, some reefs grew on the slope into adjoining basins, where little inter-reef sediment was deposited. These reefs managed to keep up with the faster subsidence rate to produce relatively tall, narrow structures termed pinnacle reefs. These frequently have an apron of reef debris on the leeward side too.

Shelf-margin carbonate sand bodies

Carbonate sands are usually generated in abundance along shelf (and platform) margins. Much of the sand is of skeletal origin, derived from shelf-break reefs (if present) and from the skeletons of organisms which live in the shelf-margin areas. The sudden barrier that the shelf-slope makes to open ocean and storm waves

ensures continuous turbulence along the shelf-margin and constant reworking of sediment and erosion of reefs. Ooids are frequently an important component of shelf margin sands since $CaCO_3$ precipitation is promoted by the active movement of suitable nuclei (fine skeletal grains) and the warming and CO_2-degassing of ocean water as it comes on to the shallow shelf. There have recently been several detailed studies of modern shelf-margin sand bodies on the Bahama Platform (Hine 1977; Hine *et al.* 1981) and from these it is clear that several types exist, dependent on the orientation of the shelf-break relative to the prevailing wind direction. Distinct differences exist between leeward and windward locations, and between those of an open and those of a protected aspect. Tide-dominated parts of the shelf-margin give a further type of sand body (see Fig. 13).

Open windward shelf margins are generally the most turbulent and sand bodies are well-developed there. Reefs are also present, usually a little seaward of the sand belt and these supply much of the sediment. The near-constant agitated conditions, however, promote the formation of ooids so that the sand is an oolitic and skeletal grainstone. The dominant bedform of the sand body is a linear sand wave with a

wavelength of 20–100 m, oriented transverse to
the flow. Superimposed upon the sand waves are
smaller-scale dunes (wavelengths 0.6–6 m) on
the sand-wave crests, and ripples (wavelengths
less than 0.6 m) on the flanks. The sand waves
are asymmetric on the seaward side, and sym-
metric on the shallower shelf-lagoon side of the
sand body, where tidal currents move the sand
backwards and forwards through the tidal cycle.
Storm-generated currents have cut channels
across the sand body and at the ends of the
channels, especially on the shelf-lagoon side,
spillover lobes are developed. These are also
covered by sand waves, migrating into the
lagoon. The internal structures to be expected
within these sands are large-scale planar cross
beds, from the sand waves, and smaller-scale
cross beds from the dunes. Orientations will be
largely lagoonward reflecting bedform movement during major storms, but some off-shelf
orientations will also occur, especially if ebb
tidal currents are important.

On windward protected shelf margins an island or barrier reef affords some protection to
the shelf lagoon. Sand is produced in abundance, from wave and storm attack on the reefs,
and this is deposited between the reefs and
island, or just off-bank from the reefs.

However, the important feature of this shelf-
margin type is that the dominant onshore wave
and storm currents are reflected offshore to
produce strong down-slope bottom currents.
Lobes of sand are thus deposited at the shelf
break, on the slope and at the toe of the
slope. Resedimentation is important as sediment
is transported off the shelf by grain flows, debris
flows and turbidity currents. The high energy
location gives a rapid turnover of sediment so
that little is micritized by algae and few ooids are
produced.

In a tide-dominated location, linear sand
ridges are produced parallel to ebb and flood
directions, usually normal to the line of the shelf
break. Sand waves and dunes are present on the
sand ridges and at the ends of the channels
between the ridges, especially at the lagoon end,
spillover lobes are developed. Sand is frequently
oolitic, as a result of the constant reworking,
and skeletal too if there are reefs nearby. Where
tidal currents are oblique to the shelf break, then
lateral development of the tidal sand ridges may
occur. Islands can be formed, especially if there
is sufficient wave and storm activity to throw up
beach ridges.

On leeward shelf margins, the dominant
direction of water movement is off-shelf, and

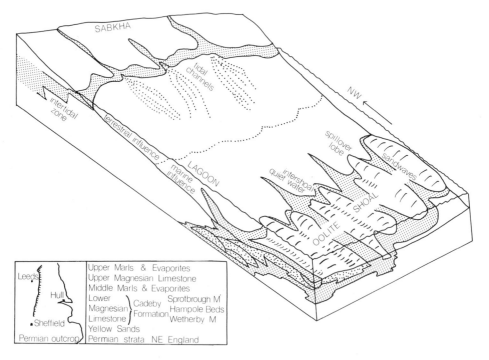

Fig. 14. An example of a facies model for a shelf-margin sand body, the upper Lower Magnesian
Limestone of Yorkshire, England. After Kaldi (1984).

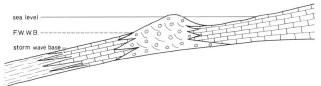

BASIN	CARBONATE RAMP deep ramp shallow ramp		PERITIDAL CARBONATE PLATFORM	
below fair weather wave base	wave dominated		protected	subaerial
shale/pelagic limestone	thin-bedded limestones	barrer island complex ooid sand shoals patch reefs	lagoonal to tidal flat carbonates	supratidal carbonates ± evaporites

FIG. 15. The carbonate ramp facies model based on Ahr (1973).

the storms and waves have moved over the shelf itself before reaching the shelf break. Sand bodies are not well developed in this situation. On open leeward margins, sand is produced by reworking of the sandy mud of the lagoon, so that grains are extensively micritized and peloids are an important constituent. Sand waves may develop at the shelf edge and there is much off-shelf transport of sediment. On protected leeward margins, islands are present and these act as a barrier to the off-shelf movement of sand. The latter thus accumulates against the islands and can cause them to enlarge. The islands themselves may be reefs or sand shoals of a former higher sea-level stand, or the result of karstic weathering of earlier carbonates during a sea-level fall. Active reefs may develop seawards of the islands since the latter will keep fine sediment from smothering them.

Carbonate sand bodies are best developed along open windward shelf margins and the sand is moved into the lagoon during major storms. The net result is a sequence of lagoonal sediments (wackestones-packstones) overlain by the carbonate sands (grainstones) with shoreward-directed cross bedding. A similar sequence can be produced along tide-dominated margins if there is much lagoonward sediment transport. Thick sequences of grainstone will be produced if there is a substantial rate of subsidence or sea-level rise and sedimentation is able to keep pace.

Ancient shelf-margin carbonate sand bodies have been described from a number of formations. One useful example from Britain occurs in the Permian Lower Magnesian Limestone (now Cadeby Formation) of Yorkshire and Nottinghamshire (Kaldi 1984). A north–south-oriented oolite shoal complex, located at a break of slope is dominated by large-scale cross bedding produced by sand waves (Fig. 14). Spillover lobes generated by major storm events

are also distinguished. Shorewards (west) of the shoal complex, a lagoon existed and tidal flats. This shelf-marginal oolite body was strongly influenced by onshore directed trade winds blowing from NE or E.

The carbonate ramp

The ramp was recognized by Ahr (1973) as a major type of depositional setting for carbonate rocks and was put forward as an alternative to the carbonate shelf. Although the ramp model has not been applied to ancient carbonates as much as the shelf and platform models, there are certainly many limestone sequences which are best understood by reference to this model of facies distribution. A carbonate ramp is a gently sloping surface, gradients of the order of a few metres per kilometre, contrasting markedly with the steep slope up to a carbonate shelf, and the relatively flat surface of the shelf itself or a platform. On a ramp, shallow-water carbonates pass gradually offshore into deeper and deeper water and then into basinal sediments (Fig. 15). There is no major break in slope which is the characteristic feature of a shelf to basin transition. In spite of this, there are often similarities between the ramp facies and nearshore open shelf facies (see earlier section).

The distinctive sediments of the inner ramp are carbonate sands formed in the agitated shallow subtidal shoreface zone (above fair-weather wave base) and low intertidal. On a ramp, wave energy is not as intense as along a shelf margin where oceanic swell and storm waves are suddenly confronted with a shallow steep slope. Nevertheless, the gradual shoaling of a ramp does result in relatively strong wave action in the shoreface-intertidal and this permits the formation of shoreline carbonate sand bodies. Storm events are generally very important on ramps and along with normal wind-wave activity give rise to beach–barrier complexes through shoreward movement of sand. Offshore storm surges are important in transporting shoreface sands to the outer, deeper ramp. Shoreward of the inner ramp sand belt, lagoons and tidal flats may develop. These will be of limited extent if the ramp continues into the supratidal, but if the ramp leads up to a platform, then a very extensive lagoon-tidal flat-supratidal area will occur behind the beach barrier.

A modern ramp

The best developed and described modern carbonate ramp is off the Trucial Coast of the Arabian Gulf (Loreau & Purser 1973; Wagner & van der Togt 1973). Here the seafloor gradually slopes down from sea level to a depth of 90 m in the axis of the Gulf. The slope is not a smooth surface; there are many local positive areas rising above the ramp surface which are structurally controlled, mostly being salt diapirs. The Trucial Coast is a mesotidal area with a tidal range of 2.1 m along the shoreline, dropping to 1.2 m within the lagoons. The NE–SW-oriented coast directly faces strong winds (the Shamal) coming from the NNW. Because of a very arid climate, and the partly enclosed nature of the Gulf, salinity (40–45‰) is a little higher than in the Indian Ocean (35–37‰) and in the lagoons it may reach 70‰.

On the deep ramp, skeletal sandy muds are extensively developed with bivalves and foraminifera the most important sediment contributors. These sediments give way to carbonate sands on the shallow ramp and these form subtidal shoals and beach-barrier–island–tidal delta complexes. The beach barriers are composed of skeletal sand reworked from the shoreface and ooids. There are aeolian dunes on the subaerial parts of the barriers. During major storms sand may be carried over the barriers to be deposited in washover fans on the back barrier. Behind the beach-barrier system there occur lagoons and then extensive intertidal flats dissected by tidal channels and partly covered by algal mats. Landwards, are the broad supratidal flats or sabkhas wherein gypsum-anhydrite is precipitating. In actual fact the back-barrier environments are particularly well developed along the Trucial Coast because this area is a tectonic depression. The lagoons are connected to the open Gulf via tidal channels through the barriers and within these inlets ooids are being precipitated. Ebb and flood tidal deltas have developed at the Gulf and lagoon ends of the tidal channels. The deltas consist of sand shoals

with distinct lobes (spillover lobes) along their margins. Sandwaves, dunes and ripples cover the surfaces of the tidal deltas.

The localized shoal areas which occur on the ramp are sites of skeletal sand accumulation where these highs extend above wave base. Reefs too may develop on the highs and these have been discussed in an earlier section (see Fig. 12). In fact small patch reefs do occur just offshore from the beach-barrier system. The slight hypersalinity of the Gulf water prevents the stenohaline corals from flourishing.

Ramp facies

The nearshore sand belt of a ramp consists of skeletal and oolitic grainstone. Peloids may be important too. Apart from compositional differences, the facies developed will be identical to that of a siliciclastic beach-barrier system (see for example Elliott 1978; Reinson 1979). The beach-barrier carbonates will show bedding dipping at a low angle offshore (surf-swash deposit) and onshore from deposition on the backsides of beach berms. On-shore directed cross-bedding will be produced by shoreface megaripples and wave-ripple cross-lamination will also occur. Aeolian cross-bedding is likely through barrier-top wind-blown dune migration (also shoreward). Rootlets and vadose diagenetic fabrics are possible in the upper barrier sediments and burrows may occur in the low intertidal-shoreface part. Tidal deltas, which could be largely oolitic, would give rise to offshore and/or onshore directed cross-bedding from sand wave and dune migration. Herring-bone cross bedding is possible from tidal current reversals. Tidal inlet migration, resulting from longshore current effects, will rework beach-barrier sediments and give rise to a sharp-based shell lag of the channel floor, overlain by variously cross-bedded, cross-laminated and flat-bedded sand of the deep to shallow channel and migrating spit.

At depths a little greater than fairweather wave base, organic productivity is still high so that bioclastic limestones will be dominant. Shoals of skeletal debris (grainstones) are likely to be formed through reworking by storm waves. These would have medium to small-scale cross-stratification, and contain winnowed horizons of coarser debris (rudstones). Hummocky cross-stratification is possible here too. Migration of these skeletal sand shoals would take place during major storms and result in coarsening-up, thickening-up units, a few metres thick, of skeletal wackestones passing up into skeletal grainstones and rudstones.

In slightly deeper areas, below storm wave base, skeletal packstones and particularly skeletal wackestones will dominate. Fine carbonate will be derived largely from shallow-water areas where fragmentation of skeletal debris takes place; some lime mud will be formed *in situ*. Thin graded bioclastic grainstones and packstones will be common in the below storm wave-base areas of the ramp, deposited from seaward flowing storm-surge currents. Scoured bases, grooves, even flutes, can be expected on the bases of these storm beds, as well as a sequence of internal structures indicating deposition from waning flow (flat bedding to ripple cross-lamination especially). Burrows may occur on the base and within the storm bed.

Sea-level changes and subsidence rates will be important in determining the thickness of the shallow ramp sand belt and also its migration direction. During a still-stand or slight sea-level fall, a thick sand body can develop through seaward progradation of the shore-line beach-barrier system, if there is a continuous supply of sand. A slow sea-level rise can lead to the shoreward migration of the sand belt, and the generation of a transgressive sequence (barrier sands overlying lagoonal sediments). In this instance the low barrier sediments will consist of washover fan sands, giving small coarsening-up cycles (lagoonal peloidal muds to the barrier sands) as the fans prograde into the lagoon. As with siliciclastic beach-barrier systems, rapid transgression may leave little record of the shoreface, other than a disconformity (a ravinement) and a basal conglomerate, produced by surf-zone erosion.

Ancient ramp facies

Ramp facies are widely developed in the geological record. Ahr (1973) interpreted the Jurassic Smackover Formation of Texas as a shallow ramp sand body. The Smackover, an important oil reservoir, consists of a seaward prograding wedge of oolite. The deeper ramp facies are peloidal wackestones and mudstones and then basinal organic-rich lime mudstones. Behind the oolitic sand belt evaporites and red beds developed in supratidal and subaerial environments. Ramp carbonates are well represented in the Upper Cambrian and Middle Ordovician of the Appalachians in Virginia (Read 1980; Markello & Read 1981). Oolites characterize the shallow ramp and ribbon carbonates the deeper ramp. The latter consist of storm-deposited shallow water carbonate sediments.

In Britain, parts of the Dinantian carbonate sequence of South Wales are ramp in character. In the southern part, deeper water argillaceous limestones give way to bioclastic-oolitic limestones farther north. Several major phases of shallowing can be recognized, where skeletal wackestones give way to skeletal-oolitic grainstones. Below fairweather wave-base skeletal wackestones contain thin skeletal packstone-grainstone beds of storm-surge origin and thicker sequences of skeletal grainstone suggest the development of sand shoals through more persistent storm reworking. Migration of these shoals gives metre-thick coarsening upward units. Oolite sand bodies were formed when above wave-base depths persisted. Then shoreface shoals and beach barriers developed, with

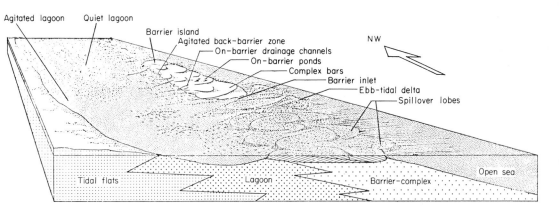

FIG. 16. An example of a facies model for a barrier-lagoon system, probably formed on a shallow ramp, the lower Bajocian (Jurassic) Lincolnshire Limestone Formation, east Midlands. From Ashton (1977). The barrier gradually builds lagoonwards in response to episodic transgression.

lagoons and tidal flats behind where lime mudstones, dolostones and cryptalgal laminites accumulated.

Another major carbonate sand body developed in the mid-Jurassic of eastern England (Ashton 1977). The Lincolnshire Limestone was deposited in a major N–S-oriented barrier island–tidal delta complex and consists of variously cross-bedded oolite. A well developed lagoon to the west was covered by the barrier as it transgressed shorewards (Fig. 16). Back barrier washover fans gave rise to coarsening up units. Supratidal barrier top, tidal delta and spillover lobe facies have all been recognized. Unfortunately the seaward facies are not exposed.

Carbonate buildups on ramps, notably mud-mounds

Since there is no major break of slope on a carbonate ramp, reef complexes of the barrier type, which are common at shelf-margins, do not develop. Small patch reefs or reef mounds are common, however, and can occur on the seaward side of the barrier, in the back-barrier lagoon, and on any topographic highs on the deep ramp. These reef mounds will be similar to those of carbonate shelves, described earlier.

One particular type of buildup which does occur on ramps is the mud mound (also called reef knoll and mud bank), composed largely of micrite, usually clotted and often pelleted. Mud mounds are particularly common in the Palaeozoic and mostly occur in offshore, deep ramp, or shelf-slope locations. They clearly had some form of seafloor expression since the flank beds generally dip off the mud-mound structure, having original depositional slopes of 5–20°. Mud mounds do not contain any metazoan frame builders and this has given rise to much discussion, posing two questions: how did the mud-mound form, and where did all the lime mud come from? Recent work by Pratt (1982) on many mud-mound limestones has revealed similarities between the clotted, rarely laminated mud-mound framework and shallow-water cryptalgal structures, particularly those of a thrombolitic character (unlaminated stromatolites). Pratt (1982) thus proposed that mud mounds were formed through organic binding by algal mats of locally produced lime mud. One particular structure that is common in many mud mounds is stromatactis. This is a cavity structure, typically occurring in families and having a flat floor of internal sediment and an irregular roof. Although variously filled, many have an initial isopachous fibrous calcite cement

and then a later drusy sparite. Stromatactis structures are now regarded as the product of early cementation of the carbonate mud, giving lithified surficial crusts, and then a degree of winnowing of uncemented sediment from beneath the crust (Bathurst 1982).

Carbonate facies through time

There are as yet no clear trends in the pattern of carbonate facies through time, but a few generalizations can be made. Ramp, platform and shelf grainstones, packstones, wackestones and lime mudstones are similar wherever they occur. There are great variations in the grain types present, skeletal and non-skeletal, but the processes operating were basically the same, so that the end-products are similar. There were certain periods when particular facies were well developed in certain areas. This is especially the case with platform carbonates, since, generally, they are best developed at times of relatively high sea-level stands, when cratons are flooded and shallow epeiric seas widespread. For example, during the lower to mid-Palaeozoic, a time of global sea-level rise, extensive platform carbonates were deposited on the North American and Russian cratons. The repetition of shallowing-upward units, so characteristic of platform sequences, also requires special conditions of periodic sea-level rises (eustatic or local tectonic), and these also only occurred at certain times in the Phanerozoic (e.g. the Carboniferous). Major variations in carbonate facies through the Phanerozoic did occur as a result of the changing fortunes of carbonate secreting organisms. The biogenic component of limestones has thus varied considerably in skeletal grainstones and packstones, but it is particularly in the formation of reefs and other carbonate buildups that these variations are most marked. The roles played by the various organisms in reef construction have varied considerably, from simple sediment trapping and baffling, through to frame building and

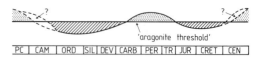

FIG. 17. The fluctuations in the mineralogy of shallow-marine carbonate precipitates, ooids and cements through time. After Sandberg (1983).

binding (see Fig. 10). As a result of this, the structure and facies distributions in ancient buildups vary tremendously. The record of reef complexes comparable to modern shelf-marginal reefs, and of reef knolls similar to modern patch reefs, has been discussed in an earlier section (see Fig. 11).

Another trend in carbonate facies which is apparent is the predominance of mud mounds in the Palaeozoic. They occur in deeper water ramp and outer shelf-slope settings but they are particularly important in Upper Ordovician to Lower Carboniferous strata. The geologic record of oolite facies also is not continuous; they are conspicuously absent in the Devonian of western Europe, although all other carbonate facies types are well developed, and Cretaceous–early Tertiary oolites are also poorly represented. Finally, the record of pelagic carbonates has not been continuous. With a few special cases, notably the Ordovician *Orthoceras* limestones of Scandinavia and the Devonian cephalopod limestones of western Europe, pelagic carbonates are poorly represented in the Palaeozoic. From the Jurassic, coccoliths and planktonic foraminifera permitted the deposition of ammonitico-rosso-type facies, as in the Alps, and chalks from the Cretaceous.

It has recently been established that there have been major fluctuations in the mineralogy of marine carbonate precipitates through the Phanerozoic (Fig. 17) (MacKenzie & Piggott 1981; Sandberg 1975, 1983). On the basis of ooid preservation and the nature of early cements, it appears that aragonite (and high-Mg calcite) was the principal precipitate in the late Precambrian to Cambrian, Upper Carboniferous to Triassic and early Tertiary to Recent, with calcite ($MgCO_3$ less than 8 mol %) the precipitate in the mid-Palaeozoic and Jurassic–Cretaceous. There is also a broad trend of more calcite-secreting organisms in the Palaeozoic and more aragonite-secreting organisms in the Mesozoic–Cainozoic. The fluctuations in mineralogy of marine carbonate precipitates are presumably a reflection of changes in seawater chemistry. The major factors controlling carbonate precipitates are Mg/Ca ratio and pCO_2, with to a lesser extent, organic geochemistry, temperature and salinity. In fact the trend of Fig. 17 coincides with the global sea-level curve; aragonite precipitates at times of low global sea-level and calcite at times of high sea-level stand. An underlying geotectonic control is thus implied. Two possibilities are: fluctuations in pCO_2 due to variations in the rate of subduction of carbonate sediment, or fluctuations in Mg/Ca ratio due to variations in the rate of seafloor spreading.

ACKNOWLEDGMENTS: I am grateful to Dave Gray and Mike Ashton for allowing me to quote from their Ph.D. theses and to use their unpublished figures.

References

AHR, W. M. 1973. The carbonate ramp; an alternative to the shelf model. *Trans. Gulf-Cst Ass. geol. Socs* 23, 221–25.

ASHTON, M. 1977. *Stratigraphy and carbonate environments of the Lincolnshire Limestone Formation, eastern England*. Unpublished Ph.D. Thesis, Hull University.

ASQUITH, G. B. 1979. *Subsurface Carbonate Depositional Models*. Pennwell Books, Tulsa. 121 pp.

BATHURST, R. G. C. 1975. *Carbonate Sediments and their Diagenesis*, Elsevier, Amsterdam. 685 pp.

—— 1982. Genesis of stromatactis cavities between submarine crusts in Palaeozoic carbonate mud buildups. *J. geol. Soc. London*, 139, 165–81.

COOGAN, A. H. 1969. Recent and ancient carbonate cyclic sequences. *In*: ELAM, J. G. & CHUBERS, S. (eds) *Symposium on Cyclic Sedimentation on Permian Basin*, 5–16. West Texas Geological Society.

DONOVAN, D. T. & JONES, E. J. W. 1979. Causes of world-wide changes in sea-level. *J. geol. Soc. London*, 136, 187–92.

ELLIOTT, T. 1978. Clastic shorelines. *In*: READING, H. G. (ed.) *Sedimentary Environments and Facies*, 143–77. Blackwell Scientific Publications, Oxford.

ENOS, P. & PERKINS, R. D. 1977. Quaternary sedimentation in South Florida. *Mem. geol. Soc. Am.* 147, 198 pp.

FISCHER, A. G. 1964. The Lofer cyclothems of the Alpine Triassic. *In*: MERRIAM, D. F. (ed.) *Symposium on Cyclic Sedimentation. Bull. geol. Surv. Kansas*, 169, 107–49.

FLÜGEL, E., 1982. *Microfacies Analysis of Limestones*. Springer-Verlag, Berlin. 633 pp.

FROST, S. H., WISS, M. P. & SAUNDERS, J. B. 1977. Reefs and related carbonates—ecology and sedimentology. *In*: *Studies in Geology*, 4, Am. Ass. Petrol. Geol. 421 pp.

GEBELEIN, C. D. *et al.* 1980. Subsurface dolomitization beneath the tidal flats of central west Andros Island, Bahamas. *Spec. Publs Soc. econ. Paleont. Miner., Tulsa*, 28, 31–49.

GEORGE, T. N. 1978. Eustasy and tectonics: sedimentary rhythms and stratigraphic units in British Dinantian correlation. *Proc. Yorks geol. Soc.* 42, 229–62.

GINSBURG, R. N. 1975. *Tidal Deposits*. Springer-Verlag, Berlin. 428 pp.

——, HARDIE, L. A., BRICKER, O. P., GARRETT, P. & WANLESS, H. R. 1977. Exposure index: a quantitative approach to defining position within the tidal zone. *In*: HARDIE, L. A. (ed.) *Sedimentation on the Modern Carbonate Tidal Flats of Northwest Andros Island, Bahamas*, 14–20. Johns Hopkins University Press, Baltimore.

—— & JAMES, N. P. 1974. Holocene carbonate sediments of continental shelves. *In*: BURK, C. A. & DRAKE, C. L. (eds) *The Geology of Continental Margins*, 137–55. Springer-Verlag, New York.

GRAY, D. I. 1981. *Lower Carboniferous shelf carbonate palaeoenvironments in North Wales*. Unpublished Ph.D. Thesis. University of Newcastle upon Tyne.

HALLOCK, P. 1981. Production of carbonate sediments by selected large benthic foraminifera on two Pacific coral reefs. *J. sedim, Petrol.* **51**, 467–74.

HARDIE, L. A. 1977. *Sedimentation on the Modern Carbonate Tidal Flats of Northwest Andros Island, Bahamas*. Johns Hopkins University Press, Baltimore. 202 pp.

HECKEL, P. H. 1972. Recognition of ancient shallow marine environments. *Spec. Publs Soc. econ. Paleont. Miner., Tulsa*, **16**, 226–96.

—— 1974. Carbonate buildups in the geologic record: a review. *Spec. Publs Soc. econ. Paleont. Miner., Tulsa*, **18**, 90–154.

HINE, A. C. 1977. Lily Bank, Bahamas; history of an active oolite sand shoal. *J. sedim. Petrol.* **47**, 1554–82.

——, WILBER, R. J. & NEUMANN, A. C. 1981. Carbonate sand bodies along contrasting shallow bank margins facing open seaways in Northern Bahamas. *Bull. Am. Ass. Petrol. Geol.* **65**, 261–90.

IRWIN, M. L. 1965. General theory of epeiric clear-water sedimentation. *Bull. Am. Ass. Petrol. Geol.* **49**, 445–59.

JAMES, N. P. 1979. Shallowing-upward sequences in carbonate reefs. *In*: WALKER, R. G. (ed.) *Facies Models*, 109–32. Geoscience Canada.

—— & GINSBURG, R. N. 1980. *The Seaward Margin of Belize Barrier and Atoll Reefs. Spec. Publs int. Ass. Sediment.* **3**. Blackwell Scientific Publications, Oxford. 191 pp.

——, ——, MARSZALEK, D. S. & CHOQUETTE, P. W. 1977. Facies and fabric specificity of early subsea cements in shallow Belize (British Honduras) Reefs. *J. sedim. Petrol.* **46**, 523–44.

JEFFERSON, D. P. 1980. Cyclic sedimentation in the Holkerian (middle Viséan) north of Settle, Yorkshire. *Proc. Yorks. geol. Soc.* **42**, 483–503.

KALDI, J. 1984. Sedimentology of sandwaves in an oolite shoal complex in the Cadeby Magnesian Limestone Formation (Upper Permian) of eastern England. *In*: HARWOOD, G. M. & SMITH, D. B. (eds) *The English Zechstein and Related Topics*. Blackwell Scientific Publications, Oxford. In press.

LAPORTE, L. F. (ed) 1974. Reefs in time and space.

Spec. Publs Soc. econ. Paleont. Miner., Tulsa, **18**, 256 pp.

LOGAN, B. W. (ed.) 1974. Evolution and diagenesis of Quaternary carbonate sequences, Shark Bay, Western Australia. *Mem. Am. Ass. Petrol. Geol.* **22**, 358 pp.

LONGMAN, M. W. 1981. A process approach to recognizing facies of reef complexes. *Spec. Publs Soc. econ. Paleont. Miner., Tulsa*, **30**, 9–40.

LOREAU, J. P. & PURSER, B. H. 1973. Distribution and ultrastructure of Holocene ooids in the Persian Gulf. *In*: PURSER, B. H. (ed.) *The Persian Gulf*, 279–328. Springer-Verlag, Berlin.

LOWENSTAM, H. A. 1950. Niagaran reefs of the Great Lakes area. *J. Geol.* **58**, 430–87.

MACKENZIE, F. T. & PIGOTT, J. D. 1981. Tectonic controls of Phanerozoic sedimentary rock cycling. *J. geol. Soc. London*, **138**, 183–96.

MARKELLO, J. R. & READ, J. F. 1981. Carbonate ramp-to-deeper shale shelf transitions of an Upper Cambrian intrashelf basin, Nolichucky Formation, Southwest Virginia Appalachians. *Sedimentology*, **28**, 573–97.

MAZZULLO, S. J. & FRIEDMAN, G. M. 1975. Conceptual model of tidally-influenced deposition on margins of epeiric seas: Lower Ordovician (Canadian) of eastern New York and Southwestern Vermont. *Bull. Am. Ass. Petrol. Geol.* **59**, 2123–41.

PERKINS, B. F. 1975. Carbonate Rocks III, organic reefs. *Bull. Am. Ass. Petrol. Geol. Reprint Series*, **15**, 190 pp.

PRATT, B. R. 1982. Stromatolitic framework of carbonate mud-mounds. *J. sedim. Petrol.* **52**, 1203–27.

PURSER, B. H. (ed.) 1973a. *The Persian Gulf.* Springer-Verlag, Berlin. 471 pp.

—— 1973b. Sedimentation around bathymetric highs in the southern Persian Gulf. *In*: PURSER, B. H. (ed.) *The Persian Gulf*, 157–78 Springer-Verlag, Berlin.

RAMSBOTTOM, W. H. C. 1977. Major cycles of transgression and regression (mesothems) in the Namurian. *Proc. Yorks. geol. Soc.* **41**, 261–91.

READ, J. F. 1980. Carbonate ramp to basin transitions and foreland basin evolution, Middle Ordovician, Virginia Appalachians. *Bull. Am. Ass. Petrol. Geol.* **64**, 1575–612.

—— & GROVER, G. A. 1977. Scalloped and planar erosion surfaces, middle Ordovician limestones, Virginia; analogues of Holocene exposed karst or tidal rock platforms. *J. sedim. Petrol.* **47**, 956–72.

REINSON, G. E. 1979. Barrier island systems. *In*: WALKER, R. G. (ed.) *Facies Models*, 57–74. Geoscience Canada.

SANDBERG, P. A. 1975. New interpretations of Great Salt Lake ooids and of ancient non-skeletal carbonate mineralogy. *Sedimentology*, **22**, 497–538.

—— 1983. An oscillating trend in Phanerozoic non-skeletal carbonate mineralogy. *Nature*, **305**, 19–22.

SCHROEDER, J. H. & ZANKL, H. 1974. Dynamic reef formation: a sedimentological concept based on studies of recent Bermuda and Bahama reefs. *Proc. 2nd int. Coral Reef Symp.* **2**, 413–28.

SELLWOOD, B. W. 1978. Shallow-water carbonate environments. *In*: READING, H. G. (ed.) *Sedimentary Environments and Facies*. 259–312. Blackwell Scientific Publications, Oxford.

SHAVER, R. H. 1977. Silurian reef geometry—new dimensions to explore. *J. sedim. Petrol.* **47**, 1409–24.

SHAW, A. B. 1964. *Time in Stratigraphy*. McGraw-Hill, New York. 365 pp.

SHINN, E. A., LLOYD, R. M. & GINSBURG, R. M. 1969. Anatomy of a modern carbonate tidal flat, Andros Island, Bahamas. *J. sedim. Petrol.* **39**, 1202–28.

SMITH, S. V. 1973. Carbon dioxide dynamics: a record of organic carbon production, respiration and calcification in the Eniwetok windward reef flat community. *Limnol. Oceanogr.* **18**, 106–20.

SOMERVILLE, I. D. 1979. Minor sedimentary cyclicity in late Asbian limestones in the Llangollen district of North Wales. *Proc. Yorks. Geol. Soc.* **42**, 317–41.

STODDART, D. R. 1969. Ecology and morphology of Recent coral reefs. *Biol. Rev.* **44**, 433–98.

TOOMEY, D. F. (ed.) 1981. European fossil reef models. *Spec. Publ. Soc. econ. Paleont. Miner.*, *Tulsa*, **30**, 544 pp.

TURMEL, R. J. & SWANSON, R. G. 1976. The development of Rodriguez Bank, a Holocene mudbank in the Florida reef tract. *J. sedim. Petrol.* **46**, 497–518.

WAGNER, C. W. & VAN DER TOGT, C. 1973. Holocene sediment types and their distribution in the southern Persian Gulf. *In*: PURSER, B. H. (ed.) *The Persian Gulf*, 123–55. Springer-Verlag, Berlin.

WALKDEN, G. M. 1974. Palaeokarstic surfaces in Upper Viséan (Carboniferous) limestones of the Derbyshire Block, England. *J. sedim. Petrol.* **44**, 1232–47.

WILSON, J. L. 1975. *Carbonate Facies in Geologic History*. Springer-Verlag, New York. 417 pp.

MAURICE E. TUCKER, Department of Geological Sciences, University of Durham, Durham DH1 3LE.

DIAGENESIS

Diagenesis of shallow-marine carbonates

J. A. D. Dickson

SUMMARY: The utility of carbonate petrography in solving diagenetic problems is often underrated. Staining and cathodoluminescence are complementary to petrographic studies; their application is compared. Interpretation of carbonate trace element concentration through the distribution coefficient, as illustrated by Sr^{2+}, was oversimplified in the 1970s and often tied to repeated recrystallization episodes in meteoric waters. Kinetic factors complicate the use of the distribution coefficient in calcite. $\delta^{13}C$ values from carbonates help in identifying the source of carbon but $\delta^{18}O$ values are difficult to interpret due to control by temperature as well as the isotopic composition of the precipitation fluids. Data on the composition of sedimentary waters, geothermometry and geobarometry of buried sedimentary rocks can be provided by fluid inclusion studies but interpretation has often been over-optimistic, because insufficient account has been taken of the difficulties which exist with fluid inclusion work.

Near-surface, carbonate diagenesis is well understood through field studies of Recent and Pleistocene sediments. Research effort in the future, it is predicted, will concentrate on burial diagenesis because it is the least understood diagenetic realm. The large variations in temperature, pressure and time which distinguish burial from near-surface environments are exemplified by the burial history of the Carboniferous Limestone of South Wales.

Recent reviews concerning carbonate diagenesis have been published by Moore (1979), Bathurst (1980a, b) and Longman (1980, 1981). Dolomites and dolomitization have been reviewed by Chilingar *et al.* (1979); they were the subject of the *SEPM Special Publication No. 28* (Zenger *et al.* 1980) and were included in the Geological Association of Canada's review series (Morrow 1982a, b). The approach and subject matter of these reviews are not reiterated here: this review considers some of the methods of investigation applied to carbonate rocks and attempts to predict where research effort may be directed in the near future.

Microscopy

At the birth of the BSRG, the polarizing microscope was probably the most important tool for investigating carbonate diagenesis. Bathurst (1958) had published his classic work on diagenetic fabrics and Folk's scheme for classifying carbonate rocks had appeared. One of the 'burning' issues of the day was the distinction between calcite cement and recrystallized spar. This distinction is difficult to make even after several decades of research, as is beautifully illustrated by Bathurst's (1983) work on a Jurassic grainstone.

Today, petrography is regarded by some as archaic. Some feel it is much more important to 'zap' minerals with an electron beam or generate gas to pass into a mass spectrometer. The sophistication of the hardware, the expense, and sometimes the difficulty in obtaining numbers from these machines, give the results more

significance than readily obtained data. The danger of this attitude is that materials are analysed in isolation without regard for field relationships or petrographic control which means much genetic information is unavailable when interpretation is attempted.

The similarity of optical properties within the trigonal carbonates makes them notoriously difficult to separate microscopically. Fortunately carbonate minerals are easily stained, which allows the common minerals to be differentiated. Some stains may be used to indicate the distribution of minor elements both between and within mineral generations. Alizarin red S/potassium ferricyanide (ARS/PF) dual stain (Evamy & Shearman 1962; Dickson 1966) distinguishes calcite from dolomite and their ferroan varieties (Fig. 1). The distinction of aragonite from calcite can be made using Feigl's solution (Friedman 1959) but this stain is unreliable because of grain size effects. High-Mg calcite (HMC) is distinguished from other carbonate minerals by Titan (Clayton) yellow (Choquette & Trusell 1978).

Staining is a fickle technique that depends on the skill of the operator; however, it is the easiest and most rapid method of carbonate identification. Stains obviate the need to prepare sections at 30 μm thickness to observe standard optical properties used for mineral identification, so ultra-thin sections may be prepared (Lindholm & Dean 1973). These have great advantages when studying micrites, for in a 30 μm section several crystals are piled one on top of the other. Even in coarsely crystalline carbonates, the positions and type of inter-

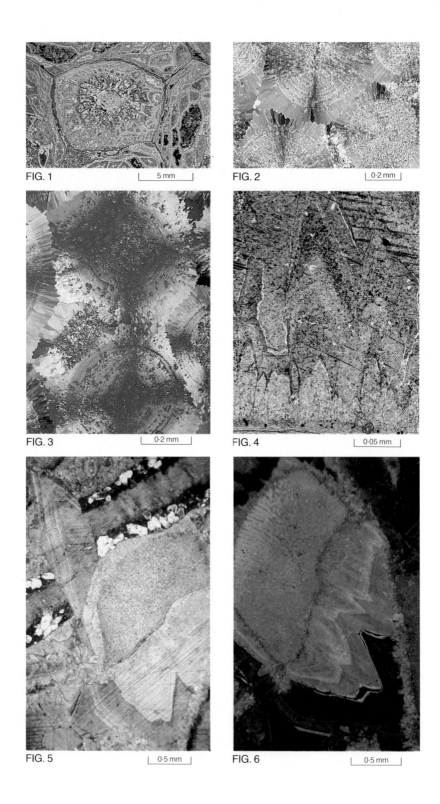

FIG. 1 |___ 5 mm ___| FIG. 2 |__ 0·2 mm __|

FIG. 3 |___ 0·2 mm ___| FIG. 4 |__ 0·05 mm __|

FIG. 5 |___ 0·5 mm ___| FIG. 6 |__ 0·5 mm __|

crystalline boundaries are much more clearly defined in an ultra-thin section (Fig. 2). The use of accessory plates in determining properties such as optical elongation in trigonal carbonates at 30 μm is difficult because of their high birefringence. However, no such difficulty is experienced if ultra-thin sections are used (Fig. 3).

Staining, in addition to differentiating a number of carbonate minerals, may display chemical variations both between generations of crystals and within individual crystals. During one episode of crystal growth, changes in the composition of the water from which crystallization is occurring may happen. These changes of fluid composition may be reflected in the crystal's minor element composition which can sometimes be shown by staining. Intra-crystalline zonation is usually assumed to be growth zonation and hence each zone represents a time surface. The succession of chemical changes or zones so identified may be dated relative to each crystal's nucleation point. Zoned cement crystals often possess a distinctive zonal pattern (Fig. 4), which can be identified and mapped through adjacent crystals. Similar zonal patterns

may be observed on stimulation of cathodoluminescence (CL).

Luminescence is generated in minerals because of their molecular distortions. Cathodoluminescence involves excitation by electron bombardment, temporary storage of energy and emission. The molecular distortions which cause CL are manifold (Nickel 1978); the emission is usually only observed in the visible wavebands although emission at other wavelengths (ultraviolet for example) is common. CL is generated on polished specimens being bombarded by an electron stream in a vacuum chamber. The vacuum chamber is placed on a microscope stage so the CL produced can be observed directly.

Staining and CL both reveal features such as minor-element intracrystalline variations not readily visible by other methods. Staining is an inexpensive, rapid technique which can be applied to large surface areas either in the field or in the laboratory but it is often difficult to replicate and depends on the skill of the stainer. Cathodoluminescence microscopy is usually studied using the commercially available, relatively inexpensive machines with cold-cathode electron guns. Machines with heated-

---

FIG. 1. Photomicrograph of *Lonsdaleia floriformis* (Fleming), stained (ARS + PF). Lower Carboniferous Limestone (Brigantian), Hotchberry Quarry, Brigham, near Cockermouth, Cumbria. Coral skeleton and oldest cement against skeleton stained pink, indicating non-ferroan calcite. Most of the intraskeletal spaces filled with ferroan calcite of variable iron content, stained mauve, purple and royal blue. Some intraskeletal spaces filled with ferroan dolomite which stains a turquoise to green colour.

FIG. 2. Photomicrograph, double-polished, ultra-thin section, 10–12 μm thick, crossed polars, Smackover oolite, Oxfordian. Pennzoil Allen Well No. 1 Lafayette County, Arkansas, 10,814' depth. Note good definition of crystal shapes in fibrous cement and optical continuity between radial structure in ooids and epitaxial cement.

FIG. 3. Photomicrograph, same rock as illustrated in Fig. 2, but section 2 μm thick. Crossed polars and sensitive tint plate inserted. Ultra-thin nature of section causes first-order, grey birefringence colours with crossed polars. Sensitive tint plate clearly indicates optical orientation of length-fast calcite which would be difficult to determine if section was 30 μm thick.

FIG. 4. Photomicrograph, stained (ARS + PF) grainstone, plane polars, Derbyhaven Beds, Arundian, L. Carboniferous, Derbyhaven foreshore, Isle of Man. Productid brachiopod (base) on which zoned calcite cement has grown. Early stage iron-free (orange); later stages variable iron (blue/purple).

FIG. 5. Photomicrograph, stained (ARS + PF) grainstone, plane polars, Derbyhaven Beds, Arundian, Lower Carboniferous, Derbyhaven foreshore, Isle of Man. Crinoid ossicle (central) displaying large zoned epitaxial overgrowth. Fractured productid (top centre to bottom right) with algal encrustations (brown) which contains many authigenic quartz crystals (clear). Fracture postdates early, non-ferroan cement (orange) but is healed by later ferroan cement (zoned, blue/purple).

FIG. 6. Photomicrograph, same subject as Fig. 5 but rotated. Cathodoluminescence. Note extra zones displayed in early, non-ferroan epitaxial overgrowth on crinoid and tiny, red/orange, microdolomite crystals both in non-ferroan overgrowth and in overgrown crinoid ossicle. These features are not apparent in Fig. 5.

filment, electron guns have been successfully used (Zinkernagel 1978) and CL attachments for the SEM are available (Grant 1978). The efficiency of cathodoluminizers differs; the image may be replicated using different machines but light intensity generated is variable.

The information produced by staining and CL is similar but different (Figs 5 and 6); CL images contain more information. The disclosure of multiple zoned crystals is perhaps the most significant feature of these images. The zones, with their time significance, allow the direction and relative speed of crystal growth to be documented, the succession of crystallographic forms and habits can be identified and the spatial distribution of zones can be plotted; a reenactment of crystal growth is possible. The correlation of particular zones has allowed Meyers (1974, 1978) to identify the dimensions of palaeoaquifer systems in the Lake Valley Formation (Mississippian), New Mexico, U.S.A. Non-zonal patterns produced by staining and CL have been used to identify neomorphic spar (Marshall 1981; Bathurst 1983). Richter & Zinkernagel (1981) list many other features of geological significance which can be deduced from CL images of carbonates.

ARS/PF staining has been related to iron concentration in calcite (Lindholm & Finkleman 1972) but this has not gained wide acceptance. Deductions made about CL dependence on composition (Meyers 1974; 1000 ppm Mn^{2+} to activate CL) have proved incorrect (Pierson 1981; Frank et al. 1982). It is tempting to use staining or CL as a quick, easy, indirect method to assess elemental concentrations in carbonates but these surrogate methods are unreliable.

The intensity of potassium ferricyanide staining depends on the reaction rate between carbonate and acid stain as well as the presence of ferrous iron. Reaction rate differs widely between carbonates; even within one carbonate the reaction rate will vary not only with composition but also with orientation of the reacting surface (calcite reacts faster along its c-axis than normally to it; Dickson 1966). In addition to these problems other elements may block the potassium ferricyanide stain: for instance, Turnbull's Blue is not produced on ferroan rhodochrosite, despite reacting with the acid, presumably because of manganous interference.

The relationship between CL response in carbonates to elemental composition is even more difficult to interpret than staining. Luminescence colour is a combination of different wavelengths (Gies 1975), yet most observations are reported as a general hue; seldom is reference to some standard, such as the Munsell colour chart, ever made. However, even if complete spectral patterns were available, it would be necessary to demonstrate that each waveband was uniquely and quantitatively related to the presence of a particular element before CL could be used as an analytical tool. At our present state of knowledge, elemental concentrations should not be assessed by staining or CL but by standard analytical methods such as atomic absorption, inductive coupled plasma, microprobe etc.

The light microscopy discussed so far may be extended with electron microscopy. The scanning electron microscope has provided exquisite images of tiny crystals developed in incompletely filled pores. The documentation of many Recent diagenetic changes in carbonates would have been poorly recorded (if at all) without the SEM. A connection has been made between crystal habit, the Mg/Ca ratio of calcite and the environment of crystallization (Lindholm 1972, Folk 1974 and others). This is an interesting connection but care must be exercised in its application. Suess (1970) has pointed out that organic molecules have a controlling effect on carbonate precipitation. Kirov & Filizova (1970) and Kirov et al. (1972) have shown how widely differing habits and forms in calcite are controlled by the anion/cation ratio and state of supersaturation of the waters. The Mg^{2+}/Ca^{2+} ratio is simply one of a number of factors which affects calcite form and habit. Accurate indexing of crystallographic form is rarely given in 'sedimentological' literature; terms such as 'steep rhombs' or 'dog tooth spar' are commonly used. The identification of form in incompletely filled pores is comparatively easy with the SEM for interfacial angles may be directly measured: calcite probably displays a greater variety of individual crystallographic forms (Palache, et al. 1951 list 19 forms and 48 less common forms), combination of forms (Whitlock 1910) and differing habits than any other natural mineral.

Transmission electron microscopy images internal structure down to the atomic scale. Reeder (1981) claims new insight into the genesis of sedimentary dolomites through TEM work but very little has been done in this field.

Minor and trace elements

Potentially the most precise way of interpreting trace element analysis of carbonates is by using the distribution coefficient. The molar ratio of trace element to dominant or carrier ion in the solid can be determined and this can be related through the distribution coefficient to the same ratio in the liquid from which the solid crystallized. The discussion that follows is chiefly concerned with strontium because it is the most thoroughly investigated and perhaps the best understood of the carbonate trace elements.

The composition of carbonate crystals may be related to two precipitation mechanisms (Katz *et al.* 1972). Homogeneous precipitation occurs when any change in solution composition causes an adjustment in the crystal's composition so the entire crystal is uniform and at equilibrium with the final solution. The distribution coefficient for homogeneous precipitation has been symbolized *D* and only applies when the system is at complete equilibrium. Heterogeneous precipitation (Doerner & Hoskins 1925) applies when the crystal's composition reflects that of the solution only at the instant of precipitation and does not readjust to later changes of solution composition. The distribution coefficient for heterogeneous precipitation has been symbolized λ (Katz *et al.* 1972). The common occurrence of multiply zoned crystals and evidence from experimental work (Kinsman & Holland 1969, Katz *et al.* 1972 and others) indicate carbonate precipitation is heterogeneous although homogeneous precipitation can be simulated when small amounts of crystal are precipitated from a large solution reservoir so the solution's composition does not significantly change. Kinsman & Holland

(1969), used the symbol k for the distribution coefficient for both homogeneous and heterogeneous precipitation: some disparity exists in the use of symbols k and *D*. k is used here in a general sense after Kinsman & Holland (1969). Holland, *et al.* (1964) and Kinsman & Holland (1969) experimentally determined k values for strontium in aragonite $k_{Sr}^{2+}{}_A = 1.12$, 25°C and calcite $k_{Sr}^{2+C} = 0.14$, 25°C. Kinsman (1969) indicated a straightforward relationship existed between k_{Sr}^{2+k} and temperature; the reaction being insensitive to changes in the solution's composition or concentration. He applied k_{Sr}^{2+A} and k_{Sr}^{2+C} values to diagenetic problems, in particular to the depletion of strontium in ancient limestones as compared to recent lime sediments. Open and closed systems were modelled and the former was held responsible for Sr^{2+} loss effected by repeated recrystallization due to the passage of meteoric water through lime sediments.

The promise offered by this work was tremendous; by analysing the trace element composition of carbonates one could determine cation chemistry of the long-departed water. However in 1972, Katz and co-workers produced a k_{Sr}^{2+} value ($\lambda_{Sr}^{2+C} = 0.05$; at 25°C) significantly different from Holland's by the recrystallization of aragonite to calcite rather than direct precipitation from a fluid of seawater composition (Fig. 7). The difference in experimental method may have guided some investigators choice as to which of the two values to adopt but uncertainty existed, and often still exists. Wigley (1973) investigated Barbadan

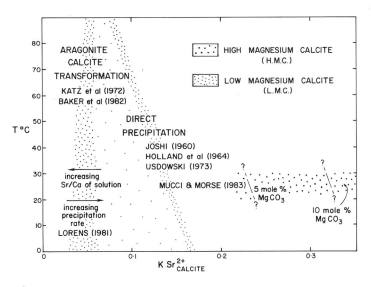

FIG. 7. k_{Sr}^{2+} calcite, plotted against temperature. Sources of data indicated; further details in text.

limestones and used Holland's value to interpret Sr^{2+} values but Morrow & Mayers (1978) using Wigley's data chose Katz *et al.*'s value to model Sr^{2+} depletion in the same Barbudan limestones. Pingitore (1978) considered Zn^{2+} and Mn^{2+} to interpret carbonate diagenesis because of the uncertainty (Pingitore 1980) which he claimed surrounded k_{Sr}^{2+C} values. The reason for the disparity in k_{Sr}^{2+C} values is that direct precipitation in seawater is not an equilibrium process, consequently k_{Sr}^{2+C} values are high.

In recent years a greater variety of trace elements have been considered for modelling changes during limestone diagenesis (Veizer 1983). Computer models (Land 1979, 1980) have been published for Sr^{2+} depletion and sometimes the Sr^{2+} depletion is cross-correlated with enrichment of other elements (e.g. Zn, Morrow & Mayers 1978) but these changes have usually been tied to the passage of vast volumes of meteoric water through the rock and repeated recrystallization episodes. The conclusion made by Katz *et al.* (1972) that Sr^{2+} depletion in calcite can occur by aragonite recrystallization in seawater has been largely ignored. However, Land (1979, 1980) reported how he was forced by the geological situation to accept that chalk from the Shatsky Rise was depleted in Sr^{2+} without the influence of meteoric waters.

A local distribution coefficient may be determined in a natural system actively undergoing precipitation where the Ca^{2+}/Sr^{2+} ratio can be determined for both solid and solution phases. Baker *et al.* (1982) investigated such a system in deep-sea carbonates. In addition to corroborating Katz *et al.*'s (1972) value for k_{Sr}^{2+C}, Baker, *et al.* used their k_{Sr}^{2+C} values, estimated temperatures and Ca^{2+}/Sr^{2+} ratio of the pore water, to predict Ca^{2+}/Sr^{2+} ratios in diagenetic calcites. The match between their predicted Ca^{2+}/Sr^{2+} ratios and that analysed in the carbonate of some DSDP cores is remarkably close (although the method does not work well with impure carbonate sediments; Elderfield, personal communication).

Experimental work by Lorens (1981) on Sr, Cd, Mn and Co in calcites indicates that both solution composition and precipitation rate are significant factors in controlling trace element partitioning. Lahann & Siebert (1982) showed that kinetic factors are even more important for Mg^{2+} distribution in calcites. Mucci & Morse (1983) found that Mg^{2+} and/or Sr^{2+} concentration in calcite is controlled by the Mg^{2+}/Ca^{2+} ratio of the solution (Fig. 7) and is influenced by surface effects close to the crystal. They also observed a correlation between increasing $MgCO_3$ and $SrCO_3$ contents of the precipitate and postulated this was due to distortion of the calcite lattice caused by the small Mg^{2+} ion (0.65 Å) creating sites suitable for a cation larger than Ca^{2+} (0.99 Å); that is Sr^{2+} (1.13 Å). This influence is most noticeable when the Mg^{2+}/Ca^{2+} ratio of the solution is below 7.5. In seawater or related solutions Mucci & Morse found precipitation rate has no effect on the amount of Mg^{2+} or Sr^{2+} incorporated into high magnesium calcite.

What state are we in?

The promise of easily derived numerical information about the long departed waters of carbonate precipitation through use of experimentally derived k values must be approached more cautiously than has been done in the past. k is not a constant. Kinetic effects are particularly important at low temperatures; at 20°C there is an order of magnitude difference in experimentally determined k_{Sr}^{2+C} values (Fig. 7). The analysed concentration of Sr^{2+} in a carbonate (whether natural or experimentally produced) must not be used in calculations involving k values unless *all* the Sr^{2+} is in solid solution in the carbonate lattice (Angus *et al.* 1979).

Skeletal carbonates are often produced with Sr values far from that predicted by an inorganic k_{Sr}^{2+} value (Bathurst 1971; Milliman 1974). Incorporation of Sr^{2+} into biogenic carbonates is compatible with k_{Sr}^{2+} work (higher k_{Sr}^{2+} with faster calcification rates) but biological partitioning is complex and poorly understood.

The use of k_{Sr}^{2+} for interpreting the origin of carbonates has concentrated on the reaction(s) which produces low magnesium calcite from aragonite and/or high magnesium calcite of shallow marine sediments. Problems exist if the reactions have gone to completion for then the Sr^{2+} content of the starting materials must be estimated. Shallow marine carbonates are predominantly biogenic so estimates of Sr^{2+} content in transformed skeletons of extinct taxa must be indirect. Many workers have assumed a Sr concentration for ancient marine sediments to be similar to that in Recent unaltered equivalents but Graham *et al.* (1982) suggest the Sr^{2+}/Ca^{2+} ratio of seawater has changed over the last 75 Ma which in turn should affect the Sr^{2+} concentration of marine precipitates!

The Sr concentration of an extant system can be determined because original aragonite is present in unaltered material. However even knowing the Sr concentration of the reactants does not eradicate misinterpretation. Part of the Sr content of modern shell carbonate is easily leached. Walls, *et al.* (1977) have suggested that the easily leached Sr^{2+} (and Mg^{2+}) is either absorbed in the conchiolin matrix or is present as solid and/or fluid inclusions in the shell; hence part of the Sr^{2+} content should not enter into k_{Sr}^{2+} calculations.

k_{Sr}^{2+} is only a true constant at equilibrium but it can be used for non-equilibrium situations if factors controlling its variability are understood. Equilibrium precipitation is claimed for deep sea carbonates by Baker *et al.* (1982) where recrystallization of low magnesium calcite is a relatively slow process. Veizer (1978) emphasizes that 'thin film' (or 'messenger-film', Brand & Veizer 1980) stabilization of aragonite to calcite in a meteoric environment is a disequilibrium process. Pingitore (1982) in discussing the same meteoric stabilization process proposes the reaction is non-steady state and, as such, Sr^{2+}/Ca^{2+} values for the solid can at best only be indirectly related to the fluid at the reaction site and not to the composition of the bulk fluid.

Awareness of the above problems will improve confidence in the interpretation of trace element partitioning. It has been traditional to use k_{Sr}^{2+} values and the tracer/carrier ratio of the solid to determine the trace concentration in the parent solution. However if the trace/carrier ratio of the solution can be determined, either by examining an extant system or through fluid inclusions, then a local value of k can be determined. This should provide information about the reaction, whether it is an equilibrium reaction or not, and this may ultimately be more significant than simply determining the trace element concentration of the parent solution!

Isotope studies

Carbon ($\delta^{13}C$) and oxygen ($\delta^{18}O$) stable isotope data have become an integral part of most modern studies of carbonate diagenesis (Hudson 1977). Information on the source of carbon used in the precipitation of carbonates can be deduced from the $\delta^{13}C$ value of the carbonate.

The oceans form a relatively well mixed carbon reservoir, so marine carbonates should typically have a $\delta^{13}C$ value around zero ($\delta^{13}C$ standard is PDB, a marine belemnite). Long-term secular variations (Veizer *et al.* 1980) and short-term variations (Scholle & Arthur 1980) in the marine carbon reservoir, although of great interest, are relatively minor (a few per mille) compared to the variations in non-marine waters. Lakes and other small reservoirs show considerable variations in $\delta^{13}C$ caused by exchange with atmospheric CO_2, input of dissolved carbonates in ground-waters, CO_2

production of organic matter decomposition and photosynthesis (Fritz & Poplawski 1974). Some other sources of carbon that contribute to primary carbonate precipitation are shown in Fig. 8. The range shown in Fig. 8 can be overlapped and extended when reactions involving CO_2 production by the oxidation of organic matter are considered. Hence the numerical value of $\delta^{13}C$ alone does not unequivocally identify a particular carbon source.

The source of carbon for diagenetic carbonates is multifarious. Any pre-existing carbonate (Fig. 8) may dissolve and be reprecipitated during diagenesis. This process may involve either a single type of primary carbonate with a constant or variable $\delta^{13}C$ value, or several types of primary carbonate with differing $\delta^{13}C$ values. Another carbon source during diagenesis is the oxidation of organic matter which produces carbon dioxide which in turn may be precipitated as carbonate. Irwin *et al.* (1977) proposed a sequence of diagenetic zones with increasing depth from the water/sediment interface for organic-rich, marine sediments. In surficial oxic sediments (Zone 1) organic matter is microbially oxidized to produce CO_2 with $\delta^{13}C$ values around $-26‰$. Zone 2, the region of bacterial sulphate reduction, also produces CO_2 with $\delta^{13}C$ values around $-26‰$ but in Zone 3 organic carbon is split by fermentation into CH_4 and CO_2. A strong fractionation occurs during fermentation, the $\delta^{12}C$ being enriched in CH_4 which means CO_2 is correspondingly heavy: $+15‰$. The final Zone 4, involves decarboxylation where CO_2 is produced with $\delta^{13}C$ ranging between -10 and $-25‰$. Carbonates precipitated from these reactions have been most conclusively recognized in concretions which formed in mudrocks. This is partly

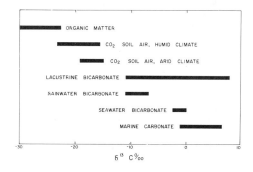

FIG. 8. Range of $\delta^{13}C$ values in various sources of CO_2 for carbonate precipitation. Data from (Degens 1967; Fritz & Poplawski, 1974; Galimov 1966; Gross 1964; Keeling 1958; Rightmire & Hanshaw 1971).

because the contribution from primary marine carbonate in mudrock is small (or better still absent) and hence does not dilute the diagenetic carbonate out of recognition and partly because the CO_2 produced by the diagenetic reaction completely overwhelms the small amount of CO_2 initially dissolved in the pore water. However many mudrock concretions do not possess $\delta^{13}C$ values which correspond to a single source. Coleman & Raiswell (1981) described Lower Jurassic concretions with $\delta^{13}C$ values of -13 to -15% which they conclude are due to a mixture of carbonate derived from sulphate reduction, fermentation and of marine origin. The marine carbonate was present in the original mudrock and was incorporated in the concretion. As the diagenetic reactions which follow one another are mutually exclusive, Coleman & Raiswell proposed the earlier, light, sulphate-reduction carbonate (Zone II) must have remained porous allowing the introduction of the later, heavy fermentation carbonate (Zone III) into the concretion so giving an intermediate $\delta^{13}C$ value. However, although the reactions are mutually exclusive, the CO_2 produced may diffuse between zones and mix, so their explanation may require modification.

All shallow marine carbonate sediments probably pass through zones of bacteriological activity but evidence for this, in terms of $\delta^{13}C$ values in carbonates, is very patchy. This is probably partly due to bulk rock sampling methods where the diagenetic addition to 'background' primary marine carbonate is so small as to go unrecognized. It may be that in some sediments bacteriologically produced CO_2 is lost before carbonate is precipitated and in well-oxygenated sediments it is possible that little metabolizable organic matter survives aerobic attack so eliminating later anaerobic bacterial processes. Analysis of the incremental additions during diagenesis (Dickson & Coleman 1980; Marshall 1981, 1982) is perhaps the most satisfactory way to identify diagenetic zones but can only be accomplished on coarsely crystalline material (millimetre size).

The most thoroughly documented diagenetic change in shallow, marine carbonates in terms of stable isotope data is exposure of marine sediments to near-surface meteoric water. Gross (1964) established this change on Bermuda where light carbon due to input of biogenic CO_2 from the soil gas zone occurs. This CO_2 is dissolved in meteoric water which also dissolves marine carbonates. The meteoric water then precipitates calcite with a $\delta^{13}C$ value intermediate between soil CO_2 and marine carbonate values, i.e. approximately -12% (Hudson

1977). The $\delta^{13}C$ value of precipitated carbonate, with increasing depth from the land surface, looses its distinctive soil CO_2 component and becomes indistinguishable from marine carbonate (Allan & Matthews 1977). The pattern of isotopic change in vertical profile associated with exposure and with vadose/phreatic and phreatic mixing zones was established on Barbados by Allan et al. (1978) and Allan & Matthews (1976, 1977) and has been recognized in Pleistocene (Videtich & Matthews 1980) and older sequences (Allen & Matthews 1982).

Distinctive isotope profiles have been shown to exist away from hardground surfaces (Marshall & Ashton 1980). In three profiles recorded by Marshall & Ashton, two showed slight $\delta^{13}C$ depletion ($+3.4$ to -0.6% and $+2.5$ to $+1.7\%$) and one showed $\delta^{13}C$ enrichment (-0.4 to $+1.3\%$), although $\delta^{18}O$ trends were much more convincing. These trends were related to varying proportions of early marine cement versus late calcite, both of which have distinctly different isotope values. The cause of these trends, is, therefore, different from that recorded by Matthews and coworkers where an overall change in isotope composition occurs in a single generation of cement away from a discontinuity surface.

Shallow marine limestones commonly show a small range of $\delta^{13}C$ values (a few $\%$ around 0) compared to a pronounced change towards lighter $\delta^{18}O$ values (from 0 to -15%; Hudson 1977). This range is due to a variable mixture of original marine carbonate and diagenetic alteration and/or addition. The enrichment in ^{16}O has been equated with the diagenetic component being added at progressively higher temperatures with increasing burial, with evolving $\delta^{18}O$ of the pore waters or a combination of both. Successive precipitates may occur which reflect conditions at the time of precipitation or pre-existing material may continuously re-equilibrate to new conditions. In the latter case the rock should be isotopically homogeneous, and in the former, heterogeneous; perhaps zoned. The presence of isotope zonation has been shown by Dickson & Coleman (1980) for Carboniferous calcite cement and by Marshall (1981, 1982) for Jurassic calcite spars. The polarity of the zonal trend towards overall lighter $\delta^{18}O$ values with time is confirmed, but some reversals occur (Marshall 1982).

The difficulty in interpreting $\delta^{18}O$ values is due to the dual control on oxygen fractionation, temperature and $\delta^{18}O$ of pore water (Arthur et al. 1982). The Smackover, a Jurassic grainstone reservoir rock which underlies many of the Gulf Coast states of USA, may be used as an example to illustrate the difficulties in stable isotope interpretation. This rock contains coarse calcite cement (calcspar) which has been analysed from Walker Creek field by both Moore & Druckman (1981) and Wagner & Matthews (1982a). These workers agree on the isotopic data ($\delta^{13}C$ + 2 to +4‰ and $\delta^{18}O$ approximately −5‰) but Moore & Druckman (1981) and Moore & Brock (1982) believe the spar was precipitated from warm, subsurface brines whilst Wagner & Matthews (1982a, b) believe the spar was precipitated in a freshwater lens; scarcely could two views be more divergent!

The numerical values of $\delta^{13}C$ and $\delta^{18}O$ for calcite are seldom sufficient to indicate the mineral's origin although certain hypotheses can often be ruled out. Detailed petrographic work before calcite is isotopically analysed and a knowledge of the sample's geological setting allows a more complete interpretation.

Do the problems uncovered in interpreting distribution coefficients have any relevance to isotope studies? The problem concerning contamination, and whether analysed concentration equals solid solution, presents little problem. Proper sample preparation can ensure all CO_2 evolved for analysis is derived from the carbonate. Kinetic factors may however be important. It is well established that many Recent skeletal carbonates are produced out of equilibrium with ambient waters. Most diagenetic studies involving stable isotope interpretation have assumed equilibrium is maintained. Turner (1982) has shown that precipitation rate affects $\delta^{14}C/\delta^{13}C$ ratio in inorganically precipitated carbonates, and Usdowski *et al.* (1980) have shown calcites precipitated from supersaturated stream waters are out of equilibrium (both $\delta^{13}C$ and $\delta^{18}O$) with respect to those waters.

Most isotope research on carbonates has concentrated on the anion radical: the cation radical is also of interest. $^{87}Sr/^{86}Sr$ ratio of seawater has varied significantly through the Phanerozoic (Burke *et al.* 1982). The same ratio has been shown to be an important indicator of diagenetic reactions in deep sea sediments (Elderfield *et al.* 1982; Elderfield & Gieskes 1982). A significant range of $^{87}Sr/^{86}Sr$ values exist in present-day, Smackover Formation water over short distances (16 km) whilst Smackover ooids and oncolites yield 'Jurassic' ratios and postcompaction calcite cement and baroque dolomite are significantly more radiogenic than the ooids in the same rock (Stueber & Pushkar 1983). The potential of $^{87}Sr/^{86}Sr$ work on shallow water carbonates is just being realized, the prospects are exciting but the work needs to be done.

Fluid inclusion studies

Sedimentologists are only now realizing the potential fluid inclusion studies hold for solving sedimentary (particularly diagenetic) problems. Fluid inclusions may yield information on temperature, pressure and the composition of the trapped fluid. Inclusions may be created: (1) contemporaneously at a crystal's growing surface; (2) penecontemporaneously within a crystal which is still growing and (3) subsequently when the grown crystal is fractured, fluid is injected and the fracture heals (Roedder 1976). It is essential to determine the origin and distribution of these various generations of fluid inclusions before full interpretation of inclusion data can be achieved. Fluid inclusion data relating to sedimentary environments have been collected from economically important deposits (evaporites, Mississippi Valley-type ores etc.), crystal lined cavities and large cement crystals (Roedder 1979). Temperature and salinity information on ancient formation waters from sedimentary basins derived from fluid inclusion studies is available (Hanor 1979) from the works of economic geologists, not sedimentologists! Speleologists have extracted the fluids from speleothem inclusions to determine D/H and $\delta^{18}O/\delta^{16}O$ ratios (Schwarcz *et al.* 1976; Thompson *et al.* 1976). The study of fluid inclusions from tectonic fracture fills (Tillman & Barnes 1983) and oil field fractures (Wayne & Burress 1982) is proving significant but geothermometry on late stage calcite and dolomite cement from fluid inclusions has poor precision; often a range of 80°–200°C for a single crystallization event (Nahnybida *et al.* 1982; Young & Jackson 1981).

Geothermometry is perhaps the commonest aim of fluid inclusion studies. The type of inclusion used is that which is completely filled by a homogeneous saline solution under the temperature and pressure conditions of trapping but at room temperature and pressure a vapour bubble exists in addition to the solution. Rehomogenization of the two phases is readily accomplished by heating on the microscope stage but this homogenization temperature will only be equal to the trapping temperature if the effects of pressure are zero, which is usually not the case, so corrections are needed to determine the trapping temperature. To make this correction it is usually necessary to know the composition of the fluid and the phase behaviour of that fluid under different $P-V-T$ conditions (Roedder & Bodnar 1980).

The commonest method of assessing an inclusion's fluid composition is by measuring the depression of the freezing point (Roedder 1963). It is usually assumed the principle solute is NaCl and salinity is expressed as equivalent weight percent NaCl. This measure is accurate but indirect. Most formation waters are much more complex than the $NaCl-H_2O$ system; not only ionic species Ca^{2+}, K^+, HCO_3^- and SO_4^{2-}, but also dissolved gases, are commonly present. The presence of methane in the bubble phase leads to significant overestimates of trapping temperature on applying pressure corrections to homogenization temperatures (Hanor 1980). Direct chemical analysis of inclusion fills often involves releasing the fluid from many inclusions with mixing and contamination problems (Roedder 1972). Partial analysis of polynuclear species (CO_3, SO_4, CH_4 etc.) from individual inclusions is possible using laser-activated, Raman spectroscopy (Rosasco & Roedder 1979; Dhamelincourt *et al.* 1979 and Beny *et al.* 1982).

The pressure correction applied to homogenization temperatures to obtain trapping temperatures can only be made if the composition of the fluid fill is known and if $P-V-T$ data are available for fluid of that composition. $P-V-T$ data are available for some aqueous fluids (Crawford 1981) but often the complex natural solutions cannot be matched by simple experimental data, extrapolation is necessary and so results are approximate.

Geothermometry and geobarometry estimations can only be made using data from fluid inclusions the internal volume of which has not been modified since trapping either in nature or the laboratory (by leakage, necking-down, decrepitation or stretching; Roedder 1981). The composition of the fill must also have remained unchanged since trapping.

The simple aqueous inclusions discussed above are just one type of inclusion to occur in sedimentary rocks; many others occur which often involve special study techniques, for instance, liquid hydrocarbons may be identified by fluorescence microscopy (Burruss 1981). It is sometimes very important to study more than one type of inclusion, for instance, if CO_2 and H_2O inclusions are separately but simultaneously trapped the common temperature and pressure of trapping may be determined (Roedder & Bodnar 1980). It is impossible to review the variety and utility of all inclusion types here; further reference may be made to COFFI (abstract volumes, editors, Roedder and Kozlowski) and the short course handbook, *Fluid Inclusions: Applications to Petrology* (Hollister & Crawford 1981).

The aim of accurately determining temperature, pressure and fluid composition during precipitation of sedimentary material through fluid inclusion study is highly desirable, but probably feasible in only a few cases. The collection of large numbers of homogenization and freezing point determinations with no regard to pressure corrections or the other prerequisites mentioned above brings this method of geothermometry and geobarometry into disrepute.

Diagenetic environments

The methods of investigation into carbonate diagenesis mentioned so far relate to the meso- and microscopic scales but our understanding of carbonate diagenesis has undoubtedly advanced most significantly through (megascopic) field studies of areas actively undergoing diagenesis. The most thoroughly investigated are those areas which have pleasant climates and are easiest to reach. One of the most fully understood diagenetic systems is reefal sediments raised to form islands in a tropical or semi-tropical climate which attract rainfall and groundwater circulation. Such a system is destined for diagenetic change because of the metastable starting mineralogy. The products of these diagenetic changes have been identified, documented and collated (reviewed by Longman 1980). Just how far this particular situation is significant in the interpretation of ancient carbonates has been queried by Hudson (1975) and Bathurst (1980b). A plethora of marine diagenetic precipitates composed of H. M. C. and/or aragonite exist in Recent shallow-water sediments (Bathurst 1974, 1980b and Longman 1980, 1981). Beach rock, hardground and reefal cements have been described but the mechanism(s) of precipitation is less clearly understood than in the meteoric situation mentioned above. The transformation of these H. M. C. and aragonite precipitates to L. M. C. whether subjected to meteoric water or not is probably the most important diagenetic change to affect shallow marine carbonates. The style and rate of this transformation has been reassessed by Walter (1983) and the importance of H. M. C. and A as primary precipitates at various times in the past has been questioned (Sandberg & Popp 1981, Wilkinson, *et al.* 1982, Tucker 1982 and Sandberg 1983). Many problems still remain to be solved in the field of near-surface, diagenetic change of shallow marine carbonates but the realm of diagenetic change least well understood is when these rocks become deeply buried. Research into this diagenetic environment is particularly difficult because of its remoteness and the difficulty of obtaining trustworthy data.

The problems of studying burial diagenesis generally become more complex with increasing age. Many Tertiary and Mesozoic sedimentary basins are still undergoing primary burial but older basins may have been involved in repeated episodes of burial and emergence. In the former case, sampling is restricted to borehole cores while in the latter, exhumed outcrop sequences may be sampled. The Carboniferous Limestone of Britain has a greater surface outcrop than any other British carbonate formation. Most studies involving diagenesis of the Carboniferous Limestone have ignored its burial history.

A time–depth reconstruction for 1000 m of Carboniferous Limestone now present in the Swansea–Gower area of South Wales is shown in Fig. 9. The near-surface conditions of sedimentation were followed by rapid burial to a depth of 7 km by the Stephanian. Beneath Swansea 3 km of compacted Namurian and Westphalian sediments now overlie the Carboniferous Limestone (Jones 1974; Thomas 1974); the Swansea Beds, close to the top of the Carboniferous sequence, contain semi-anthracites requiring a further 3 km burial to achieve rank (George 1970). The Hercynian orogeny deformed and raised the Carboniferous Limestone to different levels, so during the Trias, Gower limestones were subaerially exposed but Swansea limestones were still buried 3 km below the surface (Fig. 9). By the late Cretaceous 1½ km of Jurassic and Cretaceous sediments and the 'Chalk Sea' lay on Palaeozoic rocks but the 'Chalk Sea' had left by the end of the Cretaceous and Mesozoic sediments were

removed by the Oligocene (George 1974). The details of this reconstruction are unimportant here but the overall pattern is important. The nature of the rapid burial and uplift of the Carboniferous Limestone during the late Carboniferous/early Permian is of fundamental importance for an equilibrium temperature was not attained. This was due to the short residence time at maximum burial and the rocks' poor conductivity. If the Carboniferous Limestone had remained buried, an equilibrium temperature above 200°C would have developed and at 2 k.bar pressure, greenschist metamorphic facies would have been entered. Rapid burial may cause excess fluid pressure in confined horizons and rapid emergence may cause underpressuring. If these phenomena produce recognizable diagenetic effects then the frequency of these effects should correlate with the type and style of time–depth reconstructions.

The *P–T* conditions of the Swansea Carboniferous Limestone from Permian times until the present day are quite different from those on Gower (Fig. 9). The constant high temperature (always above 100°C) of the Swansea rocks should have affected their diagenetic 'grade' but to check this the rocks must be drilled, sampled and examined.

Consideration of time–depth curves for sedimentary rocks provides external information which helps to limit the diagenetic conditions which prevailed during the history of that rock. From these conditions it is possible to predict events which are likely to have occurred and those which are not. These curves also provide a

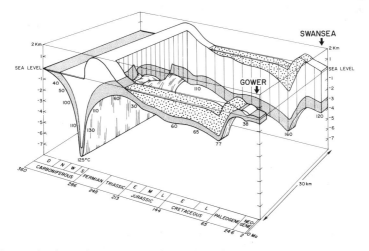

FIG. 9. Perspective time—depth reconstruction to show history of South Wales Dinantian sediments from their deposition, through burial to their present position. Dinantian sediments shown in fine stipple; Mesozoic sediments in coarse stipple. Temperatures shown at base of Dinantian in degrees celsius. Time-scale from Harland *et al.* (1982).

time base on which the diagenetic events may be fitted.

Conclusions

In the last two decades the most important advances in our understanding of diagenesis of shallow-water marine carbonates has come from studies of Recent and Pleistocene field areas. The predilection for meteoric processes has gripped many as a mental strait-jacket; virtually all diagenetic changes being explained by freshwater circulation. The new fashion appears to be burial diagenesis (Bathurst 1980a), although comments have appeared (Longman 1982) that burial diagenesis is of little significance. We are, however, ill-equipped to make general statements about the efficacy of burial diagenesis for our ignorance is great. Some evidence exists, such as the stylolite related cements (Wong & Oldershaw 1981), that points to subsurface processes being important but we do not understand how and why they occur.

Mattes & Mountjoy (1980) have shown how cements are repeatedly formed during burial and how these may be fitted on to the time–depth curve of the formation's history. Subsurface fluids have an enormous range of ionic strengths and, whilst ion activities are well understood in dilute solutions, we have problems with under-

standing the fundamental chemical behaviour of brines. Many subsurface formation waters now appear to be at equilibrium with the containing rock. Does this mean zero diagenesis? How long has this situation existed? If other waters were present where have they gone? How did they move and what drove them out and the existing ones in? One line of research with direct bearing on some of these questions is investigation into fluid inclusions (Roedder 1979). Its early exponents will probably make extravagant claims although these should be tempered by the experience of 'economic' geologists who have the greatest background in fluid inclusion work.

It seems illogical to claim that all carbonate diagenesis occurs near the surface and then nothing happens until the rocks pass into the metamorphic realm. Local changes within carbonate units are obviously important but these changes may be caused by fluids derived from or passing through adjacent sandstone or shale units.

ACKNOWLEDGMENTS: I wish to thank Professor E. R. Oxburgh for advise on geothermics and help in calculating burial temperatures for South Wales. Drs R. G. C. Bathurst, J. D. Marshall and P. F. Friend read the manuscript and made several suggestions for its improvement.

Cambridge Earth Science Series ES 426.

References

ALLAN, J. R. & MATTHEWS, R. K. 1976. C^{13}/C^{12} variations in Pleistocene reef limestones, Barbados, West Indies. *Abstr. Trans. Am. geophys. Un.* **57**, 350.

—— & —— 1977. Carbon and oxygen isotopes as diagenetic and stratigraphic tools: surface and subsurface data, Barbados, West Indies. *Geology*, **5**, 16–20.

—— & —— 1982. Isotopic signatures associated with early meteoric diagenesis. *Sedimentology*, **29**, 797–817.

——, ACHAUER, C. W. & MATTHEWS, R. K. 1978. Carbon and oxygen isotopes as indicators of meteoric and mixing zone diagenesis in limestones. *Abstr. Bull. Am. Ass. Petrol. Geol.* **62**, 489.

ANGUS, J. G., RAYNOR, J. B. & ROBSON, M. 1979. Reliability of experimental partition coefficients in carbonate systems: evidence for inhomogeneous distribution of impurity cations. *Chem. Geol.* **27**, 181–205.

ARTHUR, M. A., SCHOLLE, P. A. & HALLEY, R. B. 1982. Report on the second SEPM Research conference: "Stable-isotopes in sedimentary geology". *J. sedim. Petrol.* **52**, 1039–45.

BAKER, P. A., GIESKES, J. M. & ELDERFIELD, H. 1982. Diagenesis of carbonates in deep-sea sediments— evidence from Sr/Ca ratios and interstitial

dissolved Sr^{2+} data. *J. sedim. Petrol.* **52**, 71–82.

BATHURST, R. G. C. 1958. Diagenetic fabrics in some British Dinantian Limestones. *Lpool Manchr Geol. J.* **2**, 11–36.

—— 1971. *Carbonate Sediments and their Diagenesis.* Elsevier, Amsterdam. 620 pp.

—— 1974. Marine diagenesis of shallow water calcium carbonate sediments. *Ann. Rev. Earth planet. Sci.* **2**, 257–74.

—— 1980a. Deep crustal diagenesis in limestones. *In*: ORTI CABO, F. (ed.) *Conferencias y communicaciones del I symposium sombre diagenesis de sedimentos y Rocas sedimentarias. Revta Inst. Invest. geol., Barcelona*, **34**, 89–100.

—— 1980b. Lithification of carbonate sediments. *Sci. Prog.* **66**, 451–77.

—— 1983. Neomorphic spar versus cement in some Jurassic grainstones: significance for evaluation of porosity evolution and compaction. *J. geol. Soc. London*, **140**, 229–37.

BENY, C., GUILHAUMOV, N. & TOURAY, J. C., 1982. Native sulphur-bearing fluid inclusions in the CO_2-H_2S-H_2O-system. Microthermometry and Raman microprobe (MOLE) analysis. Thermochemical interpretations. *Chem. Geol.* **37**, 113–27.

BRAND, V. & VEIZER, J. 1980. Chemical diagenesis of a multicomponent carbonate system—1 trace elements. *J. sedim. Petrol.* **50**, 1219–36.

BURKE, W. H., DENISON, R. E., HETHERINGTON, E. A., KOEPNICK, R. B., NELSON, H. F. & OTTO, J. B. 1982. Variation of seawater $^{87}Sr/^{86}Sr$ throughout Phanerozoic time. *Geology*, **10**, 516–9.

BURRUSS, R. C. 1981. Hydrocarbon Fluid inclusions in studies of sedimentary diagenesis. *In*: HOLLISTER, L. S. & CRAWFORD, M. L. (eds) *Fluid Inclusions. Applications to Petrology*, 138–56. Short Course Handbook, Mineral Association of Canada.

CHILINGAR, G. V., ZENGER, D. H., BISSELL, H. J. & WOLF, K. H. 1979. Dolomites and dolomitization. *In*: LARSEN, G. & CHILINGARIAN, G. V. (eds) *Diagenesis in Sediments and Sedimentary Rocks. Developments in Sedimentology*, **25A**, Elsevier, Amsterdam. 423–536.

CHOQUETTE, P. W. & TRUSELL, F. C. 1978. A procedure for making the titan-yellow stain for Mg calcite permanent. *J. sedim. Petrol.* **48**, 639–41.

COLEMAN, M. L. & RAISWELL, R. 1981. Carbon, oxygen and sulphur isotope variations in concretions from the upper Lias of N.E. England. *Geochim. cosmochim. Acta*, **45**, 329–40.

CRAWFORD, M. L. 1981. Phase equilibria in aqueous fluid inclusions. *In*: HOLLISTER, L. S. & CRAWFORD, M. L. (eds) *Fluid Inclusions. Applications to Petrology*, 75–100. Short course handbook, Mineral Association of Canada.

DEGENS, E. T. 1967. Stable isotope distribution in carbonates. *In*: CHILINGAR, G. V., BISSELL, H. J. & FAIRBRIDGE, R. W. (eds) *Carbonate Rocks. Developments in Sedimentology* **9B**, 193–208. Elsevier, New York.

DHAMELINCOURT, P., BENY, J. M., DUBESSY, J. & POTY, B. 1979. Analyse d'inclusions fluides a la microsonde MOLE a effet Raman. *Bull. Soc. fr. Minér. Cristallogr.* **102**, 600.

DICKSON, J. A. D. 1966. Carbonate identification and genesis as revealed by staining. *J. sedim. Petrol.* **36**, 491–505.

—— & COLEMAN, M. L. 1980. Changes in carbon and oxygen isotope composition during limestone diagenesis. *Sedimentology*, **27**, 107–18.

DOERNER, H. A. & HOSKINS, W. M. 1925. Coprecipitation of radium and barium sulphates. *J. Am. Chem. Soc.* **47**, 662–75.

ELDERFIELD, H. & GIESKES, J. M. 1982. Sr isotopes in interstitial waters of marine sediments from Deep Sea Drilling Project cores. *Nature*, **300**, 493–7.

——, GIESKES, J. M., BAKER, P. A., OLDFIELD, R. K. HAWKESWORTH, C. J. & MILLER, R. 1982. $^{87}Sr/$ ^{86}Sr and $^{18}O/^{16}$ratios, interstitial water chemistry and diagenesis in deep-sea carbonate sediments of the Ontong Java Plateau. *Geochim. cosmochim. Acta*, **46**, 2259–68.

EVAMY, B. D. & SHEARMAN, D. J. 1962. The application of chemical staining techniques to the study of diagenesis in limestones. *Proc. geol. Soc.* **1599**, 102.

FOLK, R. L. 1959. Practical petrographic classification of limestones. *Bull. Am. Ass. Petrol. Geol.* **43**, 1–38.

—— 1974. The natural history of crystalline calcium carbonate: effect of magnesium content and salinity. *J. sedim. Petrol.* **44**, 40–53.

FRANK, J. R., CARPENTER, A. B. & OGLESBY, T. W. 1982. Cathodoluminescence and composition of calcite cement in the Taum Sauk Limestone (Upper Cambrian), Southeast Missouri. *J. sedim. Petrol.* **52**, 631–8.

FRIEDMAN, G. M. 1959. Identification of carbonate minerals by staining methods. *J. sedim. Petrol.* **29**, 87–97.

FRITZ, P. & POPLAWSKI, S. 1974. ^{18}O and ^{13}C in the shells of freshwater molluscs and their environments. *Earth planet. Sci. Lett.* **24**, 91–8.

GALIMOV, E. M. 1966. Carbon isotopes of soil CO_2 (transl.). *Geochem. Int.* **3**, 889–997.

GEORGE, T. N. 1970. *British Regional Geology: South Wales.* 3rd ed. H.M.S.O. London. 152 pp.

—— 1974. The Cenozoic evolution of Wales. *In*: OWEN, T. R. (ed.) *The Upper Palaeozoic and Post-Palaeozoic Rocks of Wales*, 341–71. University of Wales Press, Cardiff.

GIES, H. 1975. Activation possibilities and geochemical correlations of photoluminescing carbonates, particularly calcites. *Miner. Deposita*, **10**, 216–27.

GRAHAM, D. W., BENDER, M. L., WILLIAMS, D. F. & KEIGWIN, L. D. (Jr.) 1982. Strontium-calcium ratios in Cenozoic planktonic foraminifera. *Geochim. cosmochim. Acta*, **46**, 1281–92.

GRANT, P. 1978. The role of the scanning electron microscope in cathodoluminescence petrology. *In*: WHALLEY, W. B. (ed.) *Scanning Electron Microscopy in the Study of Sediments*, 1–11. Geo Abstracts, Norwich.

GROSS, M. G. 1964. Variations in the O^{18}/O^{16} and C^{13}/C^{12} ratios of diagenetically altered limestones in the Bermuda Islands. *J. Geol.* **72**, 170–94.

HANOR, J. S. 1979. The sedimentary genesis of hydrothermal fluids. *In*: BARNES, H. L. (ed) *Geochemistry of Hydrothermal Ore Deposits*, 2nd ed, 137–68. Wiley, New York.

—— 1980. Dissolved methane in sedimentary brines: potential effect on the PVT properties of fluid inclusions. *Econ. Geol.* **75**, 603–17.

HARLAND, W. B., COX, A. V., LLEWELLYN, P. G., PICKTON, C. A. G., SMITH, A. G. & WALTERS, R. 1982. *A Geologic Time Scale.* Cambridge University Press. 131 pp.

HOLLAND, H. D., HOLLAND, H. J. & MUNOZ, J. L. 1964. The coprecipitation of cations with $CaCO_3$—II. The coprecipitation of Sr^{2+} with calcite between 90 and 100°C. *Geochim. Cosmochim. Acta*, **28**, 1287–302.

HOLLISTER, L. S. & CRAWFORD, M. L. (eds) 1981. *Fluid Inclusions: Applications to Petrology.* Short course handbook, Mineralogical Association of Canada. 304 pp.

HUDSON, J. D. 1975. Carbon isotopes and limestone cement. *Geology*, **3**, 19–22.

—— 1977. Stable isotopes and limestone lithification. *J. geol. Soc. London*, **133**, 637–60.

IRWIN, H., CURTIS, C. & COLEMAN, M. 1977. Isotopic

evidence for source of diagenetic carbonates formed during burial of organic-rich sediments. *Nature, Lond.* **269**, 209–13.

JONES, D. G. 1974. The Namurian Series in South Wales. *In*: OWEN, T. R. (ed.) *The Upper Palaeozoic and Post-Palaeozoic Rocks of Wales*, 117–32. University of Wales Press, Cardiff.

JOSHI, M. S. 1960. reported by GIESKES, J. M. 1978. Strontium 38-H. Reactions controlling strontium abundance in natural water. *In*: WEDEPOHL, K. M. (ed.) *Handbook of Geochemistry*, **II/4**, 38H-1–5. Springer-Verlag, New York.

KATZ, A., SASSE, E., STARINSKY, A. & HOLLAND, H. D. 1972. Strontium behaviour in the aragonite-calcite transformation: an experimental study at 40–98°C-. *Geochim. cosmochim. Acta*, **36**, 481–96.

KEELING, C. D. 1958. The concentration and isotopic abundance of atmospheric carbon dioxide in rural areas. *Geochim. cosmochim. Acta*, **13**, 322–34.

KINSMAN, D. J. J. 1969. Interpretation of Sr^{+2} concentrations in carbonate minerals and rocks. *J. sedim. Petrol.* **39**, 486–508.

KINSMAN, D. J. J. & HOLLAND, M. D. 1969. The coprecipitation of cations with $CaCO_3$-IV. The coprecipitation of Sr^{+2} with aragonite between 16° and 96°C. *Geochim. cosmochim. Acta*, **33**, 1–17.

KIROV, G. K. & FILIZOVA, L. 1970. Über die Möglichkeiten der diffusionverfahren bei der Kristallzuchtung (II). *Krist. Tech.* **5**, 387–407.

——, VESSELINOV, & CHERNEVA, Z. 1972. Conditions of formation of calcite crystals of tabular and acute rhombohedral habits. *Krist. Tech.* **7**, 497–509.

LAHANN, R. W. & SIEBERT, R. M. 1982. A kinetic model for distribution coefficients and application to Mg-calcite. *Geochim. cosmochim. Acta*, **46**, 229–37.

LAND, L. S. 1979. Chert-chalk diagenesis: the Miocene Island slope of North Jamaica. *J. sedim. Petrol.* **49**, 223–32.

—— 1980. The isotopic and trace element geochemistry of dolomite: the state of the art. *In*: ZENGER, D. H., DUNHAM, J. B. & ETHINGTON, R. L. (eds) *Concepts and Models of Dolomitization. Spec. Publs Soc. econ. Paleont. Miner., Tulsa*, **28**, 87–110.

LINDHOLM, R. C. 1972. Magnesium content and crystal habits in ancient carbonates. *Nature, Lond.* **237**, 43–4.

—— & DEAN, D. A. 1973. Ultra-thin sections in carbonate petrology: a valuable tool. *J. sedim. Petrol.* **43**, 295–7.

—— & FINKLEMAN, R. B. 1972. Calcite staining: semi-quantitative determination of ferrous iron. *J. sedim. Petrol.* **42**, 239–42.

LONGMAN, M. W. 1980. Carbonate diagenetic textures from near-surface diagenetic environments. *Bull. Am. Ass. Petrol. Geol.* **64**, 461–87.

——1982. *Carbonate diagenesis as a control on stratigraphic traps. Education Course Notes, Am. Ass. Petrol. Geol.* **21**, 159 pp.

LORENS, R. B. 1981. Sr, Ca, Mn and Co distribution coefficients in calcite as a function of calcite precipitation rate. *Geochim. cosmochim. Acta*, **45**, 553–61.

MARSHALL, J. D. 1981. Zoned calcites in Jurassic ammonite chambers: trace elements, isotopes and neomorphic origin. *Sedimentology*, **28**, 867–87.

—— 1982. Isotopic composition of displacive fibrous calcite veins: reversals in pore-water composition trends during burial diagenesis. *J. sedim. Petrol.* **52**, 615–30.

—— & ASHTON, M. 1980. Isotopic and trace element evidence for submarine lithification of hardgrounds in the Jurassic of eastern England. *Sedimentology*, **27**, 271–89.

MATTES, B. W. & MOUNTJOY, E. W. 1980. Burial dolomitization of the Upper Devonian Miette Buildup, Jasper National Park, Alberta. *In*: ZENGER, D. H., DUNHAM, J. B. & ETHINGTON, R. L. (eds) *Concepts and Models of Dolomitization. Spec. Publs Soc. econ. Paleont. Miner., Tulsa*, **28**, 259–97.

MEYERS, W. J. 1974. Carbonate cement stratigraphy of the Lake Valley Formation (Mississippian) Sacramento Mountains, New Mexico. *J. sedim. Petrol.* **44**, 837–61.

—— 1978. Carbonate cements: their regional distribution and interpretation in Mississippian limestones of southwestern New Mexico. *Sedimentology*, **25**, 371–400.

MILLIMAN, J. D. 1974. *Marine Carbonates*. Springer-Verlag, New York. 375 pp.

MOORE, C. H. 1979. Porosity in carbonate rock sequences. *In*: *Geology of carbonate porosity: Continuing Education Course Notes, Am. Ass. Petrol. Geol.* **11**, A1–124.

—— & DRUCKMAN, Y. 1981. Burial diagenesis and porosity evolution, Upper Jurassic Smackover, Arkansas and Louisiana. *Bull. Am. Ass. Petrol. Geol.* **65**, 597–628.

—— & BROCK, F. C. (Jr.) 1982. Porosity preservation in the Upper Smackover (Jurassic) carbonate grainstones, Walker Creek field, Arkansas: response of paleophreatic lenses to burial processes—discussion. *J. sedim. Petrol.* **52**, 19–23.

MORROW, D. W. 1982a. Dolomite—part I: the chemistry of dolomitization and dolomite precipitation. *Geosci. Can.* **9**, 5–13.

—— 1982b. Dolomite—part 2: the dolomitization models and ancient dolostones. *Geosci. Can.* **9**, 95–107.

—— & MAYERS, I. R. 1978. Simulation of limestone diagenesis—a model based on strontium depletion. *Can. J. Earth Sci.* **15**, 376–96.

MUCCI, A. & MORSE, J. W. 1983. The incorporation of Mg^{2+} and Sr^{2+} into calcite overgrowths: influence of growth rate and solution composition. *Geochim. cosmochim. Acta*, **47**, 217–33.

NAHNYBIDA, C., HUTCHEON, I. & KIRKER, J. 1982. Diagenesis of the Nisku Formation and the origin of late-stage cements. *Can. Miner.* **20**, 129–40.

NICKEL, E. 1978. The present status of cathode luminescence as a tool in sedimentology. *Min.*

Sci. Eng. **10**, 73–100.

PALACHE, C., BERMAN, H. & FRONDEL, C. 1951. *Dana's System of Mineralogy volume 2.* Wiley, New York. 1124 pp.

PIERSON, B. J. 1981. The control of cathodoluminescence in dolomite by iron and manganese. *Sedimentology,* **28**, 601–10.

PINGITORE, N. E. 1978. The behaviour of Zn^{2+} and Mn^{2+} during carbonate diagenesis: theory and applications. *J. sedim. Petrol.* **48**, 799–914.

—— (Jr.) 1980. The behvaiour of Zn^{2+} and Mn^{2+} during carbonate diagenesis: theory and application—reply. *J. sedim. Petrol.* **50**, 1010–4.

—— 1982. The role of diffusion during carbonate diagenesis. *J. sedim. Petrol.* **52**, 27–39.

REEDER, R. J. 1981. Electron optical investigation of sedimentary dolomites. *Contr. Miner. Petrol.* **76**, 148–57.

RICHTER, K. & ZINKERNAGEL, V. 1981. Zur Anwendung der Kathodolumineszenz in der Karbonatpetrographie. *Geol. Rdsch.* **70**, 1276–302.

RIGHTMIRE, C. T. & HANSHAW, B. B. 1971. Relationship between the carbon isotopic composition of soil CO_2 and dissolved carbonate species in groundwater. *Abstr. Trans. Am. Geophys. Un.* **52**, 366.

ROEDDER, E. 1963. Studies of fluid inclusions II: freezing data and their interpretation. *Econ. Geol.* **58**, 167–211.

—— 1972. The composition of fluid inclusions. *Prof. Pap. U. S. geol. Surv.* **440 JJ**, 164.

—— 1976. Fluid-inclusion evidence on the genesis of ores in sedimentary and volcanic rocks. *In*: WOLF, K. H. (ed.) *Handbook of Strata-bound and stratifirm Ore Deposits,* **4** (2), 67–110. Elsevier, Amsterdam.

—— 1979. Fluid inclusion evidence on the environments of sedimentary diagenesis, a review. *In*: SCHOLLE, P. A. & SCHUGER, P. R. (eds) *Aspects of Diagenesis. Spec. Publs Soc. econ. Paleont. Miner., Tulsa,* **26**, 89–107.

—— 1981. Origin of fluid inclusions and changes that occur after trapping. *In*: HOLLISTER, S. & CRAWFORD, M. L. (eds) *Fluid Inclusions: Applications to Petrology,* 101–37. Short course handbook, Mineral Association of Canada.

—— & BODNAR, R. J. 1980. Geologic Pressure determinations from fluid inclusion studies. *Ann. Rev. Earth planet. Sci.* **8**, 263–301.

—— & KOZLOWSKI, A. *Fluid Inclusion Research: Proceedings of C.O.F.F.I.* (Commission on Ore-Forming Fluids in Inclusions). University of Michigan Press, Ann Arbor. Eleven volumes, latest 1978.

ROSASCO, G. J. & ROEDDER, E. 1979. Application of a new Raman microprobe spectrometer to non-destructive analysis of sulphate and other ions in individual phases in fluid inclusions in minerals. *Geochim. cosmochim. Acta,* **43**, 1907–15.

SANDBERG, P. A. & POPP, B. N. 1981. Pennsylvanian aragonite from south-eastern Kansas—environmental and diagenetic implications. *Abstr. Bull. Am. Ass. Petrol. Geol.* **65**, 985.

—— 1983. Evaluation of ancient aragonite cements

and their temporal distribution. *Abstr. Am. Ass. Petrol. Geol.* **67**, 544.

SCHOLLE, P. A. & ARTHUR, M. A. 1980. Carbon isotopic fluctuations in Cretaceous pelagic limestones: potential stratigraphic and petroleum exploration tool. *Bull. Am. Ass. Petrol. Geol.* 67–87.

SCHWARCZ, H. P., HARMON, R. S., THOMPSON, P. & FORD, D. C. 1976. Stable isotope studies of fluid inclusions in speleothems and their paleoclimatic significance. *Geochim. cosmochim. Acta,* **40**, 657–65.

STUEBER, A. M. & PUSHKAR, P. 1983. Application of strontium isotopes to origin of Smackover brines and diagenetic phases, Southern Arkansas. *Abstr. Am. Ass. Petrol. Geol.* **67**, 553–4.

SUESS, E. 1970. Interaction of organic compounds with calcium carbonate I, association phenomena and geochemical implications. *Geochim. cosmochim. Acta,* **34**, 157–68.

THOMAS, L. P. 1974. The Westphalian (coal measures) in South Wales. *In*: OWEN, T. R. (ed.) *The Upper Palaeozoic and Post-Palaeozoic Rocks of Wales,* 133–60. University of Wales Press, Cardiff.

THOMPSON, P., SCHWARCZ, H. P. & FORD, D. C. 1976. Stable isotope geochemistry, geothermometry, and geochronology of speleothems from West Virginia. *Bull. geol. Soc. Am.* **87**, 1730–8.

TILLMAN, J. E. & BARNES, H. L. 1983. Deciphering fracturing and fluid migration histories in Northern Appalachian Basin. *Bull. Am. Ass. Petrol. Geol.* **67**, 692–705.

TUCKER, M. E. 1982. Precambrian dolomites: petrographic and isotopic evidence that they differ from Phanerozoic dolomites. *Geology,* **10**, 7–12.

TURNER, J. V. 1982. Kinetic fractionation of carbon-13 during calcium carbonate precipitation. *Geochim. cosmochim. Acta,* **46**, 1183–91.

USDOWSKI, E. 1973. Das geochemische Berhalten des Strontiums bei der Genese und Diagenese von Ca-Karbonat- und Ca-Sulfat-Mineralen. *Contr. Miner. Petrol.* **38**, 177–95.

——, MENSCHEL, G. & HOEFS, J. 1980. Some kinetic aspects of calcite precipitation. *Abstr. int. Ass. Sediment. 1st European Regional Meeting, Bochum,* 161–3.

VEIZER, J. 1978. Simulation of limestone diagenesis—a model based on strontium depletion: discussion. *Can. J. Earth Sci.* **15**, 1683–5.

—— 1983. Chemical diagenesis of carbonates: theory and application of trace element technique. *In*: *Stable Isotopes in Sedimentary Geology. Short Course Soc. econ. Paleont. Miner. Dallas, No. 10,* 3-1–100.

——, HOLSER, W. T. & WILGUS, C. K. 1980. Correlation of $^{13}C/^{12}C$ and $^{34}S/^{32}S$ secular variations. *Geochim. cosmochim. Acta,* **44**, 579–87.

VIDETICH, P. E. & MATTHEWS, R. K. 1980. Origin of discontinuity surfaces in limestones: isotopic and petrographic data, Pleistocene of Barbados, West Indies. *J. sedim. Petrol.* **50**, 971–80.

WAGNER, P. P. & MATTHEWS, R. K. 1982a. Porosity preservation in the Upper Smackover (Jurassic) carbonate grainstone, Walker Creek Field, Arkansas: response of paleophreatic lenses to burial processes. *J. sedim. Petrol.* **52**, 3–18.

—— & —— 1982b. Porosity preservation in the Upper Smackover (Jurassic) carbonate grainstone, Walker Creek Field, Arkansas: response of paleophreatic lenses to burial processes—reply. *J. sedim. Petrol.* **52**, 24–5.

WALLS, R. A., RAGLAND, P. C. & CRISP, E. L. 1977. Experimental and natural early diagenetic mobility of Sr and Mg in biogenic carbonates. *Geochim. cosmochim. Acta*, **41**, 1731–7.

WALTER, L. M. 1983. New data on relative stability of carbonate minerals: implications for diagenesis and cementation. *Abstr. Bull. Am. Ass. Petrol. Geol.* **67**, 566.

WAYNE, N. & BURRESS, R. C. 1982. Origin of Reservoir Fracture in Little Knife Field, North Dakota. *Abstr. Bull. Am. Ass. Petrol.* **66**, 611–2.

WHITLOCK, H.P. 1910. *Calcites of New York. Mem. N.Y. St. Mus. nat. Hist.* **13**, 190. New York State Education Department.

WIGLEY, P. 1973. The distribution of strontium in limestones on Barbados, West Indies. *Sedimentology*, **20**, 295–304.

WILKINSON, B. H., JANECKE, S. U. & BRETT, C. E. 1982. Low-magnesium calcite marine cement in middle Ordovician hardgrounds from Kirkfield, Ontario. *J. sedim. Petrol.* **52**, 47–57.

WONG, P. K. & OLDERSHAW, A. 1981. Burial cementation in the Devonian Kaybob Reef Complex, Alberta, Canada. *J. sedim. Petrol.* **51**, 507–20.

YOUNG, L. M. & JACKSON, D. H. 1981. Fluid-inclusion temperature study of Paleozoic carbonates, Llano Uplift, Texas. *Trans. Gulf-Cst Ass. Geol. Socs* **31**, 421–5.

ZENGER, D. H., DUNHAM, J. B. & ETHINGTON, R. L. 1980. Concepts and models of dolomitization. *Spec. Publs Soc. econ. Paleont. Miner., Tulsa*, **28**, 320 pp.

ZINKERNAGEL, V. 1978. *Cathodoluminescence of quartz and its application to sandstone petrology. Contr. Sedimentol. Stuttgart*, **8**, 69.

J. A. D. DICKSON, Department of Earth Sciences, University of Cambridge, Downing Street, Cambridge CB2 3EQ.

Clastic diagenesis

S. D. Burley, J. D. Kantorowicz & B. Waugh

SUMMARY: Diagenesis is the sum of those processes by which originally sedimentary clastic assemblages attempt to reach equilibrium with their environments. The subject has rapidly evolved, over the last 20 years, after several decades of routine petrographical analysis, to become a discipline of sophisticated analytical geochemistry. This progression may be traced through the evolution of depositional facies models, the growth of theoretical geochemistry, and the development and application of quantitative analytical techniques. Consequently, the study of diagenesis today involves the integration of data gathered from a range of interrelated disciplines.

The publication in recent years of such integrated studies, particularly from the Texas Gulf Coast of the U.S.A., has led to a number of wide ranging models being established. These models allow an interpretation of the sequence of authigenic minerals in terms of their relationship to the depositional environment or surface chemistry (eogenesis), burial or subsurface conditions (mesogenesis) and weathering or reexposure to surface conditions (telogenesis).

More specifically, once the pre-depositional controls on diagenesis have been established, it is possible to relate the inferred eogenetic mineral assemblages to exact geochemical sedimentary environments. Sedimentary mineral assemblages chemically are characterized by relative instability and so tend to interact with interstitial pore waters. Thus, in non-marine environments, mineral authigenesis may reflect arid-oxidizing, or wet-reducing pore-water conditions, and in the marine environment, either oxidizing or reducing pore waters.

During mesogenesis different processes become important. Elevated temperatures add energy to the reacting system, lowering reaction barriers and increasing reaction rates. Furthermore, widespread pore fluid migration at depth, transporting large quantities of solute, is likely to impart major, regional diagenetic changes to sediments. That there is a remarkable degree of consistency in deep burial settings suggests that depth-related processes conform to a predictable pattern.

Additional geochemical instability is introduced into the diagenetic system during uplift and exposure to the telogenetic realm. Minerals formed during burial, at elevated temperatures and pressures, and from concentrated formation waters, may become unstable in oxidizing, meteoric waters.

Diagenetic research requires the complete dissection of sedimentary rocks and, subsequently, the quantitative chemical and mineralogical analysis of their individual components. Such an approach to diagenetic studies, when related to an assessment of their paragenesis, may eventually lead to a predictive, integrated model for the evolution of sedimentary clastic rocks.

The purpose of this review is to highlight only the more significant research of the previous 20 years which we feel has advanced diagenesis from a prolonged era of Sorbian petrographical description into a period of multidisciplinary analytical geochemistry. These advances may be discussed in several distinct categories, namely the evolution of depositional facies models, the development and application of sophisticated analytical techniques, and the generation of theoretical geochemical models. Also important in providing an impetus to the subject has been the realization by some oil companies that an understanding of diagenesis may ultimately enhance recovery from hydrocarbon reservoirs. Additionally, we describe and discuss the results of recent studies which illustrate the approach which we feel may further advance our understanding of clastic diagenesis.

In spite of the number of books and papers written on the subject in the last 20 years, there is no universally accepted definition of diagenesis. Authors vary in restricting diagenesis to processes following deposition, or to processes prior to lithification. However, a case can also be made for extending diagenesis to include weathering, and the processes involved in altering rocks exposed at the earth's surface. We favour the latter approach, and would, therefore, describe the modifications made to rocks prior to reburial beneath unconformities as well as their complete destruction during continued weathering as diagenetic.

In addition to being variously defined in an absolute sense, diagenesis is also variously defined in a relative sense—i.e. with respect to metamorphism. Heroux *et al.* (1979) provided an up-to-date appraisal of the indices which may be used to separate metamorphism from diagenesis. Wisely however, they like many other authors, and indeed ourselves, do not attempt a rigorous distinction between the two. Burial

189

diagenesis and metamorphism reflect a continuum of geological processes within which there is no single criterion for distinguishing between the two that can be used for all sediments. There is, therefore, no consensus as to what diagenesis is, when it starts, or when it stops. We do not propose a rigid definition of the subject because considerable advances can been made by the application of techniques and theories from loosely interrelated disciplines. Consequently, it is preferable perhaps to examine all aspects of all sediments, rather than to prejudice one's understanding by pre-selecting those areas on which to concentrate study.

In prefacing a review of clastic diagenesis it is, however, important to point out how vaguely defined the subject is. This is because clastic diagenesis does not lend itself to a straightforward 'State of the Art' review. Numerous subdivisions and classifications have been proposed but none adopted universally. Conflicting terminology abounds, different methods of routine investigation are pursued by those groups with a different approach, and finally, whilst the identification of authigenic minerals and diagenetic modification is now universally accepted, it is probably fair to say that we have made little progress towards understanding exactly how these processes take place (cf. Bjørlykke 1979; Blatt 1979; Boles & Franks 1979b; Land & Dutton 1979; Wood & Hewett 1982; Haszeldine *et al.* 1984).

The terminology of Schmidt & McDonald (1979a), subdividing diagenesis into eogenesis, mesogenesis, and telogenesis, is used in this review. Although this terminology is a genetic scheme and is, therefore, strictly an interpretative classification, it is preferred to the schemes which merely divide diagenetic processes into early and late stages. These often make little reference to the role of surface chemistry in diagenesis and therefore are not readily applicable to studies designed to elucidate the relationship between diagenesis and the depositional pore-water chemistry. Also, other classifications often exclude weathering by making reference to 'after deposition' or 'during burial', and we would not exclude weathering so readily.

The plates accompanying this review are intended to exhibit the range of textures and diagenetic products described in the text. Plates have also been chosen to demonstrate the application of specific techniques to the resolution of diagenetic problems. This selection is not intended to be exhaustive, it is hoped that the references herein may satisfy readers who wish further to investigate particular aspects.

Towards integrated modelling

Clastic diagenesis is low-temperature geochemistry; it is everything that contributes to making up a clastic rock, from the weathering of its source until its metamorphism. We are aware of many excellent studies on isolated aspects of a clastic sedimentary rock, but too few integrated or holistic ones. In this review we intend highlighting and promoting an integrated approach to the study of clastic diagenesis. Consequently, before briefly describing historically how the subject has developed, it is necessary to outline the philosophy and the style of investigation which is currently advancing our understanding of diagenesis.

The integrated approach involves initially establishing *all* the conditions which influenced the composition of the original clastic sedimentary assemblage, as well as its depositional pore waters. Secondly, by 'taking the rock apart', determining whether the constituents are detrital or authigenic and in what order the latter were formed. Thirdly, evaluating the entire burial history of both the sediment itself as well as any other sediments within the basin that are likely to influence pore-water composition in the subsurface. Finally, using the information above, combined with existing geochemical models, to explain how all the constituents were formed and, if possible, when. With the resulting information it may be possible to establish whether specific diagenetic modifications are related to any particular aspect of the history of the sediment, such as a distinct source area (see Stalder 1975), the depositional environment, burial temperatures and pressures, or hydrocarbon migration.

Historically, clastic rocks were originally investigated by mineralogists and petrologists. Only more recently, however, have sufficient data become available in related fields to allow some of the diagenetic processes involved to be understood more fully. These data result from research into topics as diverse as recent oceanic sediments, soils and aquifer properties, incorporating investigations of biological, chemical and hydrological processes. Results from many of these studies are included in the following sections, either specifically or within the more general reviews referenced. Unfortunately, much of this research is outside the scope of this paper and as a consequence can be mentioned only in passing.

The DSDP projects have produced a wealth of data concerning the mineralogy, biology and chemistry of surficial deep-sea sediments (see Warme *et al.* 1981, and references therein). Similarly, Berner and his co-workers in the

FOAM Group have studied shallow marine sediments intensively in recent years (see Krom & Berner 1981). The general results of these and other studies were concisely reviewed by Curtis (1978, 1980). Recently the work of hydrologists has become significant in view of our poor understanding of how material is transferred, and how large volumes of cement are produced in the subsurface. Their results suggest simple models of fluid flow through sediments may not be applicable (Wood & Hewett 1982; see also Domenico 1977 for a review). In addition the considerable volume of data collected by soil scientists in recent decades is now being utilized by sedimentologists searching for suitable analogues for their observations of the ancient sedimentary record. Dixon & Weed (1977) and Fitzpatrick (1980) exemplify pedology *per se*, whilst the investigations by Nahon, Tardy and their co-workers illustrate its possible application in a geological context (see Tardy 1971; Gac & Tardy 1980; Nahon *et al.* 1980). Wilson (1983) illustrated how study of a range of sediments may significantly contribute to analysis of the sedimentary record.

The early days

Petrography, the description of rocks, began in the 1800s with the work of Sorby (see Folk 1965). Early work though, was often mainly descriptive and interpretation of diagenetic processes did not really begin until the 1960s. Thus, by the time quantitative geochemical explanations began to appear in the literature, features as diverse as quartz overgrowths and berthierine ooliths had been recorded for up to a century (Phillips 1881).

Although our understanding of diagenesis is mostly the result of very recent advances, some notable contributions to the subject were made on the basis of detailed petrography alone. Some of the descriptions and interpretations have been superseded with time although they are significant even today in having influenced several decades of interpretation. For example, Lerbekmo (1957) described authigenic montmorillonite from andesitic Tertiary sandstones in California. Its formation was attributed to dissolution of the andesitic detritus, irrespective of the depositional environment. By contrast Carrigy & Mellon (1964) and Shelton (1964) described authigenic clays in a variety of occurrences. They concluded that there was a relationship between the clay mineralogy and depositional environment.

Other workers, however, biased future thinking towards the subsequently controversial topic of 'pressure solution'. Both Waldschmidt

(1941) and Taylor (1950) attributed the loss of porosity with depth in sandstones to solution and stylolitization. The textures they described were subsequently observed to be between overgrowths, thus demanding both a reconsideration of accepted depth/porosity gradients, and more significantly, a reevaluation of the problematical source of silica in diagenesis (see Blatt 1979).

The advent of sedimentary geochemistry

Diagenesis can be resolved into a number of distinct thermodynamic processes. The classic work of Garrels & Christ (1965) showed how mineral thermodynamics can be applied to sedimentary systems and its importance for understanding diagenetic processes. Although there are many shortcomings in the application of thermodynamics and kinetics to diagenetic studies, careful use of some of the concepts is an invaluable aid to defining and understanding some of the processes that take place during diagenesis (Berner 1971; Curtis 1978; Hutcheon 1981).

The basis for diagenetic reactions is conveniently summarized by the 'law' of mineral stability (Keller 1969), which simply states that 'minerals are in equilibrium with their surroundings only in the environment in which they form'. This fundamental principle seems to have been generally underrated by sedimentologists in the past although it has been appreciated in soil science for many years. Goldich (1938) observed that the persistence of common rock-forming minerals during weathering was generally the reverse of the sequence of minerals crystallizing as Bowen's reaction series during magmatic fractionation. Minerals stable at the highest temperatures and pressures were apparently least stable under surface conditions and rapidly underwent alteration. This relationship was considered to result from the relative cation–oxygen bond strengths in the various minerals (Keller 1954). However, Curtis (1976) demonstrated that the persistence of a particular mineral at surface temperature and pressure conditions is largely a function of total released energy liberated during breakdown into new mineral products. Providing there is a net free energy loss in a diagenetic reaction then it should proceed in the diagenetic environment. Furthermore, the greater the free energy loss, the greater the tendency to react, and hence the more unstable the mineral. Thus, not only is mineral stability a function of the diagenetic environment, but also of the rate at which the minerals react.

In diagenetic studies this general concept has widespread applications. On deposition an original detrital mineral assemblage comprises a mixture of minerals formed under widely differing temperature–pressure regimes, usually higher than those prevailing at the earth's surface. Therefore, in the sedimentary environment detrital minerals are often either metastable or unstable with respect to their pore waters. Typically, the detrital assemblage is unstable as a whole and so during diagenesis may react with the ambient environment and move towards equilibrium. Kinetic models, however, indicate

that despite favourable free energies, reactions need not proceed spontaneously because of an energy barrier which has to be overcome in the course of the reaction, so that it is possible for phases to persist metastably. Geological reaction barriers arise from several causes which may be physical, biological or chemical in origin (Berner 1980). In essence, the dominance of metastability during early diagenesis is a direct consequence of the slowness of chemical reaction rates at low temperatures (Berner 1971; Curtis 1978;

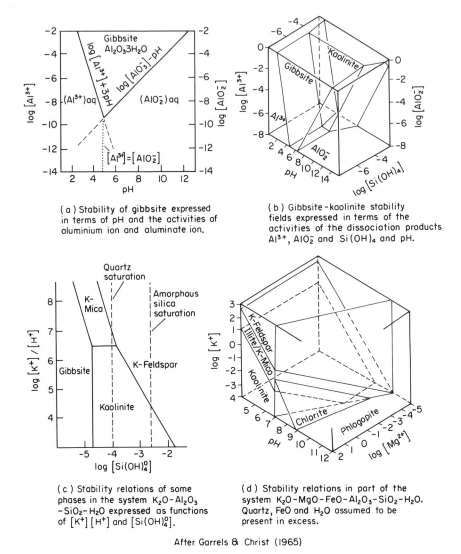

(a) Stability of gibbsite expressed in terms of pH and the activities of aluminium ion and aluminate ion.

(b) Gibbsite–kaolinite stability fields expressed in terms of the activities of the dissociation products Al^{3+}, AlO_2^- and $Si(OH)_4$ and pH.

(c) Stability relations of some phases in the system $K_2O-Al_2O_3-SiO_2-H_2O$ expressed as functions of $[K^+][H^+]$ and $[Si(OH)_4^0]$.

(d) Stability relations in part of the system $K_2O-MgO-FeO-Al_2O_3-SiO_2-H_2O$. Quartz, FeO and H_2O assumed to be present in excess.

After Garrels & Christ (1965)

FIG. 1. The development of complex stability diagrams from a simple, two-dimensional, three-component system to multicomponent, three-dimensional diagrams (after Garrels & Christ 1965). (c) is a particularly useful plot because the ratio of $(K^+)/(H^+)$ in the hydrolysis reactions which define the stability fields is unity so that the complex stability relations can be expressed in terms of a two-dimensional plot.

Lippman 1982). Thus, despite complex interaction with pore waters, it is unlikely that true thermodynamic equilibrium is ever reached during early diagenesis at near-surface temperatures and pressures (Curtis 1978), although Hutcheon (1981) suggests that authigenic minerals may reach equilibrium with their pore waters during the later, deep burial stages of diagenesis.

Useful qualitative constraints can be placed on the chemical behaviour of species in solution by plotting the stability fields of minerals as a function of the activities of the relevant ions dissolved in solution. Such activity diagrams, largely developed by Garrels & Christ (1965) and expanded by Helgeson *et al.* (1969), serve to illustrate the relationships of the various minerals in equilibrium with the pore waters that bathe them, and provide a graphic summary of the mineral sequences that might be expected to precipitate if equilibrium were attained. The stability of a mineral in aqueous solution depends upon the temperature, the total pressure and the concentration of the chemical components in the co-existing aqueous solution. Not only are stability field boundaries temperature/pressure dependent, but the limiting hydrolysis reactions depend on free energy values, many of which are poorly known. Furthermore, most theoretical thermodynamic calculations also assume that equilibrium states are approached or attained in a closed system. Activity coefficients have been determined experimentally in very dilute aqueous solutions and essentially at low, near-surface pressures. The application of theoretical thermodynamic models to open geo-logical systems at elevated temperatures and pressures in concentrated basinal oilfield brines with a dynamic flux of interstitial pore waters therefore requires extreme caution and an appreciation of the shortcomings. Despite the potential complexity, and the present lack of detailed understanding of high salinity, temperature and pressure thermodynamics, general stability relations and trends can be conveniently expressed using activity diagrams, providing they are used within their limitations. Figure 1 illustrates the development of multicomponent activity diagrams, whilst Figs 2 and 3, provide specific examples of the usefulness of such diagrams.

There are now many published studies which use thermodynamic stability models to explain diagenetic assemblages. Almon *et al.* (1976) used diagenetic cement assemblages to reconstruct the effective pore water composition which prevailed during cementation of Cretaceous sandstones from Montana. Similarly, Merino (1975) calculated the distribution of aqueous species in interstitial pore waters of the Eocene McAdams Sandstone of California and demonstrated, using activity diagrams, that the interstitial pore waters are in equilibrium with the diagenetic minerals. Also, Nesbitt (1980) showed that deep brines in the Carboniferous Illinois basin have equilibrated with the diagenetic mineralogy. Stoessell & Moore (1983) investigated the relationship between mineral diagenesis and reservoir fluid compositions in several Texas Gulf Coast reservoirs where the diagenetic sequence was already relatively well

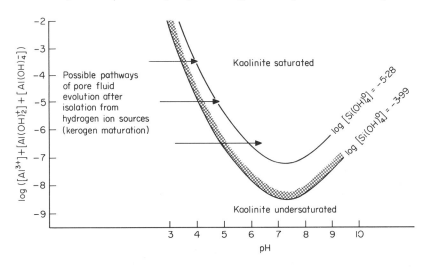

FIG. 2. The stability of kaolinite relative to the solubility of silicon and aluminium species as a function of pH (from Curtis 1983a). Pore fluids containing aluminium species must precipitate kaolinite if the pH of the solution rises.

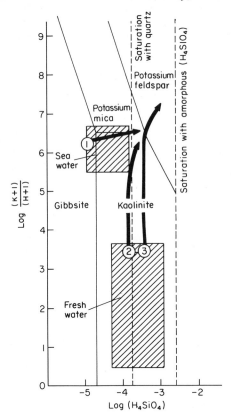

FIG. 3. Practical use of the stability diagram shown in Fig. 1c, showing the stability fields of kaolinite, illite and potassium feldspar together with approximate fields of marine pore waters and meteoric waters. Pathway 1 illustrates the evolution of marine pore waters through reaction with detrital minerals. Pathways 2 and 3 show equilibration of more acidic freshwater with detrital minerals. Pathway 3 involves more quartz precipitation.

known (see 'mesogenetic realm'). Using the thermodynamic data base of Helgeson together with several assumptions regarding stabilities of authigenic minerals they were able to propose a realistic model for the origin of present-day Gulf Coast reservoir fluids.

By contrast Curtis (1983a) used thermodynamic stability fields and kinetics to explain petrographical observations. The association of late-stage secondary porosity development and pore-filling kaolinite during burial diagenesis has been noted by several authors (Loucks *et al.* 1977; Lindquist 1977; Land & Dutton 1978; Schmidt & McDonald 1979a, b) and is readily explained in Fig. 2. Acidic pore waters generated at depth through thermal decarboxylation of

organic matter are capable of dissolving and transporting large quantities of metal solutes, including aluminium. If separated from the source of acidity during a compactional drive, the pH of these acid solutions must rise as carbonate dissolution occurs. As the solubility of kaolinite decreases rapidly with increasing pH, dissolution of calcite should cause the precipitation of kaolinite.

The scanning electron microscope can be used as a qualitative guide to determining geochemical precipitation and dissolution processes. Precipitation of a solid substance from an aqueous solution can be considered as two basic processes, nucleation and crystal growth (Berner 1980). Both the rate of crystal nucleation and crystal growth are a function of the degree of supersaturation of the solution. At low levels of supersaturation, crystal nucleation is very slow and excess dissolved material is consumed by crystal growth on a limited number of nuclei. This results in a high degree of crystallinity of the authigenic phase, and will also tend to favour growth on existing mineral phases as syntaxial overgrowths. In contrast, with high solute concentrations and a high level of supersaturation, the rate of nucleation may be so fast that virtually all the excess solute is rapidly precipitated as nuclei (in order of 10–100 Å), effectively decreasing the saturation so that little solute remains for crystal growth. The results will tend to be very poorly crystallized, often fine-grained precipitates, with a high degree of disorder.

The actual processes of dissolution of solids by an aqueous solution involves the removal of ionic species by either diffusion transport or by surface reaction. Observation of partially dissolved crystals with the scanning electron microscope helps to elucidate the dissolution rate-controlling mechanism. Slow partial dissolution of mineral phases by surface reaction will result in the formation of crystallographically controlled features such as etch pits and contrasts with non-specific corrosion resulting from more rapid transport-controlled dissolution.

Energetically, the precipitation of overgrowths can also be considered as an attempt by detrital grains to become more stable and crystal-like (Hurst & Stout 1980). Rounded or angular grains are crystallographically anhedral and characterized by high surface area to mass ratios. During diagenesis the energetic response will tend to lower both the surface free energy and surface area by precipitation of stable authigenic overgrowths, so that the detrital grain approaches the low surface free-energy state of a euhedral crystal. Similarly, dissolution is influenced by the surface free-energy properties of grain and crystal surfaces (Hurst 1981). Low-index crystal faces possess low surface free energy and are least likely to suffer corrosion or dissolution, whilst high-index crystal faces, face edges and corners, abraded detrital surfaces and fractures all have high surface free energies and are therefore susceptible to dissolution and corrosion.

The above account of sedimentary geochemistry is out of necessity, brief and simplified. To summarize, diagenesis from a geochemical viewpoint may be considered as the sum of those processes by which originally sedimentary assemblages attempt to reach equilibrium with their environment, although it is unlikely that sediments ever reach equilibrium in stages with successively changing pore-water chemistry, temperature and pressure. Diagenetic processes are controlled by the activities of dissolved ions, Eh and pH, temperature, pressure and interaction with organic systems. Not only does the original sedimentary assemblage play an active role during diagenesis but so does the interstitial fluid phase. The reaction products of diagenetic processes, released into solution may fundamentally change the pore-water composition. A detrital assemblage moving towards equilibrium with its interstitial pore-water environment may therefore become unstable and undergo further dissolution and precipitation reactions as a result of either *in situ* pore-fluid evolution, or after the introduction of formation waters. Thus, it is important to remember that diagenetic processes take place in a dynamic, evolving aqueous system, so that the availability of water within the sediment, and the ability to move water through the sediment, are as important in diagenesis as the actual pore-fluid composition.

Depositional and pre-depositional influences

Many of the variables which affect diagenesis exert an influence on the sedimentary assemblage before and during its accumulation. The complexity of isolating a single factor is discussed by Hayes (1979). Not only can the composition, texture and geometry of the original sedimentary deposit be expected to exert a controlling influence but also the gross physical and chemical environment. The depositional environment is likely, therefore, to be an important factor. As a review of this topic is to be found elsewhere in this volume we do not discuss it here in any detail. Needless to say, in order to suggest likely depositional pore-water chemistry, as well as any likely local variations in ground-water conditions, it is necessary to estab-

lish, in considerable detail, the probable depositional environment. Consequently, and especially in the field of core studies rather than outcrop-based projects little progress was made until depositional-facies models were detailed during the last 20 years (see Reading 1978; Walker 1979; and Anderton, this volume).

The depositional environment is in turn a function of other variables. Climate may play a significant role in affecting the degree of surficial weathering in a particular environment, which will determine the stability of minerals and hence the chemistry of the pore waters. Detrital mineralogy can also be affected by weathering in the source area, transport distance and provenance. These in turn may be affected by the gross tectonic setting (Siever 1979).

The influence of such variables unfortunately does not end with the eogenetic pore-water system. A specific depositional environment will determine the sediment texture, sand to mud ratio, and overall geometry. Hence, porosity-permeability characteristics, and interconnectivity of sand bodies in the subsurface may all be related to the original depositional system. Furthermore, the style and rate of burial will also be related to the sedimentation rate and especially the tectonic setting. Clearly both the tectonics and sedimentology of the depositional basin affect diagenesis during and after burial, and an appreciation of the original setting is important to any diagenetic investigation.

Techniques

Petrography (see Figs 4–7)

Thin-section studies, employing optical microscopy, still form the basis for understanding textural relationships between detrital and authigenic minerals. Furthermore, quantities derived from modal analyses provide a basis for sandstone classification and modelling of the relationship between detrital and authigenic components and porosity (see Pettijohn *et al.* 1973). Mineral identification may be refined by use of suitable staining techniques, particularly for carbonate cements (Dickson 1965) where optical properties of individual minerals are very similar (Fig. 6c), and for feldspars (Bailey & Stevens 1960) where distinguishing potassium feldspar from plagioclase is often difficult. Recently petrography has been supplemented by more sophisticated analytical techniques such as XRD and cathodoluminesence.

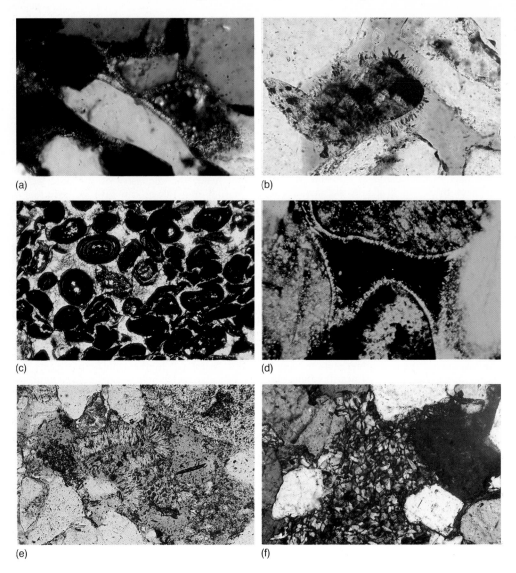

FIG. 4. Colour photomicrographs of authigenic clays. (a) Grain-coating chlorite vermiculite. Clay
inhibits overgrowths and allows preservation of intergranular porosity. Vermiculite in this
case is inferred to have formed during recent weathering of chlorite. ($\times$70.) (b) Rim of radial
authigenic smectite around an altered detrital volcanic grain. Adjacent quartz grains in contrast are
coated with mechanically infiltrated haematite-stained clay and quartz overgrowths strongly
suggesting the ions necessary for the formation of the smectite were provided by the host grain.
($\times$180.) (c) Berthierine ooliths, partially oxidized in places, and cemented with siderite. ($\times$18.)
(d) Radial illite. Thick coating of radial illite growing on a variety of host grains but inhibiting
subsequent overgrowths. ($\times$180.) (e) Large stacked crystal aggregate of authigenic vermiform
kaolinite within an oversized pore probably a result of grain dissolution. The growth of such large
kaolinite aggregates may be related to a flux of meteoric water and may be correspondingly rapid (see
Fig. 8e). ($\times$70.) (f) Hugely oversized pore filled with blocky dickite aggregates. Note the corroded
nature of quartz overgrowths suggesting that dickite postdates the removal of a carbonate cement (see
Fig. 8f). ($\times$70.)

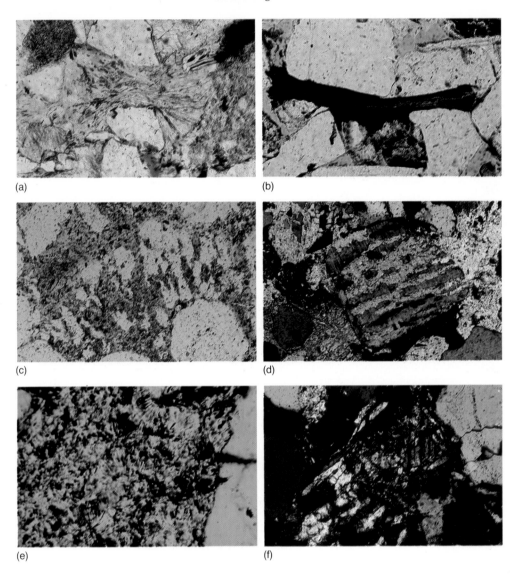

(a)

(b)

(c)

(d)

(e)

(f)

FIG. 5. Colour photomicrographs of replacement textures. (a) Mechanical and chemical alteration of a detrital muscovite grain to kaolinite. The intense basal cleavage is accompanied by extensive potassium loss clearly demonstrable by microprobe analysis. (× 200.) (b) Similar potassium loss in biotite under oxidizing conditions also results in the oxidation of octahedrally co-ordinated ferrous iron in the biotite lattice. The iron precipitates as haematite on the altering biotite which presumably acts as a nucleus. (× 180.) (c) Feldspar appears to be unstable in a variety of diagenetic environments and the alteration products depend on several factors including the type of feldspar and the ambient geochemical environment. Here a potassium feldspar grain has been intensely replaced by a pervasive authigenic kaolinite cement. (× 70.) (d) Solutions which precipitate carbonate cements are commonly highly corrosive to framework silicates, and feldspar is particularly susceptible to such alteration. This iron-rich ankerite cement has extensively corroded detrital quartz grains (left top, right top) detrital feldspar (centre) and quartz overgrowths (right top). (× 70.) (e) Earlier formed clay minerals can become unstable subsequently. Kaolinite is being replaced by strongly birefringent illite. (× 70.) (f) Authigenic lath-shaped crystals of barytes associated with and replacing an altered detrital plagioclase feldspar grain containing barium. The association suggests that whilst sulphate for the authigenic barytes was probably derived from the enclosing interstial pore waters at least some barium was provided by the altering grain which acted as a nucleus for barytes growth. (× 70.)

FIG. 6. Colour photomicrographs of carbonate and sulphate cements. (a) Siderite spheruliths, exhibiting cruciform extinction being comprised of elongate crystals of siderite radiating from an amorphous nucleus. Siderite spheruliths occur commonly in suboxic soil profiles and grow displacing and replacing surrounding clay particles. (×35.) (b) Nodular, non-ferroan calcite cement associated with calcrete development in coarse-grained alluvial sandstones. Note the displacive growth textures (sand grains concentrated outside the nodule) and the intensive grain replacement and corrosion. (×18.) (c) Complex sequential carbonate cementation in a sandstone, highlighted by staining techniques. Initial dolomite rhombs (unstained carbonate left centre, cloudy core) are enclosed by non-ferroan calcite cement (pink stain) and later ferroan calcite (blue stain). (×70.) (d) Well-developed ankerite (stained blue, identified by microprobe analysis) partially cementing late stage mesogenetic secondary porosity in hydrocarbon reservoir. (×70.) (e) Dolomite cement associated with and replacing a highly altered and subsequently oxidized ferromagnesian grain suggesting possible source of magnesium ions from altered grain. Dolomite cement was probably originally zoned (with calcite ?) and more soluble zones have been leached. (×70.) (f) Pore filling, brightly birefringent poikilotopic anhydrite cement associated with either environmentally controlled early diagenesis in arid sediments or as a late-stage cement precipitated from brines derived from dissolution of evaporite minerals. (×180.)

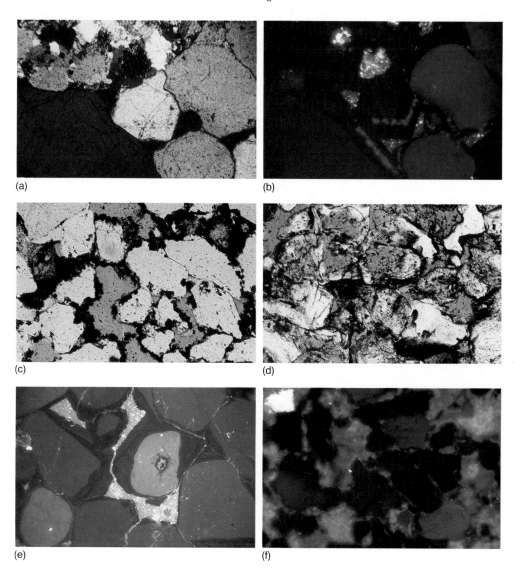

FIG. 7. Colour photomicrographs of authigenic cements and porosity. (a) and (b) Crossed polars and cathodoluminescence pair illustrating complex authigenic mineral precipitation sequence not normally visible using only optical microscopy. A thin black radial chlorite surrounds the rounded detrital grain. Two phases of quartz overgrowth cementation are separated by a phase of kaolinite precipitation (blue luminescence). (×115.) (c) Extensive intergranular and enlarged intergranular secondary porosity created by the dissolution of a replacive siderite cement. The siderite is now partially oxidised to goethite. Compare the quartz grain and overgrowth surfaces where replacement occurs, with pore walls enclosing subsequently precipitated dickite (see (f) and Fig. 4f). (×70.) (d) Intense dissolution of detrital feldspar grains producing extensive mesogenetic secondary porosity. (×70.) (e) Polyphase quartz overgrowth cementation in a quartz-rich sandstone. Note also dissolution at grain contacts. (×70.) (f) Extensive pore-filling dickite (bright blue) within occasionally enlarged intergranular porosity. Note the variety of quartz grain colours suggesting a complex source area, or sedimentary recycling. Note also the irregular nature of overgrowth surfaces (black). (×70.)

X-ray diffraction (XRD)

The introduction of this technique enabled the structural identification of minerals previously only identified on the basis of their thin-section properties—see Brindley (1951) for a contemporary review and Brindley & Brown (1980) for an up-to-date consensus. X-ray diffraction (XRD) is now widely applied to both the analysis of mudrocks and sandstones. Analysis of the powdered whole rock provides much information on the main sedimentary components of fine-grained clastics, although it is limited in its usefulness in sandstones without supportive techniques (see Almon & Davies 1978). Of much more value in diagenetic studies is XRD analysis of the separated fine fraction from both mudrocks and sandstones for determining detailed clay mineralogy. Consider the difficulty in distinguishing the range of clays exhibited in the figures on the basis of their morphology alone. The application of techniques such as XRD to diagenetic problems is well exemplified by the studies of depth-related mudrock mineralogy from the Texas Gulf Coast (Burst 1959, 1969; Perry & Hower 1970; Hower et al. 1976).

Clay-mineral studies documenting similar depth-related transformations are now recorded from other major sedimentary basins—e.g. Canadian Arctic (Foscolos & Powell 1979a, b); and North Sea (Pearson et al. 1982). X-ray diffraction has also been utilized to identify authigenic clays following their widespread recognition in sandstones (see Lerbekmo 1957; Shelton 1964; Carrigy & Mellon 1964). More recently, clay fraction XRD has supported studies distinguishing detrital and authigenic clays (see Wilson & Pittman 1977), elucidating complex clay mineral reactions (Boles & Franks 1979a) and highlighting specific clay mineral details (Almon et al. 1976; Guven et al. 1980, McHardy et al. 1982).

Scanning electron microscopy (SEM)

The SEM has revolutionized studies of sandstone diagenesis and is widely used to enhance two-dimensional (2-D) thin-section observations. The extreme depth of focus and wide range of magnification (with a resolution some 500 times greater than that of the optical microscope) make the SEM ideal for elucidating the three-dimensional (3-D) morphology of detrital and authigenic minerals and the nature of cement-porosity relationships (see Figs 8–10).

Early applications of the SEM in diagenetic studies were largely of pore throat geometry (Weinbrandt & Fatt 1969) and quartz overgrowth morphology (Waugh 1970; Pittman 1972). With the realization that authigenic clays

were widespread in sandstones, the SEM has been used extensively to characterize clays detected by XRD (Borst & Keller 1969; Wilson & Pittman 1977). Extreme caution, however, needs to be taken concerning the identification of clay minerals. A wide range of morphologies and textures may be exhibited by clay minerals of the same structural group (Keller 1976; Wilson & Pittman 1977; Tompkins 1981) and cannot therefore be identified on the basis of SEM alone. This situation has been partially resolved with the development of solid-state energy dispersive X-ray analysers (EDS) which can be attached to SEM and provide qualitative chemical identification (e.g. Guven et al. 1980). Back-scatter SEM analysis and X-ray image mapping have enabled further refinements of the distribution of elements in sandstones to be made. Critical-point drying should dramatically affect understanding of the role of clays in reservoir behaviour (McHardy et al. 1982).

Electron-microprobe analysis

Problems inherent in the SEM/EDS technique such as the takeoff angle or 3-D geometry, can be resolved largely by examining either polished thin sections or polished blocks with a perfectly flat surface, thereby overcoming absorption of X-rays, etc. This has been fully utilized in the electron microprobe (EM) which allows quantitative chemical analyses to be made of individual minerals using a fully focussed 10–0.5 μm electron beam. The EM is particularly well suited to the analysis of mineral transformation (Velde 1977; Walker et al. 1978; Boles 1981). or of zonation in cementing minerals (Keighin & Fouch 1981).

The EM has also been of particular significance in identifying specific carbonate cements such as ankerite (Boles 1978) which would have been difficult to characterize structurally (Irwin 1981). Microprobe analyses provide the only way of characterizing many phases individually (Velde 1977; Merino & Ransom 1982) although problems of mechanical mixtures may arise because of the resolution of the optical system (Curtis et al. 1984). However, this may be resolved by use of more sophisticated high-resolution electron microscopes (Huggett 1984). The electron microscope is of most value in determining the chemistry of authigenic minerals such as feldspars which can be readily optically resolved (Waugh 1978; Ali & Turner 1982).

Stable isotope analysis

It is now well known that stable isotope analysis can provide a wealth of information on

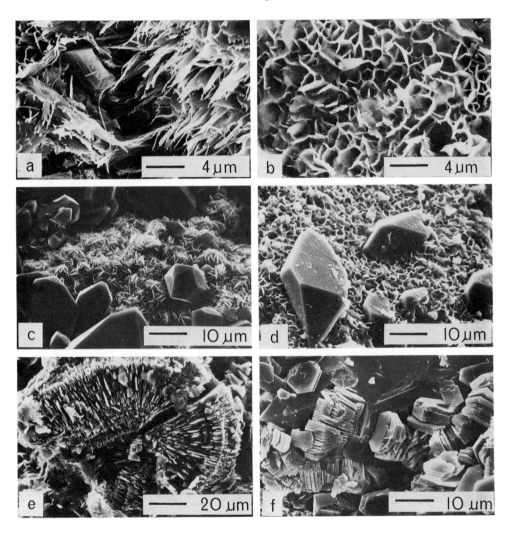

FIG. 8. Scanning electron micrographs of authigenic clays (all clays identified from $2\mu m$ clay fraction XRD analysis). (a) Elongate, lath-shaped, radially orientated illitic clay with a small expandable component (10%) and low crystallinity index. (b) Well-developed boxwork texture of ordered illite-smectite mixed layer clay, typical of eogenetic diagenesis in hot, dry red beds. (c) Iron-rich chlorite crystals (14 Å, chamosite *s.s.*) coating detrital grains and partially inhibiting development of quartz overgrowths. (d) Mixed-layer chlorite-smectite clay forming a pervasive grain coating and also inhibiting quartz overgrowth development. Note how the clay mineral morphology resembles the illite-smectite mixed-layer clay. (e) Large crystal aggregate of vermiform kaolinite. The vermiform aggregate is composed of many small stacked crystals of kaolinite. (f) Euhedral crystals of dickite with a high crystallinity partially enclosed by late quartz overgrowth development. Note the thickness of individual crystals compared with their diameter.

cementing minerals in sedimentary rocks (Hudson 1977). The majority of studies however have investigated carbonate rather than clastic sediments. Nonetheless, analysis of carbon, oxygen, and sulphur isotopes has refined our understanding of clastic diagenesis significantly in certain specific fields.

Principally, carbon- and oxygen-isotope analysis is employed to investigate the conditions

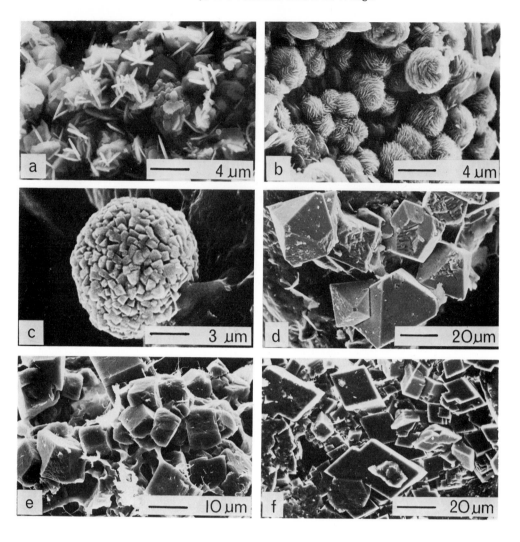

FIG. 9. Scanning electron micrographs of authigenic cements. (a) Specularite crystals of authigenic haematite mixed with authigenic clay. Such authigenic haematite is characteristically produced during eogenesis of hot, dry red beds. (b) Rosettes of highly crystalline haematite, which extensively cements a Triassic Sandstone. What controls the difference in haematite growth forms between (a) and (b)? (c) Framboid of authigenic pyrite crystals developed in a Carboniferous mudstone. (d) Euhedral octahedra of mesogenetic pyrite from a gas reservoir. Is the difference in growth form a reliable indicator of geochemical environment? (e) Non-ferroan dolomite crystals associated with illite-smectite authigenic clay interpreted to be an early pedogenic development. (f) Highly crystalline, isotopically light, calcian ankerite with well developed crystal faces. Compared with Fig. 5a is the difference in crystal form and development a function of chemistry or rate of growth?

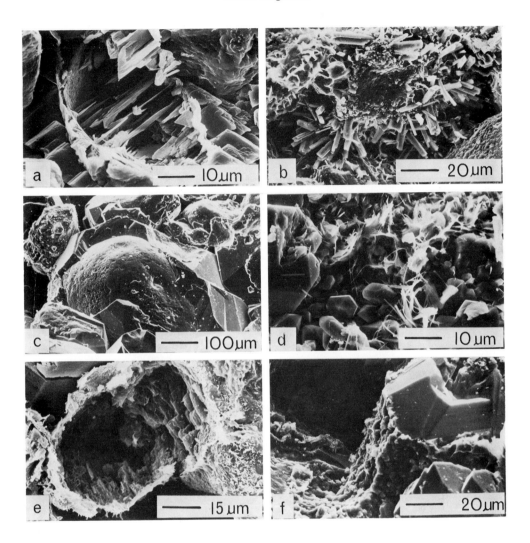

FIG. 10. Scanning electron micrographs of authigenic cements and porosity. (a) Intense dissolution of potassium feldspar leaving only thin clay rim which originally coated the detrital feldspar grain. Authigenic overgrowths and 'ingrowths' occur either side of the relic clay rim. Is the dissolution–precipitation process a response to high instability of the original high temperature feldspar with resultant precipitation of a low-temperature polymorph? (b) An unusual elongate, prismatic development of feldspar overgrowths around a detrital grain. What controls the morphology of feldspar overgrowths? (c) Typical development of euhedral quartz overgrowths (low surface free energy) enclosing rounded detrital grain (high surface free energy). (d) A cluster of small prismatic crystalls of authigenic quartz lining a dissolution void. Is crystal growth the result of precipitation from high levels of silica saturation? (e) Secondary porosity through total grain dissolution in a sandstone with a quartz grain framework preventing compaction. (f) Extensive intergranular secondary porosity through removal of a widespread carbonate phase (siderite in this example) which has previously corroded quartz overgrowths. Non-specific corrosion by transport-controlled dissolution?

under which minerals formed, and is an especially useful technique in distinguishing between non-marine and marine pore waters (e.g. Tan & Hudson 1974). However, changing pore-fluid compositions during burial also may be revealed, in both mudrocks (Irwin *et al.* 1977) and sandstones (Boles 1978; Land & Dutton 1978), the result basically of 'taking the rock apart' as originally advocated by Hudson. Irwin & Hurst (1983) have recently illustrated the significance of this technique in elucidating the nature of evolving pore-fluid compositions during diagenesis. Moreover, theoretical models allow the approximate temperature of mineral authigenesis to be deduced from stable isotope data (e.g. Milliken *et al.* 1981). Also, the study of modern calcretes has helped confirm their identification in the ancient sedimentary record (e.g. Salomons *et al.* 1978).

In addition, sulphur-isotope analysis (Coleman 1977), although applied primarily to ore deposits (e.g. Boast *et al.* 1981) may be used to assess the origin of sulphates both as cements in sandstones (e.g. Taylor 1983) and in both early and late diagenetic pyrite (e.g. Coleman & Raiswell 1981; Hudson 1982).

Cathodoluminescence (CL)

The luminoscope has evolved as a powerful petrographic tool in carbonate studies since its development 20 years ago (Smith & Stenstrom 1963; Long & Agrell 1965). Sippel (1968), Sibley & Blatt (1976) and others have applied the luminoscope in studies of silica authigenesis. However its more widespread use by clastic petrologists has been limited. Cathodoluminescence (CL) may be applied usefully in diagenetic studies to elucidate possible zonation of diagenetic minerals, and to study textures produced during development of secondary porosity. Little work has been published in these fields. This is because the majority of coarse clastic rocks are composed of quartz, which compared to non-ferroan calcites does not luminesce particularly well. Consequently, early workers were only able to record photographically their results. Sippel (1968) for example, employed 3000 ASA film with up to 20 min exposures. This problem has been overcome by recent research which has shown that quartz is more easily excited at lower temperatures and therefore emits more intense luminescence under cold stage conditions. Alternatively, Zinkernagel (1978) has developed a luminoscope with a hot cathode gun which emits more radiation than the 'cold' cathodes employed previously, excites quartz intensely and allows

observation of its luminescence. A more sophisticated hot cathode luminoscope developed at the University of Bern (Ramseyer, 1983) also produces intense luminescence in clastic rocks. Repeated zonation in quartz overgrowths and clay inclusions are easily recognized by their characteristic luminescence colours (Fig. 7a, b, e). Complex replacement textures may be interpreted (Fig. 7f) and quantification of grain-to-grain dissolution is possible.

Miscellaneous techniques

In recent years, various additional techniques have been applied to specific diagenetic problems. Studies of fluid inclusions in quartz overgrowths and carbonate cements have been used to determine the timing and conditions of cementation or mineralization (see Roedder 1979). Pittman & Duchatko (1970) describe the use of epoxy-resin casts in studying pore geometry. Age dating has been applied in both sandstone and shale studies, to pinpoint the timing of mineral authigenesis (e.g. Perry 1974; Aronson & Burtner 1982). A range of laboratory techniques derived from soil-science investigations has aided the analysis of fine-grained sediments and, moreover, of amorphous constituents in fine- and coarse-grained clastic rocks (see Jackson 1956; Dixon & Weed 1977). Finally, Ixer *et al.* (1979) illustrate the use of reflected light microscopy in an investigation of opaque cementing minerals.

Diagenetic models

A review of published studies indicates that three broad schemes of diagenetic modification occur in the eogenetic environment, whilst a variety occur during both mesogenesis and telogenesis. Although no single factor *controls* diagenesis, the three eogenetic schemes appear to be largely a function of the depositional pore-water chemistry.

Eogenetic regime

Firstly, the marine environment, with dominantly alkaline pore waters (White 1965) constitutes one end member of the scheme. Non-marine environments, however, include a wide variety of original depositional pore-water systems, in which two extremes may be recognized. Warm and wet, often sub-tropical to temperate settings, where intense weathering is typically developed in the source area, have either initially acidic or anoxic pore waters with high concentrations of dissolved species (Tardy 1971; Gibbs 1977). In contrast, semi-arid to arid environments are characterized by low precipitation and little source-area weathering. Surface depositional pore waters in such desertic settings are typically dilute and slightly alkaline (Cooke & Warren 1973). As Hurst & Irwin (1982) concluded, the environment of deposition will, therefore, tend to create a broad division of diagenetic modifications in sandstones. The stability of detrital minerals is thus a function of pore-water chemistry. Despite this pertinent observation, few recent diagenetic studies detail, or even take into account, the depositional environment.

Of equal importance in determining the stable eogenetic assemblage is the potential contribution of dissolved species from the associated mudrocks. Mudrock diagenesis is, however, predominantly controlled by the amount of organic matter present during burial (Curtis 1977). In this respect, the depositional environment also creates a broad division of diagenetic mineralogy. In most marine and non-marine environments with warm, wet conditions organic matter becomes involved in diagenetic reactions and contributes reaction products to the pore waters. However, both in semi-arid environments or in a depositional framework with a slow rate of burial, organic matter is rapidly oxidized near the sediment–water interface and effectively removed from the reacting system. This not only has important implications in the eogenetic regime but can also impart a long-lasting influence on diagenesis throughout the mesogenetic regime when depth-related mudrock reactions may release significantly concentrated solutions into the enclosed sandstones.

Marine environments

Seawater is typically slightly alkaline with a salinity of around 35 parts per thousand (White 1965). However, there is little potential for chemical reaction between the common detrital mineral phases, and normal marine pore waters (Garrels & Christ 1965). Consequently, most detrital minerals should be metastable following their deposition in marine environments. Hence, the diagenetic processes which operate in both fine- and coarse-grained clastic sediments result either from changes in pore-water chemistry during initial burial, or the reaction of amorphous material or the less stable detritus (Curtis 1980).

Diagenetic processes in marine mudrocks commence immediately following deposition. These processes are well documented in studies of various recent sediments (e.g. Curtis 1978; Froelich *et al.* 1979; Berner 1981; Krom & Berner 1981). Close to the sediment–water inter-

face the bacterial oxidation of organic matter releases bicarbonate ions into solution. These readily react with and dissolve the often amorphous or fine-grained clastic detritus, particularly weathering products such as clay minerals coated with iron sesquioxides (Greenland 1975). It is likely that this amorphous material is rapidly removed from the system by these reactions close to, or at, the sediment–water interface (Curtis 1980; cf. van Elsberg, 1978). Vertical permeability in fine-grained sediments is generally low. Consequently, whilst some reaction products may diffuse into the sea, the majority will either be fixed within the mudrock itself or during shallow burial compaction, expelled into any enclosed sandbodies (Meade 1966). Following aerobic bacterial processes sulphate-reducing bacteria liberate further bicarbonate ions into the pore waters and continue to reduce pH by the production of hydrogen sulphide. Eventually, fermentation processes dominate in the sediment column, releasing bicarbonate ions and methane. By-products of bacterial metabolic processes are thus released into the depositional pore waters and as a result authigenic minerals form rapidly at or near the sediment surface. Pyrite may precipitate by reaction between residual detrital iron oxides and hydrogen sulphide (Fig. 9c). Iron- and manganese-bearing calcite and dolomite may locally precipitate as concretions (Irwin *et al.* 1977) whilst the bicarbonate may significantly alter pore-water composition to effect the dissolution of amorphous material. Thus, the interaction of organic and inorganic detrital components with marine pore waters, modified by the reaction products of bacterial metabolism, probably causes rapid diagenetic modifications of the initial sediment (Coleman & Raiswell 1981).

Marine sandstones are characterized by an early authigenic suite of minerals which includes illite, interstratified illite-smectite, chlorite, potassium-feldspar overgrowths, quartz, pyrite and various carbonate cements (Fig. 4d; Almon *et al.* 1976; Hawkins 1978; Storey 1979; Tillman & Almon 1979; Hurst & Irwin 1982; Kantorowicz 1984). Furthermore, most workers record clay authigenesis, associated with quartz and feldspar overgrowths, followed by carbonate precipitation, as the typical order of cementation, suggesting that eogenetic marine diagenesis conforms to a predictable pattern. It is likely, however, that the paragenesis of these various minerals reflects not only the composition and chemistry of the seawater but also the original sedimentary mineralogy and the early diagenetic reactions in adjacent mudrocks.

Quartz arenites and subarkoses, with a generally stable detrital assemblage, contain authigenic illite or illite-smectite formed following grain dissolution or directly from seawater, followed by quartz overgrowths (Elverhøi & Bjørlykke 1978). Precipitation is probably the result of an increase in the ionic concentration of the seawater as silica- and alumina-bearing solutions migrate from adjacent shales to combine with potassium from the original interstitial seawater (Fig. 3; Harder 1974; Kastner & Siever 1979). With a much less stable original mineralogy, as in volcaniclastic sediments, the potential for reaction with the pore waters is increased and more ions can be added to the reacting system during grain dissolution. Thus, dissolution of volcaniclastic detritus, possibly stimulated by the reaction products of mudrock diagenesis, leads to increased activity of the iron and magnesium species in solution. Consequently chlorite or chlorite-smectite clays tend to precipitate in preference to more illitic clays (Fig. 4a; Almon et al. 1976; Dutton 1977; Galloway 1979; Wescott 1983). Illitic clays observed today are likely to have formed as disordered, more smectitic, end-members which have evolved in time. Clay mineral authigenesis is typically followed by quartz and feldspar overgrowth precipitation (Fig. 11). It is likely that this reflects the geochemical evolution of the system, with more easily formed minerals forming rapidly, whilst those requiring more ordered crystal growth forming subsequently. The diagenesis of aluminosilicate minerals in marine sediments is arrested ubiquitously by carbonate cementation (e.g. Stanton 1977). Calcite, or less commonly dolomite or siderite precipitation may involve either carbonate and bicarbonate ions originally in solution, or those released from the bacterial degradation of organic matter in the surrounding mudrocks (Coleman & Raiswell 1981).

In addition to a variety of authigenic minerals, marine clastic sedimentary assemblages are widely reported to contain a range of green sheet-silicates (Porrenga 1967; Knox 1970; James & van Houten 1979; Bhattacharyya & Kakimoto 1982). In the past, and indeed in some contemporary publications these are loosely referred to as either 'glauconite' or 'chamosite'. However, more recent research indicates that both the chemistry and the mineralogy of these sheet silicates is extremely diverse (see Odin & Matter 1981; Brindley 1982). The glauconite or glaucony minerals comprise a complex range of iron-rich, potassium-poor smectites. The complexity of recent glaucony compared with the relative purity of the mica end-member reported from ancient deposits confirms the general suggestion that clay mineral authigenesis is a relatively rapid process which, given a wide range of ions available, may initially form a chemically complex and poorly crystalline mineral. During diagenesis, this smectitic material may become more potassium-rich, and evolve or mature into an end-member which is chemically and mineralogically a potassium mica.

There appears to be a genetic relationship between glaucony and 'chamosite' (or berthierine s.s., see below), the former occurring further from river mouths and deltas (Fig. 12). However, the origin of 'chamosite' is problematical, and consequently its use as a marine indicator is in doubt. Many 'chamosites' reported from the literature have now been shown to be 7Å layer silicates, and, mineralogically therefore berthierines. In addition, chamosite is currently used to describe 2:1 14Å layer silicates, the iron-rich chlorites (Bailey 1980). Consequently, it is often difficult to establish which mineral is being described in the literature. It is also suggested that berthierine s.s. evolves during diagenesis and alters to form chamosite s.s. (Iijima & Matsumoto, 1982). However, chamosite authigenesis may take place independently causing both minerals to occur together without any genetic relationship.

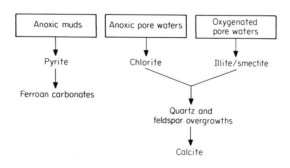

FIG. 11. Schematic diagenetic pathways in marine sediments.

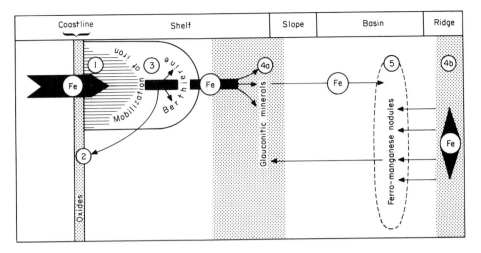

FIG. 12. Geochemical pathways of iron in the marine environment, (after Odin & Matter 1981). Iron accumulates in the coastal zone and inner shelf (1) where chemical precipitation of iron and oxidation of detrital iron (2) take place. Mobilization of iron results in authigenic growth of berthierine and (3) reorganization of detrital clay micelles. Further out on the shelf authigenic growth of glauconite (4) takes place in the deep basin. Where reducing conditions dominate, ferrous nodules (5) grow. Ooliths may be derived or accreted in place but berthierine is essentially reworked and altered lateritic or kaolinitic soil.

Furthermore, the origin of the ooliths which berthierine characteristically forms (Fig. 4c) is currently controversial. Theories proposed include replacement of carbonate ooliths, *in situ* growth on the sea floor, and alteration of derived iron-coated clay ooliths (Knox 1970; Kimberley 1979, 1980a, b; Bradshaw *et al.* 1980; Bhattacharyya & Kakimoto 1982). The association of berthierine ooliths with deltas (Fisher & McGowen 1967; James & van Houten 1979; Odin & Matter 1981), the widely reported occurrence of ferruginous ooids and clay pellets on modern tropical alluvial plains (Nahon *et al.* 1980, and references therein), and the geochemical constraint of ferrous iron requiring reducing conditions (Curtis & Spears 1968), suggest that clay authigenesis is probably related to seawards size fractionation of kaolinite and smectite.

Vegetated (warm wet) non-marine environments

Petrological studies of a variety of alluvial sediments suggest that diagenesis was influenced by initially acidic or neutral, typically anoxic, pore waters. These studies are mostly of sediments deposited in well-vegetated and hence equatorial or tropical, even temperate climates (see Hemingway 1968; Hawkins 1978; Staub & Cohen 1978; Almon & Davies 1979; Edwards 1979; Storey 1979; Keighin & Fouch 1981; Huggett 1984). The diagenesis of these

sediments is inferred to have been influenced significantly by the bacterial degradation of organic matter. Consequently, although the wide-ranging effects of grain dissolution and mineral precipitation may be seen throughout, the majority of the reactions involved probably resulted from changes in pore-fluid composition influenced strongly by bacterial activity (Curtis 1977, 1978). In addition, most authigenic minerals reflect the relative paucity of potassium, magnesium and sulphate ions in solution in freshwater (White 1965; Berner 1971).

In these warm, wet and vegetated conditions, organic matter is not completely oxidized at the sediment surface. Indeed, significant quantities survive to impart to the mudrocks present, a distinct black colour which remains throughout their burial. Subsequently, bacterially controlled processes operate, degrading this organic matter and significantly influencing ground-water chemistry immediately after deposition. These reactions initially involve aerobic bacteria, removing oxygen from the pore waters, causing the release of bicarbonate ions into solution and as a result lowering the pH of the solution. Such acidic conditions may also stimulate further organic and inorganic reactions which may subsequently affect the pore-water composition (Curtis 1977). Indeed, similar processes are likely to have operated throughout the sedimentary basin, affecting not only the composition of the original detrital clastic assemblage, but also that of the depositional groundwaters (Thomas 1974). Nonetheless, many of the studies listed above report early iron-chlorite authigenesis and hence, by contrast, suggest that pore waters were neutral and anoxic rather than acidic and oxidizing (Velde 1977; Harder 1978). It is equally possible that these requirements may also be satisfied by the continued diagenesis and interaction of organic matter with pore waters. This is because following the depletion of oxygen, anaerobic bacterial processes commence. Consequently, and in the absence of sulphate ions in solution (Berner 1971), ferric iron sesquioxides are reduced, releasing hydroxyl ions and causing a rise in pH (Curtis 1977).

Eogenetic modifications to quartz arenites and subarkoses under more acidic conditions are often similar and unspectacular (Fig. 13). Variations generally only occur where abundant rock fragments were deposited or where the detrital sediments were finer grained. Quartz overgrowths (Fig. 7c), pore-filling kaolinite (Figs 4e and 8e), altered (neomorphosed) muscovite (Fig. 5a) and dissolved feldspar grains are typically reported from quartz arenites (see Fig. 3). Both the quartz and the kaolinite may have precipitated directly from ground waters (Tardy 1971). In addition, silica and alumina released during the dissolution of feldspar grains may have contributed to these authigenic phases (Curtis & Spears 1971; Berner & Holdren 1979). This is especially true of crevasse-splay sandstones. These are often colonized by plants and as a result leached by acidic fluids, causing removal of feldspars and detrital micas whilst kaolinite precipitates in the underlying muds. The dissolution of feldspar grains and fine-grained clays, as well as alteration of muscovite may be explained by their relative instability in acidic conditions (Fanning & Keramides 1977; Huang 1977; Lin & Clemency 1981). Similarly, silica is relatively insoluble at low pH with the result that it is readily precipitated (Blatt 1966). In some sandstones diagenesis is effectively

complete by this stage, with wholesale quartz overgrowth cementation. This is perhaps best described as a wet version of Wopfner's (1978) tropical silcrete model. In more arkosic sediments more authigenic kaolinite is often reported, suggesting that a relationship may exist between detrital mineralogy and clay authigenesis. Further confirmation of this model may be found in the quartz overgrowths (Breese 1960), and large vermiform aggregates of kaolinite plates (Fitzpatrick 1980) which have been observed in modern tropical soils.

In both quartz arenites and with the introduction of more iron-rich labile grains into the detrital sedimentary assemblage, further diagenetic variations may be observed. Chlorite rather than kaolinite is often reported as the first authigenic clay. This chlorite is invariably iron-rich and technically therefore chamosite *s.s.* (Fig. 8c). In addition, siderite rather than calcite cements are often found. Here it may be inferred that bacterial processes rapidly depleted oxygen with the result that anoxic conditions prevailed before significant diagenetic modifications to the sediments had occurred. Several studies confirm the importance of the nature of particular authigenic minerals. Chlorite occurs as a pore lining, inhibiting further quartz overgrowths and enhancing the preservation potential of intergranular porosity. Kaolinite on the other hand occurs as a pore filling, creating a rock with different reservoir properties. The significance of this distinction is that only very local variations in ground-water chemistry are required to produce either of the two situations (Kantorowicz 1984). Consequently, the reservoir character of such sediments is difficult to predict (see Fig. 13).

Siderite authigenesis often marks the last diagenetic modification that may be related to the degradation of organic matter, and consequently to depositional pore-water chemistry (Fig. 13). However, the role of iron in clastic diagenesis is somewhat problematical. Many authigenic sequences include iron-bearing

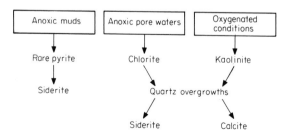

FIG. 13. Schematic diagenetic pathways in warm, wet non-marine sediments.

minerals but lack any obvious sources of iron. Nonetheless, Greenland (1975) and Picard & Felbeck (1976) suggest that iron may be transported into basins either coating clays, or complexed with organic polymers. In this context Gac & Tardy (1980) describe iron-coated clay micelles from Lake Chad, whilst Gibbs (1977) estimates that 6–7% of the material presently transported by the Amazon and Yukon Rivers is iron and that some 90% of this is transported as solid particles or as grain coatings. Gibbs (1977), Nahon *et al.* (1980) and Odin & Matter (1981) all related the mobility and derivation of iron to erosion of lateritic weathering products. This material is most likely to accumulate on flood plains as a result of overbank flooding. As a result sphaerosiderite-dominated soil horizons may develop, whilst at depth ferrous and sideritic concretions may form as detrital iron sesquioxides are mobilized (Fig. 6a; Coleman & Ho 1968; Retallack 1976; Curtis *et al.* 1972; Gould & Smith 1979).

Within the stabler sedimentary assemblages recorded in most of the above examples, diagenesis involves some dissolution of the aluminosilicate framework, as well as alteration to a variety of clay minerals, authigenesis of clays, and precipitation of both overgrowths and pore filling occasionally, replacive cements. These processes involve the interaction of the sediments themselves, particularly the organic matter present, and also the depositional pore waters. In a few extreme cases, the breakdown of abundant labile grains, such as volcaniclastics, overrides the subtler affects outlined above. Potassium, calcium, and magnesium ions may be released into solution during grain dissolution and result in the precipitation of complex diagenetic assemblages characterized by smectites and zeolites (Galloway 1974; Vessel & Davies 1981).

Although the studies reported here infer acidic or anoxic ground-water conditions they often involve essentially lateritic weathering assemblages. However, the diagenesis of similar sedimentary clastic assemblages described below is inferred to form red beds, with the eogenetic mineral assemblage dominated by haematite and illite/smectite authigenesis. The essential differences appear to reflect both the original climate and the original basin geometry and burial rate. Deposition and rapid burial often associated with tectonically active areas will create different conditions to those on stable cratons with slower sedimentation rates or in other tectonic settings with prolonged exposure to surface oxidation (Curtis 1977; Coleman *et al.* 1979; see Fig. 15). It is significant, therefore, to note that many of the rocks detailed above are from the Jurassic or Carboniferous of Western Europe (Figs 4a, c, 6a, 7c, 8c and 9c), or the Tertiary of the Gulf Coast. More significantly these studies concentrate on sediments inferred to have been deposited as basin centres and in areas with a consistently high water table.

Desertic (hot dry) non-marine environments

Red-coloured sediments have for many years been interpreted as being of continental origin. Furthermore, the cause of the red coloration has been well documented to be due to the presence of ferric oxide (see Fig. 9a; Grabau 1920; Berner 1969). The actual quantity of iron is small, often totalling less than 1% of the sediment (Walker & Honea 1969) and it is typically no greater than in non-red sediments. Clearly, high iron contents are not needed to produce red sediments and virtually all sediments would appear to have the potential to form red beds.

The origin of the haematite, however, has been the subject of considerable debate. Typically, red-bed sequences are late to post-orogenic fanglomerate deposits which grade laterally into fluvial sandstones and basin centre playa lake sediments. Mineralogically, they are often highly feldspathic, indicating a source area of crystalline granitic or metamorphic basement. Historically, two opposing theories have developed to account for this common association and the development of haematite in red beds. Earliest ideas suggested that red coloration was caused by iron liberated during the decomposition of hornblende and mica (Lyell 1852). Subsequent workers considered the haematite to form in lateritic soils in tropical, humid climates under deep weathering conditions (Krynine 1935, 1949). However, because ancient red beds are frequently associated with evaporites, Krynine (1950) suggested that red river alluvium, derived from lateritic soils, was transported from the tropical uplands and deposited as large, red alluvial fans in desertic basins. Despite widespread acceptance of this theory (see van Houten 1961) there was no proven modern analogue to such transported red soils. Indeed according to Walker (1967) the recent alluvium from Mexico quoted by Krynine was actually greyish-brown. Red coloration in the alluvium only developed on much older soils that had been exposed to prolonged oxidation.

To develop this concept of reddening increasing with age, Walker and his co-workers undertook detailed studies of Recent, Pleistocene and Pliocene deposits of the Sonoran Desert, and Baja California, and compared the results with studies of late Palaeozoic

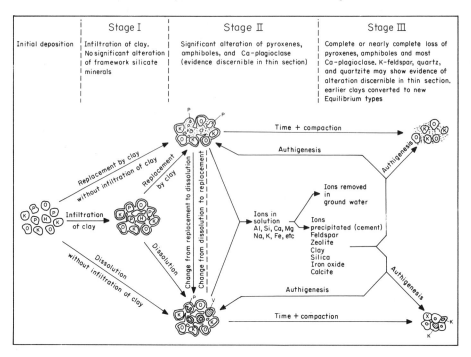

FIG. 14. Schematic diagram illustrating observed and inferred diagenetic alterations in first cycle, arkosic hot, dry (desertic) sediments (after Walker *et al.* 1978).

red-bed sediments of Colorado (see Walker 1967; Walker *et al.* 1967; Walker & Honea 1969; summarized in Walker *et al.* 1978; Fig. 14). Essentially, they were able to show that, regardless of age or location, the haematite pigment in red beds is a product of the *in situ* diagenetic alteration of iron-bearing grains.

In the modern alluvial fans of the Sonoran desert the sediments are distinctly non-red in colour, whilst the Pleistocene counterparts are brownish and the older Pliocene sediments are a characteristic brick-red colour. It appears that the reddening process takes place *in situ*, and is the result of several distinct post-depositional processes. Initially, the clay content of this coarse alluvium is negligible. However, it gradually increases with age to as much as 25% in the Pliocene sediments. Mechanical infiltration of clay contributes significantly to this increase (Fig. 4b). Clay also forms as a replacement of unstable detrital ferromagnesian grains, and is typically smectitic in composition (Fig. 8b). Some of this clay is unstable in the desert vadose environment and alters to liberate iron (Fig. 9a). During these various processes ions may be liberated into solution. In the oxidizing environment of the hot desert, iron rapidly becomes ferric and either enters the newly forming clay mineral lattice, or forms a fine coating on

individual clay particles (Fig. 9a, b). This iron is subsequently converted to an oxide, amorphous hydrated goethite in the Pleistocene sediments, which in the Pliocene deposits has aged to crystalline haematite. Selective or total dissolution of unstable silicate grains (Figs 7d and 10a, c) such as volcaniclastics, also liberates many ions into solution which again may be precipitated as a result of the local pore-water chemistry or remain in solution and be removed from the reacting system (Fig. 14). In the associated dune fields a similar series of reddening processes develops.

Walker also recognized the widespread development of red soils in sediments of the Sonoran Desert, although the youngest ones are of late Pleistocene age. Fine-grained material in the playa lakes also becomes red with time. Here the iron-oxide clay coatings age to haematite whilst the clay may undergo post-depositional alteration, yielding additional iron which ultimately forms more haematite.

In addition to iron, various other chemical species are released into solution by the replacement and dissolution of detrital grains. Although in detail the exact chemical pathways involved in these hydrolysis reactions are not fully known they may be approximated by some of the 'weathering' reactions compiled by Curtis

(1976). The dissolution of framework silicates (feldspar) rock fragments and ferromagnesian grains liberates silica, alumina, alkali and alkali-earth elements into the interstitial pore waters as hydrated ions and complexes. Much of the released silica, alumina and labile elements is precipitated as pore-lining authigenic clay (Figs 8b, d). In the Sonoran Desert the most abundant clay is randomly interstratified illite-smectite with 80–95% expandable layers.

Sediments with high detrital feldspar contents release sufficient quantities of potassium into the pore waters to precipitate potassian intermediate sanidine as authigenic overgrowths, stable in the diagenetic environment (Fig. 10b; see also Waugh 1978; Kastner & Siever 1979; Ali & Turner 1982). Any excess silica forms authigenic quartz occurring as both tabular, and prismatic, discrete, euhedral crystals in addition to incipient overgrowths (Fig. 10c). More locally, where the detrital sediments are volcaniclastic in composition, the zeolite mineral clinoptilolite is a common cement. The ions necessary for all these authigenic cements can be generated by dissolution and replacement reactions taking place within the original sediment (Fig. 14). The remaining elements may be concentrated to form caliche deposits, or as evaporites within the playa lakes of enclosed basins (Figs 6b, f and 9e).

Walker's studies were important for several reasons, particularly in the red-bed context. The occurrence in a desert region with an arid climate, of red soils formed through diagenetic processes, illustrates conclusively that red beds need not only form in areas of lateritic weathering in moist tropical climates. However, red soils may form in such tropical settings. Both in the Orinoco Basin (Walker 1974) and in Columbia (van Houten 1972) although modern alluvium is brown, older sediments have reddened due to their prolonged exposure to favourable oxidizing conditions. Walker, therefore, concluded that there is no justification for using red beds as indicators of palaeoclimate, either in the source area or depositional basin. Chemical and mineralogical analyses reveal no obvious criteria that can be used to differentiate ancient sediments with a red coloration being formed diagenetically in moist climates from those found in hot desert climates. Thus, all sediments which contain sufficient free iron are capable of being oxidized to produce red sedimentary rocks, but they will only do so if the interstitial pore-water environment favours the formation and preservation of ferric iron oxides (see Fig. 15).

There are many examples of ancient analogues to the above red-bed model. Walker (1975) considered the late Palaeozoic red beds of the Western U.S. Interior and the Triassic Moenkopi Formation of Colorado as classic examples of diagenetically produced red beds. Nagtegaal (1969) described similar red beds from post-Hercynian Carboniferous to Lower Triassic sediments in Spain, whilst Pedersen & Anderson (1980) proposed the same diagenetic origin for the red Triassic sandstones of Jutland. Here in Britain the Old Red Sandstone (Turner 1980), Permian (Waugh 1978) and Triassic (Burley, 1984) all exhibit diagenetic features typical of desertic red beds, although the cautionary remarks of Turner & Archer (1977) and Wilson & Duthie (1981) should be recalled. The Rotliegendes of the southern North Sea is also considered as a classic example of desertic red beds and a wealth of literature has been written on it (Glennie *et al.* 1978; Seeman 1979; Rossel 1982).

In general, these studies concentrate on the petrology and paragenesis of authigenic minerals. They reveal a general trend which is comparable to that observed in studies of recent sediments. Namely, detrital grains are commonly coated with haematite and either tangential or radial developments of illite/smectite and illite clays. These initial developments are succeeded by quartz and feldspar overgrowths although these are often thinly spread and the sandstones may remain friable and loosely cemented. Calcite and dolomite cementation which occurs locally is sometimes related to calcrete formation, or evaporative conditions, respectively.

In contrast, red beds considered to form in a humid, tropical climate are described from deltaic sediments in the Difunta Group of Mexico (McBride 1974) and from the late Carboniferous Coal Measures of North Staffordshire (Besly & Turner 1983) and of South Wales (Archer 1965). These red beds are typically associated with palaeosols—see Turner (1980) for a summary. Frequently the red coloration is restricted to the mudstone units which are colour mottled red, purple, green and yellow. Calcrete nodules, spherulitic siderite, pyrite and roots are present at several horizons within the mudstones. Although Walker (1974) considered that the origin of red beds in both desert and moist tropical climates was broadly similar, the above examples suggest there are distinct differences. In these clay-rich, fine-grained tropical mudrocks the *in situ* reddening of detrital ferric hydroxides appears to be the dominant process. This is brought about by the seasonal lowering of water tables and the development of well-drained oxygenated

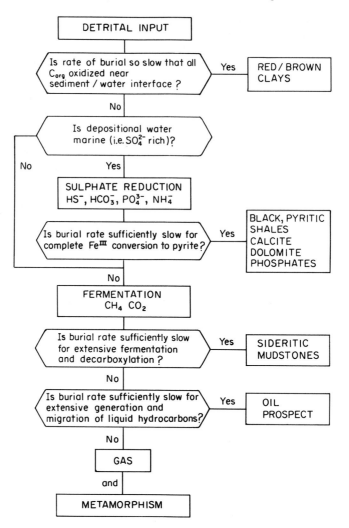

FIG. 15. Schematic diagram to illustrate the interrelations between sediment mineralogy, mud source rock potential and rate of burial (after Curtis 1977).

conditions, allowing the oxidation of organic matter and eventually development of haematite (McBride 1974; Besly & Turner 1983). A further contribution to the red coloration results from the rapid oxidation of finely disseminated pyrite producing ferric oxyhydroxides which redden with age (Berner 1969).

Mesogenetic regime

The sediment is gradually removed from the influence of depositional pore waters during burial. Although difficult to define precisely, the mesogenetic regime begins when the sediment is effectively sealed from the predominant in-

fluence of surface agents in the interstitial pore water, and persists to the onset of metamorphism or uplift and re-exposure.

Following the eogenetic modifications, regardless of original depositional environment, the sedimentary assemblage will comprise a mixture of those phases that were initially stable (or metastable) and new authigenic phases which have reached stability (or metastability) through reaction with the pore waters. To effect a further change in the diagenetic assemblage requires a change in the physicochemical conditions of temperature, pressure or pore-fluid chemistry. However, the solubility of many mineral phases increases with increasing temperature so that pore fluids will be capable of dissolving more ionic species. Also kinetically, increasing temperature adds energy to the reacting system. Reaction rates will therefore increase and reaction barriers can be more easily overcome so that metastable phases will react more readily.

This is well illustrated by the relationship between clay mineral transformations and burial temperatures which has been studied by many workers in several sedimentary basins. Probably the most extensively documented reaction is the conversion of smectites to illites through an intermediate mixed-layer illite-smectite with increasing depth and temperature, particularly well studied in the Tertiary of the Gulf Coast of the U.S.A. (e.g. Burst 1959, 1969; Perry & Hower 1970; Hower *et al.* 1976; Boles & Franks 1979a; summarized by Boles 1981 and Hower 1981). In detail the reaction of smectite to illite is complex and does not simply take place by cation exchange, but also involves aluminium substitution for silicon, reduction of octahedral iron and eventually the loss of both octahedral magnesium and iron (Hower *et al.* 1976).

Boles & Franks (1979a) suggested that the reaction does not retain the 2:1 structure of original smectite. The aluminium needed for the reaction is concentrated in the smectite causing the collapse of the 2:1 structure forming new

illitic layers and liberating silica. Any potassium present in the original smectite would contribute to the newly forming illite, the remainder coming from external sources.

However, the reaction may take place in a different manner. Hower *et al.* (1976) suggest that the original 2:1 smectite layer structure remains intact and the reaction essentially involves ionic substitution. Nadeau *et al.* (1984) reviewed the smectite-illite transformation with the aid of the TEM. They concluded that illite crystals are amalgamated 10 Å smectite layers. Solid solution recrystallization does not occur, hence this model may need to be re-assessed.

Boles (1981) considered that the smectite to illite transformation is not only significant in shale diagenesis but also has far-reaching implications for sandstone cementation (see Figs 16 and 17). Detailed petrographical studies of Gulf Coast sediments (Boles 1978; Boles & Franks 1979a; Land & Milliken 1981; Milliken *et al.* 1981) indicate that a variety of cements occur in the associated sandstones. These include quartz overgrowths, late ferroan carbonates—ferroan calcite, ferroan dolomite and ankerite (Figs 5d, 6d and 9f)—chlorite and extensive albitization of plagioclase. The breakdown of smectite to illite releases silica, calcium, sodium, iron and magnesium into the pore waters which may be capable of transporting dissolved ionic species from mudrocks into sandstones. Silica so

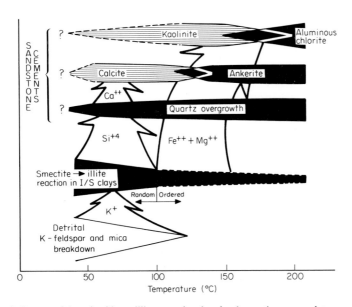

FIG. 16. The influence of the mixed layer illite-smectite clay depth reaction on sandstone cementation and mineral phases in the Wilcox Sandstones of the Texas Gulf Coast (after Boles & Franks 1979a). Vertical arrows depict ion transfer between illite-smectite reactions and mineral phases in sandstones.

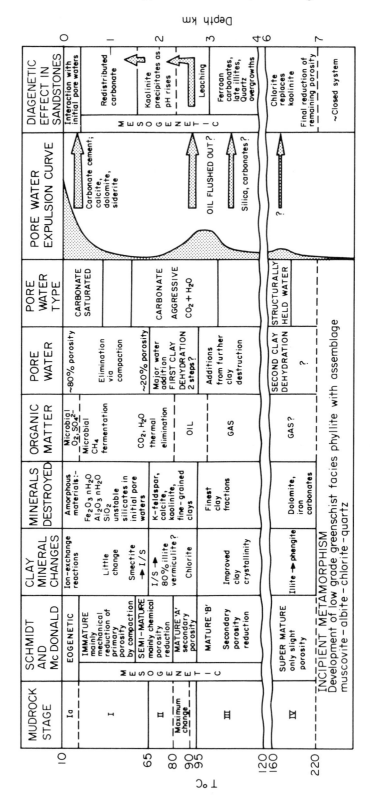

FIG. 17. Summary of depth-related diagenetic stages and important reactions in mudrocks which may contribute solutes to sandstones, via the associated trends in pore water evolution and expulsion (modified from Curtis 1983b, with additional data from Powers 1967; Perry & Hower 1970; Walker *et al.* 1978; Boles & Franks 1979a; Foscolos & Powell 1979a; Schmidt & McDonald 1979a).

released may precipitate as quartz overgrowths (Figs 7a, b, e and f), whilst iron and magnesium may contribute to the formation of chlorite from kaolinite and ferroan dolomite or ankerite from pre-existing calcites. Sodium liberated may contribute to the albitization of plagioclase.

This strongly implies that mass transfer of ionic species occurs between mudrocks and sandstones during mesogenetic diagenesis. There are other related lines of evidence to support the concept of mass transfer in the deep subsurface. Concomitant with the depth-related changes taking place in the mineralogy is the evolution and expulsion of the interstitial pore waters. Undisturbed muds, when initially deposited, have porosities in excess of 80%, whilst well sorted sands may have porosities of 50% (Burst 1969). During early burial most of the eogenetic interstitial pore waters are expelled by mechanical compaction. However, with elevated temperature, many diagenetic reactions occur which involve dehydration and will, therefore, liberate water. Hydrous or amorphous minerals rapidly dehydrate and, in smectitic clays undergoing transformation to illite, large quantities of interlayer water are expelled (Burst 1969). This smectitic clay dehydration is actually a complex process and probably takes place in at least two steps (Perry & Hower 1972). The initial water loss is associated with the collapse of smectite interlayers. Other complications arise where vermiculite is present as dehydration may be delayed until much later in diagenesis (Foscolos & Powell, 1979b). The destruction of clay minerals and organic matter also liberates structurally held water, even at considerable burial depths.

Throughout burial diagenesis therefore, in addition to evolution of the interstitial pore waters through diagenetic reactions, potentially large volumes of water may be added to the reacting system through the breakdown of the clay minerals and organic material. In addition, organic matter undergoes a series of depth-related changes following microbial degradation (Tissot *et al.* 1974) Bacterial formation may extend to depths of 1 km after which either increasing temperature or the exhaustion of suitable organic material prevents the activity of bacteria. With continued burial abiotic reactions dominate and thermal decarboxylation results in the liberation of carbon dioxide to depths of around 3 km and corresponding temperatures of 100°C. Further increase in temperature results in the generation of hydrocarbons.

All these depth-related processes may conveniently interact in the deep subsurface to produce a repeated pattern of sandstone diagenesis (Fig. 17). Throughout burial, pore waters expelled from mudrocks will tend to migrate upwards under a pressure gradient. Carbon dioxide liberated during thermal decarboxylation of organic matter produces acidic solutions. These migrating acidic pore waters can potentially carry high concentrations of aluminium and metal cations (Curtis 1983a). Water movement will tend to follow the more permeable channels in a sedimentary sequence and will therefore migrate into major sandbodies. Here, acidic pore waters will be capable of generating widespread secondary porosity by the removal of carbonate cements (see Figs 2 and 17 for explanation; and Figs 7d, f and 10e, f). These solutions may also precipitate extensive kandites (Figs 4f, 7f and 8f). Many authors now recognize deep mesogenetic secondary porosity in sandstone reservoirs (see Schmidt & McDonald 1979a, b, for a synopsis; Loucks *et al.* 1977; Lindquist 1977; Land & Dutton 1978; Milliken *et al.* 1981). An essential precursor to the development of this porosity is the presence of an early, framework-supporting cement. Given a stable silicate framework, the creation of deep mesogenetic porosity at depth clearly has important implications for petroleum reservoirs. Furthermore, smectite dehydration, supplying a water source at depth, and the release of chemical species that may precipitate higher in the section, may provide a mechanism for the migration of petroleum and formation of a 'diagenetic seal' causing hydrocarbon entrapment and overpressuring (Burst 1969; Bonham 1980).

Much of the above data has been accumulated from the Gulf Coast of the U.S.A. This is a clay-dominated sequence which has undergone continuous and active burial since the late Cretaceous and therefore has a simple burial history. Maximum burial depths, temperatures and pressure are known *in situ* (Boles 1981) and water migration is known to be up dip, largely through a compactional drive. However, the depositional environment is often ignored or given little consideration and little attention has been paid to the eogenetic processes that may have taken place. When studying depth-dependent mineralogical changes, it is important to bear in mind that the starting mineralogy in shales at different levels is likely to vary considerably (see Hower 1981).

Similar depth-related changes to those recorded from the Gulf Coast are known in other deep basins. In NE British Columbia and the NW Territories of Canada the smectite to illite transformation has been studied by Powell *et al.*

(1978). However, their results suggest that the burial depths at which clay dehydration occurs are attained prior to oil generation and that the water expelled by this process does not aid in oil migration. Foscolos & Powell (1979b) considered that the actual destruction of clay minerals (mostly expandables and kaolinite) releases crystal lattice water later during diagenesis which may assume an important role in petroleum migration. This second clay dehydration occurs after oil has become cracked to gas because of the presence of a calcium-vermiculite component in the mixed layer clays which is stable to higher temperatures (see Fig. 17).

Considerably less work has been published on deep-burial diagenesis in the U.K. Pearson et al. (1982) demonstrated that the smectite clay dehydration reaction is taking place in the early Tertiary sediments of the northern North Sea Viking Graben. The Kimmeridge Clay in the Jurassic, which is widely regarded as the main source for North Sea oil, has undergone smectite-illite ordering and Pearson et al. (1982) considered that water released during illitization would have been available at the time of hydrocarbon generation. Rossel (1982) envisaged that a similar depth-related clay mineral transformation has occurred in the Rotliegendes of the southern North Sea. Eogenetic clay mineral suites comprised a smectitic phase derived from the alteration of iron-bearing silicates. Under deep-burial conditions this smectite and kaolinite became unstable and converted to illite (e.g. Fig. 8a), although the source of potassium is envisaged to be overlying Zechstein halite. The occurrence and dominance of illite at depth in the North Sea has been noted by several authors (Fig. 5e; Sommer 1978; Hancock & Taylor 1978; McHardy et al. 1982) but there is little general agreement on its cause or relative timing.

Depth-related diagenetic assemblages are recorded from Triassic (Burley 1984) and Carboniferous sandstones (Hawkins 1978; Huggett 1984) in several basins in the U.K. Interpretation of the diagenetic history of these sediments is hindered by the complex burial history which includes tectonic inversion. However, late cements include isotopically light ankerite, quartz overgrowths and illite with associated secondary porosity (Figs 5e, f, 6d and 9d, f). That there is a remarkable degree of consistency in deep-burial diagenetic assemblages from different tectonic and environmental settings suggests that depth-related physicochemical conditions conform to a predictable pattern. Widespread pore-fluid migration at depth from compaction of mudrocks, clay mineral reactions or large-scale migration of formation water is likely to impart changes in sediments on a regional scale, resulting in a uniformity of diagenetic texture and mineral assemblages.

Although the main theme of this account of mesogenesis has dealt with chemical reactions and mineralogical changes it would be inappropriate to conclude without reference to the physical processes which may be important during burial diagenesis. Following deposition, clastic sediments are compressed and compacted by the weight of the overlying sediments which results in a reorganization of grain fabric and a general reduction in porosity.

Fine- and coarse-grained clastics react in different ways. Mud-dominated sequences may be compacted to 10% of their original thickness, coincidentally reducing intergranular porosity from 80% to perhaps 5% and hence expelling considerable quantities of water into adjacent sands. This process may be important both during the most rapid compaction at shallow depths and later during burial when temperature-related reactions release absorbed water into pore spaces. During compaction the orientation of particles in muds changes from random to orientated parallel to basal spacings. Pore water is initially easily expelled in the early stages of compaction, as both porosity and horizontal permeability are relatively high. With continued compaction, when the mudrock is reduced to around 50% of its original volume, further pore-water escape is hindered by decreasing horizontal permeabilities, whilst adsorbed water becomes more tightly bonded. At depths of up to 1 km, water expulsion in mudrocks is largely driven by gravitational displacement, but clay mineral diagenesis and aquathermal pressuring are the main contributing agencies at depth.

In coarser-grained sequences pressure brings about several porosity-reducing adjustments. During initial mechanical compaction grains move closer together. Rotation and grain slippage, brittle grain deformation and fracturing, and plastic deformation of ductile grains may all be important. Nagtegaal (1978) demonstrated the effect of primary mineralogy on the stability of various sandstone frameworks. Lithic arenites, with high contents of relatively soft and volcanic fragments, are particularly susceptible to mechanical compaction and plastic deformation. In contrast, quartz arenites and arkosic arenites typically exhibit high framework stability to compaction effects (Fig. 10e, f).

Telogenetic regime

Rocks uplifted to the earth's surface and subsequently either altered and reburied, or weathered and eroded, may exhibit a wide range of diagenetic modifications. This is because minerals formed during burial at elevated temperatures and pressures, and from concentrated formation waters, are often unstable in oxidizing and occasionally acidic conditions (e.g. Berner *et al*. 1980). Such modifications may be illustrated with reference to records of both ancient sediments exposed beneath unconformities, and a range of rocks being altered or weathered at or near the earth's surface today.

Taylor (1978) described the alteration of organic matter, ferrous minerals and sulphides in Westphalian Carboniferous sediments in the Lake District. He related these various reactions to oxidation during exposure during the early Permian. These observations are entirely consistent with theoretical models derived to explain modern weathering reactions (Curtis 1976). Hence organic matter, preserved under the originally euxinic conditions during burial, was oxidized and the carbon completely removed during early Permian exposure. In addition, iron carbonates were oxidized to iron oxides, eventually producing haematite, whilst the oxidization of sulphides resulted in the formation of sulphates such as gypsum.

It is suggested by a number of authors that telogenetic modifications occur in Middle Jurassic clastic sediments in Brent Group reservoirs from the northern North Sea. Brent Group sandstones are variously described as flushed by freshwater both in the Middle Jurassic and subsequently during uplift and subaerial exposure beneath Upper Jurassic or Lower Cretaceous unconformities (e.g. Hancock & Taylor 1978; Sommer 1978; Bjørlykke *et al*. 1979; Giles & Marshall, in press). Exposure to acidic freshwater in these conditions is proposed to explain the presence of kaolinite or dickite, occurring both as a pore filling (Fig. 8f) and as a replacement of detrital feldspars (Fig. 5c). In addition, this may explain the improved porosity that often occurs up dip beneath unconformities in these reservoirs. Nevertheless, a number of problems exist with this general theory. Its two basic premises appear to be that kaolinite can only form from the dissolution of feldspars, and freshwater is the only potential source of acidity. However, a review of the occurrences of kaolinite and dickite reveals firstly that they may form in quartz arenites and, moreover, in limestones (Shelton 1964; Dickson & Coleman 1980; Burchette & Britten, this volume). Theoretically, kaolinite may also be produced as the result of mudrock diagenesis (Curtis 1983a), whilst freshwater is

not necessarily acidic (see above). Even if such meteoric waters were available, Curtis (1983b) does not believe 'freshwater' has the chemical potential to create the secondary porosity observed. In addition, freshwater is not the only source of acidic pore waters available, as acidic solutions are believed to emanate from mudrocks during their burial, especially when they contain organic matter (e.g. Parker 1974; Tissot *et al*. 1974; Pearson *et al*. 1982; Rossel 1982). An additional point overlooked by many adherents of the 'freshwater flushing' school is that in actively forming sedimentary basins water movement is predominantly upwards driven by compactional processes (see mesogenesis section). With a change of tectonic regime to a situation in which the sedimentary pile remains static or is uplifted considerable quantities of water may be driven into and through the basin. The model of Schmidt & McDonald (1979a) was based largely on data from the former situation whilst many examples of freshwater penetration quote ancient sediments presently at or near the earth's surface. This topic is currently controversial and will probably be debated vigorously by the opposing schools of thought for some time to come.

The alteration or weathering of rocks at the earth's surface represents not only the end of one cycle but also the beginning of another in clastic diagenesis. This topic demonstrates particularly well the important contribution that has been made to our understanding of diagenetic processes by soil scientists. Pedogenic studies are particularly relevant because the alteration of detrital grains at the surface in soil profiles may take place by similar processes to those which affect grains subsequently. Hence, an understanding of the ways in which detrital minerals are modified or the mechanism by which pedogenic (authigenic) minerals form allows the conditions during sediment diagenesis to be inferred. Dixon & Weed (1977) provided comprehensive coverage of this subject, illustrating both the minerals and the processes involved as well as the variety of techniques employed. In addition, Curtis (1976) and Berner (1981 and references therein) both studied the thermodynamics of these processes. Their work and its application demonstrates the importance of an integrated approach to diagenesis.

Numerous studies of surface or near-surface alteration could be reviewed in this section. For example, Nishiyama *et al*. (1978) describe a Triassic chlorite-bearing red shale from Toyoma in Japan. At outcrop this shale is found to contain an irregularly interstratified chlorite–vermiculite rather than pure chlorite. They concluded that alteration of chlorite to vermiculite is most probably a weathering process (cf. Fig. 4a). Glasmann (1982) described the alteration of andesite from soils in the Cascades, illustrating both the more labile minerals within the rock, and their alteration products. Augite and plagioclase were both observed to alter to smectites, which grow as globules initially and act as nuclei for subsequent fibrous growth (cf. Guven *et al.* 1980). These studies are particularly useful in attempts to establish the original detrital mineralogy of greywackes and other digenetically mature sediments. Benson & Teague (1982) described the alteration of Miocene basalts from the Pasco Basin, Washington. They reported a paragenetic sequence comprising nontronite (an iron-rich smectite), clinoptilolite, and silica, with minor amounts of mordenite replacing the clinoptilolite. More significantly the alteration of basalts to these various minerals can be related to low-temperature processes, in ground-water conditions similar to those occurring today. Consequently, they proposed that zeolite formation is possible within 1000 m of the earth's surface and without elevated temperatures.

A detailed hydrogeochemical study by Walton (1981) demonstrated that dilute acidic ground waters (pH 5–6) in the Triassic Budleigh Salterton Pebble Beds aquifer of south Devon was capable of extensively leaching haematite and carbonate from the sediments. Furthermore, detrital feldspar in these sandstones has been replaced *in situ* by an intimate association of illite, kaolinite, and authigenic potassium feldspar. This telogenetic alteration is readily explained by the modern ground-water chemistry as the Pebble Bed waters are presently saturated with respect to both kaolinite and illite, containing considerable quantities of dissolved alumina. Thus, under surface conditions of temperature, pressure and modern pore-water chemistry, the Triassic mineral assemblage is unstable and has undergone extensive reaction with the ground waters in a shift towards equilibrium.

Volcaniclastic sediments

The role of volcaniclastic grains in clastic diagenesis has been mentioned briefly in the sections above. However, volcaniclastic sediments are especially significant because of the extreme results of their alteration when they dominate a clastic sedimentary assemblage. Volcaniclastic material is relatively unstable at the earth's surface. Consequently its constituents react readily with their pore waters and dissolve resulting in the precipitation of a variety

of authigenic minerals. Vessel & Davies (1981), for example, report the generation of 20% authigenic clay within 3–5 years of deposition in volcaniclastic sediments from Guatemala. The speed of diagenesis of volcaniclastic sediments is illustrated by the results of studies from the geological record. The potential reactivity of sediments of this type was originally recognized by Cummins (1962) who proposed a diagenetic rather than detrital origin for the matrix of greywackes. This proposal was subsequently confirmed experimentally by Whetten & Hawkins (1970). Galloway (1974) reported how the reaction of predominantly ferromagnesian sediments controls their diagenesis with little reference to likely variations in depositional pore-water chemistry. Moreover, whilst initial reactions formed calcite and then chlorite, reactions at higher temperatures and pressures, and involving both the remaining detritus and initial diagenetic minerals, resulted in zeolite formation (cf. Benson & Teague 1982). Because of their textural transformations during diagenesis dominantly volcaniclastic sediments have poor reservoir potential (Nagtegaal 1978). This results from the wholesale dissolution and breakdown of detrital grains, which destroys the sediment's original framework, reducing the potential for preservation of any intergranular pore space. Volcaniclastic sediments suffer perhaps the most extreme diagenetic modifications, manifesting little or no textural or mineralogical similarities to the original assemblage.

Discussion and conclusions

The past 20 years have seen significant changes in the approaches taken to studying clastic rocks. Similarly, important advances have been made in our understanding of the diagenesis of these rocks. Diagenesis has evolved from a purely petrographical microscopic-based subject into a broad discipline encompassing all aspects of sedimentary geochemistry. This change may be traced from the 1950s through the establishment of predictive depositional-facies models, and the development of sophisticated quantitative analytical techniques, to more recent integration of data from numerous related disciplines such as organic geochemistry, pedology and hydrology. As a result it is now almost possible to establish the entire history of a rock from the source of its detrital components, via the depositional environment, with the associated (possibly climatically influenced) early cementation and through deep burial and changing pore-fluid composition and eventually into weathering following uplift. During diagenesis, distinct realms may be recognized in which pore-fluid composition is influenced by the depositional environment (eogenesis), or by formation waters (mesogenesis), or finally by the reintroduction of surficial agents during uplift and exposure (telogenesis). Within each of these realms differing styles of diagenesis may be recognized. Eogenetic processes are dominated by the relatively rapid dissolution and reaction of amorphous and unstable components resulting in the formation of characteristic minerals, such as haematite. Mesogenetic processes dominantly operate within the framework established previously with most significant changes influenced by solutions introduced from elsewhere in the basin. Telogenesis, by contrast, involves the alteration of minerals often formed under drastically different conditions and may eventually lead to the wholesale destruction of the rock.

A number of important points may be emphasized by contrasting the various eogenetic assemblages, and by comparing the general style of eogenetic modifications with those occurring subsequently. Although there are similarities in the relative order of diagenetic relations during eogenesis, the diversity of products of any initial reactions suggests that truly predictive models may be extremely hard to establish (see Almon & Davies 1979; Storey 1979; Besly & Turner 1983). Nonetheless, the inherent metastability of many of these assemblages results in rocks being 'fingerprinted' by the interstitial water conditions prevailing immediately following deposition. Furthermore, because of the relative instability of the original sedimentary assemblage the changes which may occur during eogenesis are often more drastic than any subsequent modifications. Consequently, with the obvious exception of volcaniclastic sandstones, mesogenetic modifications are generally restricted to the precipitation and dissolution of authigenic mineral phases within the solid framework established during eogenesis.

In each of the three eogenetic regimes outlined above a sequence of clay authigenesis, quartz and feldspar overgrowth cementation and carbonate cementation generally occurs. Within this general pattern there is however considerable diversity. Firstly, the clay mineralogies outlined above encompass illitic and smectitic clays in dry non-marine settings, kaolinite and chlorite in wet ones, and chlorite or smectitic clays in marine depositional environments. There is clearly some considerable overlap between these various eogenetic regimes. Consequently, distinct assemblages may form, and these may be explained in terms of the known deposi-

tional history. However recognition of distinct assemblages does not allow any ready interpretation of the depositional environment.

As well as the authigenesis of clay minerals, equally rapid eogenetic precipitation, dissolution and replacement reactions, which may be observed at the sediment surface today in modern analogues, are interpreted to take place. These lead to the formation of calcretes and haematite in dry settings, lateritic and sideritic soils in wet non-marine environments, whilst in the sea glauconite and berthierine ooliths form in addition to a range of incipient ferrous and calcareous concretions within muds rich in organic matter. These various processes generally have the effect of fixing many of the previously amorphous constituents of the sedimentary assemblage. They also remove from solution many of the cations liberated during the dissolution of igneous or metamorphic rock fragments. The subsequent formation of quartz overgrowths demonstrates that pore waters at this time must be excessively supersaturated with respect to silica, more than sufficient being present to allow its continued precipitation.

During mesogenesis a range of effects may occur. These vary according to the original detrital mineralogy, with the more pronounced effects being felt where relatively unstable grains survive eogenetic reactions, and form an integral part of the rock framework during its burial. With a more stable mineralogy—e.g. in quartz arenites—early cementation with quartz overgrowths often creates a rigid framework. The stability of this framework allows the repeated precipitation and dissolution of pore-filling cements during burial without the gross composition of the rock being affected. With the introduction of more lithic components or feldspar grains into the sandstone, reactions during burial may be more effective creating significant grain-dissolution porosity (Heald & Larese 1973). Alternatively in less well-cemented rocks, feldspar or lithic grains are compressed or compacted leading to a gradual reduction in intergranular porosity (McBride 1978). With greater percentages of these labile grains, wholesale destruction of the detrital grains often occurs during eogenesis. Nonetheless, any grains surviving into mesogenesis are usually altered during this phase, with the result that the original depositional framework of the rock may cease to exist whilst the mineralogical components may have changed from those present initially (e.g. Galloway 1974). The recognition of the potential relationship between detrital mineralogy and reservoir rock potential during burial also has important implications for hydrocarbon exploration.

The relatively high initial geochemical instability of detrital minerals in the eogenetic environment will tend to produce a shift toward sediment–pore-water equilibrium during early diagenesis. To produce significant changes in the resulting stable (or metastable) assemblage would require a change in the gross physical and chemical environment. It seems probable therefore that pressures and temperature, and pore waters introduced into the sediments from elsewhere in the basin, are likely to play increasingly important roles during mesogenesis. The remarkable degree of consistency recorded in deep-burial diagenetic assemblages (crystalline illites, chlorite, isotopically light ferroan carbonates, quartz cements) and common secondary porosity, from a variety of environmental and tectonic settings, does suggest that depth-related diagenetic processes conform to a pattern that may be predictable (see Fig. 17). Clearly, large volumes of pore fluids do migrate at depth in the subsurface, and the evidence for widespread solute transport is convincing.

To date, therefore, the study of diagenesis has enabled general models for the evolution of sediments deposited in a range of environmental and tectonic settings to be developed. As a result of these investigations a wide range of diagenetic mineral associations and textures can be interpreted, and some diagenetic processes are now understood. However, there are many aspects of diagenesis that remain poorly understood and provide scope for much research. There are many situations, for example, in which the diagenetic history of a clastic sediment is too complex to unravel with standard techniques. Continued application of more sophisticated techniques, such as stable isotope analysis, fluid inclusion studies and cathodoluminescence, should help resolve this. Many diagenetic processes are not fully understood, and here the use of geochemical modelling and high-resolution quantitative analytical techniques can be expected to increase our knowledge. Hydrogeology is a field which should also make significant contributions towards our understanding of subsurface fluid flow and sediment/water interactions over the next few years.

In conclusion, diagenesis has come of age as a discipline in the last 20 years, although it still remains a long way from being a truly predictive science. We hope that this admittedly biased review will provide a useful introduction to diagenesis for workers in diverse disciplines. We also hope that we may stimulate other workers to confirm or refute the models we have described, many of which cannot be regarded as the consensus of opinion at present.

ACKNOWLEDGEMENTS: The research which forms the basis of the review was carried out whilst all the authors were in receipt of NERC financial assistance at the University of Hull. The technical and academic staff of the Department of Geology are thanked for their high-quality support and stimulating encouragement. We are also grateful to the Geological Institute at the University of Berne and AGAT Consultants Inc., for continued assistance during the preparation of the manuscript. Special mention must be made of Julia Davidson for typing and editing innumerable versions of the manuscript.

A review paper by definition utilizes the research of other workers. We should like to thank the numerous people who have listened patiently to our ideas and whose helpful discussions have clarified our thinking. Mistakes of fact and opinion are our own.

Karl Ramseyer, of the University of Berne, took the luminescence pictures, Alan Mitchell, of Britoil, kindly provided the colour slides of the Rotliegendes. Photographic plates were prepared by Andy Werthemann at the University of Berne. Finally, we extend our gratitude to our reviewers and editors and to the expert staff of Blackwell Scientific Publications and Alden Press for their patience in piecing together this multifarious contribution.

References

ALI, A. D. & TURNER, P. 1982. Authigenic K-feldspar in the Bromsgrove Sandstone Formation (Triassic) of Central England, *J. sedim. Petrol.* **52**, 187–98.

ALMON, W. R. & DAVIES, D. K. 1978. Clay technology and well stimulation *Trans. Gulf-Cst Ass. geol. Socs*, **28**, 1–6.

—— & —— 1979. Regional diagenetic trends in the Lower Cretaceous Muddy Sandstone, Powder River Basin. *In*: SCHOLLE, P. A & SCHLUGER, P. R. (eds) *Aspects of Diagenesis. Spec. Publs Soc. econ. Miner. Paleont., Tulsa*, **26**, 379–400.

——, FULLERTON, L. B. & DAVIES, D. K. 1976. Pore space reduction in Cretaceous sandstones through chemical precipitation of clay minerals. *J. sediment. Petrol.* **46**, 89–96.

ARCHER, A. A. 1965. Red beds in the Upper Coal Measures of the Western part of the South Wales Coalfield. *Bull. geol. Surv. Gt Br.* **23**, 57–64.

ARONSON, J. L. & BURTNER, R. L. 1982. K-Ar dating of illitic clay in sandstone reservoirs and timing of petroleum migration. *Am. Ass. Petrol. Geol. Bull.* **66**, 1442 (abstract).

BAILEY, E. H. & STEVENS, R. E. 1960. Selective staining of K-feldspar and plagioclase on rock slabs and thin section. *Am. Miner.* **45**, 1020–5.

BAILEY, S. W. 1980. Summary of recommendations of AIPEA nomenclature committee. *Clay Miner.* **15**, 85–93.

BENSON, L. V. & TEAGUE, L. S. 1982. Diagenesis of basalts from the Pasco Basch, Washington. I. Distribution and composition of secondary mineral phases. *J. sedim. Petrol.* **52**, 595–613.

BERNER, R. A. 1969. Goethite stability and the origin of red beds *Geochim. cosmochim. Acta*, **33**, 267–73.

—— 1971. *Principles of Chemical Sedimentology*, McGraw-Hill, New York. 240 pp.

—— 1980. *Early Diagenesis: A Theoretical Approach*, Princeton University Press, Princeton, New Jersey. 241 pp.

—— 1981. Kinetics of weathering and diagenesis. *In*: LASAGA, A. C. & KIRKPATRICK, R. J. (eds) *Kinetics of Geochemical Processes. Reviews in Mineralogy*, **8**, 111–34. Mineralogical Society of America,

—— & HOLDREN, G. J. 1979. Mechanism of feldspar weathering. II. Observations of feldspars from soils. *Geochim. cosmochim Acta*, **43**, 1173–86.

——, SJOBERG, E. L. & VELBEL, M. A. & KROM, M. D. 1980. Dissolution of pyroxenes and amphiboles during weathering. *Science*, **207**, 1205–6.

BESLY, B. M. & TURNER, P. 1983. Origin of red beds in a moist tropical climate (Etruria Formation, Upper Carboniferous, U.K.). *In*: WILSON, R. C. L. (ed.) *Residual Deposits: Surface Related Weathering Processes and Materials. Geol. Soc. London Spec. Publ.* **11**, 131–47.

BHATTACHARYYA, D. P. & KAKIMOTO, P. K. 1982. Origin of ferriferous ooids: and SEM study of ironstone ooids and bauxite pisoids. *J. sedim. Petrol.* **52**, 845–57.

BJØRLYKKE, K. 1979. Cementation of sandstone—discussion. *J. sediment. Petrol.* **49**, 1358–60.

——, ELVERHOI, K. A. & MALM, A. O. 1979. Diagenesis in Mesozoic sandstones from Spitsbergen and the North Sea, a comparison. *Geol. Rdsch.* **68**, 1152–71.

BLATT, H. 1966. Diagenesis of sandstones: processes and problems. *Wyoming Geol. Ass. Ann. Conf.* **65**, a–o.

—— 1979. Diagenetic processes in sandstones. *In*: SCHOLLE, P. A. & SCHLUGER, P. R. (eds) *Aspects of Diagenesis. Spec. Publs Soc. econ. Miner. Paleont.*, Tulsa, **26**, 141–57.

BOAST A. M., COLEMAN, M. L. & HALLS, C. 1981. Textural and stable isotopic evidence for the genesis of the Tynagh Base Metal deposit, Ireland, *Econ. Geol.* **76**, 27–55.

BOLES, J. R. 1978. Active ankerite cementation on the subsurface Eocene of south-west Texas. *Contrib. Mineral. Petrol.* **68**, 13–22.

—— 1981. Clay diagenesis and effects on sandstone cementation (case histories from the Gulf Coast Tertiary). *In*: LONGSTAFFE, F. J. (ed.) *Clays and the Resource Geologist. Min. Ass. Can., Short Course Hdbk*, **7**, 148–68.

—— 1982. Active albitisation of plagioclase, Gulf Coast Tertiary. *Am. J. Sci.* **282**, 165–80.

—— & FRANKS, S. G. 1979a. Clay mineral diagenesis in Wilcox of south-west Texas: implications of smectite diagenesis in sandstone cementation. *J. sedim. Petrol.* **49**, 55–70.

—— & —— 1979b. Cementation of sandstones—reply. *J. sedim. Petrol.* **49**, 1362.

BOHNAM, L. C. 1980. Migration of hydrocarbons in compacting basins. *Am. Ass. Petrol. Geol. Bull.* **64**, 549–67.

BORST, R. L. & KELLER, W. D. 1969. Scanning electron micrographs of API reference clay minerals and other selected samples. *Proc. Int. Clay Conf.* **1**, 871–901.

BRADSHAW, H. J., JAMES, S. J. & TURNER, P. 1980. Origin of oolitic ironstones—discussion. *J. sedimen. Petrol.* **58**, 295–9.

BREESE, D. E. 1960. Quartz overgrowths as evidence of silica deposition in soils. *Aust. J. Sci.* **23**, 18–20.

BRINDLEY, G. W. (ed.) 1951. *X-ray Identification and Crystal Structures of Clay Minerals*, Mineralogical Society, London. 345 pp.

—— 1982. Chemical composition of berthierines. *Clays Clay. Miner.* **30**, 153–5.

—— & BROWN, C. (eds) 1980. *Crystal structures of clay minerals and their x-ray identification. Mineralogical Society Monograph*, **5**, Mineralogical Society, London. 455 pp.

BURLEY, S. D. 1984. Distribution and origin of authigenic minerals in the Triassic Sherwood Sandstone Group, U.K. *Clay Miner.* **19**, 403–40.

BURST, J. F. 1959. Post diagenetic clay mineral environmental relationships in the Gulf Coast Eocene. *Clays Clay Miner.* **6**, 327–41.

—— 1969. Diagenesis of Gulf Coast clayey sediments and its possible relation to petroleum migration. *Am. Ass. Petrol. Geol. Bull.* **53**, 73–93.

CARRIGY, M. A. & MELLON, G. B. 1964. Authigenic clay mineral cements in Cretaceous and Tertiary sandstones of Alberta *J. sedim. Petrol* **34**, 461–72.

COLEMAN, J. M. & HO, C. 1968. Early diagenesis and compaction in clays. Abnormal Subsurface pressures. *Louisiana Coastal Studies Institute Technical Report 62 AD No. 687550*, pp. 23–50. Louisiana State University, Louisiana.

COLEMAN, M. L. 1977. Sulphur isotopes in petrology. *J. geol. Soc. London*, **133**, 595–608.

—— & RAISWELL, R. 1981. Carbon, oxygen and sulphur isotope variations in concretions from the Upper Lias of N.E. England. *Geochim. cosmochim. Acta*, **45**, 329–40.

——, CURTIS, C. D. & IRWIN, H. 1979. Burial rate a key to source and reservoir potential. *Wld Oil*, **5**, 83–92.

COOKE, R. U. & WARREN, A. 1973. *Geomorphology in Deserts*, Batsford Ltd London. 374 pp.

CUMMINS, W. A. 1962. The Greywacke Problem. *Lpool Manchr geol. J.* **3**, 57–72.

CURTIS, C. D. 1976. Stability of minerals in surface weathering reactions: a general thermochemical model. *Earth Surf. Proc.* **1**, 63–70.

—— 1977. Sedimentary geochemistry: environments and processes dominated by involvement in an aqueous phase. *Phil. Trans. R. Soc.* **A286**, 353–71.

—— 1978. Possible links between sandstone diagenesis and depth related geochemical reactions occurring in enclosing mudstones. *J. geol. Soc. London*, **135**, 107–17.

—— 1980. Diagenetic alteration in black shales. *J. geol. Soc. London*, **137**, 189–94.

—— 1983a. A link between aluminium mobility and destruction of secondary porosity. *Am. Ass. Petrol. Geol. Bull.* **67**, 380–4.

—— 1983b. Geochemical studies on development and destruction of secondary porosity. *Geol. Soc. London Spec. Publ.* **12**, 113–25.

—— & SPEARS, D. A. 1968. Formation of sedimentary iron minerals. II. *Econ. Geol.* **63**, 262–70.

—— & SPEARS, D. A. 1971. Diagenetic development of kaolinite. *Clays Clay Miner.* **19**, 219–27.

——, PETROWSKI, C. & OERTEL, G. 1972. Stable carbon isotopes within carbonate concretions: a clue to place and time of formation. *Nature, Lond.* **235**, 98–100.

——, IRELAND, B. J., WHITEMAN, J. A., MULLANEY, R. & COBLEY, T. 1984. Stability of authigenic clay minerals: new evidence from analytical transmission electron microscopy. *Clay Miner.* **19**, 471–81.

DICKSON, J. A. D. 1965. A modified staining technique for carbonates in thin section *Nature, Lond.* **205**, 587.

—— & COLEMAN, M. L. 1980. Changes in carbon and oxygen isotope composition during limestone diagenesis. *Sedimentology*, **27**, 107–18.

DIXON, J. E. & WEED, S. B. (eds) 1977. *Minerals in Soil Environments*, Soil Science Society of America, Madison, Wisconsin. 948 pp.

DOMENICO, P. A. 1977. Transport phenomena in chemical rate processes in sediments. *Ann. Rev. Earth planet. Sci.* **5**, 287–317.

DUTTON, S. P. 1977. Diagenesis and secondary porosity distribution in deltaic sandstone, Strawn Series (Pennsylvannian) north-central Texas. *Trans. Gulf-Cst Ass. Geol. Socs*, **27**, 272–7.

EDWARDS, M. B., 1979. Sandstone in Lower Cretaceous Helvetiafjellet Formation, Svalbard. Bearing on reservoir potential of Barents Shelf. *Am. Ass. Petrol. Geol. Bull.* **63**, 2193–203.

ELSBERG, J. N. VAN 1978. A new approach to sediment diagenesis. *Can. Pet. Geol. Bull.* **26**, 57–86.

ELVERHØI, A. & BJØRLYKKE, K. 1978. Sandstone diagenesis—Mesozoic rocks from southern Spitsbergen. *Norsk Polarinst. Arb.* **1977**, 145–57.

FANNING, D. S. & KERAMIDES, V. Z. 1977. Micas. *In*: DIXON, J. N. & WEED, S. B. (eds) *Minerals in Soil Environments*, pp. 195–258. Soil Science Society of America, Madison, Wisconsin.

FISHER, W. L. & McGOWEN, J. H. 1967. Depositional Systems in the Wilcox Group of Texas and their relationship to occurrence of oil and gas *Trans. Gulf-Cst Ass. geol. Socs*, **17**, 105–25.

FITZPATRICK, E. A. 1980. *Soils: Their Formation, Classification and Distribution*, 2nd edn, Longman, London. 353 pp.

FOLK, R. L. 1965. Hentry Clifton Sorby (1826–1908). The founder of petrology. *J. Geol. Educ.* **8**, 43–7.

FOSCOLOS, A. E. & POWELL, T. G. 1979a. Catagenesis in shales and occurrence of authigenic clay in sandstones, North Sabine Well H-49, Canadian Arctic Islands. *Can. J. Earth Sci.* **16**, 1309–14.

—— & —— 1979b. Mineralogical and geochemical transformation of clays during catagenesis and their relation to oil generation. *Can. Soc. Pet. Geol. Mem.* **6**, 153–72.

FROELICH, P. N., KLINKHAMMER, G. P., BENDER, M. L., LUEDTKE, N. A., HEATH, G. R., CULLEN, D., DAUPHIN, P., HAMMOND, D., HARTMAN, B. & MAYNARD, V. 1979. Early oxidation of organic matter in pelagic sediments of the estern equatorial Atlantic: suboxic diagenesis. *Geochim. cosmochim. Acta*, **43**, 1075–90.

GAC, J. Y. & TARDY, Y. 1980. Geochemistry of a tropical landscape on granitic rocks: the Lake Chad Basin. *In*: CAMPBELL, A. (ed.) *Proceedings 3rd International Symposium on Water–Rock Interaction*, 8–10. Alberta Research Council.

GALLOWAY, W. E. 1974. Deposition and diagenetic alteration of sandstones in northeast Pacific arc-related basins: implications of greywacke genesis. *Bull. geol. Soc. Am.* **85**, 379–390.

—— 1979. Diagenetic control of reservoir quality in arc-derived sandstones; implications for petroleum exploration. *In*: SCHOLLE, P. A. & SCHLUGER, P. R. (eds) *Aspects of Diagenesis. Spec. Publs Soc. econ. Miner. Paleont.*, Tulsa, **26**, 251–62.

GARRELS, R. M. & CHRIST, C. L. 1965. *Solutions, Minerals and Equilibria*, Harper and Row, New York, 450 pp.

GIBBS, R. J. 1977. Transport phases of transition metals in the Amazon and Yukon Rivers. *Bull. geol. Soc. Am.* **88**, 829–43.

GILES, M. R. & MARSHALL, J. D. in press. Processes controlling secondary porosity development. *Clay Miner.*

GLASMANN, J. R. 1982. Alteration of andersite in wet, unstable soils of Oregon's western Cascades *Clays Clay Miner.* **30**, 253–63.

GLENNIE, K. W., MUDD, G. C. & NAGTEGAAL, P. J. C. 1978. Depositional environment and diagenesis of Permian Rotliegendes sandstones in Leman Bank and Sole Pit areas of the U.K. Southern North Sea. *J. geol. Soc. London*, **135**, 25–34.

GOLDICH, S. S. 1938. A study in rock weathering. *J. Geol.* **46**, 17–58.

GOULD, K. W. & SMITH, J. W. 1979. The genesis and isotopic composition of carbonates associated with some Australian coals. *Chem. Geol.* **24**, 137–50.

GRABAU, A. W. 1920. *Textbook of Geology. I. General Geology*, Heath and Co., Boston. 864 pp.

GREENLAND, D. J. 1975. Charge characteristics of some kaolinite-iron hydroxide complexes. *Clay Miner.* **10**, 407–16.

GUVEN, N., HOWER, W. F. & DAVIES, D. K. 1980. Nature of authigenic illites in sandstone reservoirs. *J. sediment. Petrol.* **50**, 761–6.

HANCOCK, N. J. & TAYLOR, A. M. 1978. Clay mineral diagenesis and oil migration in the Middle Jurassic Brent Sand Formation. *J. geol. Soc. London*, **135**, 69–72.

HARDER, H. 1974. Illite mineral synthesis at surface temperatures. *Chem. Geol.* **14**, 241–53.

—— 1978. Synthesis of iron-layer silicate minerals under natural conditions. *Clays Clay Miner.* **26**, 65–72.

HASZELDINE, R. S., SAMSON, I. M. & CORNFORD, C. 1984. Quartz diagenesis and convective fluid flow: Beatrice Oilfield, U.K., North Sea. *Clay Miner.* **19**, 331–41.

HAWKINS, P. J. 1978. Relationship between diagenesis, porosity reduction and oil emplacement in late Carboniferous sandstone reservoirs: Bothamsall Oilfield, East Midlands. *J. geol. Soc. London*, **135**, 7–34.

HAYES, J. B. 1979. Sandstone diagenesis—the hole truth. *In*: SCHOLLE, P. A. & SCHLUGER, P. R. (eds) *Aspects of Diagenesis. Spec. Publs. Soc. econ. Miner. Paleont., Tulsa*, **26**, 127–9.

HEALD, M. T. & LARESE, R. E. 1973. The significance of the solution of feldspar in porosity development. *J. sediment. Petrol.* **43**, 458–60.

HELGESON, H. C., GARRELS, R. M. & MACKENZIE, F. T. 1969. Evaluation of irreversible reactions in geochemical processes involving minerals and aqueous solutions. II. Applications. *Geochim. cosmochim. Acta*, **33**, 455–81.

HEMINGWAY, J. E. 1968. Sedimentology of coal bearing strata. *In*: MURCHISON, D. & WESTOLL, T. S. (eds) *Coal and Coal Bearing Strata*, 43–69. Oliver and Boyd, Edinburgh.

HEROUK, Y., CHAGNON, A. & BERTRAND, R. 1979. Compilation and correlation of major thermal maturation indicators. *Am. Ass. Petrol. Geol. Bull.* **63**, 2128–44.

HOUTEN, F. B. VAN 1961. Climatic significance of red beds. *In*: NAIRN, A. E. M. (ed.) *Descriptive Palaeoclimatology*, 89–139. New York Interscience, New York.

HOUTEN, F. B. VAN 1972. Iron and clay in tropical savanna alluvium, North Columbia, a contribution to the origin of red beds. *Bull. geol. Soc. Am.* **83**, 2761–72.

HOWER, J. 1981. Shale diagenesis. *In*: LONGSTAFFE, F. J. (ed.) *Clays and the Resource Geologist. Min. Ass. Can., Short Course Hdbk*, **7**, 60–80.

——, ESLINGER, E. V., HOWER, M. E. & PERRY, E. A. 1976. Mechanism of burial and metamorphism of argillaceous sediment. I. Mineralogical and chemical evidence. *Bull. geol. Soc. Am.* **87**, 725–37.

HUANG, P. M. 1977. Feldspars, olivines, pyroxenes and amphiboles. *In*: DIXON, J. B. & WEED, S. B. (eds) *Minerals in Soil Environments*, 553–602. Soil Science Society of America, Madison, Wisconsin.

HUDSON, J. D. 1977. Stable isotopes and limestone lithification. *J. geol. Soc. London*, **133**, 637–60.

—— 1982. Pyrite in ammonite-bearing shales form the Jurassic of England and Germany. *Sedimentology*, **29**, 639–67.

HUGGETT, J. M. 1984. Controls of mineral authigenesis in Coal Measures Sandstones of the East Midland, U.K. *Clay Miner.* **19**, 343–57.

HURST, A. R. 1981. A scale of dissolution for quartz and its implications for diagenetic processes in sandstones. *Sedimentology*, **28**, 451–60.

—— & IRWIN, H. 1982. Geological modelling of clay diagenesis in sandstones. *Clay Miner.* **17**, 5–22.

—— & STOUT, J. H. 1980. Dissolution and growth of quartz grains from sandstone. *Int. Ass. Sedimentologists. 1st Europ. Mtg Bochum*, pp. 151–4 (abstract).

HUTCHEON, I. 1981. Application of thermodynamics to clay minerals and authigenic mineral equilibria. *In*: LONGSTAFFE, F. K. (ed.) *Clays and the Resource Geologist. Min. Ass. Can., Short Course Hdbk*, **7**, 169–92.

IIJIMA, A. & MATSUMOTO, R. 1982. Berthierine and chamosite in coal measures of Japan. *Clays Clay Miner.* **30**, 264–74.

IRWIN, H. 1981. On calcic dolomite-ankerite from the Kimmeridge clay. *Min. Mag.* **44**, 105–7.

—— & HURST, A. R., 1983. Applications of geochemistry to sandstone reservoir studies. *Geol. Soc. London Spec. Publ.* **12**, 127–46.

——, CURTIS, C. D. & COLEMAN, M. L. 1977. Isotopic evidence for source of diagenetic carbonates formed during burial of organic-rich sediments. *Nature, Lond.* **269**, 209–13.

IXER, R. A., TURNER, P. & WAUGH, B. 1979. Authigenic iron and titanium oxides in Triassic red beds (St Bees Sandstone), Cumbria, Northern England. *Geol. J.* **14**, 179–92.

JACKSON, M. L. 1956. Soil chemical analysis—advanced course. Department of Soil Science, University of Wisconsin, Madison, Wisconsin.

JAMES, H. E. & HOUTEN, F. B. VAN 1979. Miocene goethite and chamositic oolites, northeastern Columbia. *Sedimentology*, **26**, 125–34.

KANTOROWICZ, J. D. 1984. The origin, nature, and distribution of authigenic clay minerals from Middle Jurassic Ravenscar and Brent Group Sandstones. *Clay Miner.* **19**, 359–75.

KASTNER, M. & SIEVER, R. 1979. Low temperature feldspars in sedimentary rocks. *Am. J. Sci.* **279**, 435–79.

KEIGHIN, C. W. & FOUCH, T. D. 1981. Depositional environments and diagenesis of some non-marine Upper Cretaceous reservoir rocks. Uinta Basin, Utah. *In*: ETHRIDGE, F. G. & FLORES, R. M. (eds) *Recent and Ancient Non-marine Depositional Environments: Models for Exploration. Spec. Publs Soc. econ. Miner. Paleont., Tulsa*, **31**, 109–25.

KELLER, W. D. 1954. Bonding energies of some silicate minerals. *Am. Miner.* **30**, 783–93.

—— 1969. *Chemistry in Introduction Geology*, Luca Bros, Columbia, Missouri. 108 pp.

—— 1976. Scan electron micrographs of kaolins collected from diverse environments of origin. I. *Clays Clay Miner.* **24**, 107–33.

KIMBERLEY, M. M. 1979. Origin of oolitic iron formations. *J. sediment. Petrol.* **49**, 111–32.

—— 1980a. Origin of oolitic iron formations—reply.

J. sediment. Petrol. **50**, 299–302.
—— 1980b. Origin of oolitic iron formations—reply. *J. sedim. Petrol.* **50**, 1003–4.
Knox R. W. O'B. 1970. Chamositic ooliths from the Winter Gill Ironstone (Jurassic) of Yorkshire, England *J. sedim. Petrol.* **40**, 1216–25.
Krom, M. D. & Berner, R. A. 1981. The diagenesis of phosphorus in a nearshore marine environment. *Geochim. cosmochim. Acta,* **45**, 207–16.
Krynine, P. D. 1935. Arkose deposits in the humid tropics. *Am. J. Sci.* **39**, 353–63.
—— 1949. The origin of red beds. *NY Acad. Sci. Trans.* **II**, 60–68.
—— 1950. Petrology, stratigraphy and origin of the Triassic sedimentary rocks of Connecticut. *Conn. Geol. Nat. Hist. Survey Bull.* **73**, 239 pp.
Land, L. S. & Dutton, S. P. 1978. Cementation of a Pennsylvannian deltaic sandstone: isotopic data. *J. sedim. Petrol.* **48**, 1167–76.
—— & —— 1979. Cementation of sandstones—reply. *J. sedim. Petrol.* **49**, 1359–61.
—— & Milliken, K. L. 1981. Feldspar diagenesis in the Frio Formation, Brazoria County, Texas, Gulf Coast. *Geology,* **9**, 314–18.
Lerbekmo, J. F. 1957. Authigenic montmorillonoid cement in andesitic sandstones *J. sedim. Petrol.* **27**, 298–305.
Lin, F. C. & Clemency, C. V. 1981. The kinetics of dissolution of muscovite at 25°C and 1 atm. CO_2 partial pressure. *Geochim. cosmochim. Acta,* **45**, 571–6.
Lindquist, S. J. 1977. Secondary porosity development and subsequent reduction overpressed Frio Formation (Oligocene) south Texas. *Trans Gulf-Cst Ass. geol. Socs,* **27**, 99–107.
Lippman, F. 1982. The thermodynamics status of clay minerals. *Developments in Sedimentology,* **35**, 475–486.
Long, J. V. P. & Agrell, S. O. 1965. The cathodoluminescence of minerals in thin section. *Min. Mag.* **34**, 318–26.
Loucks, R. G., Bebout, D. G. & Galloway, W. E. 1977. Relationship of porosity formation and preservation to sandstone consolidation history. *Trans. Gulf-Cst Ass. geol. Socs,* **27**, 109–20.
Lyell, C. 1852. *Manual of Elementary Geology.* Murray, London.
McBride, E. F. 1974. Significance of colour in red, green, purple, olive-brown and grey beds of Difunta Group, northwestern Mexico, *J. sedim. Petrol.* **44**, 760–73.
—— 1978. Porosity loss in sandstone by ductile grain deformation during compaction. *Trans. Gulf-Cst Ass. geol. Socs,* **28**, 323–5.
McHardy, W. J., Wilson, M. J. & Tait, J. M. 1982. Electron microscope and X-ray diffraction studies of filamentous illitic clay from sandstones of the Magnus Field. *Clay Miner.* **17**f, 23–9.
Meade, R. H. 1966. Factors influencing the early stages of the compaction of clays and sands—review. *J. sedim. Petrol.* **36**, 1085–101.
Merino, E. 1975. Diagenesis in Tertiary sandstones from Kettleman, North Dome, California. I.

Diagenetic mineralogy *J. sedim. Petrol.* **45**, 320–36.
—— & Ransom, B. 1982. Free energies of formation of illite solid solutions and their compositional dependence *Clays Clay Miner.* **30**, 29–39.
Milliken, K. L., Land, L. S. & Loucks, R. G. 1981. History of burial diagenesis determined from isotopic geochemistry. Frio Formation, Brazoria Country, Texas. *Am. Ass. Petrol. Geol. Bull.* **65**, 1397–413.
Nadeau, P. H., Tait, J. M., McHardy, W. J. and Wilson, M. J. 1984. Interstratified XRD characteristics of physical mixtures of elementary clay particles. *Clay Miner.* **19**, 67–76.
Nagtegaal, P. J. C. 1969. Sedimentology, palaeoclimatology and diagenesis of post Hercynian continental deposits of the south central Pyrenees, Spain. *Leid. Geol. Meded,* **42**, 143–238.
—— 1978. Sandstone framework instability as a function of burial diagenesis. *J. geol. Soc. London,* **135**, 101–5.
Nahon, D., Carrozi, A. V. & Parron, C. 1980. Lateritic weathering as a mechanism for the generation of ferruginous ooids. *J. sedim. Petrol.* **50**, 1287–98.
Nesbitt, H. W. 1980. Characterization of mineralformation water interactions in Carboniferous sandstones and shales of the Illinois sedimentary basin. *Am. J. Sci.* **280**, 607–30.
Nishiyama, T., Oinuma, K. & Sato, M. 1978. An interstratified chlorite-vermiculite in weathered red shale near Toyoma, Japan, (Abstract) *Proc. 6th Int. Clay Conf., Oxford,* p. 157.
Odin, G. S. & Matter, A. 1981. De glauconiarum origin. *Sedimentology,* **28**, 611–41.
Parker, C. A. 1974. Geopressures and secondary porosity in the deep Jurassic of Mississippi. *Trans. Gulf-Cst Ass. geol. Socs,* **24**, 69–80.
Pearson, M. J. Watkins, D. & Small, J. S. 1982. Clay diagenesis and organic maturation in Northern North Sea sediments. *Dev. Sedimentol.* **35**, 665–75.
Pedersen, G. K. & Anderson, P. F. 1980. Depositional environments, diagenetic history and source areas of some Bunter Sandstones in Northern Jutland. *Danm. geol. Unders. Arborg,* **1979**, 69–93.
Perry, E. A. 1974. Diagenesis and the K-Ar dating of shale and clay minerals. *Bull. geol. Soc. Am.* **85**, 827–30.
—— & Hower, J. 1970. Burial diagenesis in Gulf Coast pelitic sediments. *Clays Clay Miner.* **18**, 165–178.
—— & —— 1972. Late stage dehydration in deeply buried pelitic sediments *Am. Assoc. Petrol. Geol. Bull.* **56**, 2013–21.
Pettijohn, F. J., Potter, P. E. & Siever, R. 1973. *Sand and Sandstone,* Springer Verlag, Berlin. 618 pp.
Phillips, 1881. On the constitution and history of grits and sandstones. *Q. Jl geol. Soc. London,* **37**, 6–25.

PICARD, G. L. & FELBECK, G. T. JR 1976. The complexation of iron by marine humic acid *Geochim. cosmochim Acta*, **40**, 1347–50.

PITTMAN, E. D. 1972. Diagenesis of quartz in sandstones as revealed by SEM *J. sedim. Petrol.* **42**, 507–19.

—— & DUSCHATKO, R. W. 1970. Use of pore casts and SEM to study pore geometry. *J. sedim. Petrol.* **40**, 1153–7.

PORRENGA, D. K. 1967. Glauconite and chamosite as depth indicators in the marine environment. *Mar. Geol.* **5**, 495–501.

POWELL, T. G., FOSCOLOS, A. E., GUNTHER, P. R. & SNOWDON, L. R. 1978. Diagenesis of organic matter and fine clay minerals: a comparative study. *Geochem. cosmochim. Acta*, **42**, 1121–97.

POWERS, M. C., 1967. Fluid release mechanisms in compacting marine mudrocks and their importance in oil exploration *Am. Assoc. Petrol. Geol. Bull.* **51**, 1240–54.

RAMSEYER, K. 1983. A new cathodoluminescence microscope and its application to sandstone diagenesis. *11th Int. Cong. Sed. (I.A.S.) Hamilton* p. 885 (abstract).

READING, H. G. (ed.) 1978 *Sedimentary Environments and Facies*, Blackwell Scientific Publications, Oxford. 557 pp.

RETALLACK, G. J. 1976. Triassic palaeosols in the Upper Narrabeen Group of New South Wales. I. features of the palaeosols. *J. geol. Soc. Aust.* **23**, 383–99.

ROEDDER, E. 1979. Fluid inclusion evidence on the environment of sedimentary diagenesis, a review. *In*: SCHOLLE, P. A. & SCHLUGER, P. R. (eds) *Aspects of Diagenesis. Spec. Publs Soc. econ. Miner. Paleont., Tulsa*, **26**, 89–107.

ROSSEL, N. C. 1982. Clay mineral diagenesis in Rotliegend aeolian sandstones of the Southern North Sea. *Clay Miner.* **17**, 69–77.

SALOMONS, W., GOULDIE, A. S. & MOOK, W. G. 1978. Isotopic composition of calcrete deposits from Europe, Africa and India. *Earth Surf. Proc.* **3**, 43–57.

SCHMIDT, V. & McDONALD, D. A. 1979a. The role of secondary porosity in the course of sandstone diagenesis. *In*: SCHOLLE, P. A. & SCHLUGER, P. R. (eds) *Aspects of Diagenesis. Spec. Publs Soc. econ. Miner. Paleont., Tulsa*, **26**, 175–207.

—— & —— 1979b. Texture and recognition of secondary porosity in sandstones. *In*: SCHOLLE, P. A. & SCHLUGER, P. R. (eds) *Aspects of Diagenesis. Spec. Publs Soc. econ. Miner. Paleont., Tulsa*, **26**, 209–25.

SEEMAN, U. 1979. Diagenetically formed interstitial clay minerals as a factor in Rotliegend Sandstone reservoir quality in the Dutch Sector of the North Sea. *J. Petrol. Geol.* **1**, 55–62.

SHELTON, J. W. 1964. Authigenic kaolinite in sandstone. *J. sedim. Petrol.* **34**, 102–11.

SIBLEY, D. F. & BLATT, H. 1976. Intergranular pressure solution and cementation of the Tuscorora Orthoquartzite. *J. sediment. Petrol.* **46**, 881–96.

SIEVER, R. 1979. Plate-tectonic controls on diagenesis.

J. Geol. **87**, 126–55.

SIPPEL, R. F. 1968. Sandstone petrology, evidence from luminescence petrography. *J. sedim. Petrol.* **38**, 530–54.

SMITH, J. V. & STENSTROM, R. C. 1963. Electron excited luminescence as a petrologic tool. *J. Geol.* **73**, 627–35.

SOMMER, F. 1978. Diagenesis of Jurassic sandstones in the Viking Graben. *J. geol. Soc. London*, **135**, 63–7.

STALDER, P. J. 1975. Cementation of Pliocene–Quaternary fluviatile clastic deposits in and along the Ocean Mountains. *Geologie Mijnb.* **54**, 148–56.

STANTON, G. D. 1977. Secondary porosity in sandstones of the Lower Wilcox (Eocene), Kames County, Texas. *Trans. Gulf-Cst Ass. geol. Soc.* **27**, 197–207.

STAUB, J. R. & COHEN, A. D. 1978. Kaolinite-enrichment beneath coals: a modern analog, Snuggedy Swamp, South Carolina. *J. sediment. Petrol.* **48**, 203–10.

STOESSELL, R. K. & MOORE, C. H. 1983. Chemical constraints and origins of four groups of Gulf Coast reservoir fluids. *Am. Ass. Petrol. Geol. Bull.* **67**, 896–906.

STOREY, S. R. 1979. Clay-carbonate diagenesis of deltaic sandstones—Basal Belly River Formation (Upper Cretaceous) Central Alberta, Canada (Abs) *Am. Ass. Petrol. Geol. Bull.* **63**, 534.

TAN, F. C. & HUDSON, J. D. 1974. Isotopic studies of the palaeoeology and diagenesis of the Great Estuarine Series (Jurassic) of Scotland. *Scott. J. Geol.* **10**, 91–128.

TARDY, Y. 1971. Characterisation of the principal weathering types by the geochemistry of waters from European and African crystalline massifs. *Chem. Geol.* **7**, 253–71.

TAYLOR, B. J. 1978. Westphalian. *In*: MOSELEY, F. (ed.) *The geology of the Lake District. Yorkshire Geological Society Occasional Publication*, **3**, 180–9.

TAYLOR, J. M. 1950. Pore-space reduction in sandstone. *Am. Ass. Petrol. Geol. Bull.* **34**, 701–16.

TAYLOR S. R. 1983. A stable isotope study of the Mercia Mudstones (Keuper Marl) and associated sulphate horizons in the English Midlands. *Sedimentology*, **30**, 11–31.

THOMAS, M. F. 1974. *Tropical Geomorphology. A Study of Weathering and Landform Development in Warm Climates*, MacMillan Press Ltd, London. 332 pp.

TILLMAN, R. W. & ALMON, W. R. 1979. Diagenesis of Frontier Formation offshore bar sandstones, Spearhead Ranch Field, Wyoming. *In*: SCHOLLE, P. A. & SCHLUGER, P.R. (eds) *Aspects of Diagenesis. Spec. Publs Soc. econ. Miner. Paleont., Tulsa*, **26**, 337–78.

TISSOT, B., DURAND, B., ESPITALIC, J. & COMBAZ, A. 1974. Influence of nature and diagenesis of organic matter in formation of petroleum *Am. Ass. Petrol. Geol. Bull.* **58**, 499–506.

TOMPKINS, R. E. 1981. Scanning electron microscopy

of a regular chlorite/smectite (corrensite) from a hydrocarbon reservoir sandstone *Clays Clay Miner.* **29**, 233–5.

TURNER, P. 1980. *Continental red beds. Dev. Sedimentol.* **29**, 562 pp.

—— & ARCHER, R., 1977. The role of biotite in the diagenesis of red beds from the Devonian of northern Scotland. *Sed. Geol.* **19**, 241–51.

VELDE, B. 1977. *Clays and clay minerals in natural and synthetic systems. Dev. Sedimentol.* **21**, 218 pp.

VESSEL, R. K. & DAVIES, D. K. 1981. Non-marine sedimentation in an active fore arc basin. *In*: ETHRIDGE, F. G. & FLORES, R. M. (eds) *Recent and Ancient Non-marine Depositional Environments: Models for Exploration. Spec. Publs Soc. econ. Miner. Paleont.*, Tulsa, **31**, 31–45.

WALDSCHMIDT, W. A. 1941. Cementing materials in sandstones and their probable influence on migration and accumulation of oil and gas. *Am. Ass. Petrol. Geol. Bull.* **25**, 1839–79.

WALKER, R. G. (ed.) 1979. *Facies models. Geoscience Canada Reprint Series No. 1 Geol. Assoc. Canada*, 211 pp.

WALKER, T. R. 1967. Formation of red beds in modern and ancient deserts. *Bull. geol. Soc. Am.* **79**, 353–68.

—— 1974. Formation of red beds in moist tropical climates: a hypothesis. *Bull. Geol. Soc. Am.* **85**, 633–8.

—— 1975. Diagenetic origin of red beds. *In*: FALKE, M. (ed.) *The Continental Permian in Central, West and South Europe. NATO A.S.I. Series C*, **22**, 240–82.

——, RIBBE, P. H. & HONEA, R. M. 1967. Geochemistry of hornblend alteration in Pliocene red beds, Baja California, Mexico. *Bull. geol. Soc. Am.* **78**, 1055–60.

—— & HONEA, R. M. 1969. Iron content of modern deposits in the Sonoran desert: a contribution to the origin of red beds. *Bull. geol. Soc. Am.* **80**, 534–44.

——, WAUGH, B. & CRONE, A. J. 1978. Diagenesis in first cycle desert alluvium of Cenozoic age, southwestern United States and northwestern Mexico. *Bull. geol. Soc. Am.* **89**, 19–32.

WALTON, N. R. G. 1981. A detailed hydrogeochemical study of groundwaters from the Triassic sandstone aquifer of south west England. *Rep. Inst. Geol. Sci* No. 81/5.

WARME, J. E., DOUGLAS, R. G. & WINTERER, E. L. (eds) 1981. *The Deep Sea Drilling Project: A Decade of Progress. Spec. Publs Soc. econ. Miner. Paleont.*, Tulsa, **32**, 564 pp.

WAUGH, B. 1970. Petrology, provenance, and silica diagenesis of the Penrith Sandstone (lower Permian) of northwest England. *J. sedim. Petrol.* **40**, 1226–1240.

—— 1978. Authigenic K-feldspar in British Permo-Triassic sandstones. *J. geol. Soc. London*, **135**, 51–6.

WEINBRANDT, R. M. & FATT, I. 1969. A scanning electron microscope study of the pore structure of sandstone. *J. Petrol. Tech.* **21**, 543–8.

WESCOTT, W. A. 1983. Diagenesis of Cotton Valley Sandstone (Upper Jurassic) East Texas: implications for tight gas formation pay recognition. *Am. Ass. Petrol. Geol. Bull.* **67**, 1002–13.

WHETTEN, J. T. & HAWKINS, J. W. 1970. Diagenetic origin of greywacke matrix minerals. *Sedimentology*, **15**, 347–61.

WHITE, D. E. 1965. Saline water of sedimentary rocks. *In*: YOUNG, A. & GALLEY, J. E. (eds) *Fluids in Subsurface Environments. Mem. Am. Ass. Petrol. Geol.* **4**, 342–66.

WILSON, M. D. & PITTMAN, E. D. 1977. Authigenic clays in sandstones: recognition and influence on reservoir properties and palaeoenvironment analysis. *J. sedim. Petrol*, **47**, 3–31.

WILSON, M. J. & DUTHIE, D. M. L. 1981. Some aspects of intrastratal alteration in the Old Red Sandstone. *Scott. J. Geol.* **23**, 57–64.

WILSON, R. C. L. (ed.) 1983. *Residual Deposits: Surface Related Weathering Processes and Materials. Geol. Soc. London Sp. Publ.* **11**, 258 pp.

WOOD, J. R. & HEWETT, T. A. (1982). Fluid convection and mass transfer in porous sandstones—a theoretical model. *Geochim. cosmochim. Acta*, **46**, 1707–14.

WOPFNER, H. 1978. Silcretes of Northern South Australia and adjacent regions. *In*: LANGFORD-SMITH, T. (ed.) *Silcrete in Australia*, 93–141. Department of Geography, University of New England, Australia.

ZINKERNAGEL, U. 1978. Cathodoluminescence of quartz and its applications to sandstone petrology. *Contrib. Sedimentol.* **8**, 69 pp.

S. D. BURLEY, Geologisches Institut, Universität Bern, Baltzerstrasse-1, 3012 Bern, Switzerland.

J. D. KANTOROWICZ, Koninklijke/Shell Exploratie en Produktie Laboratorium, Volmerlaan 6, Rijswijk ZH, The Netherlands.

B. WAUGH, AGAT Consultants Inc., 1215 18th Street, 260 Cable Building, Denver, Co 80202, USA.

ECONOMIC AND APPLIED ASPECTS

Sedimentary ore deposits

H. Clemmey

SUMMARY: Until the last two decades the magmato-hydrothermal theories for the origin of ores were considered to be predominant. The change in viewpoint over 20 years or so towards a general acceptance of various syngenetic and epigenetic concepts in ore genesis is outlined in the opening section of this review. In the following section the concentration of metals particularly Al, Ni and to some extent Au in the weathering crust is described. The ways in which metals are concentrated by physical sedimentation are outlined, with particular attention to gold and tin placer deposits and the importance of concepts such as geomorphic thresholds, complex response, and episodic erosion is emphasized. For these deposits it is shown that following the release of elements from bedrock sources they are cycled through the sedimentary system. Essentially, elements move from proximal to distal fluvial sites and from stratigraphically lower to higher positions. Resultant orebodies trend from high grade/low tonnage to lower grade/higher tonnage in a cycle.

Chemical sediments, and particularly ironstones are then discussed. Oolitic ironstones occur in two major phases in the Phanerozoic and although their formation in nearshore sheltered sites is becoming established the precise mode of formation, whether primary or diagenetic, is still uncertain. The mainly Precambrian 'banded iron formations' pose the problems of the source and transport mechanism for such vast volumes of iron-bearing sediment. Various hypotheses are discussed and a palaeoenvironmental approach adopted. The possibility is emphasized that the unroofing of mafic Archaean basement in the early Proterozoic may well have been a critical factor in their formation and time distribution.

Ore deposits of lead, zinc, copper and uranium are found in continental or marine sediments. The metals are derived from bedrock, pass through the weathering crust and are then incorporated into the sedimentary prism as 'fluid' or detrital oxide complexes. Migration takes place in interstratal solutions and they are finally precipitated as ores in the diagenetic environment. A broad 2-fold environmental division can be recognized, with some ores forming in fluviatile, lacustrine or paralic environments (Cu, U and to a lesser extent Pb, Zn) and secondly those in nearshore to offshore basin or platform sites (Pb, Zn). The movement of metallic fluids out of basinal shales and into basin margin facies, often up growth faults, is described and a model is outlined to account for lead-zinc deposits associated with carbonates and black shales. The formation of copper and uranium deposits in continental facies associations and above sub-redbed unconformities, the origin of uranium-rich conglomerates in the Proterozoic, the formation of sediment-hosted copper deposits and the copper–gold–uranium association in sedimentary rocks are all discussed in terms of a 'fluid unroofing' model.

In the final section it is emphasized that we need to refine 'fluid unroofing' models, and have a better understanding of the way metal is cycled through the sedimentary – diagenetic environment. The importance of sophisticated facies analysis, the relevance of 'maturity' with regard to metals as well as oils, and the significance of biological mediation in ore formation are all emphasized. Finally the critical importance of the weathering horizon in establishing the potential metal fertility of any sedimentary basin is noted.

Despite the impression that may be gained by reading the classic text books on economic geology, the bulk of the world's metal and energy reserves are contained within sedimentary rocks as either stratiform or stratabound deposits. Unfortunately in a short review it is not possible to consider all the economic aspects of sedimentary rocks so most of the bulk minerals and hydrocarbons are omitted: for reference they are given in Table 1. In most cases the formation of ore deposits of the elements shown in Table 1 is tolerably well understood. In contrast, speculation on the mode of origin of the metallic elements is vigorous and long lived.

Historical perspectives

Investigations into the origin of metal deposits must consider the four great problems:

(i) Where was the source of the metals?

(ii) What was the nature and origin of the metal bearing fluid?

(iii) How did the fluid reach the site of metal precipitation?

(iv) What caused the fluids to precipitate their load in economic quantities?

The history of developing ideas on these topics will form the basis of this section. It shows how over the past 20 years or so, sedimentary con-

TABLE 1. *Economic components of sedimentary rocks not covered in the text*

Hydrocarbons	Evaporites/brines	Industrial minerals/sediments
Petroleum	Halite	Limestone
Natural gas	Gypsum/anhydrite	Sand and gravel
Bitumen	Sodium carbonate	Stone
Oil shale	Sodium sulphate	Silica sand
Coal	Strontium	Ball clay
Peat	Magnesite	Bentonite
	Bromine	Diatomite
	Nitrates	

cepts have gradually been injected into ore genesis models. Ores of elements such as Fe, Al, Mn and Ni, found in the insoluble oxide form, are typical deposits of the residual weathering crust, and though their genesis is interesting it has long been accepted that they are formed by exogenetic processes and no great controversy prevails as to their origin. A similar situation exists for those elements present as the native metal or contained within minerals stable in the oxidizing environment. Such elements are typically concentrated in high-energy facies and form gold, diamond, tin and titanium placer deposits. No real controversy exists when, as is the usual case, the placers are found in unconsolidated Cenozoic sediments. However, if the same type of deposits are located within ancient lithified strata they then become controversial—witness the great debates on the gold palaeo-placers of the lower Proterozoic.

Elements having extreme solubility are typically found in chemical sediments and brines of evaporite origin and bear witness to the great efficiency of the weathering/diagenetic couple in fractionation and concentration. Again the basic ideas on economic concentration are generally accepted. However when the great deposits of the base metal sulphides are discussed the spectrum of ideas is wide and rigorous sedimentology has been late in making a contribution to the debate. So much so that at the beginning of the period in question that great student of ore deposits, Professor David Williams, could state in the conclusions to his presidential address to the Geological Association (1960): 'Current dissatisfaction in certain quarters with the magmatic/hydrothermal theory of the origin of mineral deposits seems in the main, to be without justification' and 'among hydrothermal solutions the role of metamorphic and ground-water types seems subordinate to that of magmatic water' and finally 'sedimentary sulphide ores derived from

the erosion of pre-existing mineralization are of minor importance, except possibly in the case of iron pyrites'. It is interesting to see how these statements survived the ensuing 20 years.

Metal sources

Depth, heat and high pressure are factors which have dominated ore-deposit modelling so that early ideas on metal sources were concerned with lower crust/mantle sources. Spurr (1923) invoked metal magma periodically injected to higher levels and Brown (1950) envisaged a series of metal belts encircling the deep earth and periodically releasing metal as 'vapours' able to move into upper crustal sites of deposition. Powerful as these models were at the time, many geologists, especially those in industry, expressed increasing unease with magmato-thermal models for all orebodies. They realized the strong stratigraphic control on many strat-abound and stratiform ores noting that many of the ore minerals were either disseminated within normal sediments faithfully following sedimentary structures, or were themselves bedded. The great copper deposits of Zambia (formerly Northern Rhodesia) and the historically important lead-zinc deposits of Meggen and Rammelsberg in Devonian black shales of Germany and of course the lead-zinc and minor copper of the European Kupferscheifer, provide good examples. To account for these observations the syngenetic theory was developed. According to this theory the metal contained in deposits was precipitated from seawater and deposited with the sediments containing it; thus ultimately the metal source was derived by erosion of primary sources in the hinterland. During the late 1940s and early 1950s E. G. Garlick developed the syngenetic theory, based on evidence from the Copperbelt and his ideas had a profound effect on economic geologists who became acquainted with them. He was amongst the first geologists to use modern principles of facies analysis. Garlick's work first reached a wider audience with the publication of *'The Geology of the Northern Rhodesian Copperbelt* (Mendelsohn 1961) and later in a series of powerful papers (Garlick 1967, 1972, 1981). He drew attention to the existence of a zoning in metal sulphide ores,

Chalcocite   Bornite   Chalcopyrite   Pyrite
$CuS \longrightarrow Cu_5FeS_4$ $CuFeS_2 \longrightarrow FeS_2$

reflecting a passage from nearshore to offshore environments, interpreting this to mean a decreasing copper content of the seawaters away from the land (his metal source). At a similar time syngenetic principles were incorporated into the source-bed concept (Knight 1957). Low background concentrations of metal were envisaged as precipitating syngenetically from seawater and were later concentrated into ore by lateral movements. Metamorphism was regarded as the major influence and the theory helped to reconcile the obvious epigenetic features of many stratiform ores with the syngenetic theory. Knight's theory was the forerunner of many developed since, involving more modern concepts of diagenesis caused by interstratal fluid migration.

The implications of the source-bed concept were soon apparent and workers began to look to the sediments themselves as metal sources. Barnes (1959), for example, was amongst the first to draw attention to the possibility of black shales as metal donors and also indicated the significance of the black shale–carbonate facies front in metallogenesis.

Gradually the concepts of sedimentary prisms as metal sources and the significance of burial diagenesis became established leading Beales & Jackson (1966), and Jackson & Beales (1967) to propose an integrated sedimentary model for the epigenetic, so-called 'Mississippi Valley Type' lead-zinc ores. Beales emphasized the location of such orebodies in platform marginal facies, using the Pine Point district in NW Canada as an example and predicted the metal source migrating out of the 'offshore' basinal shales to the basin margin during dewatering.

In effect the metal source was derived from erosion of a hinterland but not by the simple one-step process envisaged by Garlick. In reality, any sedimentary prism inherits the geochemistry of its source area via erosion and solution and many workers in the U.S.S.R. and on the continent held this as a central concept when interpreting ore deposits in sediments (Strakhov 1970; Samama 1973).

Ore fluids

During the last two or three decades the vapour and magmatic models for the composition of ore fluids gradually became discredited except in a few exceptional cases. More attention has been given to the possibility that metals could travel as suspensions, colloids and above all as complex ions in hydrothermal solutions. Two major influences developed. The first was the idea that ore fluids need not be of high temperature but that ordinary ground water could suffice. Such ideas were prompted by work on the Colorado Plateau-type uranium deposits in red-bed sandstones and given substance by work on the low-temperature geochemistry of uranium (Hostetler & Garrels 1962; Garrels & Christ 1965). The second major influence was the work of the Russians on their deep geothermal wells, and the reports in the West of the Salton Sea well (White *et al.* 1963, etc.) and of the Red Sea brine pools (e.g. Miller 1966). Such brines were shown to be chloride-rich and known to precipitate metal oxides and sulphides. The low sulphur contents of the brines surprised many workers. Reports of oilfield brines confirmed the chloride-rich nature and high trace-metal content of deep brines, and the concept of formation-waters scavenging metals deep in sedimentary prisms and carrying their metals as a chloride complex became well established (Dunham 1970). Fluid inclusion work (see Roedder 1976, 1979 for reviews), theoretical modelling (Anderson 1978) and work on the relations between dewatering of shale basins and the association between oil and ore deposits (see Garrard 1977) has shown the role of dense (1.1 g cm^{-3}), slow moving (m s^{-1} year^{-1}), low temperature (100–160°C), high salinity (15 wt% NaCl equiv.), low-sulphur brines, in ore genesis. True magmato-hydrothermal connate brines and meteoric ground water were all now established as possible ore fluids.

Modelling of flow characteristics of those brines, which may reach the sea floor and 'exhale' (see later section on plumbing) is currently in vogue (Turner & Gustafson 1978) following the impetus provided by the discovery of mid-ocean-ridge 'smokers' and other brine pools with associated ores (Francheteau *et al.* 1979)

Precipitation of ore metals

Temperature fall, pressure release and the geochemical environment (e.g. Eh/pH changes, chemically receptive hostrock) have always been invoked to explain metal precipitation, and are still considered to be important factors. However, the development of the syngenetic theory proposed by Garlick brought with it the concept of biologically mediated precipitation of metals. Initially, models were concerned with direct precipitation from marine waters (Garlick 1961, etc.; Temple 1964). Although proven as a process, it became discredited as a means of obtaining vast tonnages of metal, and attention switched to the diagenetic environment of H_2S-rich fetid muds as a biologically produced reducing environment in which metal precipitation could take place. The work on sedimentary pyrite formation provided impetus, especially when many workers began to describe early copper replacements of diagenetic pyrite (Bartholome 1974).

It has commonly been suggested that the bacterial reduction of evaporite sulphate (Beales & Jackson 1966; Beales 1975; Annels 1974), explains the association between evaporites and many base metal deposits. This must provide a way around the dilemma that in low-temperature fluids metal and sulphur can not travel together, as witnessed by the low sulphur contents of brines. Similar concepts of mixing two fluids to create precipitation are now extended into the ground-water realm (Mann & Deutscher 1978). The use of sulphur isotopes in 'proving' the role of bacterial reduction showed great promise (Jensen 1959) in being able to indicate the existence or not of bacterial involvement in metal precipitation.

Plumbing systems for migrating ore fluids

Deep-seated faults and fractures are still regarded as the prime conduits for migrating fluids of any origin. Probably the most important development has been the recognition of the importance of growth faults, particularly at basin margins. Many orebodies (e.g. the Irish Pb/Zn deposits in Carboniferous rocks) are known to be related to these synsedimentary structures. The influential 'submarine exhalite concept' (Schneiderhohn 1955; Oftedahl 1958) has now been firmly wedded to the growing documentation of growth faulting. Primary porosity/permeability in sediments is also

Element	Weathering horizon	Physical sedimentation	Chemical sediments	Brines	Diagenetic-epigenetic	By-product
Fe	■ ■ ■		■ ■ ■			V
Al	■ ■ ■					
Mn	■ ■ ■		■ ■ ■			
Ni	■ ■ ■					
C		■ ■ ■				
Zr		■ ■ ■				
Sn		■ ■ ■				
RE		■ ■ ■			●	
Au		■ ■ ■				
Pt		■ ■ ■				
Th		■ ■ · ■				
Ba			■ ■ ■		■ ■ ■	
B			■ ■ ■			
I,			■ ■ ■	■ ■ ■		N
Li				■ ■ ■		
K				■ ■ ■		
Cu	●		●		■ ■ ■	Ag, Co
U		●	■ ■ ■		■ ■ ■	V
Pb					■ ■ ■	Bi, Ag, Sb
Zn					■ ■ ■	Hg, Cd, Ag, Sb
P						V

FIG. 1. Economic elements for which sedimentary rocks provide a major, and in some cases the total, world resource. Grouping is by environment or process of formation. Major source (■); minor source (●).

regarded as important for fluid migration and the role of secondary porosity, particularly in carbonates, is becoming well established in the ore-deposits literature.

Summary

At the start of the era under review epigenetic concepts were dominant, magma-derived hydrothermal fluids were regarded as of pre-eminent importance and if diagenesis was recognized at all, it was termed metamorphism. Only residual ores and placer deposits were accepted as sedimentary. Most of the principal texts on ore deposits still reflect this thinking.

Gradually the concept of a syngenetic origin for metal was accepted and followed by developing ideas on syngenetic origins and diagenetic precipitation so that now we may refer to diagenetic intra-sediment sources and diagenetic precipitation mechanisms for late obviously epigenetic ores. We accept that many faults can be synsedimentary, tapping basinal not hydrothermal fluids, leading to the concept of diagenetic sources and syngenetic precipitation (exhalites). This progress has been documented firstly by the publication of thematic volumes devoted to sedimentary ores (James 1967; Amstutz & Bernard 1973) and many metal-specific thematic sets (e.g. uranium: Kimberley 1978b; lead-zinc: Sangster 1983), while Badham (1981a) reviewed the origins of metal deposits, more generally, in sedimentary rocks. As a basis for this review of the 'present state of the art' the compilation presented in Fig. 1 is used.

Sedimentary ore deposits

Ore deposits of the weathering crust

The origins of bauxite (the principal ore of aluminium) and of nickel in laterites are tolerably well understood. Such ores are a product of:-

> Heavy rainfall: 1500–4000 mm year^{-1}.
> Hot climate: 25°C annual average.
> Long wet season: 10 months + .
> Mature geomorphology.

Silicate minerals are hydrolysed and high oxidation potentials together with an overall pH of around 4/5 result in the eluviation of silica and most of the bases. Elements such as Fe and Al, whose oxides are stable, remain behind; in this sense the ores are residual. Generally the best residual ores are developed on a bedrock containing little free quartz and enhanced concentrations of the element sought. Thus economic concentrations of bauxite are preferentially developed on syenites and nepheline syenites and nickel ores over ultrabasic rocks.

Important advances have been concerned with the subtleties of mineral breakdown and inter-horizon conditions of element separation (Loughnan 1969).

Bauxitic laterites provide over 95% of the world's Al and are developed throughout the tropical world in Australia, Jamaica, South America, India, North America, Africa and Asia. Cenozoic and Cretaceous rocks provide the best potential, whilst older bauxite ores are found in Europe and the U.S.S.R. in the Palaeozoic and in the lower Precambrian of Swaziland. The principal hydroxide minerals show an interesting generalized diagenetic ageing scheme:

Cenozoic	Mesozoic	Palaeozoic
Gibbsite,	Boehmite,	Diaspore,
Al(OH)$_3$	AlO(OH)	AlO(OH)

A key question of bauxite genesis relates to the separation of iron and aluminium, for the oxides are both insoluble over the pH range of most laterites. However a small field exists around Eh 0.6 in which iron is still soluble as Fe^{++} whilst aluminium is not. Perhaps significantly bauxitic laterites tend to form beneath areas having a denser vegetation cover than ferritic laterites. In such areas soil profiles should carry slightly higher soil moisture regimes and have reduced ground-water Eh, reflecting the higher organic supply. Magnetite, a common accessory of bauxite ore, reflects the lower redox state. Further separation can be effected by detrital transport—e.g. bedded lacustrine bauxites of the U.S.S.R. and, more commonly those in China, may grade 70% Al$_2$O$_3$.

Nickel laterites provide the principal future reserve of this strategic metal. They are common on ophiolites and the basic/ultrabasic rocks in back-arc terrains throughout the Pacific region, and also in Indonesia, Philippines, Dominican Republic and Queensland whilst vast reserves exist in Cuba and New Caledonia. The ore is concentrated at the base of the laterite profile in the saprolite zone of the bedrock, which is commonly serpentinite, peridotite or harzburgite (Zeissnik 1969). Enrichment factors are usually in the range ×6 to × 10 and clearly Ni is leached down from the upper levels of the soil profile. A typical situation would be: whole profile grade 1.5% Ni, base of profile 2.85% Ni, bedrock serpentine 0.24% Ni (Esson & Surcon dos Santos 1978). The bucket term 'garnerite' is used to name a series of nickel silicates approximating to $(Ni\ Mg)_3Si_4O_{10}(OH)_2H_2O$ which can hold 10–25% Ni and are often regarded as the ore mineral assemblage. However, it seems that the bulk of the Ni is associated with goethite (Fe:OH).

Other elements are commonly enriched in the weathering crust over orebodies and contribute greatly to their economics. Gold is often enriched in the ferric gossan produced during oxidation of sulphide orebodies, as for example the gossan associated with the massive sulphides ores at Rio Tinto on the Spanish Pyrite Belt. Low values of less than 1 ppm are brought up to values of 3 ppm. Locally at the base of the gossans values up to 24 ppm are reported. Controversy surrounds the subject of gold mobility in the weathering zone for it is traditionally regarded as insoluble. However the concentration towards the base of many gossan profiles, as in the case of nickel, must indicate some movement.

Enrichment of gold in Australian vein deposits, associated with weathering profiles, seems to confirm this (McIlveen & Stevens 1979). It appears that in the presence of abundant Fe^{+++}, MnO_2 and Cu^{+++} and acid conditions, gold may pass into 'solution' and move as the chloride complex, i.e.:

$$2Au + 12H + 3MnO_2 + 8Cl =$$
$$3Mn^2 + 2AuCl_4 ? 6H_2O.$$

Such conditions are met above an oxidizing orebody.

The process of low background metal contents in an orebody being enriched by oxidation/solution/reprecipitation reactions is termed supergene enrichment and plays an important role in determing the economics of many low-grade, large-tonnage, base-metal deposits in arid regions. It must also be regarded as a powerful mechanism for removing economic elements from primary, bedrock sources and transferring them to the surrounding sediment; we may term this process 'fluid unroofing' and it is important in the genesis of many copper and uranium deposits in sediments. It provides an efficient mechanism for concentrating low background metal contents of whole source areas before releasing them to the sedimentary prism.

We may conclude that the weathering crust is not only an important source of economic ores but that it also exerts a strong influence on the concentration, timing and form (detrital bedload, complex, colloid, adsorbed or in suspension) that economic elements reach the sedimentary system. In this way the weathering crust determines the potential fertility of the sedimentary system with regard to the formation of ore deposits.

Metals concentrated by physical sedimentation

Ores formed by the mechanical action of streams and waves are termed 'placer deposits' and the minerals must fulfil the requirements of chemical stability, fracture and abrasion resistance and ideally have a high specific gravity to aid their separation. The most common placer minerals are the heavy metal phosphates oxides and silicates, monazite $CeLaThPO_4$, zircon $ZrSiO_4$, rutile TiO_2, ilmenite $FeTiO_3$, magnetite Fe_2O_3, cassiterite SnO_2 and chromite $FeCr_2O_4$, forming the so-called black-sand assemblage. Economically, they form the major reserve of titanium zirconium and the rare earths. Rare but far more valuable placers are diamond and gold. Reworked coastal dune sands provide most of the black-sand and diamond reserves, and 'pay streaks' preserved along fossil strand lines are usually the target. Reworking is the key to a good placer concentration and the combination of cyclonic storm surf and longshore drift is an ideal one. Lower-energy storms may sometimes destroy the placers but fair-weather build-ups preserve them. A classic concentration could occupy the top 10 m of a beach, be around 30 cm thick and grade 70% ore constituent. The great placer fields of Australia along the coasts of New South Wales (Welch et al. 1975; Lissman & Oxenford 1975) and Queensland (McKeller 1975) for black sand, and of Namibia (Hallam 1964) for diamond, are the principal reserves.

Fluvial placer deposits

Gold is the classic example of fluvial placer deposits and the giant gold placer fields (Henley & Adams 1979) of Colombia/Peru, California, British Columbia, Yukon/Fairbanks, Lena River (U.S.S.R) and South Island, New Zealand together with the Proterozoic Witwatersrand sequence, have provided most of the gold ever mined. The offshore tin fields of Malaysia also contain fluviatile sediments of late Cenozoic age, and provide the bulk of the world tin supply (Hutchinson & Taylor 1978; Batchelor 1979).

The basic principles of placer formation are well understood. Any factor creating a reduction in flow power may result in a concentration in such features as sediment bedforms, meander bends, gradient changes, stream junctions and bedrock irregularities.

As with marine placers, reworking is important in increasing grade and size of reserve. Thus best grades are obtained where older gravels have been reworked during incision/aggradation cycles. Schumm (1978) has applied his stimulating concepts of geomorphic thresholds, complex response and episodic erosion to placer problems. His contention that many erosion/aggradation cycles are part of normal geomorphological evolution has removed the need for repeatedly invoking fault movements, particularly when interpreting ancient placers. In experiments designed to simulate placer formation, Schumm found that in Zone I (drainage basin and upper part of transfer zone) bedrock-lag, channel-lag and scour-generated pay streaks will dominate. In Zone II in the main valley fills, channel lags, flow-separation lags and floodplain lags provide the main pay streaks. Simulation experiments on Zone III (piedmont) recorded differences between arid and humid conditions. Under the influence of an arid climate, placer formation was discouraged because rivers discharging on to the piedmont are characterized by mudflow deposition and little reworking of sediment. Wet alluvial fans—such as would be expected to predominate in pre-vegetation times (cf. Witwatersrand)—are characterized by extensive reworking and placer formation. Fan aggradation favours storage in the valleys of zone II and at the fan apex. Incision creates rich pay streaks at the fan head and within the trench valley and lateral migration of the fan valley forms placer sheets. Continued reworking cycles result in several stratigraphic sheets and the movement of most of the placers to upper levels in the mid-fan region.

These models are helpful in understanding the Witwatersrand deposits.

Malaysian tin deposits

The Malaysian tin deposits provide a good example of the fate of detrital elements released from primary bedrock sources and then cycled through the sedimentary system. Essentially the element moves from proximal to distal fluvial sites (Zone I to Zone III) and from stratigraphically low to stratigraphically high positions. The orebodies change from early, proximal, high-grade, small-tonnage deposits of Zone I to the giant, Padang-type, thick, high-tonnage but lower-grade and finer-grain-sized deposits of Zone III braid sheets. The data are summarized in Fig. 2.

Special problems of gold placers

Once it is in the fluvial system, gold obeys the laws of the other heavy minerals—cf. the comments on placer formation and the Malaysian tin (e.g. Hester 1970). However, the problem of the gold source is often controversial. Basically the problem is that the vast amounts of gold yielded by placer fields cannot have been supplied by vein-type deposits of the hinterlands in the same way that tin has. Many workers have referred to the apparent gold budget for giant placer fields in the Cenozoic (Henley & Adams 1979) and the Archaean (Reimer 1984) and many old prospectors knew of the problem for smaller fields. They conclude that some form of pre-sedimentation concentration was needed.

Earlier we approached the problem of gold

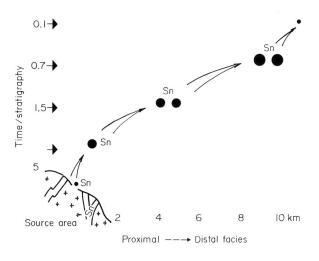

FIG. 2. Schematic representation of the clastic unroofing cycle. The economic element is seen to move through the fluvial system. The size of the dots indicates the probable increasing size (though lower grade) of an orebody. Numbers on the horizontal axis refer to distance in kilometres from primary bedrock source and those on the vertical axis to the age in millions of years of reworking events in the Malaysian tin province where the primary source is Triassic. Thus with increasing reworking the metal content migrates to more distal and to stratigraphically younger sediments.

mobility and the efficiency of deep weathering crusts in concentrating metal. Notably, all the Cenozoic fields, and apparently the Witwatersrand, form at the end of a geomorphological cycle and weathering profiles are present on the subdued topography of the source areas. Perhaps the gold disseminated through the source regions associated with pyrite and arsenopyrite is released during oxidation and concentrates in the regolith to be stripped off during erosional cycles. Large nuggets could be detrital but could just as well form by amalgamation in the weathering crust (Clemmey 1981a). In the case of the Witwatersrand this could be enhanced by the occurrence of auriferous iron formations (Fripp 1976) which presumably were present in the Archaean source areas and provide ideal environments for gold mobility in the weathering crust.

Witwatersrand ores

The quartz-chert-pebble conglomerates of the Witwatersrand and others formed around the Archaean/Proterozoic boundary are the subject of a voluminous literature and it would be impossible to attempt a full review here. The reader is referred to the excellent review by Pretorious (1975).

Interpretations of the sedimentary environment are not without dispute but a fluviatile piedmont fan model (Schumm 1977) seems to satisfy most of the observations. The pay sheets form thin gravel beds within a thick quartzite sequence and within them pyrite is the heavy mineral. Gold is found along basal scour planes, on foresets and within the coarse fraction of graded beds and is aligned in pay streaks often along the palaeocurrent direction, and a fossil placer origin is generally accepted (Minter 1976). Epigenetic features are ascribed to diagenesis and metamorphism.

The pyrite of the Rand is usually viewed as a heavy mineral assemblage, stable under the supposed anoxic Lower Proterozoic atmosphere. However Clemmey & Badham (1982) question the presence of a reducing atmosphere and Clemmey (1981a) indicates that sulphidation of a normal black-sand assemblage probably accounts for the bulk of the pyrite and also the enigma of missing magnetite. Dimroth (1979) also describes abundant pyrite clasts of apparent syngenetic origin. Probably the most exciting development regarding the sedimentology of the Rand is the realization that the situation fits the predictions of Schumm's fluvial models. The gold is found within an alluvial piedmont wet fan; it is at the higher stratigraphic levels of the sequence; it is found in

preferred stratigraphic horizons (reefs); and, moreover, it is located preferentially in the mid-fan position in all the goldfields.

Chemical sediments: brines and metal deposits

Here must be included the normal suite of evaporite minerals (Table 1) and the rare elements included on Figure 1. Perhaps the greatest advances have involved increasingly elegant facies analysis, particularly the sabkha model, and recognition of the role of diagenesis in evaporite formation (Shearman 1966). Although the concentration of metallic elements in evaporites is generally understood, some specific problems remain. For example, why is lithium concentrated in such great quantities in Chilean calcrete and nowhere else?

The great controversies concerning chemical sediments are reserved for the ironstones, and in particular the so-called 'banded iron formations' (BIF) so well developed in the Precambrian. It is however the oolitic ironstones of the Phanerozoic which are most familiar to the sedimentologist.

Oolitic iron formations

Two major phases of ironstone development are present in the Phanerozoic. The earlier are the Ordovician/Silurian ironstones formed at Clinton in the U.S.A. and also in Normandy, Brittany, Bohemia, Sardinia, Wales and Morocco. The later phase is in the Jurassic/Cretaceous when they formed in Lorraine, Germany, England and Chile. The ironstones have been subject to concentrated attention from sedimentologists (e.g. Talbot 1974) who interpret their environment of formation as nearshore lagoonal, in embayments or behind offshore bars. However, it is an open question as to how they formed in that environment and whether they are sedimentary or diagenetic.

Typically the ironstones are chamosite (Fe_4Al_2) (Si_2Al_2) $O_{10}(OH)_8$ oolites in a sideritic matrix. Kimberley (1980) argues that they result from diagenetic replacement and cites many examples where this is happening now or has happened in the recent past. The basic model involves reducing ground waters derived from a continent carrying Fe^{++} in solution. Replacement can take place by downward percolation or ground waters upwelling into overlying carbonates. Proponents of a primary origin suggest that the metal is derived from surface or ground waters where it might be in solution, or as colloids, or adsorbed on other minerals. On entering the 'lagoon' the metal coagulates and precipitates a gel from which the ooids form.

Evaporation may be involved, for Curtis & Spears (1968) indicate that hypersalinity enhances iron precipitation. There is a problem concerning the valency state of the iron, for chamosite and siderite are reduced (ferrous) iron minerals yet the environment of the ooids was in all probability oxidizing—witness the common high-energy indicators (rolled fossils, cross lamination, grading). Perhaps algal coats, mediating the precipitation of the iron, created a micro-reducing environment with the ooids. Another alternative is that they formed as oxide ooids and were reduced during diagenesis. The debate will continue for all models have modern examples (perhaps there lies the answer?).

Banded iron formation (BIF)

Perhaps as much has been written concerning the origins of the BIF as about the origins of the Witwatersrand (see UNESCO 1973). The banded iron formations are generally regarded as being predominantly of Lower Proterozoic age, and there is indeed a great tonnage of iron of this age. However they are found throughout the complete range of geological history and the oldest dated rock is in fact the Isua BIF (Moorbath *et al.* 1973).

James (1954) was the first to recognize the different facies of BIF, viz.:

oxide (haematite, Fe_2O_3; magnetite $FeOFe_2O_3$)
$\longrightarrow$ carbonate (siderite $FeCO_3$)
$\longrightarrow$ sulphide (pyrite FeS_2).

He interpreted this as an onshore to offshore change. The extensive deposits of the Canadian Archaean greenstone belts were seen to be different in geological setting and trace-element content to the vast Proterozoic ores of the Superior region and a complementary division was made into the Algoma Type (volcano-sedimentary-associated, often interbedded with turbidites) and the Superior Type (shelf-sediment assemblage associated).

The great problem of their genesis concerns the transport of such vast quantities of iron to the depository. In many instances the depository may have been huge and long lived. Iron formations can comprise 10–20% of the Proterozoic stratigraphic thickness in some areas. The largest of the 'basins' are of regional extent, viz. the Superior Basin and Labrador Trough (lower to middle Proterozoic) and the Rapitan Basin (late Proterozoic) of the North American/Canadian craton, the Hammersley Basin of Western Australia, the Krivoy Rog of the U.S.S.R. the Transvaal of South Africa, and others in India and Brazil. The iron is principally in the ferric state and yet as we have seen it is

practically insoluble in this state. Detrital transport is not favoured on account of the delicate, sometimes micro-scale, interlamination with the other principal constituent—silica. Nonetheless, Clemmey & Badham (1982) suggest detrital introduction should not be rejected totally.

Chemical precipitation, the favoured mechanism, would involve the introduction of the iron in solution in the ferrous form and its subsequent oxidation and precipitation. Holland (1973) has suggested upwelling ocean bottom waters saturated with respect to $FeCO_3$. Borchert (1960) envisages the sub-sea exhalation of waters derived from leaching out the iron from reduced clastics, while Govett (1966) has suggested seasonal convective overturn in lakes with a developed hypolimnion, the metal being derived from sediments in the reduced bottom layer. Many others have suggested that they form from submarine volcanic-related exhalations. As with the Phanerozoic problem they are all mechanisms that may work and certainly can be seen to operate on a modest scale at present.

Cloud (1973) explains the BIF as a response to the supposed anoxic Proterozoic atmosphere. His argument is based on the premise that under an oxygen-bearing atmosphere iron is locked up in continental sediments, particularly in weathering horizons and red beds. He suggests that under the proposed anoxic atmosphere vast quantities of iron are in solution in surface waters and precipitation is in response to an oxygen-producing, algal bloom in the oceans. Many arguments can be brought to bear against this model, not the least of which being the existence of vast Rapitan Formation in the latest Proterozoic, by when an oxygenic environment had most probably developed and of red beds of Archaean age (Clemmey & Badham 1982).

Dimroth (1977a, b) in his classic work on the vast Sokoman Iron Formation in the Labrador Trough brought the techniques of modern sedimentology to the problem; he indicates the range of sedimentary facies present in the iron formations and suggests a diagenetic origin. He recognized oolites, cross-bedded former carbonates, classic BIF, etc., (Chauvel & Dimroth 1974) and suggests that the common view of BIF as continuous delicate laminae of chert and iron is too simplistic.

A major problem in iron-formation geology is the question of why so much iron exists in the geological record in the Lower Proterozoic. Perhaps it is a function of their special position in crustal evolution occurring at the end of the cycle of sedimentation and tectonics that closed the Archaean era and opened the Proterozoic (Button 1976; Badham 1981a; Clemmey & Badham 1982). The cycle of different ages on different cratons commenced with clastic deposition and ended with carbonate/iron formation sedimentation, and marks the initial unroofing of the Archaean basement. Such an unroofing event must have stripped off a vast amount of mafic material from greenstone belts bringing with it an unprecedented influx of iron, silica, magnesium and manganese to the sedimentary system. Whatever their origin, the iron–silica–dolomite–manganese chemical sediments of the early Proterozoic probably reflect this influx.

In view of all the possible mechanisms for precipitating the iron formations, perhaps the most stimulating advance of the past few years is the palaeoenvironmental approach of Kimberley (1978c, 1979). In a comprehensive literature survey he recognized that whilst all types of iron formation have existed through most of geological time certain types are more prevalent at certain times than others. His classification uses the criteria of associated environments in order to determine the environment of formation. Thus he finds that turbidite-hosted and volcanogenic-associated types are most prevalent in the Archaean possibly reflecting the lack of stable shelves and steep land-to-ocean transitions of the greenstone belt era.

The Proterozoic saw the first stable shelves, massive early cratonization and a long initial period of tectonic stability. This was the era (enhanced by the iron-supply factors discussed earlier) of the extensive, shelf-sediment-associated BIF.

Figure 3 shows a model relating BIF to environment of formation and geologic time.

Ore deposits of the diagenetic environment

We are concerned here with the massive deposits of the principal base metals—lead, zinc, copper and uranium. Whilst there is an environmental continuum of these deposits, a 2-fold division can be made. First, there are those ores associated with proximal to distal continental fluviatile and lacustrine sediments and the paralic/nearshore zone. Typically, the ores are of copper and uranium which may be associated with lead and zinc in the distal fluviatile or nearshore marine zones. The second group comprises the ores associated with nearshore-marine through offshore-basin to platform-marginal sites. These are the ores of lead, zinc and barium. Copper is only associated with these ores when a continent connection is recognizable in the underlying rocks.

Lead-zinc ores: the carbonate–black-shale association

Stratabound and stratiform lead-zinc ores (± Ba, F) form some of the largest known metal accumulations. They have a special place in the history of work on sedimentary ores, for the Pine Point district, as mentioned earlier, is the first area where ores in sediments were recognized as belonging to the normal evolution of a

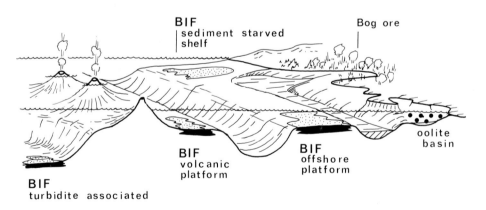

FIG. 3. Palaeoenvironmental facies of iron formations. Whilst all types of facies association are recognized throughout geological time, the lower scale indicates the dominance of certain environmental types at specific stages of earth history, reflecting the Earth's geotectonic development.

sedimentary basin in much the same way as oil does. This concept developed for these so called Mississippi-Valley-Type mineralizations allowed geologically 'late' and clearly crosscutting ores to be considered under the mantle of diagenesis. The basics of the model are outlined in Fig. 4. The model also reconciles the low fluid-inclusion temperatures and hydrocarbon associations of the ore (see the opening discussion on fluid and metal sources).

It was a short step from the ideas involved in Fig. 4 to realize that slope-marginal growth faults (or for that matter the facies mosaic of a non-depositional margin) could intercept fluids migrating out of the shale basin and be the focus of epigenetic and exhalative mineralization. Thus a new variation on the exhalite model was born (Badham 1975; Morganti 1981) and was used to explain the extensive mineralization in the base of slope/basinal facies of the Howards Pass area in the Mackenzie shale basin (off-palaeoshore of the Pine Point area.) Ensuing debate led to the development of the concept of a continuum between the epigenetic Mississippi Valley Type on the one extreme and the syngenetic black-shale-hosted mineralization on the other. Badham suggests that the Irish lead-zinc deposits, showing characteristics of epigenetic and syngenetic mineralization and related to Carboniferous growth faults (Russell 1975, 1983) are transitional between the two.

The exhalative theory is well presented by both Badham (1981b) and Russell (1983). Badham derives the metal and fluids from dewatering host shales; Russell on the other hand derives the fluid from downward percolating contemporary seawater and the metals from warm basement greywacke prisms. Associated syngenetic manganese haloes and distal oxide-facies, chemical sediments of Ba, Fe, Mn, are regarded as products of the spent brine.

'Black shale', used in the ore genesis sense, carries the connotation of a deep-water facies but this is manifestly not always so. For whilst the black-shale ores in the Howards Pass area and Sullivan may be deep-water black shales (front of major carbonate platform; turbidite-associated), others are not, as recent detailed facies work on the huge Mt Isa (Neudert 1981) and HYC (Here's yer chance) (Muir 1983) deposits in Australia has shown. An association of evaporite minerals in lacustrine and other ephemeral water bodies has led Muir to conclude that a rift-valley, alkaline, playa-lake is a likely type of setting for the deposits. This conclusion is in marked contrast to the assumed (though never defended) deep-water settings for Mt Isa (Finlow-Bates 1979) and HYC deposits (Croxford & Jephcott 1972). A Red-Sea-type model has been offered by Croxford & Jephcott (1972) for the last deposit.

There is evidence to suggest that the blanket application of the exhalite model is sometimes inappropriate. At the HYC deposit, Williams (1978a, b, 1979) and Rye & Williams (1981) have found critical evidence that the mineralization is indeed epigenetic but occurred during very early diagenesis whilst the sediment was still 'wet'. They document mineralization ranging from hot, crosscutting phases near the assumed feeder fault zone to the cool, stratabound deposits of the main mineralized 'shale' in the HYC basin. Muir (1983) interprets the faults as synsedimentary, strike-slip features bounding the rift.

Synthesizing my own observations in the field and the laboratory and using evidence from the literature, I have produced the 'Inhalite' model shown in Fig. 5. Importantly, fluids which are

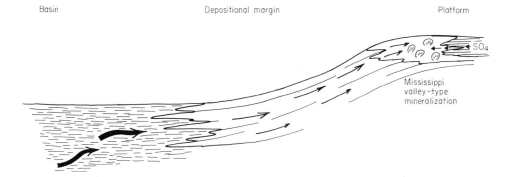

FIG. 4. Model for epigenetic formation of water-derived mineralization in carbonates (Mississippi Valley Type). Non-fault-controlled, depositional margins are envisaged as providing the best plumbing possibilities. Metals are derived from basinal shale and $SO_4^=$ from back-reef deposits. The host is commonly lithified, porous, dolomitized, barrier (reefal) carbonate. Scale: c. 100 km.

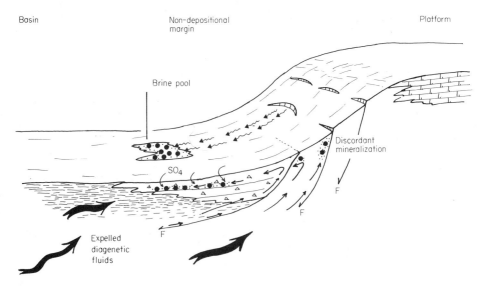

Fig. 5. Model for the 'inhalite' hypothesis (with a reconciliation with the exhalite school). A fault bounded basin with a non-depositional margin is envisaged as providing the best plumbing. Thus depending on the nature of margin control, both Mississippi Valley (cf. Fig. 4) and inhalite mineralization may be present along the same margin. Discordant mineralization can result, associated with the conduits but the host should be a slope facies carbonate not a marginal shelf facies. Fluids forced up the faults will encounter the first porous horizon(s) and move along them; base of slope debris flows often provide a conduit. Seawater $SO_4^=$ diffuses into the porous horizon which will be reducing below the sediment/seawater interface. If tapped by slope-failure faults the brines may exhale. The model reconciles observations of early diagenetic emplacement of Pb/Zn and the fact that many of the host beds are not reduzate facies sediments.

moving up growth faults are likely to be channelled back down the first, porous, slope facies encountered or the first horizon in which the fluids may generate their own porosity. Debris flows are an obvious lower-slope candidate and it is surprising how many lead-zinc mineralizations are emplaced at the base of debris-flow units or in their lateral equivalent. The sulphur source can be from contemporary seawater diffusing into the reducing lower levels of the porous beds, and this possibility removes the need for black fetid shales as a universal source. This explains many mineralizations not in reduced sediments and not related to evaporite sulphur sources.

A final point concerns the plumbing geometry of Figs 4 and 5. This geometry will survive burial, which in some cases may be needed to mature the shale prism. Thus both models can explain mineralization taking place long after sedimentation and under a considerable thickness of sediment. The vein- and flat-type deposits of the English Pennines may be Mississippi-Valley-Type deposits of this kind (Ford 1976).

Copper and uranium ores: the continental-facies association

Stratabound and stratiform copper and uranium ores are typically associated with the red-bed environment and other arid-zone lithofacies. Clemmey (1980) has argued that this link stems from the importance of having an arid climate to establish optimum metal fertility (the potential of any sedimentary prism to produce ore during diagenetic evolution or physical reworking of any unroofing sequence) (Clemmey 1981b).

Uranium ores

World uranium ore reserve is located predominantly in the following:

5% ground-water calcrete
   (W Australia, Namibia);
35% red-bed fluviatile sediments
   (SW U.S.A.);
25% Lower Proterozoic conglomerates
   (Canada, South Africa);
25% sub-red-bed unconformity associated
   (Canada, Australia); and
10% igneous and volcanic.

Calcrete mineralization occurs in areas of hyper-arid soil-moisture regimes and the bulk of the mineralization is held in the gravels below the main calcrete. Precipitation is in the form of the mineral carnotite ($K_2 (UO_2)_2 (VO_4)H_2O$ from alkaline ground waters which carry the uranyl ion predominantly in association with the carbonate complex. Precipitation follows the de-stabilization of the carbonate complex and the addition of vanadium in the V^{5+} form. Thus a source of vanadium is the key to economic calcrete mineralization. Vanadium in the five-plus valency state cannot travel in the same oxidizing ground waters as the uranyl ion or else precipitation would occur constantly. Also V^{5+} is not very soluble and vanadium travels better in the V^{4+} state.

Calcrete deposits are known to occur most frequently over bedrock constrictions when ground waters are forced to near-surface. Ion precipitation and equilibration with atmospheric CO_2 are enhanced by evaporation from the surface (Carlisle 1978), but the principal mechanism probably involves the forced mixing of stratified ground waters (Mann & Deutscher 1978). The lower, more reducing phase, probably carries the vanadium in the lower valency state and the upper, more oxidized layers, the uranyl complexes. On forced mixing the vanadium will oxidize with the resulting precipitation of the carnotite. The vanadium may be obtained from magnetite of metamorphic origin or possibly complexed with organic matter in the underlying sediments dating from an earlier pluvial episode (Clemmey 1981c). Significantly, lacustrine greywacke underlies both of the potentially economic calcrete

deposits at Langer Heinrich in Namibia and Yeelirie in Western Australia indicating an early pluvial and a lower Eh environment from which to derive ground waters carrying V^{4+}.

Uranium deposits in sandstones, typified by those of the Colorado Plateau area in the U.S.A., are usually found associated with fluviatile red beds, though some may be found in littoral marine sediments. The commonly arcuate form of the deposits, situated at the interface between oxidized (on the concave side) and reduced facies has given rise to the term 'roll-front' ore (Shawe 1956). The roll is thought to represent the chemically active front of an oxidized cell created by ground waters carrying the uranyl ion which is precipitated by reaction with reduced species. Organic remains or sulphidized sediment can cause the reduction. Other deposits formed at the flanks of channel sands or in their floors are termed 'concordant' and 'penconcordant'.

The roll-front cells are active and represent uranium en route through the system. Concordant ores are a more permanent phase. Roll fronts are common in Mesozoic sediments and concordant ores in the Palaeozoic.

The source of the metal is disputed but is in all probability related to inter-formational leaching. Low-temperature solution geochemistry has been successfully applied to sandstone bearing ore deposits (Hostetler & Garrels 1962; Langmuir 1978). Other main advances are concerned with understanding the ore in terms of ground-water/aquifer dynamics (e.g. Huang 1978; Ortiz *et al.* 1980), and increasingly elegant facies work in the hope of understanding the control of facies on the location of ore bodies

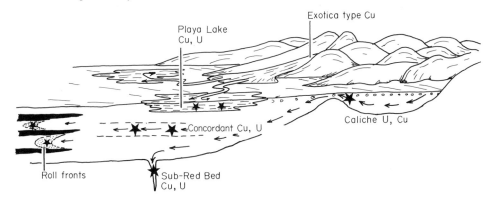

FIG. 6. Schematic presentation of the facies associations during the early/proximal part of a fluid unroofing cycle, applicable to Cu, U, ores. The stratigraphic factor (cf. Fig. 2) is also important. Thus playa lake host ore will normally be found at a higher level than fluviatile type. In the case of uranium the factor of geotectonic evolution (cf. Fig. 3) should also be applied. See text. As with the clastic unroofing model of Fig. 2 we are seeing the metal cycled through the fluvial system.

(e.g. Tyler & Ethridge 1983).

Unconformity deposits

Unconformity deposits—often referred to as 'vein types'—are related to some Proterozoic sub-red-bed unconformities particularly those of the Martin and Athabasca sandstones in Saskatchewan and the sandstones of the Pine Creek geosyncline area near Darwin, Australia. Mineralization is related to fissures in the underlying basement cutting graphitic schist and is usually found as the pitchblende form of uraninite. Whilst this is the low-valency form of uranium, oxidizing alteration is common (especially to haematite) and the depositional system fluctuated between a reducing and oxidizing nature. Mineralization may be found in the overlying regolith and basal sandstone.

The origin of unconformity deposits was traditionally regarded as hydrothermal with fluids moving up from the basement along fissures and depositing the ore on encountering the pressure drop at the unconformity. However, it would be strange if this only happened below fluviatile red beds. Langford (1974) inspired by the then recent descriptions of calcrete deposits, proposed a *perdescensum* origin from downward-percolating surface waters charged with the uranyl ion. Hoeve & Sibbald (1978) support a *perdescensum* model but prefer to derive the fluid from interstratal leaching. In effect they are envisaging a kind of unconfined roll front when the main aquiclude is the gravel bedrock surface. Given time and interstratal oxidizing conditions (especially typical of Precambrian red beds with no well-developed land vegetation) all the uranium finds its way down to the base of the pile. The essentials of the uranium continum model are summarized in Fig. 6.

Uranium conglomerates of the Proterozoic

The reader is referred to the excellent review of this problem in Kimberley (1978a) and to the paper by Clemmey (1981a). The deposits of the Witwatersrand are a major source of uranium but it is the conglomerates of the Elliot Lake region in Canada that constitute the primary reserve. They do not carry exploitable gold and this may reflect their lack of reworking compared to the Rand sediments. A simple anoxic atmosphere placer model has been advanced to explain the existence of apparently water-worn uraninite grains. However, many workers now recognize considerable interstratal mobility of uranium (Simpson & Bowles 1981), and doubts about the role of the anoxic atmosphere have been raised earlier (Clemmey & Badham 1982). A new model proposed by Clemmey involving

interstratal oxidation of primary detrital phases and diagenetic precipitation may account for the interstratal mobility observed by Simpson & Bowles (1981).

Nevertheless, the question remains open and Kimberley argues for a pre-sedimentation concentration of uranium caused by reaction of downward-leaching uranyl solutions and reduced H_2S (of volcanic origin) saturated ground waters. It seems the two models are quite compatible.

Copper in sedimentary rocks

Copper more than any other metal has figured in the syngenesis debate and the influence of Garlick (1961, etc.), in drawing the attention of the exploration geologist to the significance of nearshore redox reactions, was described earlier. Copper is readily observed in arid regions moving out of bedrock sources into the sediments (Clemmey 1981b) and from Garlick onward many workers have retained the concept of direct erosion of bedrock to supply the copper metal.

Copper, like uranium, is often associated with red-bed sequences and other arid-zone lithofacies. Davidson (1965) recognized this and argued that the association had genetic significance. He envisaged brines of evaporite parentage descending to deep levels in the crust and returning to deposit metals leached from deep magmatic sources. Copper deposits do show many 'epigenetic' features and Davidson was an opponent of the Garlick syngenesis school. Many workers have since supported Davidson's thesis of late copper introduction, though early diagenetic brines are also invoked (Annels 1974; Bartholome 1974; Brown 1974; Clemmey 1976). True syngenesis was also adjudged to be untenable because, with increasing sophistication of facies analysis of many of the units ascribed to a reducing marine nearshore origin, they were shown to be high-energy, fluviatile sediments. Clemmey (1976) recognized that in the Zambian Copperbelt, as indeed in all major copper provinces in the U.S.S.R. (e.g. in Udokan: Bakun *et al.* 1966; and in Dzhezkazgan: Feoktistov & Kochin 1972) and China (Li Hsi-chi *et al.* 1964), copper occurred in a range of facies from proximal to distal fluviatile and nearshore playa-lake/marine facies. In general there is a progradation from stratigraphically low to high positions, going from proximal to distal, similar to that shown by clastic unroofing and sedimentary reworking of tin (Fig. 2).

It is unlikely that copper has travelled through the fluvial system as a major detrital phase though brief periods of detrital transport coupled with *in situ* oxidation are possible (Clemmey 1978, 1982). Oxidizing interstratal brines are thought to be most probable (Rose 1976). Thus a form of 'fluid unroofing' model is preferred and the essence is presented in Fig. 6.

The copper–uranium–gold association

As indicated in Fig. 6 and in the text, copper and uranium share the same paths through fluvial, red-bed sediments and indeed they are often found together. Both metals can be observed actively moving out of bedrock sources at present.

Gold and uranium are found together in sub-unconformity deposits (e.g. Jaibaluka, Australia) and stratiform deposits (cf. Witwatersrand) and in red-bed copper deposits. Gold mobility in ferric-iron-rich environments is described earlier.

The recently described giant ore body at Olympic Dam on the Stuart Shelf in southern Australia (Roberts & Hudson 1983) is perhaps the classic red-bed Cu/U/Au orebody. It is located in a probable rift-valley fill of Late Proterozoic age. From the descriptions, the host facies are red-bed sequences of fan origin and facies changes seem to be telescoped. The best ore is found in the most distal (playa sheetwash?) facies. Interformational reworking is evident, and haematite (often diagenetically aged to specularite) is abundant as matrix and clasts.

Conclusions: possible way forward

In 25 years we have advanced from magmato-thermal dominated models to a position where we can now accept:

(i) Low-temperature ore fluids of meteoric or connate origin.

(ii) Metal sources from within sedimentary prisms inherited from source area.

(iii) Diagenesis, both early and late, provides the middle ground between epigenetic and syngenetic models.

(iv) Facies analysis and concepts of sedimentary basin evolution as exploration tools.

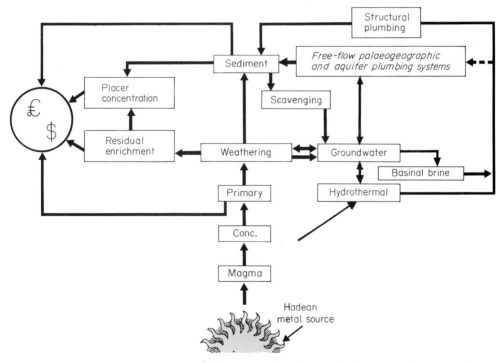

FIG. 7. Flow sheet for ore deposits in sedimentary rocks. The central role of the weathering crust is emphasized. The weathering crust (i) provides primary ores by residual or supergene concentration; (ii) releases stable elements to be concentrated in placers; (iii) eluviates the soluble elements to ground waters which may produce mineralization and evaporites; and (iv) releases elements to the sediments in a form which may yield the metal during diagenesis (detrital sulphides, adsorbed on clays or ferric sponges, contained in interlayer water, humic complexes). The possible interaction of hydrothermal and ground-water/connate fluids is indicated. Palaeogeographic plumbing includes aquifers whose characteristics are established during deposition and the irregularities of the sediment—bedrock interface. Faults are envisaged as being synsedimentary.

However the geology of metal deposits still lags behind the oil industry in the use and acceptance of sedimentary principles. Perhaps this reflects the 400+ years of academic research on ore-bodies and over 6000 years in the application of geological principles to exploration and extraction of metals. By comparison the oil industry had a clean start (almost!).

What is the way forward? Perhaps of immediate application to the search for new deposits and the definition of new types of target would be 'fine tuning' of the fluid-unroofing models. How metal is cycled through the sedimentary and diagenetic environments is a crucial question and is not well understood. Models such as those presented in Fig. 6 could be developed to aid in exploration target selection and to maximize effort during an exploration programme.

What is also needed is increasing sophistication of facies analysis in mineralized areas and the careful relating of mineralizing events to the total diagenetic picture.

Concepts of sediment maturity should be developed with regard to metals as well as oil. To do this we need to know in what form the metal arrives in the sediment, its diagenetic history and what conditions persuade the sediment to yield that metal. Stable isotopes, as used in determining burial histories (e.g. Milliken et al. 1981), together with established concepts of shale donor diagenesis, could be developed in ore-deposit geology.

More attention should be paid to the geomicrobiologist's concepts of biological mediation in metal cycles. The evaluation of the role of bacteria in surface and interstratal metal mobility, and particularly their role in producing specific metal association, should prove to be of great value.

Finally, of critical importance is the weathering horizon (Fig. 7): all elements entering the sedimentary system in fluid or detrital form must first pass through the screen of the weathering crust. We have established the concept of geochemical inheritance, but the weathering crust determines the timing of element release from bedrock and the form in which it arrives in the basin. In this way climate and the stage of geomorphological evolution determine the potential metal fertility of a sedimentary basin. We need a greater understanding here.

In this short and inevitably highly personal review the theme has been that ore deposits are abnormal products of normal geological events; we need to establish what enhancement factors are at work.

References

AMSTUTZ, G. C. & BERNARD, A. J. (eds) 1973. *Ores in Sediments Int. Union Geol. Sci. Ser. A*, **3**, 350 pp. Springer Verlag, Berlin.

ANDERSON, G. M. 1978. Basinal brines and Mississippi Valley Type ore deposits. *Episodes*, **2**, 15–9.

ANNELS, A. E. 1974. Some aspects of the stratiform ore deposits of the Zambian Copperbelt and their genetic significance. *In*: BARTHOLOME, P. (ed.) *Gisements Stratiformes et Provinces Cuprifères*, pp. 235–54. Soc. Geol. Belg. Liege.

BADHAM, J. P. N. 1975. A new Pb-Zn deposit in the Yukon, *Abstr. Mineral Deposits Studies Group Meeting, Leicester*.

—— 1981a. Origins of ore deposits in sedimentary rocks. *In*: TARLING, D. (ed.) *Economic Geology and Geotectonics*, pp. 149–91. Blackwell Scientific Publications, Oxford.

—— 1981b. Shale hosted lead-zinc deposits; products of exhalation of formation waters? *Trans. Instn Min. Metall.* **B90**, 70–6.

BAKUN, N. N., VOLODIN, R. N. & KRENDELEV, F. P. 1966. Genesis of the Udokansk cupriferous sandstone deposit. *Int. Geol. Rev.* **8**, 455–66.

BARNES, H. L. 1959. The effect of metamorphism on metal distribution near base metal deposits. *Econ. Geol.* **54**, 919.

BARTHOLOME, P. (ed.) 1974. *Gisements Stratiformes et Provinces Cuprifères*. Soc. Geol. Belg., Liege.

BATCHELOR, B. C. 1979. Geological characteristics of certain coastal and offshore placers as essential guides for tin exploration in Sundaland, Southeast Asia. *Geol. Soc. Malaysia Bull.* **11**, 283–313.

BEALES, F. W. 1975. Precipitation mechanisms for Mississippi Valley Type ore deposits. *Econ. Geol.* **70**, 943–8.

—— & JACKSON, S. A. 1966. Precipitation of lead zinc ores in carbonate reservoirs as illustrated by the Pine Point ore field, Canada. *Trans. Instn Min. Metall.* **B75**, 278–85.

BORCHERT, H. 1960. Genesis of marine sedimentary iron ores. *Instn Min. Metall. Bull.* **640**, 261–79.

BROWN, A. C. 1974. The copper province of northern Michigan, U.S.A. *In*: BARTHOLOME, P. (ed.) *Gisements Stratiformes et Provinces Cuprifères*. Geol. Soc. Belg., Liège.

BROWN, J. S. 1950. *Ore Genesis. A Metallurgical Interpretation: An Alternative to the Hydrothermal Theory*. Thos. Murby, London.

BUTTON, A. 1976. Iron formation as an end member in carbonate sedimentary cycles in the Transvaal Supergroup, South Africa. *Econ. Geol.* **71**, 193–201.

CARLISLE, D. 1978. The distribution of calcretes and gypcretes in the southwestern U.S.A. and their uranium favourability based on a study of deposits in Western Australia and South West Africa. *U.S. Dept. Energy—open file report*.

CHARTRAND, F. M. & BROWN, A. C. 1984. A

preliminary comparison of diagenetic stratiform copper mineralisation at Redstone, N.W. Territories, Canada and Komoto, Shaban Copperbelt, Zaire. *J. geol. Soc. London* **141**, 291–7.

CHAUVEL, J. & DIMROTH, E. 1974. Facies types and depositional environment of the Sokoman iron formation in the Labrador Trough, Quebec, Canada. *J. sedim. Petrol.* **44**, 299–327.

CLEMMEY, H. 1976. Aspects of stratigraphy, sedimentology and ore genesis in the Zambian Copperbelt with special reference to Rokana mines. Unpublished Ph.D. Thesis, Univ. Leeds.

—— 1978. The implications of a recent copper sulphide placer concentration from Chile. *Trans. Instn Min. Metall.* **B33**.

—— 1980. Copper and Uranium ore: the red-bed association. *Br. Ass. Adv. Sci. Salford*, **C65**, 10 pp.

—— 1981a. Some aspects of the genesis of heavy mineral assemblages in Lower Proterozoic uranium-gold conglomerates. *Min. Mag.* **44**, 399–408

—— 1981b. Copper mineralisation in Andean gravels; first stage in stratiform ore formation. *Trans. Instn Min. Metall.* **B90**, 63.

—— 1981c. Palaeoclimatic control on uranium mineralisation in Namibian calcrete. *Trans. Instn Min. Metall.* **B90**, 63.

—— 1982. Copper in Precambrian alluvial systems: the importance of interstratal oxidation. *J. geol. Soc. London Newsletter*, **II** 20–1.

—— & BADHAM, J. P. N. 1982. Oxygen in the Precambrian atmosphere; an evaluation of the geological evidence. *Geology*, **10**, 141–6.

CLOUD, P. 1973. Paleoecological significance of the banded iron formations. *Econ. Geol.* **68**, 1135–43.

CROXFORD, N. J. W. & JEPHCOTT, S. 1972. The McArthur River lead zinc silver deposit. Northern Territories, Australia. *Aust. Instn Min. Metall. Proc.* **243**, 1–26.

CURTIS, C. D. & SPEARS, D. A. 1968. The formation of sedimentary iron minerals. *Econ. Geol.* **63**, 257–70.

DAVIDSON, C. F. 1965. A possible mode of origin of stratabound copper ores. *Econ. Geol.* **60**, 942–54.

DIMROTH, E. 1977a. Models of physical sedimentation of iron formations. *Geosci. Can.* **4**, 23–30.

—— 1977b. Diagenetic facies of iron formation. *Geosci. Can.* **4**, 83–8.

—— 1979. Significance of diagenesis for the origin of Witwatersrand type uraniferous conglomerates. *Phil. Trans. R. Soc.* **A291**, 277–87.

DUNHAM, K. C. 1970. Mineralisation by deep formation waters: a review. *Trans. Instn Min. Metall.* **B79**, 127–36.

ESSON, J. & SURCAN DOS SANTOS, L. C. 1978. Chemistry and mineralogy of a section through a lateritic nickel deposit at Liberdade Brazil. *Trans. Instn Min. Metall.* **B87**, 53–60.

FEOKTISTOV, V. P. & KOCHIN, G. G. 1972. Certain distinctions in the localisation of stratified deposits of copper. *Int. Geol. Rev.* **14**, 1138–46.

FINLOW-BATES, T. 1979. Cyclicity in the lead-zinc-silver bearing sediments at Mount Isa Mine Queensland, Australia, and rates of sulphide accumulation. *Econ. Geol.* **74**, 1408–19.

FORD, T. D. 1976. The ores of the south Pennines and the Mendip Hills, England—a comparative study. *In*: WOLF, K. (ed.) *Handbook of Stratiform and Stratabound Ore Deposits*, Vol. 5, pp. 161–95. Elsevier, Amsterdam.

FRANCHETEAU, J., NEEDHAM, H., CHOUKROUNE, P. *et al.* 1979. Massive deep sea sulphide ore deposits discovered on the East Pacific Rise. *Nature, Lond.* **277**, 523–8.

FRIPP, R. 1976. Stratabound gold deposits in Archean banded iron formations, Rhodesia. *Econ. Geol.* **71**, 58–75.

GARLICK, E. G. 1961. The syngenetic theory. *In*: MENDELSOHN, F. (ed.) *The Geology of the Northern Rhodesian Copperbelt*, pp. 146–65. Macdonald, London.

—— 1967. Special features and sedimentary facies of stratiform sulphide deposits in arenites. *In*: JAMES, C. (ed.) *Sedimentary Ores Ancient and Modern (Revised)*, pp. 107–69.

—— 1972. Sedimentary environment of Zambian copper deposition. *Geol. Mijnbouw*, **51**, 277–98.

—— 1981. Sabkhas, slumping and compaction at Mufulira, Zambia. *Econ. Geol.* **76**, 1817–47.

GARRARD, P. (ed.) 1977. *Forum on Oil and Ore in Sediments*, Imperial College, London. 202 pp.

GARRELS, R. M. & CHRIST, C. L. 1965. *Solutions Minerals and Equilibria*, Harper Row, New York. 450 pp.

GOVETT, G. J. S. 1966. Origin of banded iron formations. *Bull. geol. Soc. Am.* **77**, 1191–212.

HALLAM, C. D., 1964. The geology of the coastal diamond deposits of Southern Africa. *In*: HAUGHTON, S. (ed.) *The Geology of Some Ore Deposits in Southern Africa*, Vol. 2, pp. 671–728. Geol. Soc. South Africa, Johannesburg.

HENLEY, R. W. & ADAMS, J. 1979. On the evolution of giant gold placers. *Trans. Instn Min. Metall.* **B88**, 41–50.

HESTER, B. W. 1970. Geology and evaluation of placer gold deposits in the Klondyke area, Yukon Territory. *Trans. Instn Min. Metall.* **B79**, 60–7.

HOEVE, J. & SIBBALD, T. 1978. Uranium concentrations related to the sub-Athabasca unconformity, Northern Saskatchewan Canada. *In*: KIMBERLEY, M. (ed.) *Short Course in Uranium Deposits: Their Mineralogy and Origin.* pp. 475–83. Min. Ass. Canada, Toronto.

HOLLAND, H. D. 1973. The oceans, a possible source of iron in iron formations. *Econ. Geol.* **68**, 1169–72.

HOSTETLER, P. B. & GARRELS, R. M. 1962. Transportation and precipitation of uranium and vanadium at low temperatures with special reference to sandstone type uranium deposits. *Econ. Geol.* **57**, 137–67.

HUANG, W. H. 1978. Geochemical and sedimentological problems of uranium deposits of Texas Gulf coastal plain. *Bull. Am. Ass. Petrol Geol.* **62**, 1049–62.

HUTCHINSON, C. S. & TAYLOR, D. 1978. Metallogenesis in South East Asia. *J. geol. Soc. London*, **135**, 407–28.

JACKSON, S. H. & BEALES, F. W. 1967. An aspect of sedimentary basin evolution: the concentration of Mississippi Valley type ores during late stage of diagenesis. *Bull. Can. Ass. Pet. Geol.* **15**, 383–433.

JAMES, C. H. (ed) 1967. Sedimentary ores ancient and modern (revised). *Proc. 15th Int. Univ. Geolg. Cong., Leicester.*

JAMES, H. L. 1954. Sedimentary facies of iron formation *Econ. Geol.* **49**, 235–85.

JENSEN, M. L. 1959. Sulphur isotopes and hydrothermal mineral deposits *Econ. Geol.* **54**, 374.

KIMBERLEY, M. M. 1978a. Origin of stratiform uranium deposits in sandstone, conglomerate and pyroclastic rocks. *In*: KIMBERLEY, M. M. (ed.) *Short Course in Uranium Deposits: Their Mineralogy and Origin. Short Course Handbook*, Vol. 3, pp. 339–81. Min. Ass. Can., Toronto.

—— 1978b (ed.) *Short Course in uranium deposits: Their Mineralogy and Origin. Short Course Handbook*, Vol. 3, Min. Ass. Can., Toronto. 521 pp.

—— 1978c. Paleoenvironmental classification of iron formations. *Econ. Geol.* **73**, 215–29.

—— 1979. Geochemical distinctions among environmental types of iron formations. *Chem. Geol.* **25**, 185–212.

—— 1980. Origin of oolitic iron formations. *J. sedim. Petrol.*

KNIGHT, C. L. (ed.) 1957. Ore genesis—the source bed concept *Econ. Geol.* **52**, 808.

—— 1975. *Economic Geology of Australia and Papua New Guinea. I. Metals.* Aust. Instn Min. Metall., Victoria.

LANGFORD, F. F. 1974. A supergene origin for vein type uranium ores in the light of the Western Australian calcrete carnotite deposits. *Econ. Geol.* **69**, 516–26.

LANGMUIR, D. 1978. Uranium solution-mineral equilibria at low temperatures with applications to sedimentary ore deposits. *In*: KIMBERLEY, M. M. (ed.) *Short Course in Uranium Deposits: Their Mineralogy and Origin. Short Course Handbook*, Vol. 3, pp. 17–55. Min. Ass. Can., Toronto.

LI HSI-CHI, PAN KAI-WEN, YANG CH'ENG-FUNG, TS'AI CHIEN MING 1964. Copper bearing sandstone (shale) deposits in Yunnan. *Int. Geol. Rev.* **10**, 870–82.

LISSMAN, J. C. & OXENFORD, R. J. 1975. Eneabba rutile-zircon-ilmenite sand deposits. *In*: KNIGHT, C. L. (ed.) *Economic Geology of Australia and Papua New Guinea. I. Metals*, pp. 1062–70. Aust. Instn Min. Metall., Victoria.

LOUGHNAN, F. C. 1969. *Chemical Weathering of the Silicate Minerals*, Elsevier, Amsterdam. 154 pp.

MANN, A. W. & DEUTSCHER, R. L. 1978. Genesis principles for the precipitation of carnotite in calcrete drainages in Western Australia. *Econ. Geol.* **73**, 1724–37.

MCILVEEN, G. R. & STEVENS, B. P. J. 1979. Supergene enrichment an important process in the development of economic grades in gold-quartz

veins. *Abstr. Geol. Soc. Australia, Economic Group Special Meeting, Sydney.*

MCKELLAR, J. B. 1975. The eastern Australian rutile province. *In*: KNIGHT, C. L. (ed.) *Economic Geology of Australia and Papua New Guinea. I. Metals*, pp. 1055–62. Aust. Inst. Min. Metall., Victoria.

MENDELSOHN, F. (ed.) 1961. *The Geology of the Northern Rhodesian Copperbelt*, Macdonald, London. 523 pp.

MILLER, A. R. 1966. Hot brines and recent iron deposits in deeps of the Red Sea. *Geochim. Cosmochim. Acta*, **30**, 341–59.

MILLIKEN, K. L., LAND, L. S. & LOUCKS, R. G. 1981. History of burial diagenesis determined from isotopic geochemistry, Frio Formation, Brazoria County, Texas. *Bull. Am. Ass. Petrol. Geol.* **65**, 1397–413.

MINTER, W. E. L. 1976. Detrital gold, uranium and pyrite concentrations relating to sedimentology in the Precambrian Vaal Reef placer, Witwatersrand, South Africa. *Econ. Geol.* **71**, 157–76.

MOORBATH, S., O'NIONS, R. K. & PANKHURST, R. J. 1973. Early Archaean age for the Isua iron formation, West Greenland. *Nature, Lond.* **245**, 138–9.

MORGANTI, J. M. 1981. Sedimentary stratiform ore deposits; some models and a new classification. *Geosci. Can.* **8**, 65–75.

MUIR, M. D. 1983. Depositional environment of North Australian lead zinc deposits, with special reference to McArthur River. *In*: SANGSTER, D. F. (ed.) *Sediment Hosted Stratiform Lead Zinc Deposits. Short course handbook, Vol. 8*, pp. 251–82. Min. Assoc. Can., Toronto.

NEUDERT, M. K. 1981. Shallow water and hypersaline features from the Middle Proterozoic Mount Isa sequence, Northern Australia. *Nature, Lond.* **203**, 284–8.

OFTEDAHL, C. 1958. A theory of exhalative sedimentary ores. *Forh. Geol. Foren. Stockholm*, **80**, 1.

ORTIZ, N. V., FERENTCHAK, J. A., ETHRIDGE, F. E., GRANGER, H. C. & SUNADA, D. K. 1980. Groundwater flow and uranium in Colorado plateau. *Groundwater*, **18**, 596–605.

PRETORIOUS, D. A. 1975. The depositional environment of the Witwatersrand goldfields—a chronological review of speculations and observations. *Min. Sci. Eng.* **7**, 18–47.

ROBERTS, D. E. & HUDSON, G. R. T. 1983. The Olympic Dam copper uranium gold deposit. Roxby Downs, South Australia. *Econ. Geol.* **78**, 799–822.

ROEDDER, E. 1976. Fluid inclusion evidence on the genesis of ores in sedimentary and volcanic rocks. *In*: WOLF, K. H. (ed.) *Handbook of Stratabound and Stratiform Ore Deposits*, Vol. 5, pp. 67–110. Elsevier, Amsterdam.

—— 1979. Fluid inclusion evidence on the environments of sedimentary diagenesis; a review. *In*: SCHOLLE, P. A. & SCHLUGER, P. R. (eds) *Aspects of Diagenesis. Spec. Publ. Soc. econ. Paleontol. Mineral., Tulsa*, **26**, 89–107.

ROSE, A. W. 1976. The effect of cupric chloride complexes in the origin of red bed copper and related deposits. *Econ. Geol.* **71**, 1036–48.

RUSSELL, M. J. 1975. Lithogeochemical environment of the Tynagh base metal deposits, Ireland, and its bearing on ore deposition. *Trans. Instn Min. Metall.* **B84**, 128–33.

—— 1983. Major sediment hosted exhalative zinc + lead deposits: formation from hydrothermal convection cells that deepen during crustal extension. *In*: SANGSTER, D. F. (ed.) *Sediment Hosted Stratiform Lead Zinc Deposits. Short Course Handbook*. Vol. 8, pp. 251–92. Min. Ass. Canada.

RYE, D. M. & WILLIAMS, N. 1981. Studies of base metal sulphide deposits at McArthur River, Northern Territory, Australia. III. The stable isotope geochemistry of the HYC, Ridge and Cooley Deposits. *Econ. Geol.* **76**, 1–26.

SAMAMA, J. C. 1973. Ore deposits and continental weathering. A contribution to the problem of geochemical inheritance of heavy metal contents of basement areas of sedimentary basins. *In*: AMSTUTZ, G. C. & BERNARD, A. J. (eds.) *Ores in Sediments. Int. Union Geol. Sci. Ser. A.* **3**, 247–65. Springer Verlag, Berlin.

SANGSTER, D. F. 1983. *Sediment Hosted Stratiform Lead Zinc Deposits. Short course Handbook*, **8**, Min. Ass. Can., Toronto.

SCHNEIDERHOHN, H. 1955. *Erzlagerstatten*. Stuttgart.

SCHUMM, S. A. 1977. *The Fluvial System*, John Wiley, New York. 338 pp.

SHAWE, D. R. 1956. Significance of roll orebodies in genesis of uranium-v anadium deposits on the Colorado Plateau. Contribution to the geology of uranium and thorium. *Prof. Pap. U.S. Geol. Surv.* **300**, 239–41.

SHEARMAN, D. J. 1966. Origin of marine evaporites by diagenesis. *Trans. Instn Min. Metall.* **B75**, 208–15.

SIMPSON, P. R. & BOWLES, J. F. W. 1981. Detrital uraninite and pyrite: are they evidence for a reducing atmosphere? *In*: ARMSTRONG, F. C. (ed.) *Genesis of Uranium and Gold Bearing Precambrian Quartz Pebble Conglomerates. Prof. Pap. U.S. Geol. Surv.* **1161**(A-BB), 1–26.

SPURR, J. C. 1923. *The Ore Magmas.*

STRAKHOV, N. M. 1970. *Principles of Lithogenesis*, Vol. 3. Oliver and Boyd, Edinburgh. 577 pp.

TALBOT, M. R. 1974. Ironstones in the Upper Oxfordian of southern England. *Sedimentology*, **71**, 433–450.

TEMPLE, K. L. 1964. Syngenesis of sulphide ores: an evaluation of biochemical aspects. *Econ. Geol.* **59**, 1473–91.

TURNER, J. S. & GUSTAFSON, L. B. 1978. The flow of hot saline solutions from vents in the sea floor—some implications for exhalative massive sulphide and other deposits. *Econ. Geol.* **73**, 1082–100.

TYLER, N. & ETHRIDGE, F. G. 1983. Depositional setting of the Salt Wash Member of the Morrison Formation, southwest Colorado. *J. sedim. Petrol.* **53**, 67–82.

UNESCO 1973. *Genesis of Precambrian Iron and Manganese Deposits. Proc. Kiev Symposium, 1970*, 382 pp. UNESCO, Paris.

WELCH, B. K., SOFOULIS, J. FITZGERALD, A. C. F. 1975. Mineral sand deposits of the Capel Area, Western Australia, pp 1070 and 1088. *In*: KNIGHT, C. L. (ed.) *Economic Geology of Australia and Papua New Guinea. I. Metals*, pp. 1070–88. Aust. Inst Min. Metall., Victoria.

WHITE, D. E., ANDERSON, E. T. & GRUBBS, D. K. 1963. Geothermal brine well: mile deep drill hole may tap ore bearing magmatic water and rocks undergoing metamorphism. *Science*, **139**, 919–22.

WILLIAMS, D. 1960. Genesis of sulphide ores. *Proc. Geol. Ass.* **71**, 245–84.

WILLIAMS, N. 1978. Studies of the base metal sulphide deposits at the McArthur River, Northern Territory, Australia. I. The Cooley and Ridge deposits (*Econ. Geol.* **73**, 1005–35). II. The sulphide-S and organic-C relationships of the Concordant deposits and their significance. *Econ. Geol.* **73**, 1036–56.

—— 1979. The timing and mechanisms of the formation of the Proterozoic stratiform Pb-Zn and related Mississippi Valley type deposits at McArthur River, Northern Territories, Australia. *Soc. Econ. Geol./A.I.M.E. Joint Meeting, New Orleans.* Reprint 79/51, 15 pp.

ZEISSNIK, E. H. 1969. The mineralogy and geochemistry of a nickeliferous laterite profile. *Min. Deposits*, **4**, 132–52.

H. CLEMMEY, Economic Sedimentology Laboratory, Earth Sciences Department, Leeds University, Leeds LS2 9JT.

Role of clastic sedimentology in the exploration and production of oil and gas in the North Sea

H. D. Johnson & D. J. Stewart

SUMMARY: The nature and quality of clastic reservoir rocks is the result of a complex interplay of tectonic setting, provenance, depositional environment and diagenesis. In particular, it is now well established that the reconstruction of depositional environments in clastic successions provides the optimum framework for describing and predicting reservoir development and reservoir quality distribution on both regional (exploration) and field (production) scales.

This review, in the context of the 21st anniversary of the British Sedimentological Research Group (BSRG), expands the above-mentioned theme in relation to the development of depositional models of each of the major oil- and/or gas-bearing clastic reservoirs in the U.K. sector of the North Sea, which has been the subject of a surge in geological interest following the first gas discoveries approximately 21 years ago.

Each major reservoir interval is used to illustrate specific themes, with emphasis placed on how the application of depositional models has contributed to a better understanding of the exploration for, and the production from, the following reservoirs: (i) Permian aeolian and fluvial sands (Rotliegendes Sandstones, southern North Sea), (ii) Jurassic fluvial, deltaic and marginal marine sands (Lower Jurassic Statfjord Formation and Middle Jurassic Brent Group, northern North Sea), (iii) Jurassic shallow marine and deeper marine (mass flow) sands and conglomerates (Upper Jurassic Fulmar/Piper-type sands and Brae/Magnus-type deposits; central and northern North Sea), and (iv) Tertiary deep marine, turbidite fan deposits (Palaeocene and Eocene sands, central and northern North Sea).

Sedimentology has established itself over the past 20–25 years as an integral part of the multi-disciplinary studies that are involved in the exploration and production of oil and gas. It is now well known, for example, that the reconstruction of depositional environments in clastic sequences provides the optimum framework for describing and predicting reservoir development and reservoir quality distribution on both regional-(exploration) and field-(production) scales. Nowhere has this involvement been more apparent than in the North Sea, where the phenomenal success of oil and gas exploration over the past 20 years has transformed the area into one of the most well-known hydrocarbon basins in the world, containing around 50 commercially proven oil and gas fields (Fig. 1). The current geological understanding of the North Sea area has been achieved through a broad range of multidisciplinary and multiorganizational studies involving geophysics, structural geology, stratigraphy, geochemistry, sedimentology and the many branches of petroleum engineering (e.g. production/development geology, petrophysics, reservoir engineering and production technology).

Sedimentology is only one part of the North Sea story but, in the context of the British Sedimentological Research Group's 21st anniversary, an extremely relevant subject for this meeting for several reasons. Firstly, the birth of the BSRG 21 years ago occurred very shortly after the initial realization by the oil industry that the North Sea could be a hydrocarbon-bearing basin, following the discovery of the giant Groningen gas field in the northern part of the Netherlands in 1959. Secondly, over the past 21 years, but particularly during the last 10 years, many members of the BSRG have been involved in North Sea studies in a wide range of different capacities, both directly and indirectly. Finally, the North Sea remains a vitally important and exceptionally interesting geological region which will be the subject of considerable sedimentological attention for at least the next 20–30 years.

The principal aim of this review is to highlight the contribution of sedimentological studies in the exploration of oil and gas specifically in the North Sea. This is achieved by reviewing the sedimentological characteristics of the major clastic reservoir intervals encountered in the U.K. sector within the context of their respective depositional models. The following reservoirs are considered:

(i) Permian aeolian and associated fluvial sands;
(ii) Lower–Middle Jurassic fluvial and deltaic sands;
(iii) Upper Jurassic shallow and deeper marine sands; and
(iv) Tertiary deep marine sands.

The objective of this review is not to give a detailed regional synthesis of these various rock

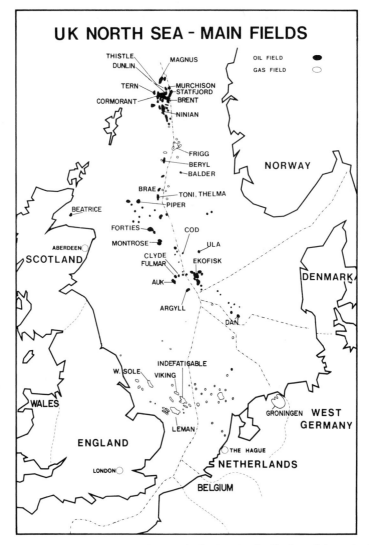

Fig. 1. Geological sketch map of the British Isles indicating the distribution of the main oil and gas fields.

units, since this has been amply covered elsewhere. Instead, emphasis is placed on how facies models and the interpretation of clastic depositional environments have influenced exploration and production activities.

The examples used partly comprise published material and also in-house reports prepared by Shell on behalf of the Shell/Esso North Sea partnership.

Sedimentology in hydrocarbon exploration and production

The nature and quality of clastic reservoirs are determined by a complex interplay of processes involving source area, depositional environment, tectonic setting and diagenesis.

Essentially the composition, climate and geomorphology of the source area determine the primary detrital composition of sediment deposited in the receiving basin which, in turn, will affect any subsequent diagenetic trends (Nagtegaal 1978). Depositional environment exerts the most basic control on clastic facies distribution, notably by determining the size, shape and trend of sand bodies, and it therefore controls primary reservoir quality (porosity-permeability) distribution (e.g. LeBlanc 1972; Shelton 1973; Pryor 1973). Reservoir development (thickness distribution) reflects the relative

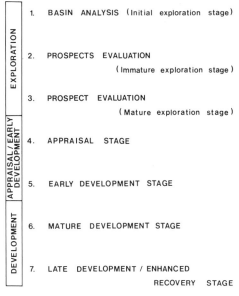

FIG. 2. Tabulation illustrating the various stages of exploration and production in which sedimentological studies are applied.

roles of tectonically induced subsidence, base-level changes and sedimentation rates. Tectonics exerts a fundamental control on all clastic sedimentation by determining the size and location of depositional (sub-) basins, rates of uplift and subsidence, directions of sediment transport, and the thickness and orientation of facies belts. Subsurface reservoir quality is finally determined by diagenesis, initially involving variable burial, compaction and cementation histories, and subsequently including any uplift, dissolution and differential diagenesis associated with hydrocarbon emplacement.

Unravelling the significance and interrelationships of these processes is fundamental to understanding the nature and quality of clastic reservoirs and to predicting their development and distribution in the subsurface. Sedimentological studies contribute to all stages of exploration and production (development) activities, ranging from the initial basin analysis stage, through the prospect evaluation of increasingly mature plays, and continuing into field appraisal and throughout the various stages of field development (Fig. 2). The typical role and application of sedimentological studies to the exploration and production of oil and gas are outlined in Fig. 3. In practice there are two distinct areas of interest, exploration (regional scale) and production (field scale), with each having its own specific objectives. The basic sedimentological data are common to both but

the aims and applications of the depositional models are different (Fig. 3). Exploration studies are primarily concerned with locating optimum areas of reservoir development with suitable source rock, seal and trapping potential to allow hydrocarbon accumulation. In production, sedimentological studies are essentially required to provide an accurate reservoir description to ensure optimum recovery of hydrocarbon reserves.

This review expands on all these aforementioned themes in the context of North Sea clastic reservoirs.

Permian, rotliegend sandstones, southern North Sea

In the context of both exploration and production, the southern North Sea was the first major European offshore hydrocarbon province to be developed. Interest in the southern North Sea as a potential hydrocarbon province was triggered by the discovery of the huge Groningen gas field in the northern Netherlands in 1959, which spurred the drive to acquire seismic data over the adjacent offshore areas in the early 1960s. This exploration effort has culminated in the discovery of several similar hydrocarbon accumulations with reserves totalling some 646 billion m^3 (22.8 trillion ft^3) of proven recoverable gas in the U.K. sector (Department of Energy 1983, p.5). Exploration and production in this offshore area were particularly attractive because of the shallow water depths ideal for the use of small jack-up rigs. Nearly 20 years later this province is experiencing a further resurgence owing to a combination of higher gas prices and lower oil prices which have renewed interest in some of the previously marginal commercial prospects.

Geological framework

The main gas accumulations are found in Permian sandstones of the Rotliegendes Group underlying a tight seal of Zechstein dolomites and evaporites. The Rotliegendes Group has a wide distribution in three intracratonic basins (Glennie 1981), the largest being the South Permian Basin which contains the southern North Sea gas fields (Fig. 4).

Although the Rotliegendes Group has a fairly wide distribution, it was quickly realized that it was composed of markedly different lithologies with varying geographical distributions. In particular, two main types of reservoir sandstone were recognized: (1) large scale cross-bedded sandstones with well developed bimodal foreset

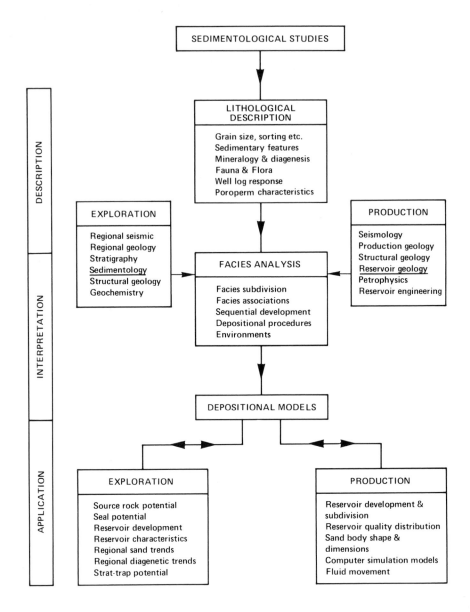

FIG. 3. Tabulation illustrating the nature and application of sedimentological studies in exploration and production.

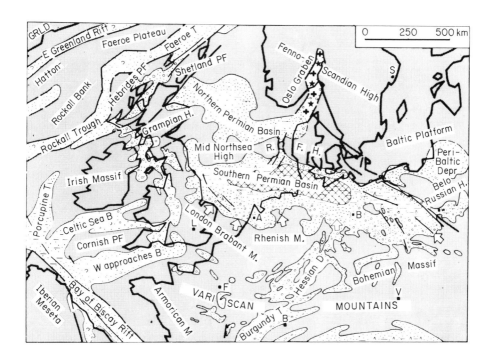

Fig. 4. Early Permian, Rotliegendes palaeogeography (from Ziegler 1981).

laminae believed to be aeolian in origin, and (2) poorly sorted, pebbly sandstones deposited in fluviatile wadis. The association of these sandstones with fine-grained evaporitic deposits and the frequent reddening of the sandstones indicated deposition in an arid, desert basin.

Whilst considerable knowledge was available in the early 1960s on the geomorphology of desert landscapes and on the external morphology of aeolian sand dunes, there was little detailed knowledge of the internal structure of desert deposits. However, this information was necessary to identify facies associations in cores and from other well-log data. In addition, a basin-scale depositional model was required to aid the prediction of sand thickness trends and the geographical distribution of facies belts for exploration purposes.

Facies-related studies

It was against this background that renewed studies of desert environments were initiated, notably by Glennie (1972), to assist the broader interpretation of ancient desert deposits and, more specifically, to assist the search for hydrocarbons in the South Permian Basin.

Glennie's work on arid environments of the present-day Persian Gulf adjacent to the Oman Mountains of the United Arab Emirates (Glennie 1972) illustrates how studies of modern sedimentary environments can aid the search for hydrocarbons in comparable ancient deposits. The depositional model for the Rotliegendes Group consists of facies belts of fluviatile, aeolian and evaporite/lacustrine deposits, roughly paralleling the London–Brabant massif/Variscan highlands which lay to the south of the South Permian Basin (Figs 5 and 6). The facies belts pass northward from wadi/alluvial fan facies, best developed in proximity to the mountain chain, through dune and inter-dune aeolian sands further to the north-east, and finally into sabkha desert lake facies at the basin centre. In building up a basin-scale depositional model, regional facies mapping has proved to be an essential part of exploration studies, largely because the depositional en-

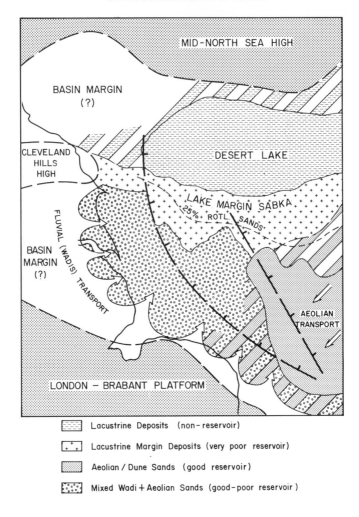

FIG. 5. Rotliegendes facies distribution (based on Glennie *et al.* 1978).

vironment of the sandstones has a fundamental influence on reservoir quality. Hence, detailed sand quality studies have attempted to distinguish the high quality aeolian sandstones from the poorer quality fluviatile sandstones. This type of study is most clearly illustrated in the work of Marie (1975; see also Fig. 7), who used a variety of mappable parameters to identify areas of optimum reservoir development based on the previously outlined facies model. The early stages of basin analysis typically involve the construction of thickness and/or percentage maps of sandstone and claystone, based initially on widely spaced well data. This is followed up by more detailed core studies, which in the case of the Rotliegendes have en-

abled the two main sandstone facies associations (fluviatile and aeolian) to be identified and mapped. The combination of these maps provide an excellent quantitative framework for evaluating reservoir quality distribution, and for recognizing the relative importance of additional factors such as tectonics and diagenesis.

Facies analysis of the Rotliegendes Group of the U.K. sector shows that it consists of a lower interval of mainly variable alternations of fluviatile/wadi sands and gravels, and an upper interval of aeolian dune and interdune sands (Fig. 8).

From the above it is apparent that core studies are essential in order to build up depositional models. However, many exploration and pro-

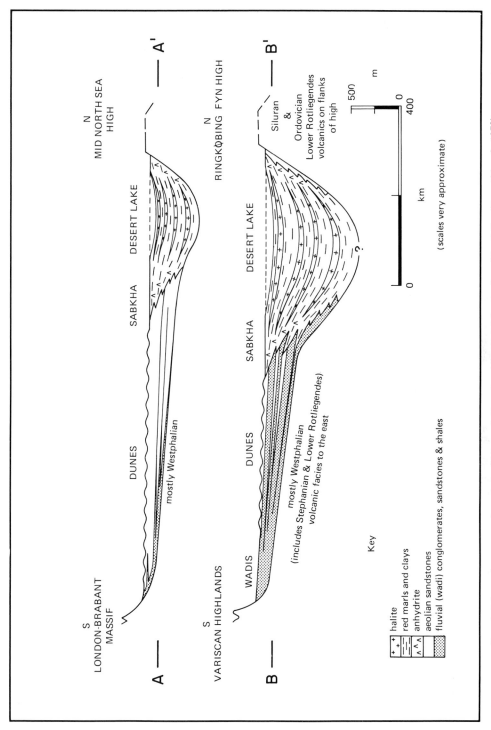

Fig. 6. Schematic cross-sections through the early Permian deposits of the southern North Sea (from Glennie 1972).

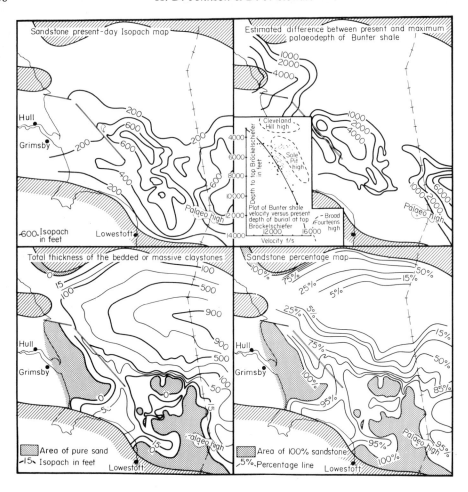

FIG. 7. Various mapping parameters used to reconstruct the depositional framework and geological burial history of the Rotliegendes Group (from Marie 1975).

duction wells are not extensively cored owing to the higher cost. In this context, it proved possible to use wireline log data not only to identify the two main facies associations for work on an exploration scale, but also to identify smaller facies variations for reservoir correlation on a field scale (Van Veen 1975). For example, the large-scale cross-sets of the aeolian dunes and the interposed finer grained, more argillaceous interdune or bottom set (including desert lag and adhesion ripple sands) deposits can be detected accurately using a combination of gamma ray, bulk density (FDC) and microlaterolog records (Fig. 9). This technique depends on the higher clay and cement contents of the dune bottom set beds. The clay causes a peak on the gamma ray log and the cement is detected on the FDC log. The combination of

clay and cement causes a similar but large peak on the microlaterolog (resistivity).

The additional use of dipmeter logs (Fig. 10) enables the measurement of dune thickness, by plotting the gradual upwards increase in foreset inclination, and the deduction of wind directions by measuring the direction of foreset inclination. This information can help predict field-scale variations in the quality and thickness of smaller-scale reservoir units necessary for field development studies. Further development information can be gained by studies of bedding characteristics of the dune sequences in cores, with the ultimate aim of identifying the size, geometry and orientation of individual cross-bed sets. The prediction of the size and orientation of dune cross-bedding is important for the construction of reservoir geological

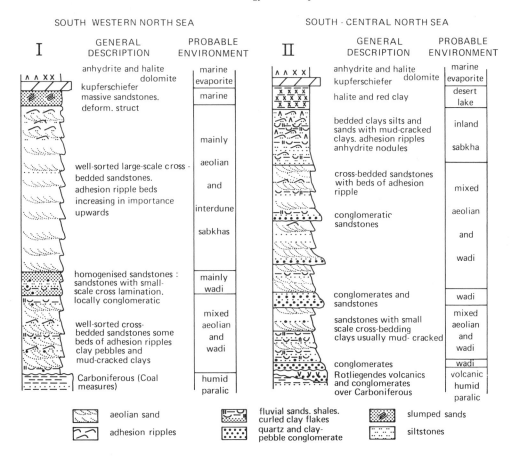

FIG. 8. Generalized sedimentology and vertical lithological profiles through the Rotliegendes of the southern North Sea (from Glennie 1972).

models because permeabilities are significantly greater parallel to the dune bedding as compared to the perpendicular direction (van Veen 1975; Weber 1984). It becomes vital, therefore, to know the internal geometry of the dune deposits when designing the field development programme, since this allows optimum location of production wells with regard to maximum well productivity.

Studies of dune bedding in the Rotliegendes have been carried out by Glennie (1982), who, with the aid of core and dip-meter data, has inferred the presence of two main types of dune: (1) barchan or transverse dunes with the bulk of the bedding dipping downwind (unimodal), and (2) longitudinal seif dunes with bimodal foreset directions at right angles to the main wind direction (Fig. 11). This type of analysis may enhance the interpretation of palaeocurrent data

in terms of transport directions and dune bedding permeability characteristics.

Diagenetic studies

The relatively simple picture of reservoir quality based directly on depositional facies distributions is complicated by the intricate burial history of the Rotliegendes Group in the southern North Sea. This has resulted in a diagenetic overprint which has a important detrimental effect on porosity/permeability characteristics of all facies associations. It was Stalder (1973) who, with the aid of the scanning electron microscope, demonstrated that the poorer reservoir quality resulted from the presence of authigenic clay minerals and that the type and morphology of these clay minerals have an important effect on porosity/permeability relationships.

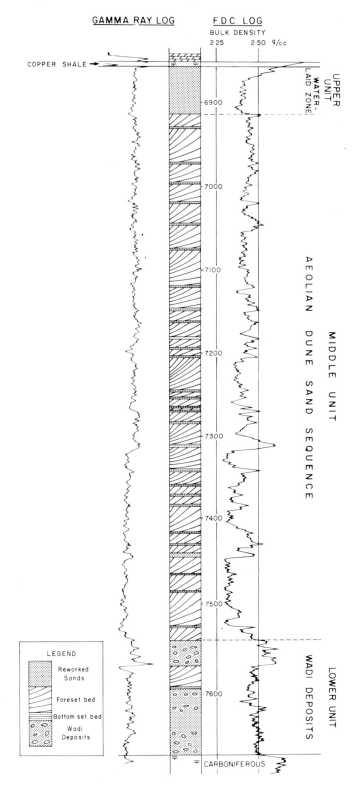

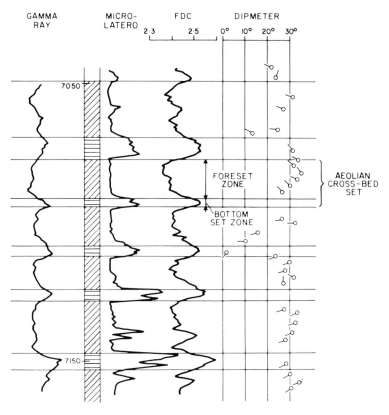

WELL 49/26-2 (VERTICAL BOREHOLE)

FIG. 10. Dipmeter response through dune-bedded Rotliegendes sandstones (from van Veen 1975).

It is now well known that illite, in particular, is a common cause of lower permeabilities in Rotliegendes reservoirs. The explanation for this is the fibrous bridging habit of the illite crystals (Fig. 12), which reduces the size of pore throats and increases the tortuosity in the interstitial pore spaces, whilst having little effect on porosity (Stalder 1973). Conversely, the more blocky kaolinite, also a common authigenic clay mineral in the Rotliegendes has a lesser effect on permeability owing to its pore-filling, rather than pore-bridging, habit (Fig. 13). It is, therefore, essential to be able to predict the distribution of the various authigenic clays, notably those areas most susceptible to kaolinite or illite development.

Regional mapping of authigenic clays initially showed that poor reservoir quality, illitic sandstones often occur at present-day shallower depths than the kaolinitic sandstones (Seeman 1979). At first sight this appears to suggest that kaolinite is formed at the expense of illite with increasing depths of burial. However, by using a combination of geological well data and sonic velocities (Glennie *et al.* 1978), it can be deduced that the illitic sandstones were in fact, buried more deeply than the kaolinitic sandstones and, in most cases, in excess of 11,000 ft (Fig. 14). In demonstrating this relationship, sonic velocities in the Bunter Shale overlying the gas reservoirs are important since the illitic sandstones are associated with shales of higher sonic velocities than those associated with the kaolinitic sandstones (Glennie & Boegner 1981), suggesting that the former have undergone greater compaction and, therefore, greater burial. The present-day inverse relationship is the result of later uplift of a previously deeply buried sequence (Glennie &

FIG. 9. Example of gamma ray and FDC (density) log responses to various facies types in the Rotliegendes Group of the Leman Field (from Van Veen 1975).

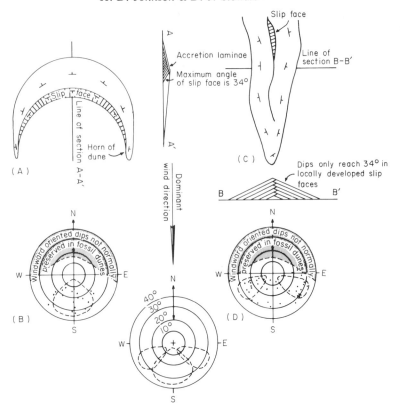

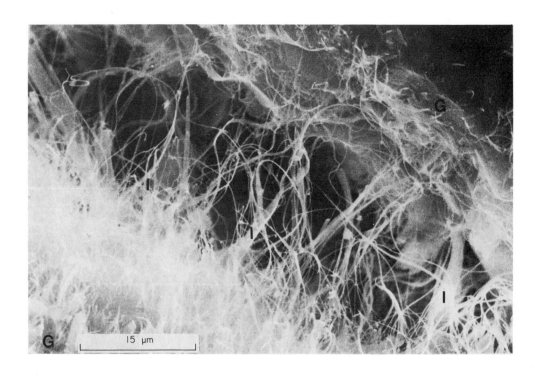

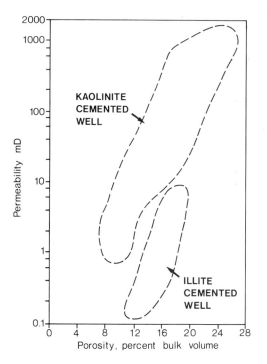

FIG. 13. Different effects of kaolinite and illite on porosity/permeability characteristics in the Rotliegendes sandstones (from Stalder 1973).

Boegner 1981).

Further studies have culminated in the recognition of a complex diagenetic sequence (Fig. 14) ranging from early environment-related features to late, post-uplift features (Glennie *et al.* 1978). These studies also show that depth is not alone in controlling diagenesis. It has been suggested, for example, that feldspar may be a source of kaolinite and, therefore, variations in detrital feldspar may control the abundance of kaolinite. In addition, the degree of feldspar dissolution at a location may be related to the thickness of underlying organic-rich Carboniferous strata available to generate CO_2 gases (Rossel 1982). The situation is further complicated by the possibility of cements such as carbonate or anhydrite being introduced from the overlying Zechstein strata. Nevertheless, porosity distribution at depth is still largely a

function of variations in sorting, as reflected in the dune, adhesion and wadi sands (Glennie *et al.* 1978).

Lower and Middle Jurassic Sands, northern North Sea

The Lower Jurassic Statfjord Sands, and particularly the Middle Jurassic Brent Sands, represent the main reservoirs of the prolific North Viking Graben oil province in the northern North Sea. In terms of the U.K. Sector of the North Sea, the discovery of major hydrocarbon reserves in these reservoirs and their subsequent development represented the second *major* phase of hydrocarbon exploration and production, following on from the successes of the southern North Sea gas province (although oil was first discovered in Palaeocene sands of the Forties and Montrose fields). However, the discovery of the Brent Sands in 1971 in well 211/29-1 (Brent field), and the Statfjord Sands in later wells, revealed a more complex and regionally less predictable reservoir sequence compared to the Rotliegend Sandstones (e.g. Bowen 1975).

One early investigation by C. Kruit (Shell E & P Laboratory, Rijswijk, personal communication) recognized the large-scale regressive/transgressive character of the Brent Sands (Fig. 15). Based on limited data (fourteen wells and 2500 ft core) the Brent Sands were interpreted as the product of a wave-dominated delta to interdeltaic system which prograded northwards along the axis of an active rift system (Fig. 16). This interpretation remains essentially correct and it illustrates the value of a conceptual depositional model, particularly during the early stages of basin analysis, in providing a framework for interpreting stratigraphic relationships and predicting regional trends in reservoir distribution.

Increased subsurface control, provided by substantially more wells (with more abundant cores and well logs) and better seismic coverage, has provided a wealth of new sedimentological data. These data, when integrated with the increased knowledge of the regional tectonic evolution and improved stratigraphic control, have resulted in the development of a more comprehensive depositional model. The

---

FIG. 11 (facing, top). Comparison of bedding architecture in ideal barchan and transverse dunes (a) and seif dunes (c) (adapted from Glennie 1982).

FIG. 12 (facing, bottom). Scanning electron photomicrograph 'hairy' illite crystals.

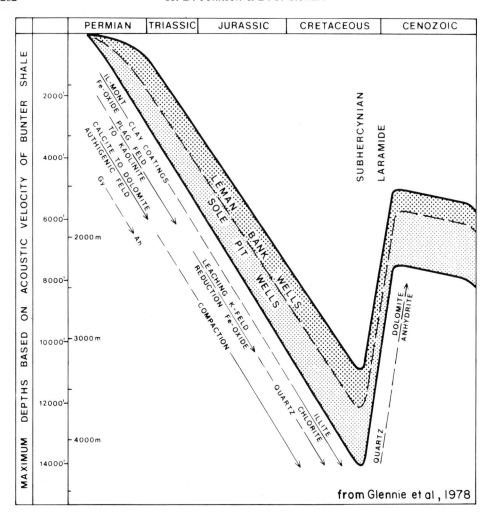

from Glennie et al, 1978

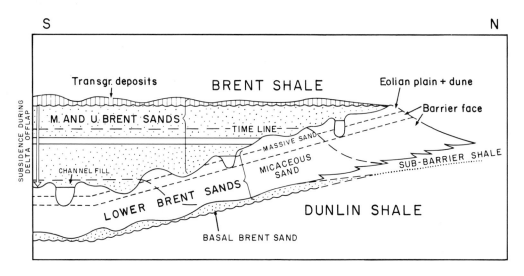

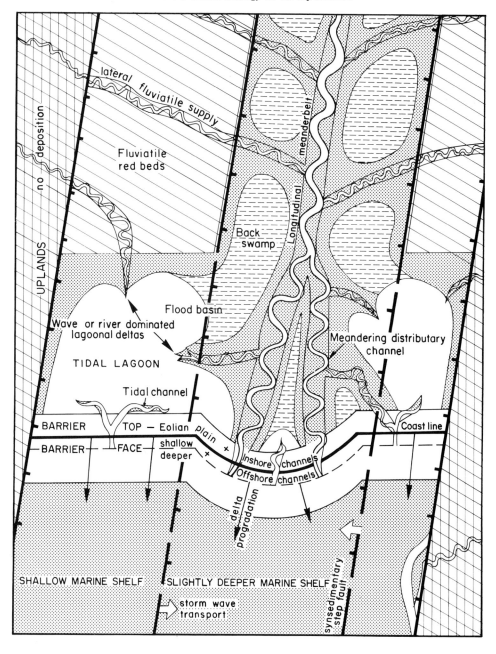

FIG. 14 (facing, top). Rotliegende burial history and sequence of diagenetic events in Sole Pit and Leman Bank areas (from Glennie *et al.* 1978).

FIG. 15 (facing, bottom). Basic depositional framework of the Middle Jurassic, Brent Sand Viking Graben, northern North Sea.

FIG. 16 (above). Model of axial fill of a shallow marine rift valley by a wave-dominated delta.

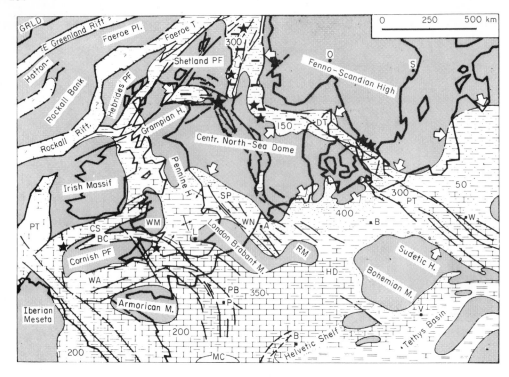

Fɪɢ. 17. Bajocian-Bathonian palaeogeography (from Ziegler 1981).

regional framework, for example, indicates that the Brent Sands were deposited following the widespread mid-Cimmerian tectonic phase and a eustatic lowering of sea-level, when a large rift dome was uplifted in the central North Sea which became a major sediment source area (Fig. 17; Ziegler 1982). Clastic sediments were shed radially off this dome and sediment transport paths developed through the continuously subsiding Viking Graben (e.g. Eynon 1981), as predicted in the earlier conceptual depositional model.

The increased well control has also provided positive definition of facies trends, which confirm the south to north progradation of the coastal/deltaic clastic wedge. Detailed facies distribution patterns in part of the Ness Formation, for example, indicate that fluvial deposits gradually merge northwards into lagoonal deposits, which are themselves replaced by coastal barrier and offshore marine sediments (Fig. 18). Furthermore, the convergence of progradational and transgressive shallow marine sands at the northern pinch-out of the clastic wedge, demonstrates that the time-equivalent coastal and delta plain complex (Ness Formation) was located to the south behind a contemporaneous coastal barrier complex

(Rannoch–Etive Formations).

The important interplay between sedimentological and graben tectonic processes in influencing the thickness and lateral distribution of the Brent Sands, together with the main facies associations, is illustrated in a series of S–N-trending time-slices (Fig. 19). The sedimentological relationships of the main types of sediment body (facies associations) have been outlined in a later conceptual depositional model (Fig. 20; Budding & Inglin 1981).

The model of a wave-dominated delta to coastal (interdeltaic) coastal barrier complex enables comparison with several modern wave-dominated and fluvial-wave interaction deltas and their associated delta and marginal coastal plain environments (Galloway 1975; Coleman *et al.* 1981; Elliott 1978; cf. Fig. 21).

Reservoir characteristics of the Brent Sands and their relation to exploration and production

The reservoir properties of the Brent Sands are, for the most part, controlled by the depositional environments within which the various facies associations were deposited (Simpson & Whitley 1981; Proctor 1981). Diagenesis is

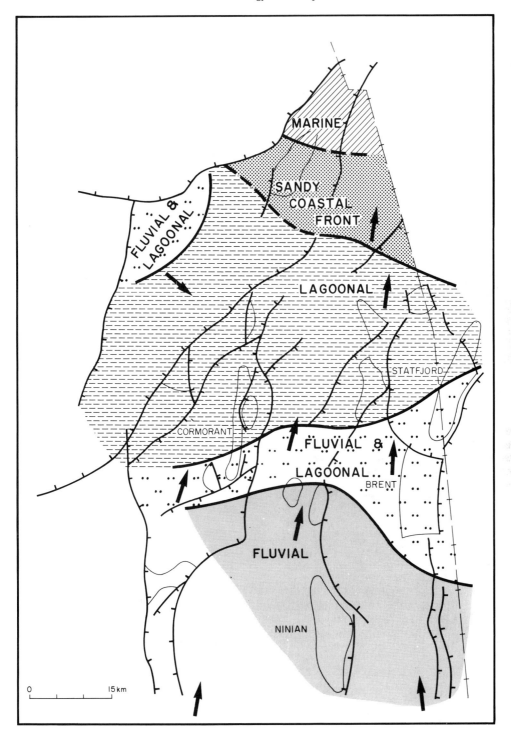

FIG. 18. Lateral facies relationships in late Brent Sands times in the North Viking Graben.

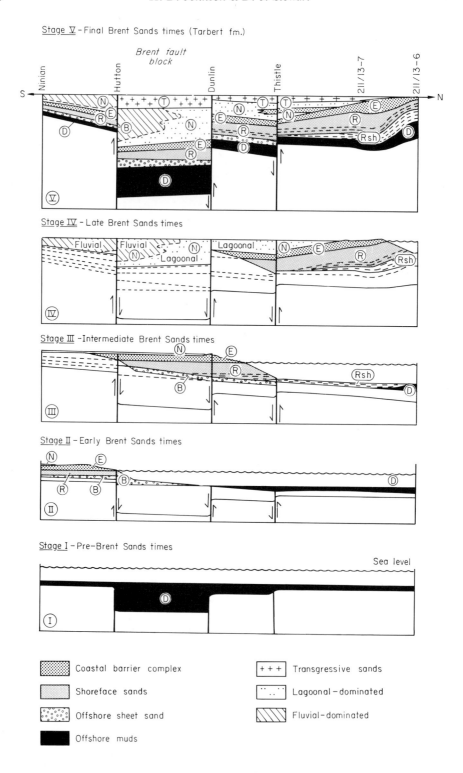

FIG. 19. Conceptual model of the evolution of the Brent Sands depositional system.

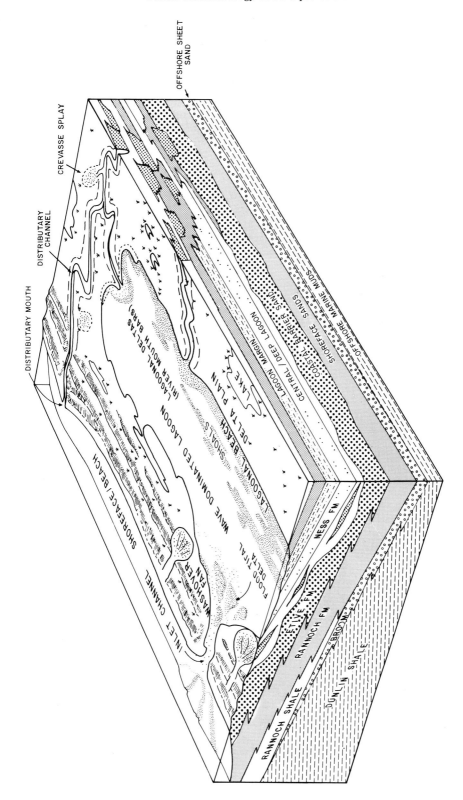

OFFSHORE SHEET SAND

CREVASSE SPLAY

DISTRIBUTARY CHANNEL

DISTRIBUTARY MOUTH

LAGOONAL / DEEP LAGOON MARGIN SAND

CENTRAL / BARRIER COASTAL SANDS

SHOREFACE MARINE MUDS

OFF SHORE MARINE MUDS

LAGOONAL DELTAS (RIVER MOUTH BARS)

SHOALS

WAVE DOMINATED LAGOON

DELTA PLAIN

LAKE

SHOREFACE / BEACH

WASHOVER FAN

FLOOD TIDAL DELTA

INLET CHANNEL

NESS FM

ETIVE FM

RANNOCH FM

BROOM

RANNOCH SHALE

DUNLIN SHALE

Fig. 20. Conceptual model of deposition in the Brent delta (from Budding & Inglin 1981).

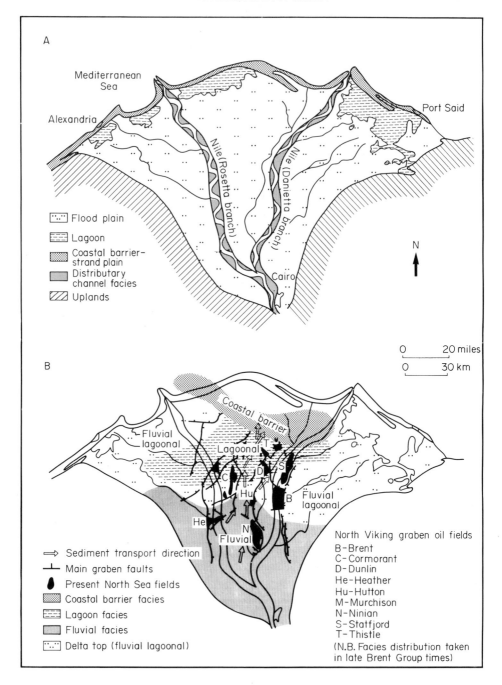

FIG. 21. (a) A possible modern analogue of the Brent Group provided by the Nile delta. (b) Time-slice in late Brent Group times illustrating the observed facies distributions, which are superimposed on the Nile delta to allow comparison of scales.

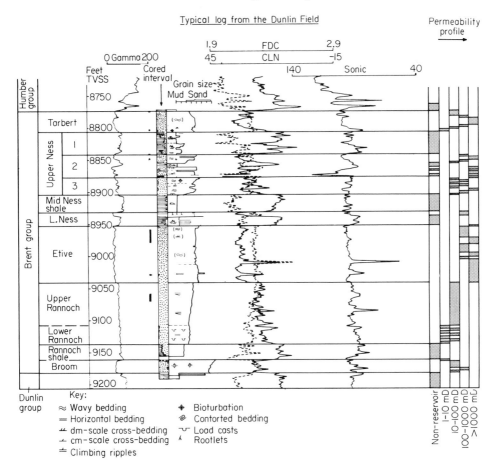

FIG. 22. Typical sedimentary sequences and corresponding wireline log and permeability profiles for Brent Group reservoir sands.

widespread but, in most cases, of secondary importance since it tends to enhance primary variations in reservoir quality rather than overprint them. It has, therefore, been necessary to have a clear understanding of the type and nature of the different facies associations, their reservoir properties, and their vertical and lateral relationships in order to interpret and predict reservoir development and reservoir quality distribution on both regional and field scales.

From a reservoir quality/development point of view, the Brent Sands comprise three main elements: (1) a sand-dominated coastal barrier-shoreface complex (Rannoch and Etive Formations), (2) a variable, alternating sequence of sands and shales deposited in coastal and delta plain environments (Ness Formation), and (3) a relatively thin, transgressive sand sheet (Tarbert Formation).

The *coastal barrier-shoreface complex* forms as essentially tabular-shaped sand sheet characterized by an upward transition from offshore shales and fine grained, micaceous shoreface sands (Rannoch Formation), into coarse grained, well-sorted and non-micaceous sands deposited in upper shoreface/foreshore, distributary channel and beach-ridge environments (Etive Formation). Permeability profiles closely follow the grain size and sorting profiles with the abrupt jump at the Rannoch–Etive boundaries (e.g. Fig. 22) commonly associated with distributary and barrier inlet channels cutting into finer grained shoreface sands.

The *coastal and delta plain complex* comprises a wide range of coal-bearing and rootletted

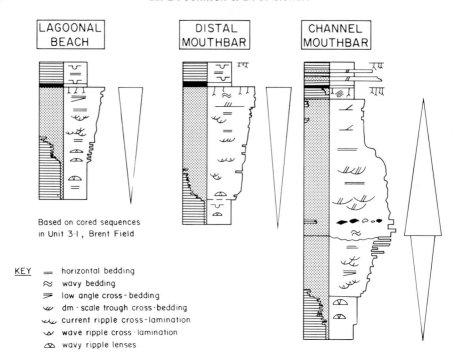

KEY
= horizontal bedding
≈ wavy bedding
⇛ low angle cross-bedding
⌣ dm-scale trough cross-bedding
⌣⌣ current ripple cross-lamination
⌣ wave ripple cross-lamination
⌔ wavy ripple lenses

Based on cored sequences
in Unit 3·1, Brent Field

FIG. 23. Examples of lithological sequences through the three main sand body types of the Ness
Formation.

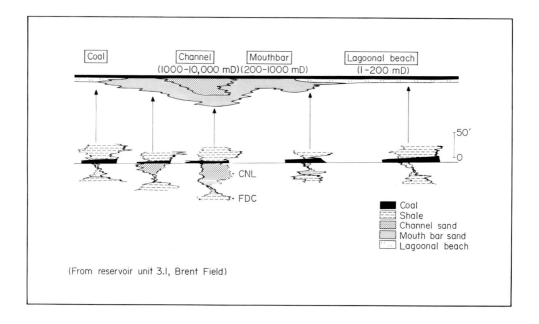

FIG. 24. Examples of FDC/CNL log response through the three main sand body types in the Ness
Formation.

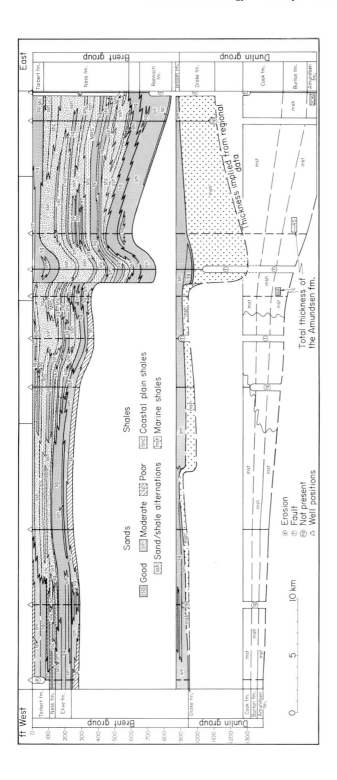

FIG. 25. East to west panel highlighting the influence of syndepositional graben faults on reservoir development in the Brent Group.

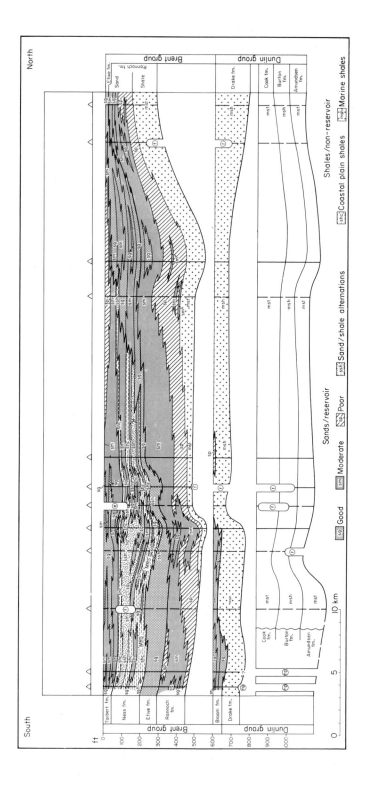

FIG. 26. North to south panel illustrating the northward termination of coastal barrier and concomitant wedge-out of lagoonal shales into barrier and back barrier sands.

sands and shales. Detailed studies of cores and well logs have identified the following facies associations: (1) fluvio-distributary channels, (2) crevasse/sheet splays, (3) river mouth bars/lagoonal deltas, (4) lagoonal beaches, (5) washover fans, (6) floodplain shales, and (7) lagoonal/lacustrine shales (e.g. Fig. 23). Although complex, these deposits occur in vertically and laterally predictable relationships (Figs 23 and 24). Distinguishing between the different types of sand bodies requires careful core/well log calibration and interpretation, which can then enable similar deposits to be distinguished by well logs alone (e.g. Fig. 24). This also represents an essential step in predicting the shape, orientation and porosity/permeability distribution within individual or multistorey sand bodies.

The *transgressive sand sheet* consists of an irregular blanket of reworked, shallow marine sands which were deposited during the final retreat of the Brent shoreline/delta system. Reservoir quality tends to be variable, but in general, porosity and permeabilities are good. In detail this interval appears to comprise several elements of transgressive/onlapping coastal barrier deposits (e.g. shoreface, inlet channels, washover fans and offshore sands).

At the common depth of many North Viking Graben oil fields of around 8000–10,000 ft, and with gross thicknesses of approximately 300–900 ft, the internal characteristics of the Brent Sands have not been able to be resolved by seismic characteristics. This has only been achieved by using the facies characteristics and depositional model, as outlined earlier on the basis of cores and well logs, to provide a framework for predicting reservoir development and reservoir quality distribution on both regional and field scales

Two arbitrary examples of E–W- and N–S-trending correlation panels illustrate typical regional-scale patterns of reservoir distribution. The E–W panel (Fig. 25) highlights the importance of syndepositional graben faults in influencing reservoir development, while the N–S panel (Fig. 26) illustrates the northward termination of the coastal barrier complex and the northward wedge-out of lagoonal shales into barrier and back barrier sands. This type of information enables prediction of anticipated reservoir development and porosity characteristics ahead of exploratory drilling once a basin has reached a relatively mature stage, as in the North Viking Graben. At the same time it provides a framework for more detailed studies associated with appraisal and development drilling.

One application of sedimentological studies to field development is provided by the Brent field in which the Brent Sands are approximately 900 ft thick. The lagoonal/lacustrine deposits noted earlier form four important laterally extensive shale layers which divide the reservoir into five separate drainage units (Fig. 27). Different sedimentological and reservoir characteristics divide the Brent Sands into 15 distinct units which can be correlated over the whole field. Within this framework it is possible to identify the different types of sand body and shale layers and, thereby, reconstruct in some detail their vertical and lateral distribution (Fig. 28). This enables the accurate mapping of a whole suite of reservoir quality parameters such as net sand, porosity, permeability and fluid saturations, amongst others.

A final example of how a sedimentological model can help predict reservoir development trends is provided by the Cormorant field. Here Budding & Inglin (1981) demonstrated how large thickness changes in the Rannoch, Etive and Lower Ness Formations can be interpreted in terms of varying rates of shoreline progradation (Fig. 29). Rapid rates of progradation lead to relatively thin sequences of coastal barrier deposits whereas slow rates of progradation allow accumulation of much thicker sequences. Thickness patterns combined with a knowledge of both regional- and field-scale facies relationships and depositional models were, therefore, able to locate areas of optimum reservoir development, particularly of the high-quality Etive Formation, ahead of extensive development drilling. From a sedimentological point of view the results of this Cormorant field study also gave independent support to the regional model, particularly in terms of shoreline orientation and the direction of progradation (to the NNE).

Geological reservoir modelling of the Statfjord Formation

The Lower Jurassic Statfjord Formation is discussed here in the context of the development of a geological model of the Brent field, which was constructed for a reservoir engineering computer simulation study utilizing a large capacity CRAY computer.

Although economically less important than the Brent Group, the Statfjord Formation contains substantial hydrocarbon accumulations in both the Brent and Statfjord fields (Bowen 1975; Chauvin & Valachi 1980). Regionally, the Statfjord Formation comprises a lower fluviatile succession, which is restricted to the central part

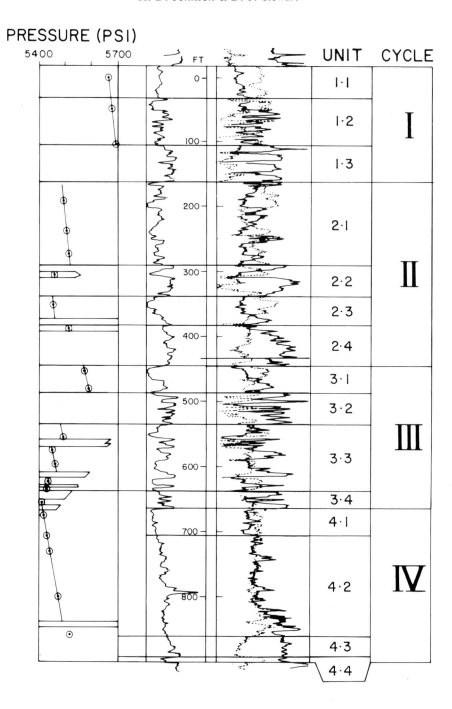

FIG. 27. Subdivision of the Brent Sands in the Brent field illustrating the subdivision of the reservoir into four drainage units and 15 distinct correlation units.

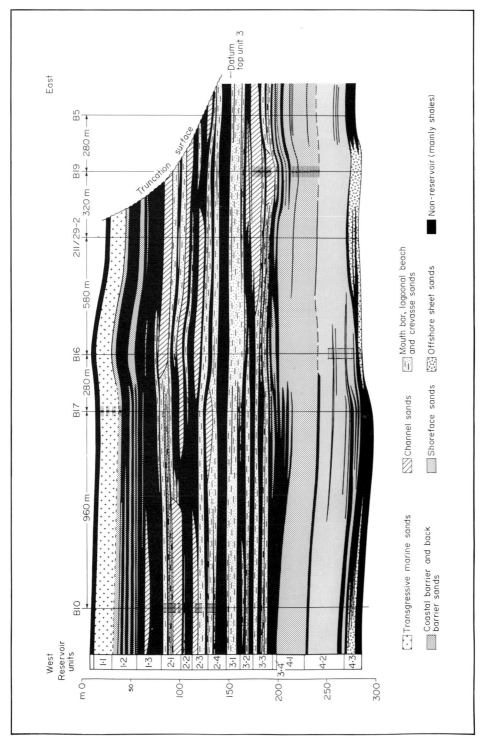

FIG. 28. East to west panel illustrating detailed reconstruction of vertical and lateral distributions of sand and shale bodies in the Brent field.

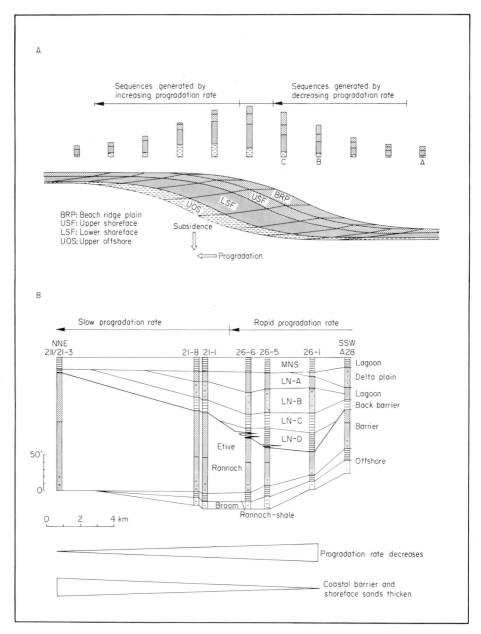

Fig. 29. (A, B) Thickness variations of coastal barrier sediments as a function of progradation rate (from Budding & Inglin 1981). (A) model for thickness variations, (B) example from South Cormorant.

of the graben east of the Hutton–Dunlin fault zone, and an upper and laterally more extensive shallow marine sandstone sequence (Fig. 30). The large-scale depositional history of the Statfjord Formation reflects two main phases of sedimentation: (1) *fluviatile sedimentation* was probably associated with a possibly eustatically induced lowering of the erosional base level, and a gradual increase in surface water run-off accompanying a change from a semi-arid (early Triassic) to a humid/sub-tropical (early Jurassic) climate, and (2) *shallow marine sedimentation* occurred in response to the early Sinemurian (*c.* 189 Ma) eustatic sea-level rise (Vail & Todd 1981) which ultimately led to the deposition of offshore/shelf muds throughout

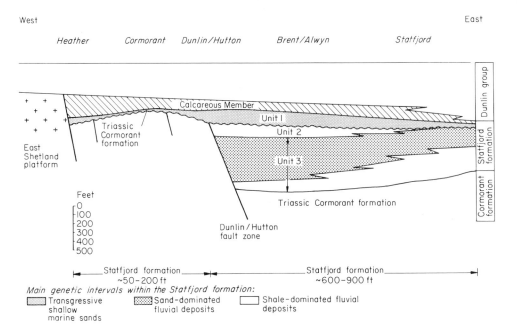

West   East

Heather   Cormorant   Dunlin/Hutton   Brent/Alwyn   Statfjord

FIG. 30. Cross-section through the Statfjord Formation in the North Viking Graben.

the North Sea area.

In the Brent field, the Statfjord Formation is represented by a large-scale offlapping-onlapping couplet. The offlapping sequence occurs in the lower part of the reservoir (Cormorant Formation → Unit 3.2 → Unit 3.1) and comprises an upward increase in sand content, pebble size and frequency, channel sand body density and channel sand body inter-connectedness. The onlapping/transgressive sequence begins with an abrupt change from the fluvial channel sand-dominated Unit 3.1 into the floodplain shale-dominated Unit 2 (Fig. 31), and this is followed by a succession of increasingly lower energy/offshore shallow marine deposits (Unit 1 → Calcareous Member → Amundsen Formation), with the high-quality sands of Unit 1 forming the important, uppermost reservoir unit in the Brent field. A conceptual reservoir geological model of the Brent field area, illustrated in Fig. 32, schematically outlines the inferred distribution of sand bodies and shale layers.

Of particular importance to the development of the Statfjord Formation in the Brent field was the potential for fluid communication between the various reservoir units, in particular between Units 1 and 3.1 across the shaley Unit 2. This stems from an investigation into the possibility of producing the reservoir by injecting gas under miscible conditions into Unit 1 and injecting water into Unit 3 (Fig. 33); such a scheme requires Unit 2 to behave as a seal. In order to test the significance and sealing potential of Unit 2 a detailed reservoir geological model was constructed from the B-platform area of the field where vertical fluid communication is a possibility for the following reasons: (1) the basal shale layer is locally cut through by channel sands, (2) sand/shale ratios are abnormally high for Unit 2, ranging between 0.4 and 0.65 in this area, (3) multistorey channel sands are locally well-developed, and (4) theoretical studies indicate substantial increases in channel interconnectedness when sand/shale ratios exceed 0.5 (Allen 1978).

The geological model was constructed within the framework of both the B-platform area in particular (Fig. 34) and the field as a whole, and to the limitations of the computer. The model comprised 30 layers with each layer being 25 ft thick, which approximates to the thickness of individual channel sand bodies. Areally the model comprised 10 × 11 grid blocks, measuring 5000 × 5500 ft (1.5 × 1.7 km), and included three wells within the model area plus seven wells in the immediate vicinity. The model was constructed in a series of panels depicting both observed and inferred distributions of sand and shale layers (Fig. 35).

In order to model the geological heterogeneity of this type of reservoir realistically it was

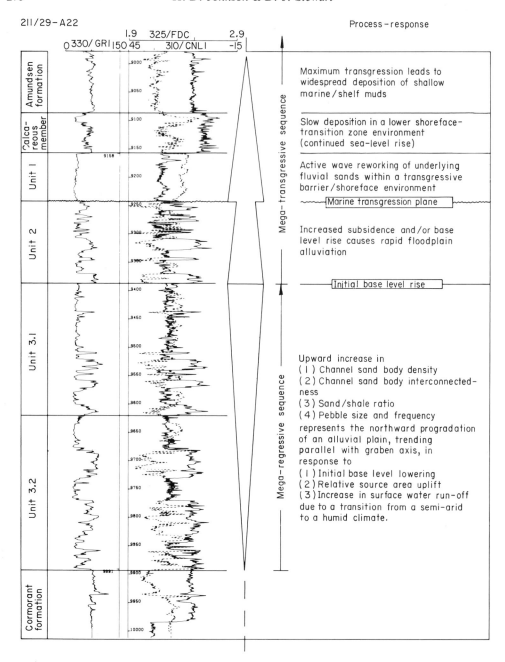

FIG. 31 (above). Interpretation of the large-scale depositional history of the Statfjord Formation in the Brent Field.

FIG. 32 (facing). Conceptual reservoir geological model through the Statfjord Formation in the Brent Field.

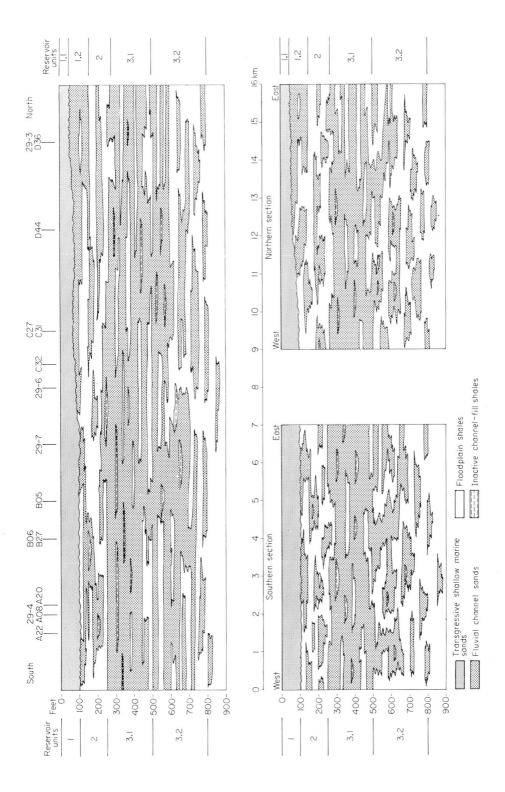

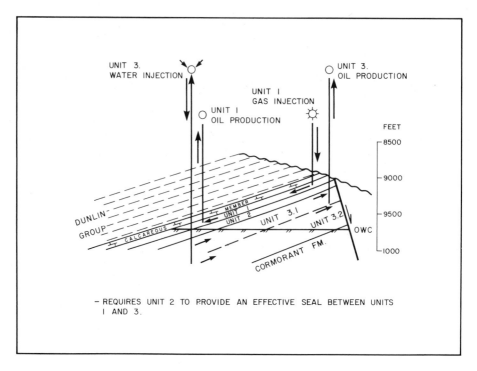

FIG. 33. Statfjord reservoir, Brent field, indicating a possible recovery scheme involving combined miscible gas/water injection.

necessary to utilize a CRAY computer, which has the capacity for handling complex three-dimensional models. Indeed, the increasing use of the new breed of supercomputers in reservoir simulation studies is an important step forward because it allows the use of more realistic geological models than has been previously possible. In all cases, however, the type of model, particularly its complexity, should be developed in the context of the objectives of the simulation exercise and it requires close liaison between the geologist and reservoir engineer. In the case cited above, it is possible to test the sensitivity of the system to different geological conditions (e.g. varying degrees of channel interconnectedness) which can given important insight into the behaviour of fluid movements in complex reservoirs.

Upper Jurassic sandstones, northern and central North Sea

Regionally the Lower and Middle Jurassic reservoirs of the Northern Viking Graben present a relatively simple 'layer cake' geological framework, and the oil accumulates in rather simple, truncated structural traps (tilted fault blocks). In contrast, Upper Jurassic oil accumulations occur in reservoirs which are often of more restricted lateral extent, owing to rapid lateral changes in facies type and reservoir thickness. They therefore confront both the explorationist and production geologist with a somewhat different set of problems.

Geological framework

Four main facies associations have been recognized within the Upper Jurassic, Humber Group (Oxfordian to Portlandian) of the Central and Northern North Sea:

   (i) shallow marine sand association (reservoir);
  (ii) shallow marine shale association, characterized by low levels of radioactivity (seal);
 (iii) gravity/mass flow sands and conglomerates (reservoir); and
  (iv) basinal shales (highly radioactive; gamma ray readings approximately over 100 API), deposited in anoxic conditions (seal/source rock).

During the Upper Jurassic much of the northern and central North Sea was the site of argillaceous sedimentation, including the widespread development of organic shales, thought

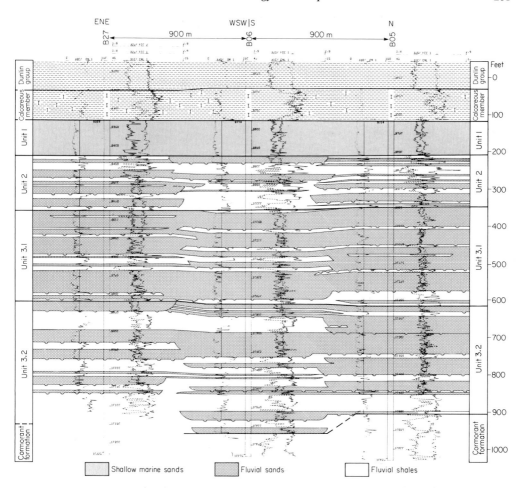

FIG. 34. Well log correlations in the Statfjord Formation in the southern part of the Brent field.

to be the main source rock for North Sea oil (Fig. 36). However, tectonic factors, in particular the continued movement of major faults and the associated development of half grabens, have caused the local accumulation of clastic deposits. In the Central Graben salt movements have had additional influences on sedimentation. Furthermore, the Humber Group forms an overall transgressive sequence, reflecting a relative rise in sea level which may be eustatic in nature (Vail *et al.* & 1977; Hallam 1978).

Two main styles of clastic reservoir are recognized in the Humber Group; shallow marine sand and mass flow sand/conglomerate accumulations (Fig. 37).

The mass flow sand/conglomerate sequences (e.g. Brae, Toni, Thelma, Magnus) are enclosed in, or interfinger with, the organic-rich Kimmeridge Clay, which frequently acts as source rock and seal. Shallow marine sand reservoirs (example Piper, Fulmar, Ula) are also generally capped by Kimmeridge Clay. Syndepositional tectonics appear to have played an important role during the accumulation of sand bodies of both facies types. Geographically, the Upper Jurassic shallow marine sandstones are restricted mainly to the Moray

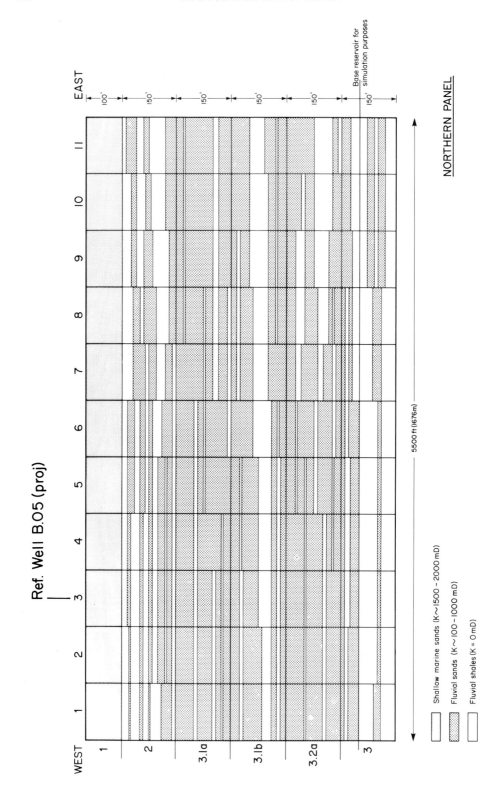

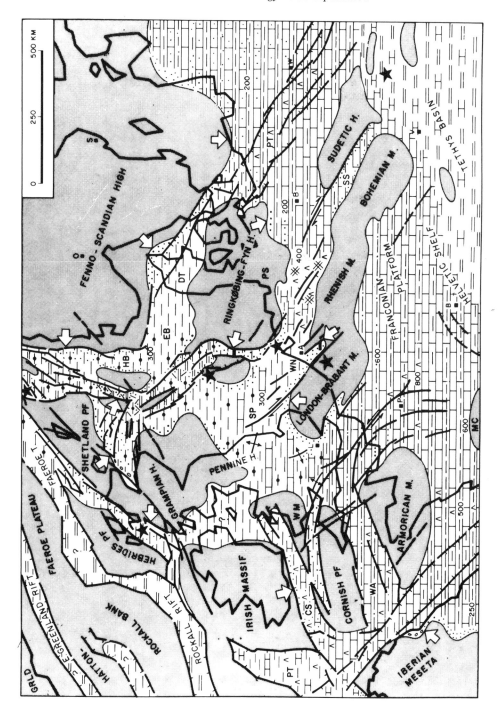

FIG. 35 (facing). An example of a panel used as input for a reservoir simulation model of the Statfjord Formation in the Brent field using a CRAY computer.

FIG. 36 (above). Oxfordian–Portlandian palaeogeography (from Ziegler 1981).

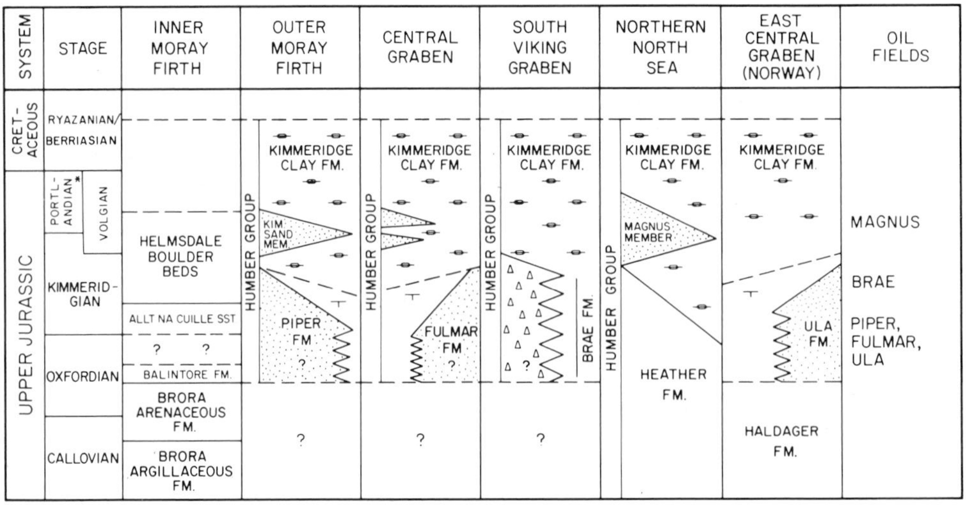

SYSTEM	STAGE	INNER MORAY FIRTH	OUTER MORAY FIRTH	CENTRAL GRABEN	SOUTH VIKING GRABEN	NORTHERN NORTH SEA	EAST CENTRAL GRABEN (NORWAY)	OIL FIELDS
CRET-ACEOUS	RYAZANIAN/ BERRIASIAN		KIMMERIDGE CLAY FM.	KIMMERIDGE CLAY FM.	KIMMERIDGE CLAY FM.	KIMMERIDGE CLAY FM.	KIMMERIDGE CLAY FM.	
UPPER JURASSIC	VOLGIAN PORTL-ANDIAN	HELMSDALE BOULDER BEDS	KIM. SAND MEM.			MAGNUS MEMBER		MAGNUS
	KIMMERID-GIAN	ALLT NA CUILLE SST	PIPER FM	FULMAR FM		HEATHER FM.	ULA FM.	BRAE PIPER, FULMAR, ULA
	OXFORDIAN	BALINTORE FM. BRORA ARENACEOUS F.M.					HALDAGER FM.	
	CALLOVIAN	BRORA ARGILLACEOUS FM.						

FIG. 37. Stratigraphic framework of the Upper Jurassic of the central and northern North Sea. Age equivalents are approximate.

Firth and central North Sea. Mass flow sands/ conglomerates occur over a larger area, but are most common in the South Viking Graben and Moray Firth.

Mass flow sand/conglomerate reservoirs

Reservoirs comprising this reservoir type are formed by two end members:
(i) conglomerate-dominated facies associations; and
(ii) sand-dominated facies associations.
In neither case do we attach any depth connotations to the deposition of these mass flow sands, but in our opinion they were deposited by gravity-driven processes.

Conglomerate-dominated facies associations

The western edge of the South Viking Graben and the Helmsdale region of the Inner Moray Firth are characterized by prominent fault zones with an overall present-day throw on a scale of thousands of feet. In the case of the South Viking Graben the respective fault system delimits the margin of the stable East Shetland Platform from the rapidly subsiding Viking Graben (Fig. 38); Jurassic movement along this fault dramatically affected the type and thickness of sediments deposited in the graben. The sediments were accumulated on the down-thrown side of the fault as a thick syntectonic clastic wedge (Fig. 39). This wedge comprises thick scree-like deposits of angular cobbles and boulders (mainly of Devonian sandstones), amalgamated sandstone units and interbedded sandstones and shales. A simplified facies break-

down is outlined on Fig. 40.

The gross facies types, recognized in cores, can also be recognized in uncored wells with fair success by using a combination of gamma ray and FDC/CNL logs, as demonstrated on Fig. 40.

The best known accumulation of this type is the Brae field. Facies changes in Brae are abrupt (Fig. 39) and the conglomeratic facies lies close to the fault scarp, wedging out in 2–5 km into thin sandstones, laminated siltstones and sandy siltstones (Harms *et al.* 1981).

Harms *et al.* (1981) considered two interpretations for these deposits (but favoured the former):
(i) fan delta systems, and
(ii) submarine fan systems.
In their preferred model the breccia and conglomerate is considered to be subaerial and the finer grained deposits submarine.

Other authors (Stow *et al.* 1982), including ourselves, favour the submarine fan interpretation. In this respect, much emphasis is placed on outcrop studies of the Helmsdale Boulder beds (Neves & Selley 1975), where it can be demonstrated that ammonite- and belemnite-bearing marine shales are intercalated with the boulder beds. The model for these strata has been greatly influenced by the outcrop studies of closely comparable deposits by Surlyk (1978) on the Lower Cretaceous of East Greenland (Fig. 41), and thus serves to emphasize the importance of relevant outcrop data. It cannot be proven whether the apices of the fans were submarine or subaerial. Furthermore, sediment

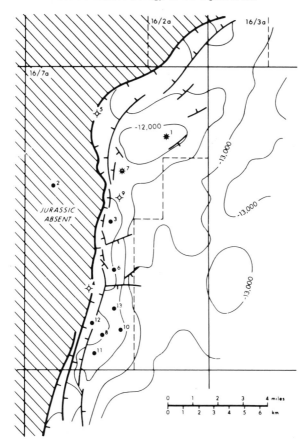

FIG. 38. Structure at top Jurassic level in Block 16/7a with faults cutting through the uppermost Jurassic. The closures approximate to coarse clastic fan deposits which wedge out basinwards (from Harms *et al.* 1981).

may have been transported to the fault edge by a combination of shallow marine, fluvial and mass emplacement processes. Whichever model is used the resultant reservoir distribution model will be very similar.

The main features can be summarized as follows:

(i) The sizes of individual fans are small, the radius of the main reservoir section (i.e. conglomerates and pebbly sands) being 4–5 km. These pass basinward into thinly bedded (mainly non-commercial) sands and silts and then into shales.

(ii) The fans may coalesce laterally to form continuous facies belts parallel to the fault zone but, judging from the location of dry wells in Brae, interfan areas are composed either of tight siltstones or similarly tight breccias.

(iii) The hydrocarbon column in the Brae Field is in the order of 1500 ft. To

account for this extreme column length a stratigraphic trapping element probably involving a seal against the East Shetland boundary fault is required. This stratigraphic trapping element is probably provided by the basinward passage into siltstones and shales (Fig. 39). Since the structural closure is only a fraction of the oil column (Fig. 38) it is extremely difficult to locate such traps and once located it is difficult to estimate their reserves. The latter problem is further hampered because it is difficult to calculate net oil sand from wireline logs in some of the more heterogeneous facies due to a lack of resolution in very thinly bedded sequences. In these cases cores become very important for such estimates.

(iv) In the Viking Graben, Brae-type reservoirs are found between 11,000 and 13,000 ft depth, which results in rather low

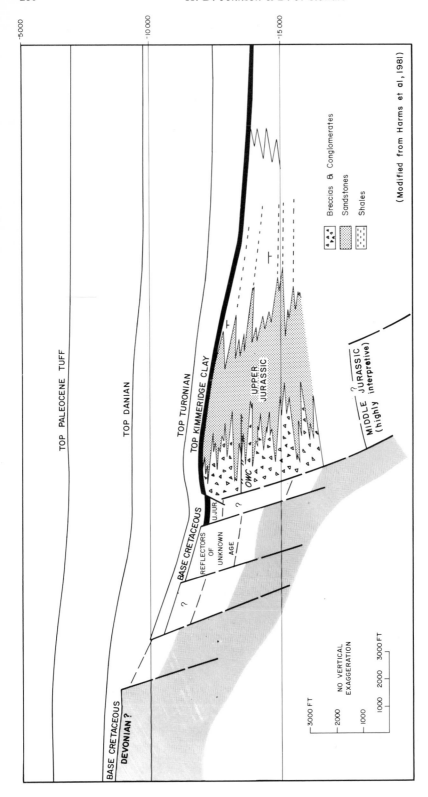

(Modified from Harms et al, 1981)

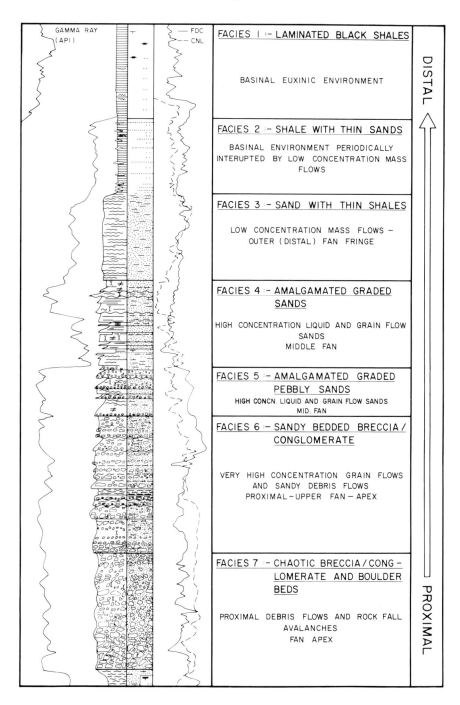

GAMMA RAY
(API)

— FDC
-- CNL

FACIES I :- LAMINATED BLACK SHALES

BASINAL EUXINIC ENVIRONMENT

FACIES 2 :- SHALE WITH THIN SANDS

BASINAL ENVIRONMENT PERIODICALLY
INTERUPTED BY LOW CONCENTRATION MASS
FLOWS

FACIES 3 :- SAND WITH THIN SHALES

LOW CONCENTRATION MASS FLOWS –
OUTER (DISTAL) FAN FRINGE

FACIES 4 :- AMALGAMATED GRADED
SANDS

HIGH CONCENTRATION LIQUID AND GRAIN FLOW
SANDS
MIDDLE FAN

FACIES 5 :- AMALGAMATED GRADED
PEBBLY SANDS

HIGH CONCN. LIQUID AND GRAIN FLOW SANDS
MID. FAN

FACIES 6 :- SANDY BEDDED BRECCIA /
CONGLOMERATE

VERY HIGH CONCENTRATION GRAIN FLOWS
AND SANDY DEBRIS FLOWS
PROXIMAL – UPPER FAN – APEX

FACIES 7 :- CHAOTIC BRECCIA / CONG –
LOMERATE AND BOULDER
BEDS

PROXIMAL DEBRIS FLOWS AND ROCK FALL
AVALANCHES
FAN APEX

DISTAL

PROXIMAL

FIG. 39 (facing). East-west cross-section through the southern part of the Brae area (from Harms *et al.* 1981).

FIG. 40 (above). Schematic vertical profiles with inferred well log response through a typical Brae-type sequence.

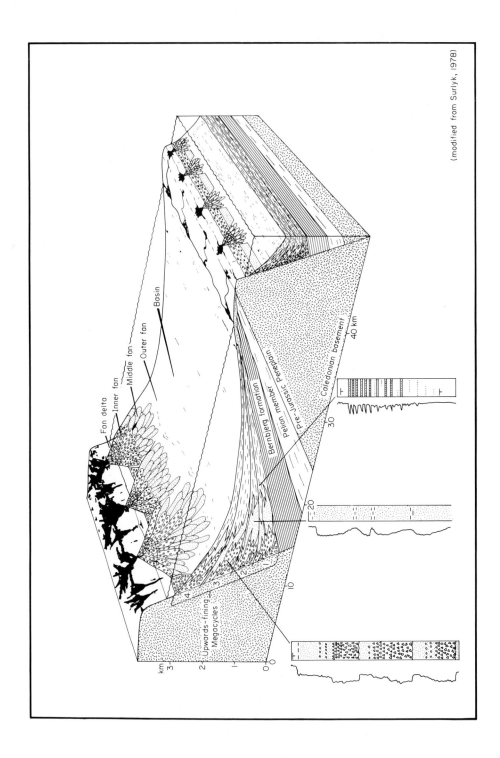

(modified from Surlyk, 1978)

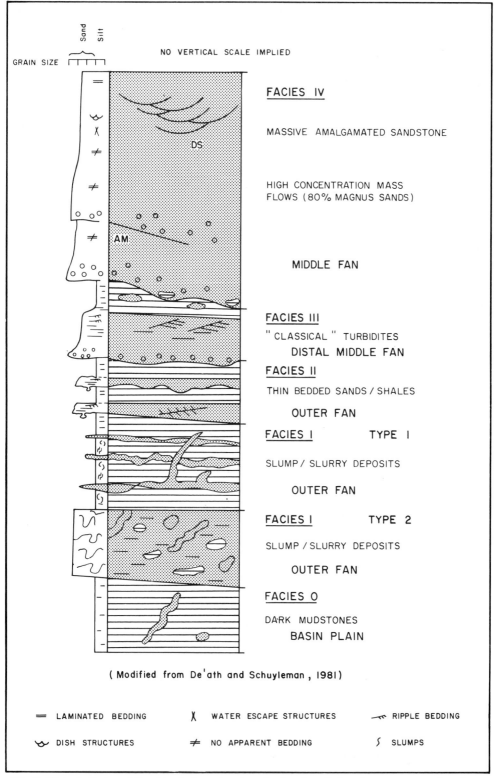

FIG. 41 (facing). Facies relationships in Jurassic submarine fan deposits based on outcrop studies of the Wollaston Forland Group of East Greenland (modified form Surlyk 1978). Vertical profiles depict schematic gamma ray profiles.

FIG. 42 (above). Facies scheme for sandy fan Magnus-type deposits (modified from De'Ath & Schuyleman 1981).

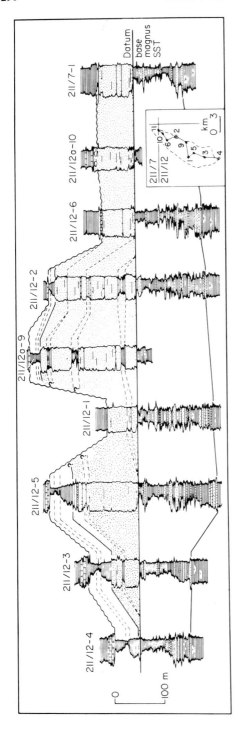

FIG. 43. Cross-section through the Magnus field illustrating typical sand distribution patterns (from De 'Ath & Schuyleman 1981).

average porosities and permeabilities. However, the reservoir rocks interdigitate with basinal shale source rocks, which are sufficiently mature to generate oil. Consequently there is a high probability that oil will migrate into the reservoir.

(v) A further point to consider is the possibility of enhancement in reservoir quality at these depths by the leaching of detrital feldspar (Harms *et al.* 1981).

It is apparent from the above that conglomerate-dominated reservoirs of this type are complex and difficult to develop. This is clearly demonstrated in the drastic reduction in reserve estimates, from 800 to 300 mb, that has taken place between initial discovery and the subsequent drilling of a further twelve exploratory and appraisal wells in the first three years of appraisal of the Brae field (Harms *et al.* 1981).

Sand-dominated facies association

Upper Jurassic sandy mass flows occur throughout the Viking Graben, Central Graben and Moray Firth, but the only field of this reservoir type under development is BP's Magnus field in the North Viking Graben (De'Ath & Schuyleman 1981). Sand-dominated submarine fans appear to be slope-related rather than fault-related, although it cannot be ruled out that small-throw faults have influenced deposition. Like the conglomerate-dominated fans, sand-dominated fans have a patchy distribution, making them difficult exploration targets. They contrast with conglomerate facies associations in tending to be of greater aerial extent, whilst individual fans are isolated rather than coalescent.

The Magnus Sands comprise five basic facies (Fig. 42) as originally described by De 'Ath & Schuyleman (1981). However, the majority of the reservoir comprises massive, amalgamated sandstone units, which represents their facies IV. These sandstones have been interpreted as the Middle Fan deposits of an eastward prograding submarine fan complex (De 'Ath & Schuyleman 1981). This interpretation is based on sedimentological studies which have shown an eastward fining in grain size, an increase in detrital clay content and decrease in bed thickness as the outer fan is approached. In addition, the sands wedge out into shales to the south (Fig. 43).

The main reservoir characteristics of sandy mass flow facies associations are outlined below:

(i) Fans of this type appear to extend over considerable areas. The Magnus fan covers an area of approximately 12 ×

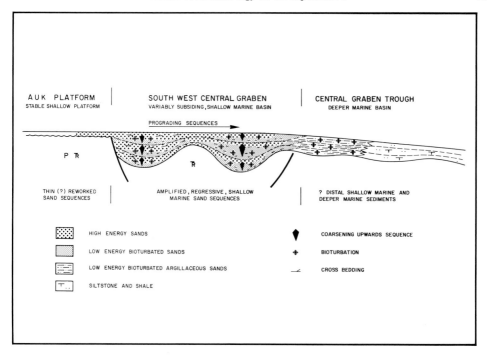

AUK PLATFORM
STABLE SHALLOW PLATFORM

SOUTH WEST CENTRAL GRABEN
VARIABLY SUBSIDING, SHALLOW MARINE BASIN

CENTRAL GRABEN TROUGH
DEEPER MARINE BASIN

PROGRADING SEQUENCES

P R

R

THIN (?) REWORKED
SAND SEQUENCES

AMPLIFIED, REGRESSIVE, SHALLOW
MARINE SAND SEQUENCES

? DISTAL SHALLOW MARINE AND
DEEPER MARINE SEDIMENTS

HIGH ENERGY SANDS

LOW ENERGY BIOTURBATED SANDS

LOW ENERGY BIOTURBATED ARGILLACEOUS SANDS

SILTSTONE AND SHALE

COARSENING UPWARDS SEQUENCE

BIOTURBATION

CROSS BEDDING

FIG. 44. Schematic cross-section through the Fulmar Sands in the Southwest Central Graben area.

12 km, but thin turbiditic sands could penetrate into the basin to considerable distances. However, these latter sands are unlikely to be of commercial significance.

(ii) Although porosities and permeabilities of these sands are fairly good, average permeabilities are, on the whole, lower than in shallow marine sands at corresponding depths.

(iii) Lateral facies changes are common. However, owing to the sandy nature of most facies the reservoir appears to be homogeneous enough to act as a single producing unit (De 'Ath & Schuyleman 1981). Appraisal of the Magnus field has therefore, been more straightforward than that of the Brae field, with fewer and more predictable appraisal wells required (eight appraisal wells).

(iv) The massive amalgamated sandstones that form the main reservoir of the sandy mass flow facies association can be recognized on wireline logs by their blocky, cylindrical shapes (cf. conglomerate-dominated mass flow association and shallow marine associations).

(v) These reservoirs possess good stratigraphic trap possibilities since they are generally encased in organic shales. These shales may act both as the source for oil and the reservoir seal. The close proximity of the source rock and reservoir rock is an obvious advantage, especially, as in Magnus, when the source rock is mature to generate oil.

Shallow marine sandstone reservoirs

In the U.K. sector of the North Sea, Upper Jurassic, shallow marine reservoirs are mainly located in the Moray Firth and Central Graben; they also occur in adjacent areas of the Norwegian sector. Reservoirs of this type are represented by the Piper (Moray Firth), Fulmar (Central Graben) and Ula (Norwegian sector near U.K./Norway boundary) fields. In comparison to mass flow sand/conglomerate reservoirs, the shallow marine reservoirs show greater lateral continuity, but as a consequence there is a reduced potential for stratigraphic trapping. Nevertheless, these shallow marine sands were deposited over the irregular mid-Kimmerian erosional surface which has contributed to their variable distribution.

For the purpose of this discussion, the Upper Jurassic shallow marine sandstone reservoirs have been divided into two types: one showing strong tectonic influence on sedimentation and the other moderate tectonic influence on

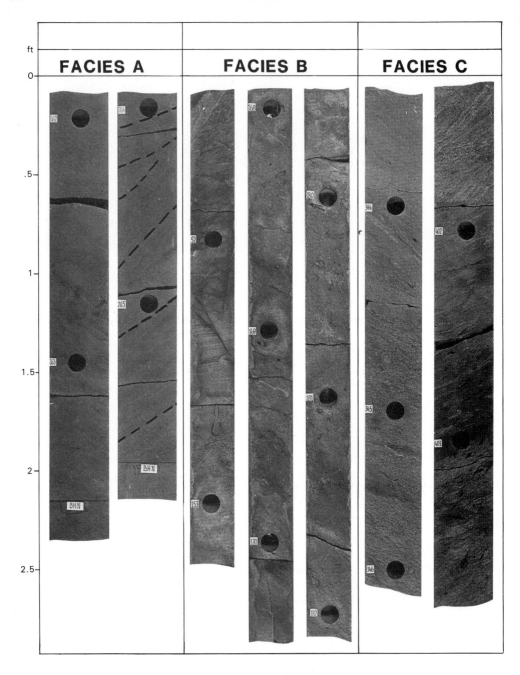

FIG. 45. Core photographs illustrating a typical Fulmar-type regressive sequence.

sedimentation.

Shallow marine sandstone reservoirs belonging to the first category are typified by the Fulmar Sandstones. The main Fulmar reservoir sandstones were deposited in a strongly subsiding basin with an abrupt faulted western margin abutting against the Auk Platform (Fig. 44). Facies identification in the Fulmar Field has required extensive core coverage and the use of X-radiography to aid interpretation, since in normal light much of the core is structureless (Fig. 45). Consequently slabs of

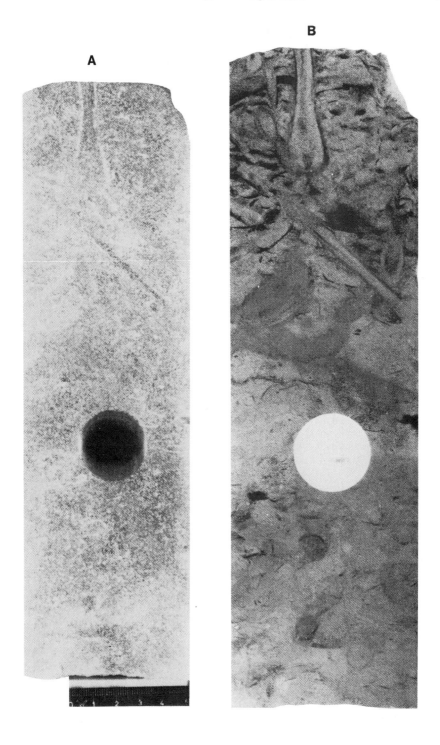

FIG. 46. Normal photograph and X-radiography of the 'massive' Fulmar Sands.

cores subjected to X-radiography reveal extensive bioturbation and burrowing (Fig. 46) and occasional cross-bedding etc., suggesting an overall shallow marine depositional environment. The sandstones form thick packages, up to 1000 ft thick, with the most characteristic feature being 'amplified' coarsening-upwards sequences up to 600 ft thick which display an upward passage from argillaceous fine grained sandstones to clean, medium grained sandstones (i.e. facies A–C, Fig. 47). In the area of the Fulmar field the highest energy sandstones (i.e. facies A) are more common near the Auk platform, with finer grained, more argillaceous and more bioturbated sandstones (facies B and C) encountered in more basinward (?deeper water) locations (Fig. 44). The 'amplified' coarsening upwards sequences can be accounted for by strong subsidence along the Auk boundary fault

and/or localized salt withdrawal from the underlying Zechstein salt.

A delicate balance between subsidence and sedimentation resulted in the substantial accumulation of good quality reservoir sands. In general the problems of field appraisal and development are less acute for Fulmar-type reservoirs than for the Brae- and Magnus-type sands. However, extensive core coverage and a good biostratigraphic framework is required. Correlation between wells can be most effectively carried out using core-derived depositional facies sequences, since the use of log markers, which often includes diagenetic features, can be misleading in these relatively homogeneous sandstones.

In parallel with Brae-type reservoirs, facies belts are strongly controlled by the tectonic grain. Consequently, the good reservoir quality

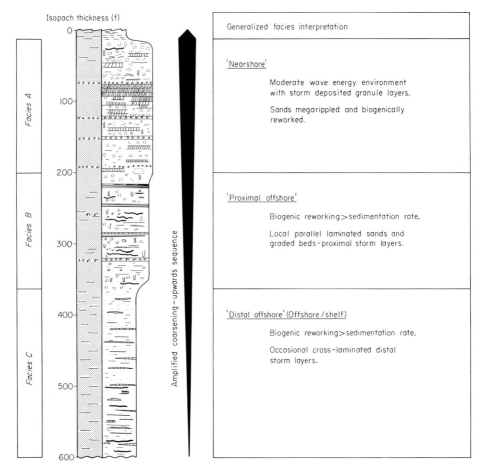

Fig. 47. Schematic vertical profile through Upper Jurassic shallow marine sands (based on the Fulmar Sands in the Fulmar field).

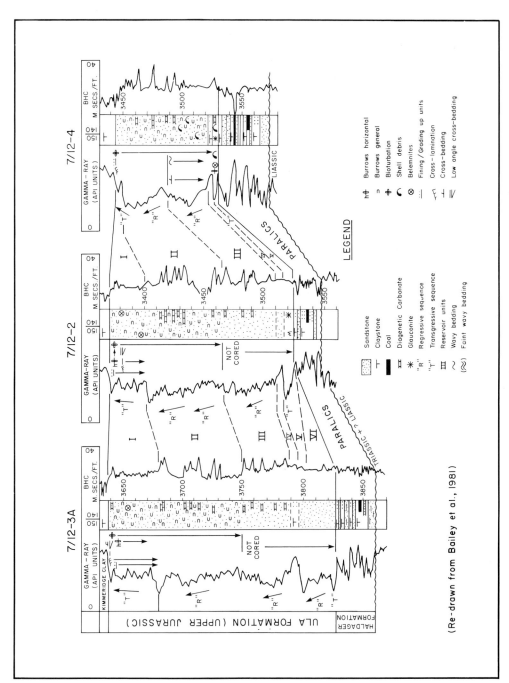

FIG. 48. Sedimentology and reservoir correlation of the Ula Field.

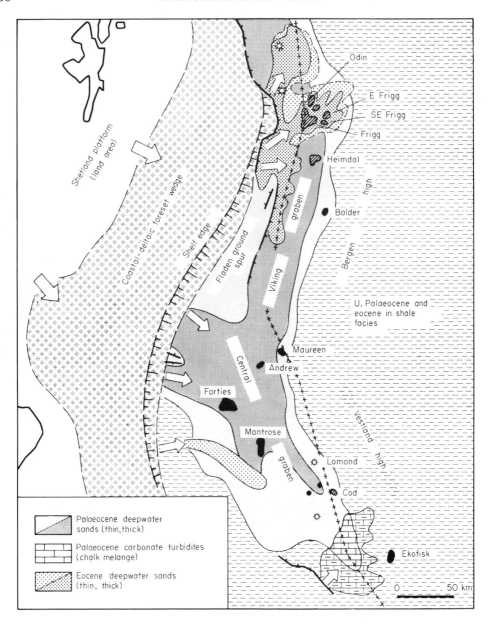

FIG. 49. Regional facies distribution of Lower Tertiary sediments in the central and northern North Sea (from Ziegler 1982).

facies belts are generally elongate and narrow, and parallel major graben faults. However, syndepositional salt diapirism has, in some places, played an equally important role in controlling reservoir development. The above factors indicate why good quality reservoir sands can be difficult to locate on a regional scale.

Reservoirs of the second, moderate tectonic control, category differ from the Fulmar example in only a few respects. The best examples of shallow marine sandstones deposited under these conditions are found in the Ula field (Bailey *et al.* 1981) and Piper field (Williams *et al.* 1975; Maher 1980, 1981). The reservoirs of both these fields, like the Fulmar field, are composed of a stacked sequence of predominantly bioturbated coarsening upwards units. However, in contrast to Fulmar, the coarsening upwards sequences of the Ula and Piper reservoirs are much thinner (90–170 ft). Furthermore, particularly in the Piper field, the coarsening upwards units show greater variations in reservoir quality, ranging from non-reservoir offshore shales at the base and passing upwards into high-energy, cross-bedded sandstones (with Darcy permeabilities) at the top. In both Ula and Piper there is evidence that the reservoir sands are thicker in the vicinity of faults, but faulting has had less impact that it has on Fulmar. Therefore, in the Ula and Piper Fields depositional facies are laterally more homogeneous than in the Fulmar Field (Fig. 48). Owing to this lateral homogeneity of facies and to the thinner and more sharply defined coarsening-upwards units in both Ula and Piper, these reservoirs have the potential to exhibit strong layering of reservoir properties. This could lead to extensive field wide permeability barriers, which could in some cases hamper attempts to maintain reservoir pressure by water injection.

The shallow marine sandstones of both of the above categories can be recognized on wireline logs by funnel-shaped gamma ray profiles and increasing upwards positive separation of FDC/CNL logs. This contrasts with the more blocky (mid-fan) or serrate (thin bedded mass flows/basinal shales) profiles of mass flow deposits discussed earlier in this section. However, extensive core coverage is essential in order to identify positively the various types of Upper Jurassic sandstone reservoirs.

Tertiary sands, northern and central North Sea

The Tertiary deposits of the northern and central North Sea areas include several important oil and gas accumulations, including the Montrose field (Fowler 1975) which was the first commercial oil discovery in the U.K. sector of the North Sea (1969). Tertiary sands also form the reservoir of the giant Forties oil field (Thomas *et al.* 1974), and the giant Frigg gas field (Heritier *et al.* 1979). Other important commercially proven hydrocarbon accumulations, occurring in both the U.K. and Norwegian sectors, include Andrew, Maureen, Lomond, Cod, Odin, Heimdal and Balder (Fig. 49).

The reservoir sands of all these fields were deposited in a variety of submarine fan complexes of Palaeocene and Eocene age. A major similarity between many of these Tertiary sand plays and the previously discussed Upper Jurassic fan bodies is the importance of, and

considerable potential for, stratigraphic trapping. Furthermore, all these fan bodies display substantial variability in sand distribution, particularly on a field scale. One major difference between these Tertiary sand plays and other, generally older North Sea sand plays, is the potential for applying seismostratigraphy to both exploration and production studies, which will be discussed more fully later. In addition, this section also highlights the application of a submarine fan facies model to the development of a reservoir geological model of the Forties field.

The regional geological framework of the Cenozoic period has been covered amply elsewhere (e.g. Ziegler 1981, 1982; Rochow 1981; Knox *et al.* 1981; Morton 1979) and is not discussed specifically here.

Seismostratigraphic studies

The regional geological picture of the Cenozoic of the North Sea has been compiled from a combination of sedimentology, palaeontology and seismostratigraphy. The latter has proved particularly valuable in both exploration and production studies for several reasons. Particularly important is the fact that the Cenozoic represents a period of relatively rapid sedimentation in which up to 3.5 km of sands and shales were deposited, mainly in the form of discrete clastic wedges which built out from the basin margins (Fig. 49). This, combined with only minor tectonism and moderate burial/compaction, has resulted in the partial preservation of the primary, large-scale depositional geometry and internal structure of individual clastic bodies, which can, in some instances, be identified on seismic records.

The Palaeocene sands of the central North Sea have proved particularly receptive to seismostratigraphic studies. This is largely due to their relatively shallow burial (c. 3000–9000 ft), and the commensurate high resolution of reflection seismic data. Furthermore, the top of the Danian chalks and the early Eocene volcanic ash marker provide regionally correlative seismic reflectors of considerable lithostratigraphic value (Rochow 1981). The first published seismostratigraphic study of the Palaeocene interval was by Parker (1975), who identified on seismic records the wedge-shaped geometry of the unit and its internal organisation which comprised large-scale topset, foreset and bottomset reflectors (Fig. 50). Sedimentological and palaeontological data established that this geometry and seismic character represents a sequence of prograding coastal/deltaic deposits

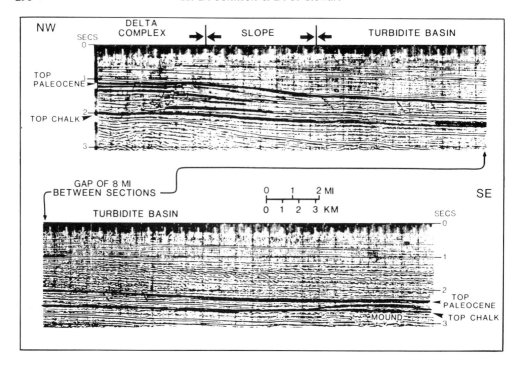

FIG. 50. Seismic sections illustrating seismic reflection patterns of Palaeocene seismic facies in the central North Sea (Parker 1975).

(topset) which pass laterally and downdip into a partly channelized slope complex (foreset), and beyond into a base of slope, deep water turbidite fan system (bottom set; Fig. 51).

The North Sea hydrocarbon productive Palaeocene sands display some of the general seismic attributes considered characteristic of turbidite sequences associated with prograding deltas (Fig. 52; Berg 1982). In the context of recent seismostratigraphic terminology (Vail *et al.* 1977), the slope or pro-delta facies are represented in dip sections by prograding clinoforms, which in this case are of a very low angle and continue far out into the basin. However, the characteristic mounding ('hummocky' to chaotic internal seismic reflection pattern) of turbidite fan systems (Sangree & Widmier 1977) is only faintly developed in Parker's Palaeocene example (observed by Berg 1982) but elsewhere they are more striking and contain hydrocarbons which are partially stratigraphically trapped (e.g. Frigg).

The Frigg field is a particularly striking example of a turbidite fan reservoir complex in which the primary fan geometry is clearly visible from the top structure map. This map displays a radial body with discrete 'fingers' which correspond to sand-rich channel/fan lobe deposits (Fig. 53). The mounded or 'hummocky' character of in-

ternal seismic reflectors can be shown from detailed well log correlations to reflect the large number of channel/fan lobe sand bodies encased in interchannel and distal fan lobe shales (Figs 53 and 54). Within the framework of the regional depositional model, as outlined earlier, the recognition on seismic of a fan-shaped body with internal mounded reflectors can be confidently predicted as representing a submarine fan complex.

The Palaeocene turbidite fan reservoirs of the Balder and Cod fields display complex patterns of reservoir distribution which presented considerable problems during appraisal and early development stages (Skjold 1981; Kessler *et al.* 1981). In both examples the complex sand body distribution patterns have resulted in large variations in hydrocarbon distribution, local stratigraphic trapping and even substantial variations in hydrocarbon-water contacts. These complexities are clearly not unusual and represent an integral feature of submarine fan reservoirs. Early recognition of this type of sand play from seismic studies should help in the planning of appraisal and development drilling campaigns. In practice, this needs supplementing by a thorough facies analysis of each reservoir complex as outlined in the following section.

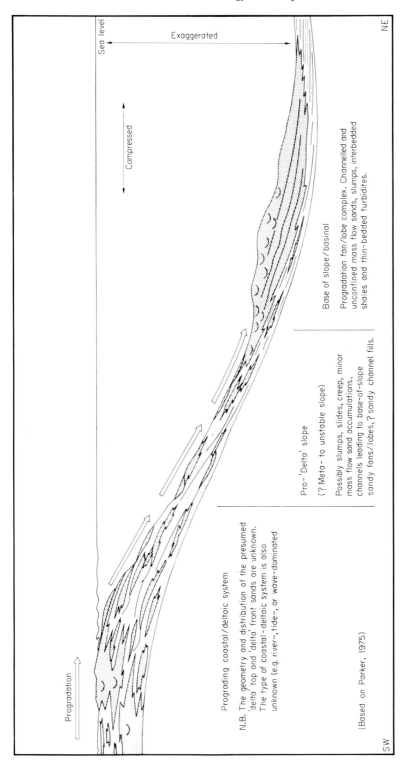

FIG. 51. Facies model for Tertiary clastic deposits illustrating the main depositional elements (Parker 1975).

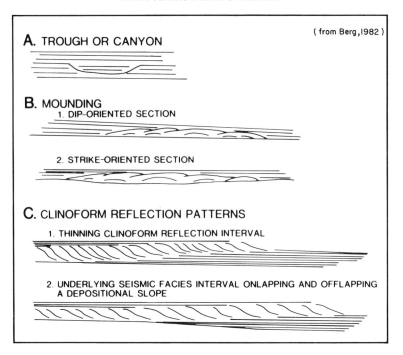

(from Berg,1982)

A. TROUGH OR CANYON

B. MOUNDING
1. DIP-ORIENTED SECTION

2. STRIKE-ORIENTED SECTION

C. CLINOFORM REFLECTION PATTERNS
1. THINNING CLINOFORM REFLECTION INTERVAL

2. UNDERLYING SEISMIC FACIES INTERVAL ONLAPPING AND OFFLAPPING A DEPOSITIONAL SLOPE

FIG. 52. Seismic indicators of turbidite sequences (from Berg 1982).

Deep sea fan sedimentation and reservoir modelling in the Forties field

The Tertiary sand reservoirs of the Forties field are illustrated here as an example of how the development of a general facies model, in this case the deep sea turbidite fan model, has proved beneficial to the understanding and prediction of subsurface reservoir distribution.

During the appraisal drilling of the Forties field, in the early 1970s, it was recognized on the basis of five wells that reservoir distribution was highly variable. Although the average sand content was generally high, it appeared that optimum sand development occurred in the north and north-western parts of the field, and that these good sands passed laterally into thin sands and shales to the SW (Fig. 55). In the context of the deep sea turbidite facies models available at that time, this sand distribution pattern could have represented a transition from proximal to distal turbidites, moving in a basinward direction towards the SE (Fig. 55). In such a model the sands could have been expected to have tabular, sheet-like geometries, which over the field would have allowed lateral and, possibly, vertical fluid communication. It was therefore anticipated that development wells would be located on a regular drainage pattern,

and that the irregular sand distribution patterns noted earlier would not influence the location of such wells nor substantially affect reservoir performance (Walmsley 1975, and 'Discussion' by R. C. Selley, p. 485).

By the mid-1970s, however, submarine fan facies models were being widely applied to ancient turbidite sequences (e.g. Walker & Mutti 1973). Furthermore, and as indicated earlier, seismostratigraphic studies demonstrated that the reservoir sands of the Forties Formation accumulated as deep water submarine fans at the base of actively prograding delta slopes (Parker 1975, predicted water depths of 3000–8000 ft). This regional depositional model of submarine fan sedimentation provided an important framework for the interpretation of current development well drilling results and of reservoir performance data in the Forties field. In addition, it was noted that 'the very variable sand characteristics and geometry discovered did cause numerous changes to the early development drilling and production schedules' (Hillier *et al.* 1978, p. 327). In particular, it was recognized that the main sand bodies were extensively channelized (Fig. 56), with one major channel sand complex (the Charlie Sand on the western flank of the field) isolated from the main reservoir unit (the multistorey Main

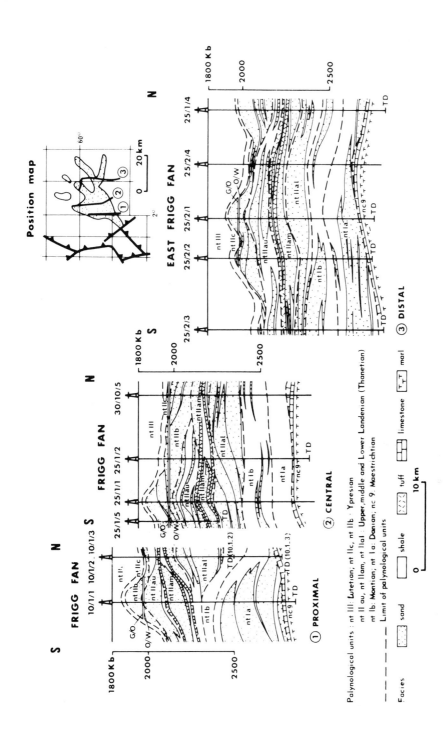

FIG. 53. Top structure map cross-sections of the Frigg Field showing the complex mounded or hummocky character of reservoir sands. (from Heritier *et al.* 1979).

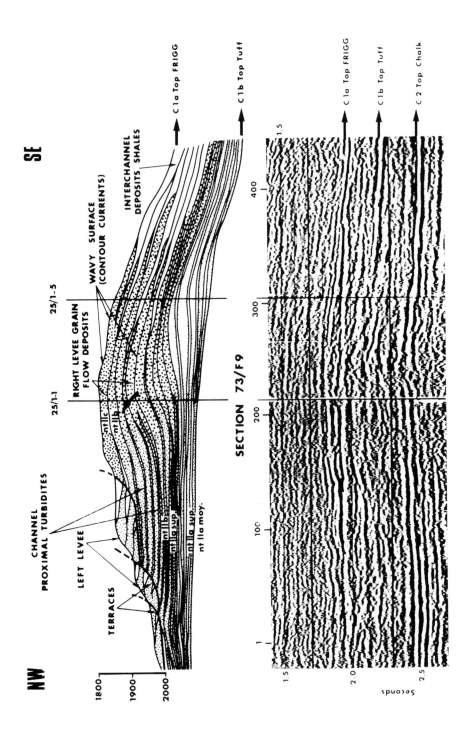

FIG. 54. Seismic section and sedimentological interpretation of a typical Frigg mound (from Heritier *et al.* 1979).

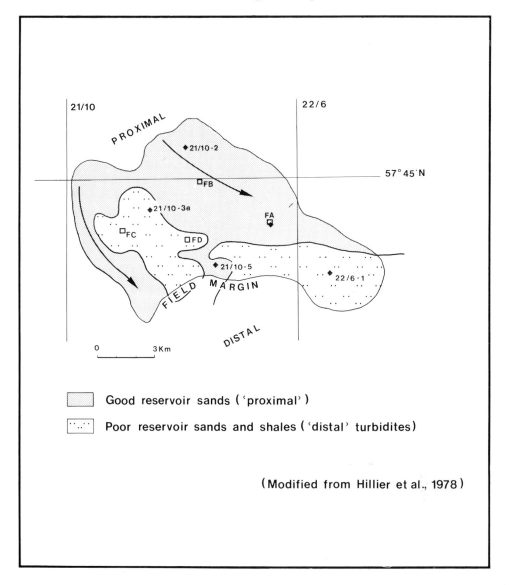

FIG. 55. Pre-development facies distribution model in the Forties field (from Hillier *et al.* 1978).

Sands). Clearly, the accurate prediction of the distribution of channel and interchannel facies will significantly affect all aspects of reservoir management. A prerequisite to reservoir prediction is the correct identification of the different sand and shale body types as seen in cores and well logs.

A facies analysis of the Forties Sands in the Forties field by Carman & Young (1981) outlines the various sedimentological elements of this complex reservoir which comprises four main lithofacies: facies A = mainly medium to thickly bedded sands displaying well-developed Bouma sequences and analogous to the proximal 'classic turbidite' group of Walker (1967, 1978); facies B = mainly massive, thickly bedded sands with subordinate pebbly sands, interpreted as the deposits of fluidized flows (= massive sandstones of Walker 1978); facies C = laminated grey mudstone with 1–5 mm thick, graded siltstone-mudstone couplets and subordinate fine grained sandstone beds probably deposited by relatively distal turbidity currents; facies D = green mudstones with abundant microfauna and microflora, which are interpreted as hemipelagic deposits.

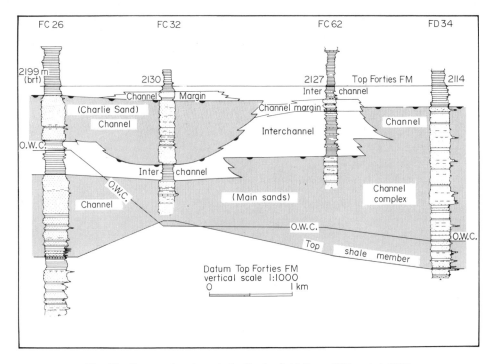

FIG. 56. Cross-section through the Forties field (from Hillier *et al.* 1978).

These four lithofacies are arranged in several distinctive facies associations and sequences which represent deposition in five environments belonging to the middle and lower regions of a submarine fan system (Fig. 57). Particularly important is the fact that all five facies associations have distinctive well log patterns which ensures correct facies interpretation in uncored wells. The present facies distribution model of the Forties field, based on the above facies analysis, identifies a series of mainly channelized sand bodies trending in a NW–SE direction (Fig. 58), which resulted from deposition in a middle fan environment. This model is substantially different in many important respects to the initial sheet-like distribution model outlined earlier. As seen in the Shell/Esso portion of the field (Block 22/6a), there are marked contrasts in reservoir development between the high-quality, multistorey channel sands and the low-quality interchannel sands and shales (Fig. 59). This type of detailed facies mapping provides a three-dimensional model capable of accurately monitoring reservoir performance, and which should also ensure the optimum location of infill wells and provide an essential framework for the application of any enhanced oil recovery techniques.

Concluding remarks

The 21 year history of the British Sedimentological Research Group has been paralleled by the exploration and production of the North Sea's oil and gas resources, and both have experienced a continual expansion in activity during this period.

Sedimentological studies have already made a significant contribution in assisting in the location and development of North Sea oil and gas accumulations, principally by increasing our understanding of the origin, nature and distribution of all the major clastic reservoirs. In particular, the reconstruction of depositional environments in clastic sequences has been applied extensively to provide the essential framework for describing and understanding the qualitative and quantitative reservoir distribution on both regional (exploration) and field (production) scales.

The continuing search for hydrocarbons in the relatively mature sub-basins within the North Sea will require the development of further imaginative models, particularly since many future discoveries are likely to include traps with an increasing stratigraphic component. The continued development of predictive deposition-

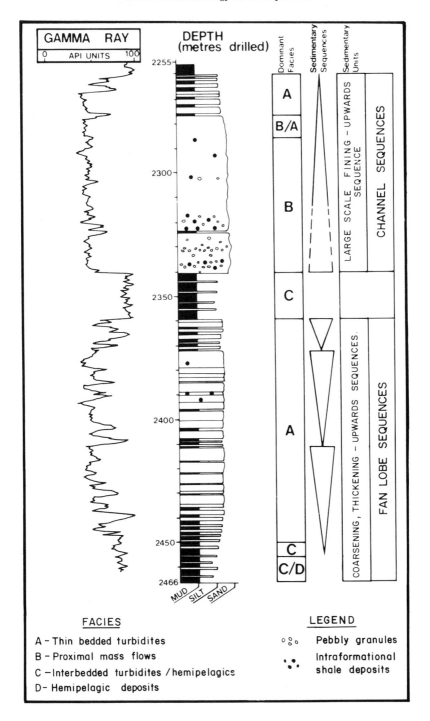

FIG. 57. Vertical facies profile through the Forties Sands (from Carman & Young 1981).

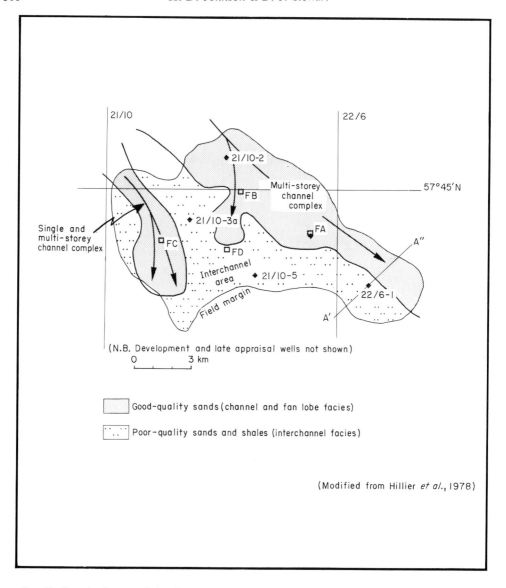

Fig. 58. Post-development facies distribution model in the Forties field (modified from Hillier *et al.* 1978).

al models is required, therefore, to help locate these stratigraphic traps. Furthermore, it remains essential to be able to predict the reservoir potential of undrilled structures in terms of reservoir characteristics (lithology, mineralogy, diagenesis etc.), thickness and quality (porosity-permeability), and to assess the likely extent of such reservoirs.

The appraisal and development of oil and gas accumulations require an accurate geological description of reservoir characteristics, particularly in terms of thickness and quality (porosity-permeability) distribution (e.g. Harris & Hewitt 1977). In clastic reservoirs this is best achieved through a thorough understanding of depositional environments, particularly the size, shape, trend and quality of sand bodies and the thickness and continuity of shale layers (e.g. LeBlanc 1977) and the tectonic framework of the basin and/or hydrocarbon accumulation. The resulting reservoir geological model provides the optimum framework for reservoir subdivision and correlation, reservoir quality mapping (e.g. net sand, porosity, permeability,

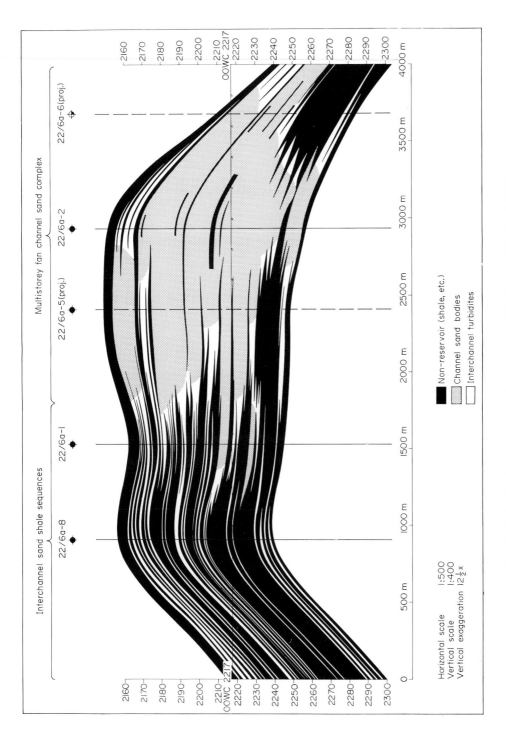

FIG. 59. Facies distribution in the south-eastern part of the Forties field, Block 22/6a.

saturation maps etc; e.g. Sneider *et al.* 1976) estimating reserves, developing reservoir simulation models (e.g. Harris 1975; Weber *et al.* 1978) and monitoring and predicting reservoir performance (e.g. fluid movements; cf. Van Veen 1977; Hartman & Paynter 1979). The relatively long productive life of a field (e.g. 20–30 years) usually requires a continual update of reservoir geological models in order to ensure optimum reservoir management and, thereby, to maximize ultimate recoveries. The quality of the reservoir geological model can prove a particularly significant factor during the latest stages of development, particularly when decisions are taken on the positions of infill wells and the possible application of enhanced oil recovery techniques. Indeed, the success of these latter projects can hinge on the accuracy of the geological description of the reservoir (e.g. Basan *et al.* 1978; Cook 1977; Kossack & Bilhartz 1976; Webb 1978).

In conclusion, it is apparent that sedimentology will continue to play an important role in the multidisciplinary studies being employed in the exploration and production of hydrocarbons from the North Sea.

ACKNOWLEDGMENTS: The authors are indebted to Shell U.K. Ltd and Esso Petroleum Co. Ltd for permission to publish this paper. We have drawn freely from the large amount of data available in existing publications, in-house reports and the personal knowledge and experience of our colleagues. The development of exploration and production sedimentological models requires a multidisciplinary approach and, as a result, the ideas and concepts outlined in this paper have crystallized through discussions between many specialists in various fields (e.g. stratigraphy, geophysics, geochemistry, reservoir engineering, petrophysics etc). Ideas have also evolved over several 'generations' of specialist activities and, consequently, it is not possible to single out all the individuals who originated the different contributions to this text, but we are indebted to all our colleagues. Specifically, however, the authors wish to acknowledge their predecessors in Shell Expro, M. C. Budding (Reservoir Geology/H. D. J.) and A. T. Buller (Sedimentology/D. J. S.), who made significant contributions in their specialist activities, particularly in relation to their separate studies on the Brent Sands. Comments by Drs P. J. C. Nagtegaal and P. Ziegler improved the final manuscript.

Special thanks are also due to Catherine Senior, Yvonne Jensen and Martin Lewis for valuable help with preparing the illustrations for our presentation of this paper at the 21st anniversary meeting of the BSRG in December 1982 at Liverpool University. Finally, we are indebted to Mrs Diane Symmons for all her work on the published figures.

References

ALLEN, J. R. L. 1978. Studies in fluviatile sedimentation: an exploratory quantitative model for the architecture of avulsion-controlled alluvial sites. *Sedim. Geol.* **21**, 129–47.

BAILEY, C. C., PRICE, I. & SPENCER, A. M. 1981. The Ula oil field, block 7/12, Norway. *In*: *The Sedimentation of the North Sea Reservoir rocks.* Norsk Petroleumsforening (NPF), Geilo.

BASAN, P. B., McCALEB, J. A. & BUXTON, T. S. 1978. Important geological factors affecting the Sloss field micellar pilot project. *Soc. Petrol. Engrs, Pap. no. 7047, 5th Symp. on Improved Oil Recovery*, Tulsa, Oklahoma.

BERG, O. R. 1982. Seismic detection and evaluation of delta and turbidite sequences: their application to exploration for the subtle trap. *Bull. Am. Ass. Petrol. Geol.* **66**, 1271–88.

BOWEN, J. M. 1975. The Brent Oil-field. *In*: WOODLAND, A. W. (ed.) *Petroleum and the Continental Shelf of Northwest Europe*, pp. 353–60. Applied Science Publishers, London.

BUDDING, M. & INGLIN, H. 1981. A reservoir geological model of the Brent Sands in Southern Cormorant. *In*: ILLING, L. V. & HOBSON, G. D. (eds) *Petroleum Geology of the Continental Shelf of North-West Europe*, 326–34. Heyden, London.

CARMAN, C. J. & YOUNG, R. 1981. Reservoir geology of the Forties oil field. *In*: ILLING, L. V. & HOBSON, G. D. (eds) *Petroleum Geology of the Continental Shelf of North-West Europe*, 371–9. Heyden, London.

CHAUVIN, A. & VALACHI, L. 1980. Sedimentology of the Brent and Statfjord Formations of Statfjord field. *In*: *The Sedimentation of the North Sea Reservoir Rocks.* Norsk Petroleumsforening (NPF) Geilo.

COLEMAN, J. M., ROBERTS, H. H., MURRAY, S. P. & SALAMA, M. 1981. Morphology and dynamic sedimentology of the eastern Nile delta shelf. *Mar. Geol.* **42**, 301–26.

COOK, D. L. 1977. Influence of silt zones on steam drive performance—Upper Conglomerate zone, Yorba Linda field, California. *J. Petrol. Technol.* **29**, 1397–404.

DE 'ATH, N. G. & SCHUYLEMAN, S. F. 1981. The geology of the Magnus Oilfield. *In*: ILLING, L. V. & HOBSON, G. P. (eds) *Petroleum Geology of the Continental Shelf of North-West Europe*, 342–51. Heyden, London.

DEEGAN, C. E. & SCULL, B. J. 1977. A proposed standard lithostratigraphic nomenclature for the Central and Northern North Sea. *Rep. Inst. Geol. Sci.* 77/25.

Department of Energy 1983. *Development of the Oil and Gas Resources of the United Kingdom 1983.* HMSO, London. 65 pp.

ELLIOTT, T. 1978. Deltas. *In*: READING, H. G. (ed.) *Sedimentary Environments and Facies*, 97–142. Blackwell Scientific Publications, Oxford.

EYNON, G. 1981. Basin development and sedimentation in the Middle Jurassic of the North Sea. *In*: ILLING, L. V. & HOBSON, G. D. (eds) *Petroleum Geology of the Continental Shelf of North-West Europe*, 196–204. Heyden, London.

FOWLER, C. 1975. The geology of the Montrose Field. *In*: WOODLAND, A. W. (ed.) *Petroleum and the Continental Shelf of North West Europe*, pp. 467–76. Applied Science Publishers, London.

GALLOWAY, W. E. 1975. Process framework for describing the morphologic and stratigraphic evolution of the deltaic depositional systems. *In*: BROUSSARD, M. L. (ed.) *Deltas, Models for Exploration*, pp. 87–98. Houston Geological Society.

GLENNIE, K. W. 1972. Permian Rotliegendes of North-West Europe interpreted in the light of modern desert sedimentation studies. *Bull. Am. Ass. Petrol. Geol.* **56**, 1048–71.

—— 1981. Early Permain Rotliegendes. *In*: *Introduction to the Petroleum Geology of the North Sea.* Joint Association of Petroleum Exploration Courses (U.K.). *Course Notes No. 1.* Burlington House, London.

—— 1982. Early Permian (Rotliegendes) palaeowinds of the North Sea. *Sedim. Geol.* **34**, 245–65.

—— & BOEGNER, P. L. E. 1981. Sole Pit inversion tectonics. *In*: ILLING, L. V. & HOBSON, G. D. (eds) *Petroleum geology of the Continental Shelf of North-West Europe*, 110–20. Heyden, London.

——, MUDD, G. C. & NAGTEGAAL, P. J. C. 1978. Depositional environment and diagenesis of Permian Rotliegendes sandstones in the Leman Bank and Sole Pit areas of the U.K. Southern North Sea. *J. geol. Soc. London*, **135**, 25–34.

HALLAM, A. 1978. Eustatic cycles in the Jurassic. *Palaeogeogr. Palaeoclim. Palaeoecol.* **12**, 1–32.

HARMS, J. C., TACKENBERG, P., POLLOCK, R. E. & PICKLES, E. 1981. The Brae Field area. *In*: ILLING, L. V. & HOBSON, G. D. (eds) *Petroleum Geology of the Continental Shelf of North-West Europe*, 352–7, Heyden, London.

HARRIS, D. G. 1975. The role of geology in reservoir simulation studies. *J. Petrol. Technol.* May, 625–32.

—— & HEWITT, C. H. 1977. Synergism in reservoir management—the geologic perspective. *J. Petrol. Technol.* July, 761–70.

HARTMAN, J. A. & PAYNTER, D. D. 1979. Drainage anomalies in Gulf Coast Tertiary Sandstones. *J. Petrol. Technol.* October, 1313–22.

HERITIER, F. E., LOSSELL, P. & WATHNE, E. 1979. Frigg Field: large submarine fan trap in lower Eocene rocks of North Sea Viking Graben. *Bull. Am. Ass. Petrol. Geol.* **63**, 1999–2020.

HILLIER, G. R. K., COBB, R. M. & DIMMOCK, P. A. 1978. Reservoir development planning for the Forties Field. *Proceedings of the European Offshore Conference*, Vol. II, pp. 325–35. Soc. Petrol. Eng., Dallas.

KESSLER, L. G., ZANG, R. D., ENGLEHORN, J. A. & EAGER, J. D. 1981. Stratigraphy and sedimentology of a Paleocene submarine fan complex, Cod Field, Norwegian North Sea. *In*: *The Sedimentation of North Sea Reservoir Rocks.* Norsk Petroleums forening (NPF), Geilo.

KNOX, R. W. O'B, MORTON, A. C. & HARLAND, R. 1981. Stratigraphical relationships of Palaeocene sands in the UK sector of the Central North Sea. *In*: ILLING, L. V. & HOBSON, G. D. (eds) *Petroleum Geology of the Continental Shelf of North-West Europe*, 267–81. Heyden, London.

KOSSACK, C. A. & BILHARTZ, H. L. 1976. The sensitivity of micellar flooding to reservoir heterogeneities. *Soc. Petrol. Engrs, Pap. no. 5808, 'Improved Oil Recovery Symposium'*, March, Tulsa, Oklahoma.

LEBLANC, R. J. 1972. Geometry of sandstone reservoir bodies. *Mem. Am. Ass. Petrol. Geol.* **18**, 133–90.

—— 1977. Distribution and continuity of sandstone reservoirs—Parts 1 and 2. *J. Petrol. Technol.* July, 776–804.

MAHER, C. E. 1980. Development geology of the Piper Field. *Oceanol. Int. 80*, 13–24.

—— 1981. The Piper Oilfield. *In*: ILLING, L. V. & HOBSON, G. D. (eds) *Petroleum Geology of the Continental Shelf of North-West Europe*, 358–70. Heyden, London.

MARIE, J. P. P. 1975. Rotliegendes stratigraphy and diagenesis. *In*: WOODLAND, A. W. (ed.) *Petroleum and the Continental Shelf of North-West Europe*, 205–12. Applied Science Publishers, London.

MORTON, A. C. 1979. The provenance and distribution of the Palaeocene sands of the Central North Sea. *J. Petrol. Geol.* **2**, 11–21.

NAGTEGAAL, P. J. C. 1978. Sandstone-framework instability as a function of burial diagenesis. *J. geol. Soc. London*, **135**, 101–5.

NEVES, R. & SELLEY, R. C. 1975. A review of the Jurassic rocks of the North-east Scotland. *In*: FINSTEAD, K. S. & SELLEY, R. C. (eds) *Jurassic Northern North Sea Symposium, Stavanger*, JNNSS/5, 1–29 Norwegian Petroleum Society.

PARKER, J. R. 1975. Lower Tertiary sand development in the Central North Sea. *In*: WOODLAND, A. W. (ed.) *Petroleum and the Continental Shelf of North west Europe*, 447–51, Applied Science Publishers, London.

PROCTOR, C. V. 1981. Distribution of Middle Jurassic facies in the East Shetlands basin and their control of reservoir capability. *In*: *The Sedimentation of the North Sea Reservoir Rocks.* Norsk Petroleumsforening (NPF) Geilo.

PRYOR, W. A. 1973. Permeability-porosity patterns and variations in some Holocene sand bodies. *Bull. Am. Ass. Petrol Geol.* **57**, 162–89.

ROCHOW, K. A. 1981. Seismic stratigraphy of the North Sea "Palaeocene" deposits. *In*: ILLING, L. V. & HOBSON, G. D. (eds) *Petroleum Geology of the Continental Shelf of North-West Europe*, 255–66. Heyden, London.

ROSSEL, N. C. 1982. Clay mineral diagenesis in Rotliegend aeolian sandstones of the southern North Sea. *Clay Miner.* **17**, 69–77.

SANGREE, J. B. & WIDMIER, J. M. 1977. Seismic stratigraphy and global changes of sea level, part 9: seismic interpretation of clastic depositional facies. *In*: PAYTON, C. E. (ed.) *Seismic Stratigraphy—applications to hydrocarbon exploration. Mem. Am. Ass. Petrol. Geol.* **26**, 165–84.

SEEMAN, U. 1979. Diagenetically formed interstitial clay minerals as a factor in Rotliegende sandstone reservoir quality in the Dutch sector of the North Sea. *J. Petrol. Geol.* **1**, 55–62.

SHELTON, J. W. 1973. Models of sand and sandstone deposits: a methodology for determing sand genesis and trend. *Bull. Oklahoma geol. Surv.* **118**, 122 pp.

SIMPSON, R. D. H. & WHITLEY, P. K. J. 1981. Geological input to reservoir simulation of the Brent Formation. *In*: ILLING, L. V. & HOBSON, G. D. (eds) *Petroleum Geology of the Continental Shelf of North-West Europe*, 310–09. Heyden, London.

SKJOLD, L. J. 1981. Paleocene sands of the Balder Field. *In*: *The Sedimentation of the North Sea Reservoir Rocks*. Norsk Petroleums forening (NPF), Geilo.

SNEIDER, R. M., RICHARDSON, F. H., PAYNTER, D. D., EDDY, R. E. & WYANT, I. A. 1976. Predicting reservoir-rock geometry and continuity in Pennsylvanian reservoirs, Elk City field, Oklahoma. *Soc. Petrol. Engrs, Pap. no 6138, 51st Ann Tech. Conf.* New Orleans.

STALDER, P. J. 1973. Influence of crystallographic habit and aggregate structure of authigenic clay minerals on sandstone permeability. *Geologie Mijnb* **52**, 217–9.

STOW, D. A. V., BISHOP, C. D. & MILLS, S. I. 1982. Sedimentology of the Brae oilfield, North Sea: fan models and controls. *J. Petrol. Geol.* **5**, 129–48.

SURLYK, F. 1978. Submarine fan sedimentation along fault scarps on tilted fault blocks (Jurassic/Cretaceous boundary, East Greenland) *Bull. Grønlands Geologiske Undersøgelse*, **128**, 108 pp.

THOMAS, A. N., WALMSLEY, P. J. & JENKINS, D. A. L. 1974. Forties Field North Sea. *Bull. Am. Ass. Petrol. Geol.* **58**, 396–405.

VAIL, P. R., MITCHUM, R. M. & TODD, R. G. 1977. Eustatic model for the North Sea during the Mesozoic. *In*: *Proc. Mesozoic Northern North Sea Symposium*. MNNSS/12, 1–35. Norwegian Petroleum Society.

—— & TODD, R. G. 1981. Northern North Sea Jurassic unconformities, chronostratigraphy and sea level changes from seismic stratigraphy. *In*: ILLING, L. V. & HOBSON, G. D. *Petroleum Geology of the Continental Shelf of North-West Europe* 216–35. Heyden, London.

VEEN, F. R. VAN 1975. Geology of the Leman field. *In*: WOODLAND, A. W. (ed.) *Petroleum and the Continental Shelf of North-West Europe*, 223–32. Applied Science Publishers, London.

—— 1977. Prediction of permeability trends for water injection in a channel-type reservoir, Lake Maracaibo, Venezuala. *Soc. Petrol. Engrs, Pap. no. 6703, 52nd Ann. Techn. Conf.* Denver, Colorado.

WALKER, R. G. 1967. Turbidite sedimentary structures and their relationship to proximal and distal depositional environments. *J. sedim. Petrol.* **37**, 25–43.

—— 1978. Deep-water sandstone facies and ancient submarine fans: models for exploration for stratigraphic traps. *Bull. Am. Ass. Petrol. Geol.* **62**, 932–66.

WALKER, R. G. & MUTTI, E. 1973. Turbidite facies and facies associations. *In*: MIDDLETON, G. V. & BOUMA, A. H. (eds) *Turbidites and Deep-Water Sedimentation*, 119–57 Society of Economic Paleontologists and Mineralogists (Pacific Section), Anaheim.

WALMSLEY, P. J. 1975. The Forties field. *In*: WOODLAND, A. W. (ed.) *Petroleum and the Continental Shelf of North-West Europe*, 477–85. Applied Science Publishers, London.

WEBB, M. G. 1978. Reservoir description, Kf Sandstone, Redwash field, Utah. *Soc. Petrol. Engrs, Pap. no. 7046, 5th Symp., Improved Methods for Oil Recovery*, Tulsa, Oklahoma.

WEBER, K. J. 1984. Computation of initial well productivities in aeolian sandstone on the basis of a geological model, Leman gas field, U.K. *In*: TILLMAN, R. W. & WEBER, K. J. (eds) *Reservoir Sedimentology. Spec. Publ. Soc. econ. Paleont. Mineral.* (in press).

WEBER, K. J., KLOOTWIJK, P. H., KONIECZEK, J. & VLUGT, W. R. VAN DER 1978. Simulation of water injection in a barrier-bar type, oil rim reservoir in Nigeria. *J. Petrol. Technol.* November, 1555–65.

WILLIAMS, J. J., CONNER, D. C. & PETERSON, K. E. 1975. Piper oilfield, North sea: fault block structure with Upper Jurassic beach/bar reservoir sands. *Bull. Am. Ass. Petrol. Geol.* **59**, 1581–601.

ZIEGLER, P. 1981. Evolution of sedimentary basins in Northwest Europe. *In*: ILLING, L. V. & HOBSON, G. D. (eds) *Petroleum Geology of the Continental Shelf of North-West Europe*, 3–39, Heyden, London.

—— 1982. *Geological Atlas of Western and Central Europe*. Shell Internationale Petroleum Maatschappij. B. V., 130 pp.

H. D. JOHNSON & D. J. STEWART, Shell U.K. Exploration and Production, London.

Carbonate facies analysis in the exploration for hydrocarbons: a case-study from the Cretaceous of the Middle East

T. P. Burchette & S. R. Britton

SUMMARY: This paper presents a case-study of an investigation into a Middle Eastern carbonate reservoir, the Mishrif Formation of eastern offshore Abu Dhabi, and its lateral equivalent, the Shilaif Member of the Salabikh Formation. The work represents a major part of a sedimentological/seismic stratigraphic study, the aims of which were to model reservoir facies and geometry, and to predict areas of best reservoir potential. The study is based on interpretation of cores, cuttings and wireline logs from 30 wells distributed over some 200 km^2.

The Mishrif of the study area forms part of an extensive Cenomanian carbonate platform. It represents deposition on a local leeward (i.e. shelfward) margin which prograded westwards into an intrashelf (the Shilaif) basin. Six lithofacies associations were recognized in the Shilaif and Mishrif, representing lateral and vertical transitions from silt-grade basinal pelagic-foraminiferal wackestones (I), through a platform margin sequence (II) which coarsens upwards to coarse rudist/ostreiid shoal (III) and biostromal (IV) packstones and grainstones. Shoal and biostromal facies are overlain in the east by bedded back-shoal packstones (V) and nodular micritic platform-lagoonal (VI) sequences. A significant unconformity exists at the top of the Mishrif.

Three stages of diagenesis have been recognized in Mishrif limestones: early (micritization, submarine cementation); unconformity-related (extensive leaching, freshwater cementation); and burial diagenesis (stylolitization, fracturing, burial cementation, neomorphism). Leaching and burial cementation were the most important diagenetic events and significantly overprint the generally simple variations in reservoir character due to original depositional facies distribution. Most favourable porosity and permeability values occur in coarse-grained shoal, upper-slope, and biostromal sediments (upper II, III, IV) where exposed and leached at the final platform margin, whereas to the east, in the platform interior, these facies have suffered severe burial cementation. The slight regional eastwards dip of the Mishrif reservoir, its westwards pinch-out into basinal facies, and the presence of a proved top-seal fulfil the requirements for stratigraphic trapping of hydrocarbons, with the best reservoir facies advantageously situated. A well drilled to test this play encountered the predicted sequence but found only residual oil. This may be due to imperfect lateral or bottom reservoir sealing.

Carbonate sediments form important hydrocarbon reservoirs in many parts of the world, and contain some of the world's largest oil and gas fields. Around 40% of known hydrocarbon reserves occur in carbonates, with 15–20% held in the giant Middle Eastern carbonate reservoirs alone (Klemme 1977; Taylor 1977). The present trend in exploration strategies towards increasing emphasis on the possibility of stratigraphic trapping of hydrocarbons has necessitated the development of refined reservoir models calling upon expertise in a range of disciplines. Among the most important of these in initially locating potential oil or gas reservoirs and traps are sedimentology and geophysics.

This paper presents a project case history illustrating the application of some techniques available to carbonate sedimentologists engaged in the exploration for hydrocarbons and demonstrates how these can be integrated with geophysical studies to produce a predictive reservoir model. The work reported is an analysis of a Middle Eastern Cretaceous carbonate succes-sion, the Mishrif Formation of eastern offshore Abu Dhabi, carried out as part of a major inter-disciplinary reservoir study for the Abu Dhabi Marine Areas Operating Company (ADMA-OPCO).

The mid-Cretaceous (Cenomanian) Mishrif Formation (Fig. 1) has an extensive distribution in the Middle East. It has been a major oil-producing horizon in offshore Dubai since 1970, but testing of Mishrif structural plays in the adjacent eastern offshore area of Abu Dhabi (Fig. 2) has yielded disappointing results. The only exception was the discovery of the Umm Addalkh field (Fig. 3), which forms a large combined structural/stratigraphic trap in an area of anomalously thick Mishrif. In the light of the latter occurrence further stratigraphic trapping of oil at the westward (basinward) edge of the Mishrif appeared likely, and was thus considered to form a valid exploration objective. The disappointing results of a number of wells drilled in eastern offshore Abu Dhabi in the pursuit of this concept led ADMA-OPCO to

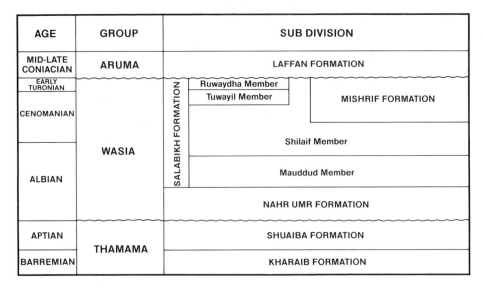

AGE	GROUP	SUB DIVISION		
MID-LATE CONIACIAN	ARUMA	LAFFAN FORMATION		
EARLY TURONIAN			Ruwaydha Member	MISHRIF FORMATION
CENOMANIAN		SALABIKH FORMATION	Tuwayil Member	
	WASIA		Shilaif Member	
ALBIAN			Mauddud Member	
		NAHR UMR FORMATION		
APTIAN	THAMAMA	SHUAIBA FORMATION		
BARREMIAN		KHARAIB FORMATION		

FIG. 1. Stratigraphic table for the mid-Cretaceous, southern Arabian Gulf.

initiate the combined seismic stratigraphic/ sedimentological study, sedimentological aspects of which are reported here. The objectives of this study were to:

(i) develop a depositional/diagenetic model for Mishrif sediments, and identify the distribution of favourable reservoir facies;

(ii) apply geophysical modelling techniques to mapping the westwards 'pinch out' of the Mishrif, ensuring compatibility between sedimentological and seismic stratigraphic models; and

(iii) evaluate the potential for stratigraphic trapping within the Mishrif reservoir, and so identify further drillable Mishrif prospects in the eastern ADMA-OPCO concession area.

Stratigraphic and palaeogeographic background

The Wasia Group (Albian-L.Turonian) in the southern Arabian Gulf represents a depositional megacycle starting with the Nahr Umr unconformity and ending with the Laffan unconformity (Fig. 1). The Shilaif Member of the Salabikh Formation, and the Mishrif Formation, form a transgressive-regressive subcycle up to 200 m thick in the upper part of this Wasia megasequence. The Shilaif comprises basinal and deeper 'shelf' pelagic limestones and occurs over the whole study area. It represents deposition in an arm of the Rub al Khali basin (Fig. 2), one of a series of intracratonic depressions

which developed along the eastern margin of the Arabo–Nubian massif during the late Mesozoic (Murris 1980). The drowned margin of the latter feature formed a shallow-water shelf (Fig. 2), bounded on its eastern side by the Tethys Ocean. The Mishrif Formation represents deposition in an extensive carbonate platform environment which covered most of the shelf. In the study area the Mishrif formed a partially attached platform (Fig. 2), the margin of which prograded actively westwards into the Shilaif 'basin': the character of the eastern, windward Tethys Mishrif limit is unknown, since a significant part of the section has been removed by erosion at the post-Mishrif unconformity. Recent data suggest that this part of the platform may have initiated around cores formed by previous carbonate platform highs (T. Thackeray, BP, personal communication). Centripetal progradation of Mishrif carbonates largely filled the Shilaif basin by the end of the Cenomanian.

On a more local scale, Cretaceous deposition was modified by diapiric movement of the deeply buried Cambrian Hormuz salt, which caused the development of broad domal uplifts throughout the southern Arabian Gulf. Movement was particularly pronounced towards the end of the Cenomanian, and in places this has had a profound effect on Mishrif diagenesis and thickness distributions.

Methods and materials utilized

The work utilized data from 30 wells distributed

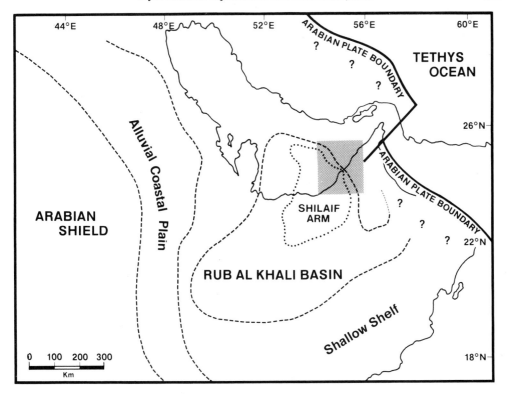

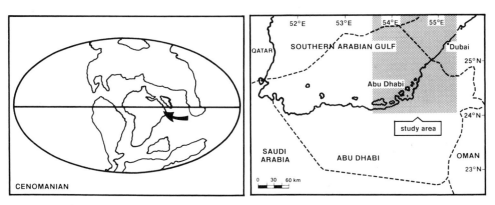

FIG. 2. Location maps and mid-Cretaceous palaeogeography of the study area.

over the whole of the study area (Fig. 3). Full suites of wireline logs were available from most of the wells studied. From seven of these wells, cores totalling 500 m were logged and examined in detail; a further 580 m of core were available for purposes of comparison and correlation. Some 2600 thin sections from core samples were examined. Ditch cuttings samples were available from uncored intervals, and selected samples of these were also thin-sectioned. All sectioned samples were pressure-impregnated with blue-

dyed resin to permit careful study of the types of porosity.

Detailed examination of fine-grained fabrics in 30 selected samples was undertaken using scanning electron microscopy. Results were integrated with conventional optical microscopic thin section studies. Insoluble residue values were determined for 15 samples, composition of the residues being analysed by means of X-ray diffraction. XRD analysis was also carried out on a limited number of whole rock samples.

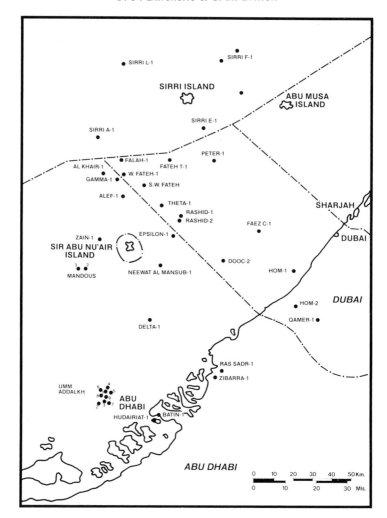

FIG. 3. Eastern offshore Abu Dhabi and Dubai, showing well locations.

Stable isotope and cathodoluminescence studies were also made of selected samples.

Close liaison was maintained throughout the study with geophysicists involved in seismic interpretation and modelling of Mishrif Formation distribution.

Lithofacies description and interpretation

Lithofacies

In the following section the lithofacies encountered are described using the textural scheme of Dunham (1962); a brief interpretation is given following each description. Sediments of the Mishrif and Shilaif have been grouped into six *lithofacies associations*, each of which re-

presents a distinct subenvironment of deposition (Fig. 4). Although facies identifications were made in cored intervals, each association exhibits a distinctive wireline log character which, combined with examination of cuttings samples, facilitated extrapolation of facies identifications into the uncored intervals of all wells used. It has been possible in this way to prove the consistency of both vertical and lateral facies transitions over the whole of the study area, and construct a depositional model for the Mishrif carbonates (Fig. 5).

Lithofacies association I. Basin (Figs 6 and 8)

Description

Shilaif Member. Hard, dark brown to dark tan, silt to very fine sand-grade organic-rich lime

LITHOFACIES ASSOCIATION	LITHOLOGY	BIOTA
I BASIN	Well-bedded, locally bioturbated, stylolitized, silt-grade bioclastic packstones and wackestones.	*Oligostegina*; pelagic foraminifera; rare *Placunopsis* and *Exogyra*; uncommon burrowing organisms.
II PLATFORM-MARGIN SLOPE	Coarsens upwards from well-bedded, bioturbated silt-grade bioclastic packstones to poorly-bedded medium-grained bioclastic packstones.	Bivalves: *Lima semiornata, Lithophaga* sp., *Plagiostoma* sp., *Agerostrea* sp.; burrowers.
III PLATFORM-MARGIN SHOAL	Unbedded. medium to very coarse grained bioclastic packstones and grainstones.	Rudists: *Praeradiolites* sp., *Radiolites* sp., *Sauvagesia* sp., uncommon caprinids and monopleurids. Other molluscs: *Chondrodonta, Tylostoma*. Corals: *Cladophyllia* (rare). Uncommon benthonic foraminifera and echinoids.
IV BIOSTROME/ BIOHERM	Unbedded. extremely coarse. shelly bioclastic packstones.	
V BACK—SHOAL	Interbedded. fine to very coarse grained bioclastic packstones. wackestones and grainstones.	Abundant *Chondrodonta*, scattered radiolitid rudists. Burrowing organisms.
VI PLATFORM-LAGOON	Indistinctly-bedded. burrow-mottled. benthonic-foraminiferal and peloidal mudstones and wackestones.	Benthonic foraminifera: *Dicyclina, Orbitolina, Ovalveolina, Praealveolina, Valvulammina picardi, Nezzazata conica, Pseudochrysalidina*. Ostracodes. uncommon molluscs. Burrowers.

FIG. 4. Lithofacies associations recognized in the Shilaif and Mishrif with their principal lithological and biotic characteristics.

packstones and wackestones comprising a pelagic biota (*Oligostegina* and hedbergellid foraminifera), set in a well-cemented lime-mudstone matrix (Figs 6 and 8A, B). Comminuted bioclastic debris and peloids are major constituents in places. The sediments are well-bedded, but intensely compacted, locally with a nodular fabric. Discrete low-amplitude and wispy 'horsetail' microstylolites are abundant. Shalier intervals are present in the lower and middle parts of the sequence; in the lower interval *Exogyra*-like bivalves occur locally.

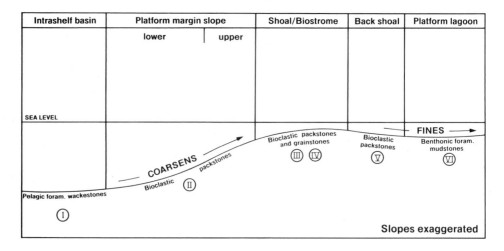

FIG. 5. Schematic cross-section of low-gradient carbonate platform margin showing distribution of Shilaif and Mishrif sediments inferred from lithofacies analysis.

Where overlain by the shallower-water Mishrif, the Shilaif grades up into this formation through a bioturbated, interbedded transition zone in which thin (up to 5 cm) graded beds locally occur. In places these are amalgamated into small tiered packets.

Interpretation

The predominantly pelagic microbiota (Fig. 8A, B) and the absence of a significant pelagic macrofauna in these sediments indicates an open-marine, probably basinal, but somewhat restricted depositional environment (Fig. 5); the fine grain size and the high content of organic material (the Shilaif is a prolific source rock) suggests that this was a regime of low turbulence and poor oxygenation. Burrowing in the Shilaif/Mishrif transition zone implies better oxygenation in basin-marginal situations. Sparsely preserved thin graded beds in this zone in some areas represent deposition from occasional basinward-flowing density currents, possibly storm-ebb currents generated in shallower water.

Lithofacies association II. Platform-margin slope (Figs 7 and 8)

Description

These sediments form the lower part of the coarsening-upwards Mishrif succession, grading up from moderately well-bedded (5–15 cm units) grey, well-cemented, silt to very fine sand-grade bioclastic packstones, to poorly bedded, friable, cream-coloured medium-grained bioclastic

packstones at the top (Figs 7A–E and 8C, D). Grains are commonly highly micritized, but predominantly molluscan in origin; echinoderm fragments are locally common. Middle and upper parts of the sequence possess a diverse molluscan fauna. Small radiolitid rudists occur rarely in the uppermost beds. Pelagic organisms are common only in the basal part of the sequence and decrease in abundance upwards. There is also a significant upwards decrease in the degree of stylolitization, with discrete microstylolites disappearing before wispy horse-tail types, and the coarser sediments near the top of the sequence virtually stylolite free (Fig. 7A–E). Bioturbation is prominent in the lower zones but less obvious in the upper.

Interpretation

The sequence, coarsening upwards from basinal limestones to shallow-marine pack-stones, represents progradation of a carbonate platform slope into the basinal environment (Fig. 5). Sediment grain size decreased distally across this zone (Fig. 8C, D). Bioturbation has destroyed sedimentary structures, and may be responsible for the absence of bedding in the upper part of the sequence. Using the features described above, fine-grained lower and coarser-grained upper shelf-slope depositional environments can be recognized (Fig. 5). They are comparable with the 'shallow' and 'outer-barrier foreslope' environments recognized in rudist 'reefs' of the Caribbean area (Kauffman & Soul 1974).

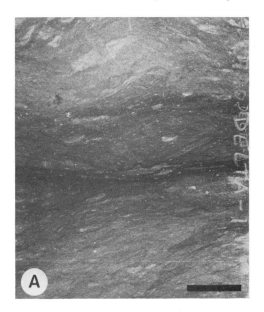

FIG. 6. Lithofacies association I; basin. Shilaif Member, just below the Mishrif. Showing its compacted, microstylolitized and nodular character. Note presence of scattered compressed burrows and uncommon thin-shelled bivalves. Scale bars represent 2 cm.

Lithofacies association III. Shoal
(Figs 9 and 12)

Description

Medium to very coarse-grained bioclastic packstones and grainstones (Fig. 9A, B). These white-cream, friable sediments gradationally overlie the shelf slope deposits and locally contain finer-grained intercalations. Bioclasts are almost exclusively molluscan, the majority probably rudistid (Fig. 12A, B). The fauna, rarely *in situ*, is a low diversity one dominated by small radiolitid rudists. Bedding is ill-defined and no other sedimentary structures are present.

Interpretation

These sediments represent the coarsest widespread facies within the Mishrif, and in most places the culmination of the coarsening-upwards sequence. They are interpreted in view of character and context as the deposits of low-energy shoals and banks at the platform margin (Fig. 5). Sedimentary structures are poorly seen in this friable facies; their absence may be attributable in part to bioturbation. Well-sorted grainstone intervals present in places at the top of the succession may represent beach or shoal calcarenites.

Lithofacies association IV. Rudist biostrome
(Fig. 9)

Description

Extremely coarse, cream-coloured shelly bioclastic packstones containing a more diverse, intact fauna than association III (Fig. 9C, D). This is dominated by radiolitid rudists, which are commonly *in situ*, with scattered caprinids and monopleurids. Other molluscs, including the ostreiid *Chondrodonta* are common. In places, small branching corals, such as *Cladophyllia*, are present (Jordan *et al.* 1981). Interskeletal matrices comprise finer molluscan wackestones and packstones.

Interpretation

These sediments occur in scattered patches up to 15 m thick and a few square kilometres in area at the top of the Mishrif coarsening-upwards succession, one example forms the Fateh oil field of northern offshore Dubai (Jordan *et al.* 1981) (Fig. 3). The more diverse fauna and the presence of *in situ* rudists indicate that these deposits originated as biostromes or small domal bioherms near the prograding shelf margin (Fig. 5).

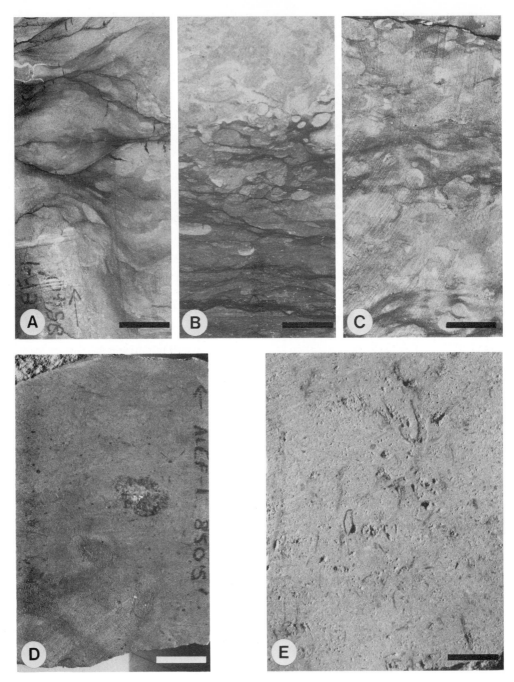

FIG. 7. Lithofacies association II; platform margin slope, Mishrif Formation. (A–E) forms the platform margin slope coarsening-upwards sequence. (A) silt-grade lowest slope packstones showing fairly bioturbated, nodular character, with abundant 'horsetail' and common discrete microstylolites. (B, C) Coarse silt-grade middle and upper-slope packstones. Note extensive bioturbation and upwards-decreasing intensity of stylolitization. Bedding is less nodular, with prominent areas of pre-compaction cementation. (D, E) Fine to medium sand-grade upper slope packstones. Note the absence of stylolitization and the presence of large molluscan bioclasts. Bioturbation decreases in prominence upwards. Scale bars represent 2 cm in all cases.

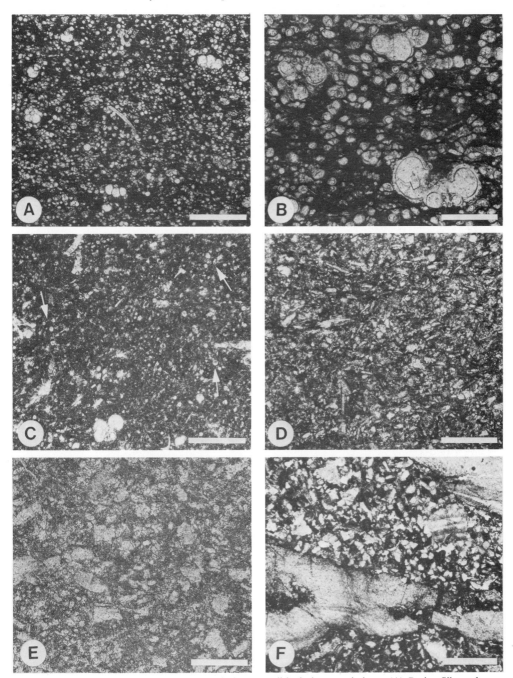

FIG. 8. Photomicrographs; basin (I) and slope (II) lithofacies associations. (A) Basin. Silt-grade pelagic packstone comprising abundant *Oligostegina* (a calcisphere) and scattered hedbergellid foraminifera. Scale bar = 0.5 mm. (B) Close-up of above. Scale bar = 250 μm. (C–F) Lithofacies association II, slope. (C) Lower slope. Silt-grade packstone comprising largely indeterminate bioclasts with common *Oligostegina* (examples arrowed) and rarer pelagic foraminifera. Scale bar = 0.5 mm. (D) Mid-slope. Coarse silt-grade packstone comprising molluscan and echinoderm fragments. Scale bar = 0.5 mm. (E) Mid-slope. Close up, showing micrite matrix and abraided nature of grains. Scale bar = 250 μm. (F) Upper slope. Fine to medium-grained packstone consisting largely of molluscan fragments. Scale bar = 1 mm.

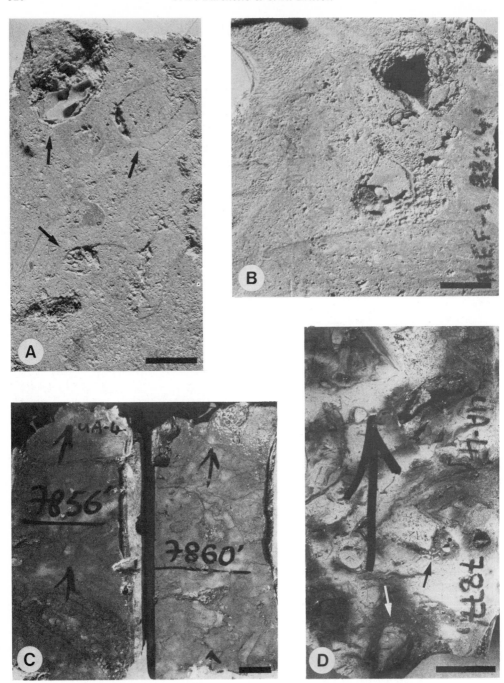

Fig. 9. Lithofacies associations III and IV; shoal and biostrome. Mishrif Formation. (A) Shoal. Coarse-grained packstone/grainstone containing broken and abraded radiolitid rudist shells (arrowed). Note extensive vuggy biomouldic porosity. (B) Shoal. Large fragments of radiolitid rudists in coarse-grained packstone, showing typical boxwork shell wall structure. (C) Biostrome. Large radiolitid rudists (arrows) and other bivalves in a fine packstone matrix. (D) Biostrome. Radiolitid (small black arrow) and caprinid (large black and small white arrows) rudist shells in a fine-grained packstone/wackestone matrix. Note dark patchy oil staining. Scale bars represent 2 cm in all cases.

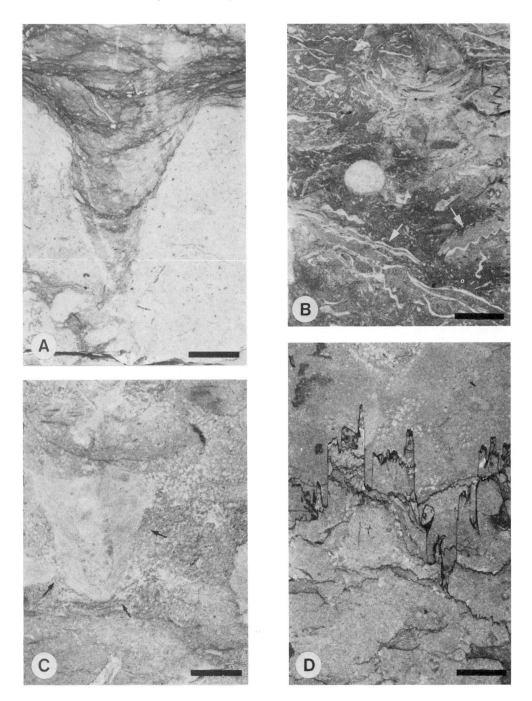

FIG. 10. Lithofacies association V; back shoal. Mishrif Formation. (A) Large ?crustacean burrow in white medium-grained molluscan packstone, infilled by overlying fine-grained, darker wackestone. (B) Large articulated, sediment-filled *Chondrodonta* shells (examples arrowed) in bioclastic packstone/wackestone matrix. Indeterminate molluscan bioclast at centre left. (C) Possibly *in situ* radiolitid rudist from thin back-shoal biostrome. (D) Large-amplitude castellate horizontal stylolite along bedding plane between different lithologies. Scale bars represent 2 cm in all cases.

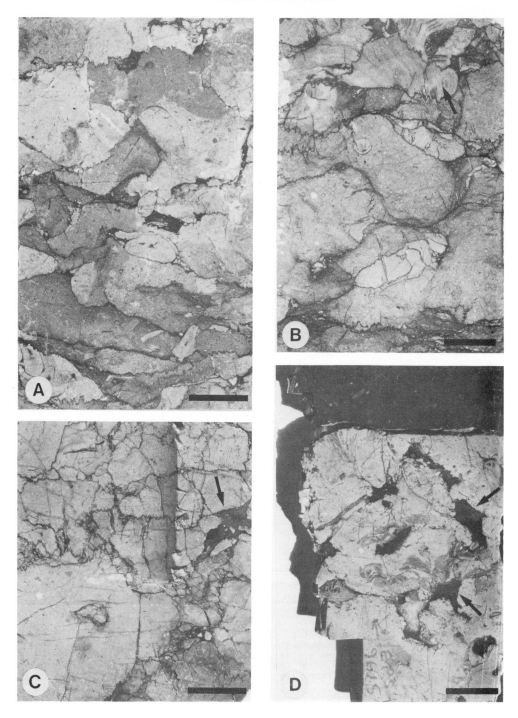

Fig. 11. Lithofacies association VI; platform lagoon. Mishrif Formation. (A, B) Highly nodular, stylolitized and fractured lagoonal mudstones and wackestones. Some nodules are irregularly bleached. Note rudist fragments at top right of (B). (C) Similar to (B), showing sediment-filled cavity. (D) Lagoonal micrite showing tubular solution cavities infilled with several generations of internal sediment. Scale bars represent 2 cm in all cases.

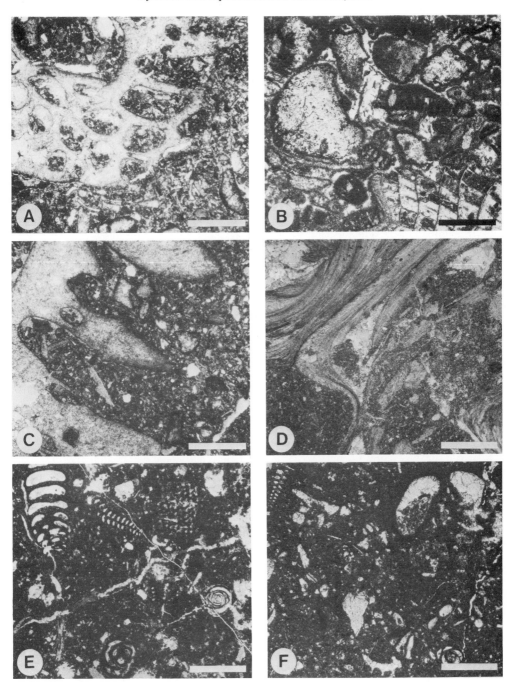

FIG. 12. Photomicrographs; shoal (III), back-shoal (V) and lagoonal (VI) lithofacies associations. (A) Shoal. Large rudist fragment in a sand-grade molluscan packstone matrix. Scale bar = 1 mm. (B) Shoal. Medium-grained molluscan grainstone, showing rounded and abraded clasts. Large radiolitid rudist fragment in lower right. Scale bar = 0.5 mm (C, D) Back-shoal. Large molluscan shells (*Chondrodonta* in D) in a fine to medium sand-grade molluscan/echinoderm packstone matrix. Scale bar = 1 mm. (E, F) Platform lagoon. Benthonic-foraminiferal wackestone. Diverse foraminifera, including large agglutinating forms, in a fossiliferous micrite matrix. Scattered open-space structures may be due to burrowing or desiccation. Scale bars = 0.5 mm.

Lithofacies association V. Back-shoal
(Figs 10 and 12)

Description

Thinly to medium-interbedded fine to very coarse-grained, fawn bioclastic packstones, wackestones and grainstones (Fig. 10A–C & 12C, D). The sediments are characterized by more varied lithologies than the shoal or biostromal facies; bedding is discrete and the sediments more indurated, and less chalky than other coarse facies. Boundaries between beds are commonly stylolitized (Fig. 10D). The coarser sediments in this association contain a fauna dominated by intact *Chondrodonta*, gastropods, and small, locally *in situ*, radiolitid rudists. Bioturbation is common and includes recognisable ?*Ophiomorpha* up to 3 cm in diameter (Fig. 10A). Ripple cross-lamination occurs locally.

Interpretation

These deposits reflect deposition in an immediate back-shoal environment, a transition zone in which mixing of sediments from the platform-margin shoals and the platform-interior lagoon occurred (Fig. 5). They resemble the 'back-barrier slope' deposits of Caribbean rudist complexes (Kauffman & Soul 1974). They overlie the Mishrif coarsening-upwards succession and are intercalated packet-wise with lagoonal lime mudstones in the base of the lagoonal section. *Chondrodonta* and radiolitid-rich beds probably represent thin (*c.* 1 m) biostromes; the interbedded, bioturbated packstones and grainstones are attributed to shoal-derived carbonate sand sheets and washovers.

Lithofacies association VI. Platform-lagoon
(Figs 11 and 12)

Description

Hard, greenish-grey, light brown or fawn, indistinctly bedded benthonic-foraminiferal and peloidal mudstones and wackestones (Fig. 11A–C). These sediments are extensively burrow-mottled and have a characteristic stylolitized, nodular fabric. They contain abundant, diverse benthonic foraminifera (Fig. 12E, F); groups identified include miliolids, orbitolinids, dicyclinids and textulariids. Molluscan (including rudist) and echinoderm debris, calcispheres and ostracods are locally important. The lime mudstone matrix is well cemented and locally slightly argillaceous. Rare laminated algal intraclasts are present.

In the top 15–20 m of the succession these sediments contain breccia zones and vertical tubular cavities up to several centimetres across filled with argillaceous carbonate silt (Fig. 11D). In places the walls to these cavities are bleached.

Interpretation

These sediments form the uppermost unit of the Mishrif succession. They represent the deposits of a broad quiet-water lagoon sheltered by the platform-margin shoals (Fig. 5). Intercalations of coarser association V sediments represent episodes during which back-shoal skeletal sand flats were extended, as areas of intra-lagoonal biostromal growth occurred.

The vertical cavities and conglomeratic units in lagoonal micrites in the zone below the sub-Laffan unconformity represent vadose solution pipes and karst breccias containing 'vadose silts' (Dunham 1969). Bleached zones around these cavities may represent contemporary oxidation aureoles.

Depositional model

Prograding platform margin

Well studies show a consistent regional upwards transition from lithofacies association I to VI (Fig. 13). The lagoonal association (VI) thus forms the uppermost part of any Mishrif succession in which it occurs, always overlying the back-shoal facies association (IV), but restricted to the east of the study area. Towards the west both lagoonal and back-shoal successions grade laterally into shoal calcarenites (Fig. 13). The latter form the uppermost lithofacies division near the western Mishrif limit. These relationships imply progradation of the platform westwards or southwestwards into the Shilaif basin (Figs 14 and 15), a deduction additionally supported by the fact that the Mishrif/Shilaif transition occurs at a progressively higher level in the basinal succession from the east to west.

The Shilaif succession exhibits a number of characteristic log markers (Fig. 13). These are attributable to subtle changes in lithology, and represent discrete events of sedimentation correlatable throughout the Shilaif basin. In the east of the study area Mishrif deposition (lower slope facies) began at the marker 2 stage, whereas the Mishrif/Shilaif transition occurs at the marker 6 stage near the final Mishrif platform margin (Figs 13 and 16).

Shoal and slope facies exhibit an increase in thickness from the east towards the west (Figs 13 and 14). This is considered to reflect the effects of: (1) steadily rising sea-level, causing the platform to aggrade as well as build out, combined with (2) progradation of this platform

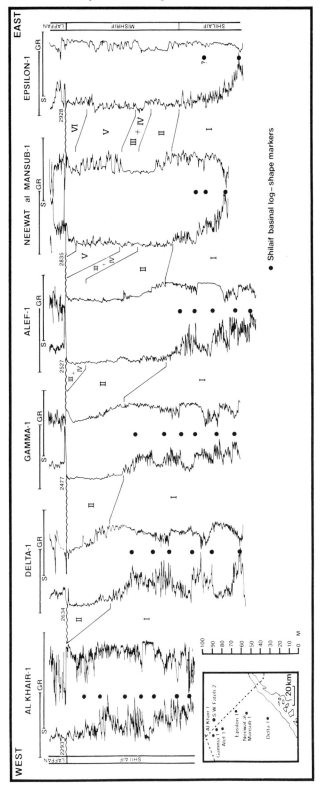

FIG. 13. Correlation of gamma ray (GR: 0–100 api units) and sonic (S: 45–140 s ft⁻¹) wireline logs from representative wells across the Mishrif platform margin, showing log responses characteristic of Shilaif and Mishrif lithofacies (I–IV). Datum: post-Mishrif unconformity. Note that the number of isochronous Shilaif log-shape markers (black dots) increases from east to west, suggesting progradation of the carbonate platform margin. Seven markers are present in Al Khair-1, the most basinal well; these are numbered in correlations from bottom to top.

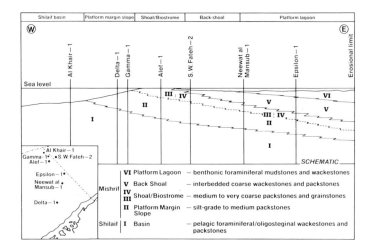

FIG. 14. Schematic cross-section based on wells in Fig. 13 effectively normal to depositional strike, showing prograding platform margin model. Wells projected along strike to line of section.

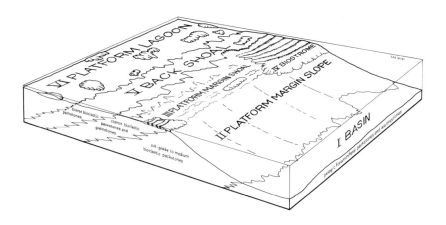

FIG. 15. Block diagram showing model proposed for Mishrif and Shilaif depositional environments. Horizontal scale is several tens of kilometres, vertical scale several hundreds of metres. Roman numerals indicate lithofacies as in Fig. 5.

margin into increasingly deeper water of the Shilaif basin. Both processes would result in expansion of the lateral and vertical extent of the shoal to slope transition with the most marked effect apparent in the thickness of the coarsening-upwards sequence.

Platform morphology

The consistent gradual upwards-coarsening transition from Shilaif basinal limestones (I) to the coarser calcarenites of association III in the lower part of the Mishrif (Fig. 13) indicates a very low-gradient ramp-like platform margin (Ahr 1973) over which shoal-derived sediment dispersed and fined basinwards in response to ebb-tidal or ebbing storm-surge currents. The absence of significant buildups, or slumps and gravity flows in the slope sediments implies a uniformly shallow profile, conforming in character to a 'homoclinal ramp' (Read 1982). The Mishrif of the study area, however, represents deposition on the shallow lee-side of a broad, shelf-edge carbonate platform (Fig. 2), and it is

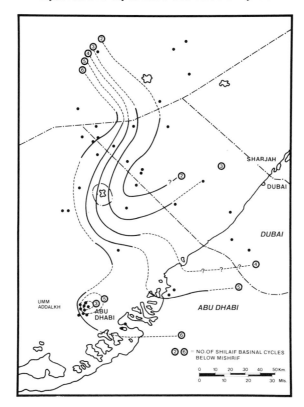

FIG. 16. Number of Shilaif log markers up to the onset of Mishrif deposition, contoured to illustrate progradation of the Mishrif platform margin with time. The Umm Addalkh field represents an isolated complex which amalgamated with the platform at the marker 5 or 6 stage.

possible that the oceanic (Tethyan) eastern margin of the complex was steeper. This margin may have been reef-rimmed ('reef-rimmed platform', Ginsburg & James 1974; Read 1982), comparable with the extensive early and mid-Cretaceous rudist barrier reef tracts which fringed the ancestral Gulf of Mexico (Coogan, Bebout & Maggio 1972; Bebout & Loucks 1974; Bay 1977).

Mishrif palaeoecology and depositional environment

Four broad assemblages of organisms have been identified in the Mishrif sediments. These are closely related to the depositional environments outlined above (Fig. 4). The lower portion of the platform margin slope (lower II) possesses an essentially basinal pelagic biota, whereas the upper is characterized by a varied assemblage of epifaunal and semi-infaunal bivalves. Shoal and biostromal facies (III, IV) are dominated by rudists, and the bedded back-shoal (V) by a mixture of the ostreiid *Chondrodonta* and

rudists. The lagoonal lime muds (VI) possess a rich biota of benthonic foraminifera and algae (Fig. 4).

The wide extent of Mishrif platform progradation in the study area demonstrates the prolific carbonate production by the organisms of these environments. Nevertheless, several features suggest that environmental conditions were not optimal. This is particularly well seen within the shoal and biostromal facies. Rudists are abundant here, but comprise a low-diversity assemblage dominated by small radiolitids, with only subordinate caprinids. Even within biostromes, where diversity is greatest, *in situ* individuals are not densely packed. Rudists in themselves seldom appear to have constructed truly wave-resistant reefs (Barnetch & Illing 1956; Philip 1972). In the Mishrif, however, accumulations nowhere developed beyond the 'dense association' or 'thicket' foundation stages observed in many rudist-bank complexes of the Caribbean realm (Kauffman & Soul 1974), or the 'loose networks' and 'tabular beds' described by Philip (1972) from the mid- and late Cretaceous of southern France.

The dominance of radiolitids, the sparse associated fauna, and restriction of their development largely to thin biostromes and low-amplitude domal bioherms in platform-margin sequences all imply that the Mishrif rudist accumulations represent rather 'immature' examples. There may be a number of reasons for this. Radiolitids appear to have been one of the rudist families most tolerant of environmental stresses such as elevated salinity, turbidity or temperature fluctuations (Wilson 1975; Bein 1976). They commonly form a principal element in rudist pioneer communities (Kauffman & Soul 1974), and their dominance in rudist faunas of the Mishrif suggests that one or more of these adverse factors operated during deposition.

Evaporation over the broad, shallow Mishrif platform undoubtedly resulted in elevated lagoonal salinities, though definite evidence for extreme hypersalinity and associated evaporite deposition is absent. Elevated salinities (up to 45 ppm) occur during summer months over parts of the comparable modern Great Bahama Bank shelf-lagoon adjacent to Andros Island (Cloud 1962; Bathurst 1975). In view of its isolated situation (Fig. 2), tidal exchange with the Shilaif basin across the leeward Mishrif platform margin would also have been restricted. However, even limited movement of hypersaline platform-interior waters across the shoals may have sufficed to influence faunal diversity in these rudist-dominated environments (cf. Wilson 1975, p. 325).

The prograding Mishrif shoals are composed predominantly of packstones with only thin grainstone intervals (Fig. 5). Reef-like developments are restricted in area to a few square kilometres individually, and form only a small part of the total sediment volume (Fig. 15). This paucity of cross-stratified grainstones and reefal facies suggests that Mishrif calcarenites (facies III) represent deposition in a relatively low turbulence environment. The resultant large quantities of primary lime mud present in Mishrif shoal and biostromal lithofacies point to the possibility of at least periodic turbidity within this depositional environment. This also may have been true of many other rudist complexes (Wilson 1975, p. 325), and lends further support to the view that rudists as a group were tolerant of muddier environments than their bank- or reef-constructing contemporaries.

Controls on buildup development

The scale of buildup development within the Mishrif was probably limited independently of the above environmental conditions by two factors: (a) the uniformly ramp-like basinal slope at the platform edge (Fig. 16); and (b) the flat-topped, actively prograding character of the platform ('keep-up type', Kendal & Schlager 1981), over the whole of which deposition kept pace with, or exceeded, rises in relative sea-level. The effects of these relationships were to limit the extent of favourable initial substrates and largely eliminate situations of optimum balance between organic growth and deepening water conditions. Large thicknesses of reefal facies did not therefore develop anywhere within the Mishrif platform margin.

The anomalous Umm Addalkh complex in the south of the study area represents a small 'buildup' initiated in front of the prograding Mishrif platform margin. Facies differ little from those described above, and imply no great relief of this feature during growth. It was engulfed and fully incorporated into the platform at about the Shilaif basinal marker 5 or 6 stage (Figs 16 and 17). Causes for the development of this complex are unknown at present, but associated slope facies are highly attenuated in the early stages, suggesting initiation in shallow water, with subsequent shoaling and then rapid vertical accretion, possibly in response to local subsidence.

Post-depositional platform evolution

Mishrif deposition in the study area was terminated in the late Cenomanian when much of the platform became emergent, creating a regional unconformity. Mature flint clays and kaolinite conglomerates rich in terrestrial palynomorphs and woody plant debris are distributed widely across the platform surface, particularly overlying the marginal shoal facies, and point to the development of extensive swamps or shallow lakes. These events resulted in local, marked karstification of Mishrif carbonates, with clay-filled fissures penetrating in places to several tens of metres.

The final phases of Mishrif platform evolution and the subsequent pattern of erosion were strongly affected by the development of intra-platform flexures derived through halo-kinesis in the deeply buried Cambrian Hormuz evaporites. Many of the oil fields of this area are structures formed in Cretaceous and older sediments due to updoming around deep-seated salt diapirs.

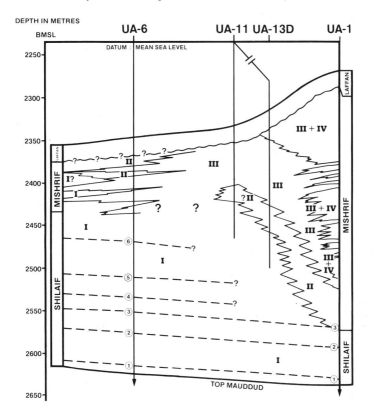

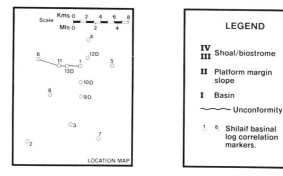

FIG. 17. Cross-section of the western margin of the Umm Addalkh field showing its isolated development. Development was largely through vertical accretion until the Shilaif log marker 5 or 6 stage, following which rapid westwards lateral progradation occurred after amalgamation with the prograding platform margin. Mishrif relief in the UA-1 area is largely apparent, due to differential compaction between basinal and shoal/biostrome lithofacies. Lithofacies notation as in Fig. 14.

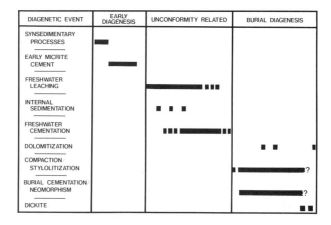

FIG. 18. Mishrif and Shilaif diagenetic processes and their relative distribution in time.

Diagenesis

Three broad stages of diagenesis have been recognized in the Shilaif and Mishrif: early burial, unconformity-related and deep burial (Fig. 18). These were of highly unequal durations, the first two together lasting for a period of *c.* 1 Ma, compared with the *c.* 88 Ma which elapsed until the sediments reached their present burial depths (2400 m in the west, dipping at 2–3° to 2900 m in the east of the study area). Diagenetic processes within each stage, and their relative importance, vary both from one lithofacies association to another as well as geographically.

Early burial diagenesis

This includes both synsedimentary processes and early post-depositional cementation of the sediments.

Synsedimentary diagenetic effects are abundant in all but the basinal facies. They comprise comminution of bioclastic debris by deposit feeders (e.g. Fig. 7D–F), and micritiz-

ation of grains through the action of endolithic algae, fungi and boring ?sponges (Fig. 19A).

Micritic cementation is prominent in proximal basinal, lower to mid-slope and lagoonal facies (i.e. in all but the platform-margin sediments). It demonstrably predates significant burial compaction and its fabric and distribution suggest that it represents a submarine cement: it was probably precipitated at shallow depth beneath the sediment surface. In basinal and lower-slope facies cementation occurred preferentially in more permeable zones, enhancing the original bedding, or highlighting and protecting burrows from compaction (cf. Fürsich 1973; Abed & Schneider 1980). This selectivity commonly resulted in the development of a nodular texture during burial (Figs 6A, B, 7A–C and 11A, B), the lithified zones being fractured but little affected by the stylolitization imposed on the surrounding rock.

In the coarsest, upper-slope and shoal, sediments this micritic cementation has not been observed, neither is there any evidence for

FIG. 19 (facing). Diagenesis. (A) Lithofacies association III, shoal. Grainstone showing the effects of synsedimentary micritization, by boring organisms (algae, fungi, ?sponges) and the development of micrite envelopes after leaching. Note extensive biomouldic and intergranular porosity (white areas). Scale bar = 1 mm. (B) Hand specimen of upper-slope sediments, immediately beneath sub-Laffan unconformity, showing probable solution brecciation. Scale bar = 2 cm. (C) Solution cavity in lagoonal lime mudstones showing several layers of finely laminated internal (?vadose) calcite silt overlain by equant calcite spar. Scale bar = 1 mm. (D, E) Early freshwater cement in shoal sediments. Fine rhombs of polyhedral non-ferroan calcite forming isopachous fringes in secondary and intergranular pores and delicate pore-filling aggregates in biomoulds. Scale bars = 0.5 mm. (F) Syntaxial overgrowths upon echinoderm fragments. These are of same cement phase as that in (D) and (E). They tend to grow poikilotopically and account for a large proportion of the cement in lower- and mid-slope sediments. Scale bar = 0.5 mm.

fibrous, fringing, submarine cement. The platform margin shoals were thus largely unlithified during early diagenesis.

Unconformity-related diagenesis

Freshwater invasion during exposure at the post-Mishrif unconformity resulted in extensive dissolution, minor internal sedimentation and local calcite cementation (Fig. 18).

Dissolution was particularly pronounced directly beneath the unconformity, although its precise effects are variable in character, depending upon lithofacies. Grain leaching and

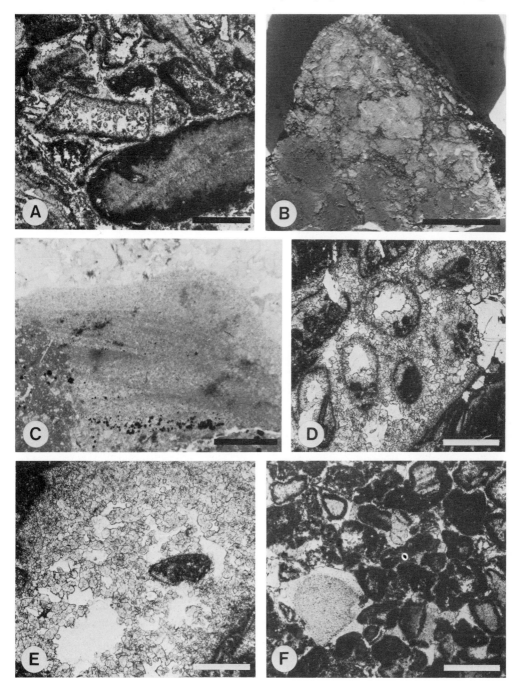

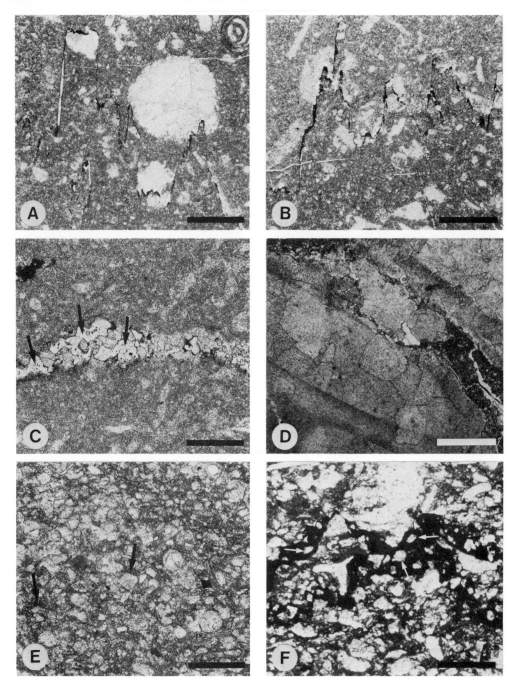

Fig. 20. Photomicrographs; diagenesis. (A, B) Examples of stylolites in lagoonal sediments from platform interior, showing how they commonly affect blocky burial cements. Scale bars = 1 mm. (C) Small fracture adjacent to stylolite, lagoonal lime mudstone. Partially bridged by non-ferroan calcite; remaining porosity filled by authigenic dickite (arrowed). Scale bar 250 μm. (D) Neomorphic calcite. Molluscan bioclast partly obliterated by neomorphic size increase in cement or matrix crystals. Scale bar = 1 mm. (E, F) Dolomite. Scattered silt-grade euhedral non-ferroan dolomite rhombs in lower slope sediments (examples arrowed). Rhombs have remained as insoluble residues in a microstylolite seam (arrowed). Scale bars = 250 μm.

porosity creation were most pronounced in the coarser-grained, biostrome, shoal and upper slope facies, these sediments contain the most abundant and coarsest molluscan bioclasts. Here dissolution of aragonitic components produced biomouldic porosity preserved within micrite envelopes (Fig. 19A), and in places resulted in enlargement of biomoulds to create non-fabric-selective, vuggy porosity. In contrast, exposed lagoonal sediments developed an interconnecting network of tubular solution vugs (Fig. 11D) to a depth of *c.* 20 m beneath the unconformity (Fig. 10).

Locally the uppermost few metres of platform margin sediments, directly beneath the unconformity, have suffered intense solution brecciation (Fig. 19B). In the Umm Addalkh area, where the most severe examples of cavernous leaching occur, karstic solution produced polymict breccias and flint clay-filled dolines of several hundred metres diameter within the Mishrif. Lateral variations in the intensity of these dissolution effects may reflect palaeogeographic relief or relative durations of exposure at the platform surface.

Internal sedimentation Immediately beneath the unconformity both solution vugs and finer-scale mouldic and primary pores may be partially infilled with overlying sediment or vadose silt (cf. Dunham 1969). This is best developed in the lagoonal facies (Figs 11D & 19C).

Calcite cementation was locally important during this diagenetic stage. Two fabrics are distinguished—fine fringing and overgrowth—each being characteristic of different lithofacies.

In the shoal facies directly subcropping the unconformity the cement consists of a fine (*c.* 20 μm), rhombic to polyhedral, non-ferroan calcite, partly occluding, and forming delicate isopachous linings within pores (Fig. 19D, E). This fabric is atypical of marine cements. It post-dates bioclast dissolution since it coats both internal and external surfaces of micrite envelopes. Furthermore, it pre-dates compactional grain-breakage, being absent from fractured grain surfaces. It was therefore precipitated prior to significant burial of these—at that stage highly friable—shoal sediments. Cathodoluminescence data tentatively indicate that down-dip, away from the shoal facies-belt subcrop, this fringing cement is in fact still present. It is overlain, but in plane-polarized light completely masked, by a blocky, burial calcite cement (described below). This petrographic relationship suggests relatively early precipitation of the fine fringing cement. The above paragenetic constraints point to a fresh-water origin for this cement. Meniscus and drip-stone features are absent, however, suggesting a phreatic rather than vadose environment of precipitation.

In the lower to mid-slope facies, where echinoderm debris is locally an important constituent, similar calcite cement occurs as poikilotopically extended syntaxial outgrowths around echinoderm fragments (Fig. 19F).

Deep burial diagenesis

Following exposure, the Mishrif and Shilaif were progressively buried to their present depth. Burial resulted in a variety of diagenetic pro-

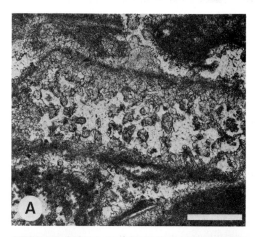

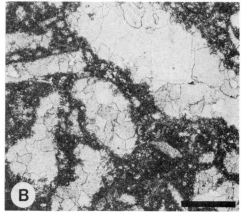

FIG. 21. Photomicrographs; diagenesis. Delicate rhombic to polyhedral early fresh-water cement (A) from a platform margin situation contrasted with the equant pore-occluding burial cement (B) from a platform interior well. Both examples from lithofacies association III shoal sediments. Scale bars = 0.5 mm.

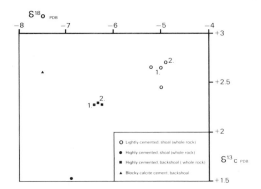

FIG. 22. $\delta^{13}C/\delta^{18}O$ cross-plot for selected Mishrif samples showing possible trend towards lighter $\delta^{18}O$ values in more deeply buried limestones. This may reflect the elevated temperature experienced at the depths where burial cements were precipitated.

cesses (Fig. 18), most important of which were compaction, expressed principally as stylolitization but also as fracturing, and calcite cementation. Less important processes were calcite neomorphism, dolomitization and dickite cementation.

Stylolitization

The distribution and character of stylolites in the Mishrif and Shilaif are strongly related to grain size and therefore to lithofacies, although bedding inhomogeneities also exercise an important influence in certain facies. Ubiquitous 'horsetail' microstylolites are developed throughout the basinal and lower- to mid-slope deposits, with discrete microstylolites (amplitude *c.* 0.5 cm) also occurring in the lower part of this gradational sequence (Figs 6 & 7A–C). Here, microstylolitization contributes to the nodular character of the sediments and their early cementation-enhanced bedded fabric. The highly nodular fabric of the lagoonal facies is accentuated in the same way (Fig. 11). Discrete, horizontal, large-amplitude (up to 5 cm) stylolites are present locally in the backshoal facies. They have commonly developed along bedding planes (Fig. 10D).

In contrast to all other facies, shoal and upper-slope sediments are devoid of stylolites (Figs 7C, D & 9A, B). This is attributed to their unbedded and poorly cemented, friable nature. Compaction here is relatively minor, being expressed by fractured micrite envelopes and locally by grain interpenetration.

Fracturing

Macroscopic fracturing is of relatively little importance in the Mishrif and Shilaif sediments. Early cemented burrow-nodules and beds tend to have suffered fracturing while uncemented sediments deformed plastically around them. Small (up to 1 cm long) fractures, developed normally to stylolites (cf. Nelson 1981), are characteristic of the lower slope and particularly the lagoonal sediments (Figs 7A & 11). Fractures are commonly open, but may be calcite- or dickite-cemented (Fig. 20C).

Calcite cementation

Blocky, equant, non-ferroan calcite forms an extensive pore-occluding cement in the backshoal facies and in those shoal and biostromal sediments down-dip from the unconformity surface. It thus displays an antithetic relationship with the freshwater, fine fringing calcite cement described above (cf. Fig. 21A with B).

Preliminary carbon and oxygen stable isotope analyses on whole-rock samples (Fig. 22) suggest a positive relationship between lighter oxygen values and increasing proportions of blocky calcite cement; the single separated-cement analysis shows the most strongly negative $\delta^{18}O$ value and thus continues this trend. Carbon isotope values show little variation. These results are consistent with the blocky calcite being of burial-diagenetic origin (cf. Hudson 1977; Dickson & Coleman 1980). Furthermore, results from fluid-inclusion studies on blocky calcite cements from the Mishrif of the Fateh Field (McLimans 1981) are consistent with a Miocene age for cement precipitation.

Stylolitization is an obvious potential source for this burial cement. However, cementation ceased before stylolitization, at least locally, as demonstrated by stylolitic cross-cutting of the cement (Fig. 20A, B).

Calcite neomorphism proceeded during the period of burial diagenesis. Neomorphic crystal enlargement (Fig. 20D) can be demonstrated from all but the shoal facies.

Dolomite and dickite

Both of these minerals are represented as accessory diagenetic phases. The former occurs as replacive, non-ferroan rhombs up to 100 μm across, sparsely scattered throughout the basinal and lower- to mid-slope sediments. The rhombs are also present in these facies within microstylolitic seams, indicating an origin prior to or during stylolitization (Fig. 20E, F). Dickite occurs in small quantities, partially filling fractures or other remnant porosity. It is the

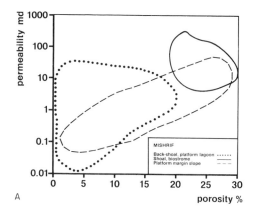

A

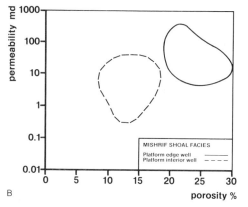

B

ranges from poor to good. Porosity occurs principally within the lime mudstone matrix, with small contributions from intraparticle pores within micritized grains and remnant bio-mouldic pores. Porosity increases markedly upwards, a direct result of the coarsening-upwards character of the slope sediments. Both back-shoal and lagoonal facies associations exhibit poor reservoir qualities; porosity is predominantly of matrix type. Although bio-mouldic porosity was created during leaching in the backshoal sediments, this was subsequently cemented during deep burial by blocky calcite. The wide variability in lagoonal facies poroperm values (Fig. 23A) reflects their characteristic nodular, micro-fractured character.

The best reservoir quality is present in those shoal, biostromal, and uppermost-slope facies directly subcropping the top Mishrif unconformity: the result of extensive leaching and only fine fringing calcite cementation in these coarse and relatively matrix-poor sediments. In contrast, the same facies associations down-dip, towards the east, form very poor reservoirs (Fig. 23B), due to extensive burial-diagenetic blocky calcite cementation. This crucial difference, where late calcite cement has destroyed porosity in the down-dip, but not in the up-dip position, is attributed to synchronous migration and up-dip accumulation of hydrocarbons: that is, a palaeo-oil-water contact effect. Reported immature oil within fluid inclusion in the late calcite cement (McLimans 1981) supports this conclusion.

Application to exploration

This sedimentological study has demonstrated that the most favourable reservoir facies occur within uppermost slope, shoal and biostromal sediments only where these subcrop the post-Mishrif unconformity. This structural position is ideal with respect to the Mishrif stratigraphic hydrocarbon trap concept: a good reservoir pinching out up-dip (westwards) against the overlying Laffan Formation (a regionally proven seal) and the underlying Shilaif Member (Fig. 24A). The Shilaif is a proven hydrocarbon source rock.

Seismic modelling, utilizing results from the sedimentological study, has permitted delineation and mapping of the westwards Mishrif pinch-out (Fig. 25), as well as the 60 m Mishrif isopach. The latter is considered to represent the ideal condition where the maximum shoal-facies thickness subcrops the unconformity (Fig. 24A). This concept was tested by a well which en-

FIG. 23. (A) Porosity/permeability cross-plots for cored Mishrif lithofacies. Highest values occur in shoal and upper-slope facies, with low values in lower-slope sediments. The range in poroperm values for slope sediments reflects the coarsening-upwards sequence. Variable, but generally very low, poroperm values occur in back-shoal and lagoonal sediments. Variability here is due to the presence of stylolites and fractures in largely tight rocks. (B) Cross-plot of porosity/permeability values for cored lithofacies association III (shoal) sediments in a platform-margin situation, where they were subjected to leaching at the post-Mishrif unconformity, and in a platform-interior situation where burial cementation has occluded much porosity.

youngest authigenic phase identified.

Reservoir quality

Mishrif reservoir porosity and permeability are strongly related to lithofacies (Fig. 23A).

The reservoir quality of the slope sediments

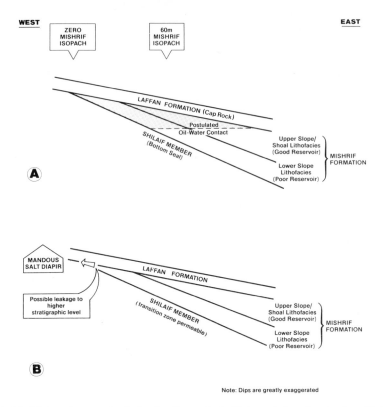

Note: Dips are greatly exaggerated

FIG. 24. Models based on geophysical and sedimentological interpretations for stratigraphic trapping of oil at the Mishrif platform margin. (A) Exploration working model. Top seal for Mishrif reservoir is formed by micritic carbonates and shales of the Laffan Formation, bottom and lateral seals by basinal sediments of the Shilaif Member. (B) Model modified after testing prospect by drilling a well. Abundant oil shows proved the validity of the concept, but water saturations were high. The lower slope facies may extend into the area affected by the Mandous salt diapir, possibly allowing further migration from the Mishrif platform-margin trap to a higher stratigraphic level.

countered the predicted Mishrif lithofacies, formation thickness and high-quality reservoir. However, only residual oil was present in the formation. This is considered to have been due to imperfect lateral-(bottom-) sealing by the Shilaif Member, possibly an effect of the anomalously microporous nature of the uppermost Shilaif (i.e. transition zone) at this locality. It is possible that oil has migrated laterally through the Mishrif and further up-dip into the large salt-induced Mandous structure immediately to the west (Fig. 24B).

Conclusions

1 The Cenomanian Mishrif reservoir carbonates (Fig. 1) in eastern offshore Abu Dhabi represent the progradational low-energy leeward margin of a partially attached shelf-edge carbonate platform (Fig. 2). This built out into the relatively shallow, Shilaif intrashelf basin, in which pelagic foraminiferal limestones rich in organic material were deposited.

2 Six lithofacies associations, representing distinct depositional subenvironments, have been recognised in Shilaif and Mishrif carbonates (Fig. 4):

    I Basin—pelagic-foraminiferal packstones and wackestones.

    II Platform-margin slope—succession coarsening upwards from silt-grade to medium-grained bioclastic packstones.

    III Platform-margin shoal—medium to very coarse-grained molluscan packstones and grainstones.

    IV Biostrome—extemely coarse, shelly molluscan packstones.

    V Back-shoal—fine to very coarse-grained molluscan packstones, wackestones and grainstones.

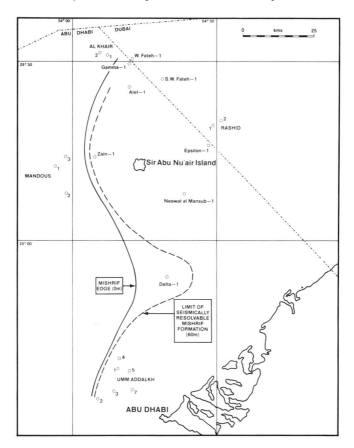

FIG. 25. The eastern Abu Dhabi concession area (ADMA) showing the westwards Mishrif limit (0 and 60 m contours) mapped using seismic stratigraphic techniques.

VI Platform lagoon—benthonic-foraminiferal mudstones and wackestones.

3 Mishrif carbonates were subaerially exposed during the Late Cenomanian and Turonian and in many places are overlain by plant-rich freshwater mudstones.

4 Several phases of diagenesis have been recognized in Mishrif and Shilaif carbonate sediments. These correspond broadly to: early diagenesis, unconformity-related processes, and burial diagenesis. The processes most significantly affecting reservoir properties were freshwater leaching at the unconformity surface and compaction and burial cementation.

5 Porosity and permeability distributions in the Mishrif reservoir are functions of original facies character (grain size, homogeneity) and subsequent diagenesis. The coarsest-grained, lightly cemented upper-slope (upper II), shoal (III) and biostromal (IV) lithofacies associations possess the most favourable reservoir characteristics. In situations of deeper burial, these facies also exhibit severely reduced poroperm values due to extensive burial cementation.

6 The validity of a Mishrif platform-margin stratigraphic play, as defined by a combined sedimentological and seismic stratigraphic study, has been proved. However, testing with a well at a selected location demonstrated that hydrocarbons had migrated, through faulty bottom-sealing, from the trap to a higher stratigraphic level in the area affected by the Mandous salt diapir.

ACKNOWLEDGMENTS: We gratefully acknowledge the permission to publish this work granted by Abu Dhabi Marine Operating Company (ADMA-OPCO) and The British Petroleum Company. Most of the data presented are extracted from the proprietary material of the former company. We thank The British Petroleum Company for technical support during preparation of the manuscript. Special thanks are due to Ron Walters (BP Research Centre, Sunbury) and Dick Moody (Kingston Polytechnic) for assistance in identifying the micro- and macrofaunas respectively, and to Phil Johnson (BP, Sunbury) for isotopic analyses. We are indebted to Katy Frankiel for preparation of many of the diagrams. The seismic stratigraphic interpretations which formed an integral part of this study were carried out for ADMA-OPCO by Patrick Lantigner (Compagnie Francaise de Petroles).

References

ABED, A. M. & SCHNEIDER, W. 1980. A general aspect in the genesis of nodular limestones documented by the Upper Cretaceous limestones of Jordan. *Sedim. Geol.* **26**, 329–35.

AHR, W. M. 1973. The carbonate ramp: an alternative to the shelf model. *Trans. Gulf-Cst Ass. Geol.* **23**, 221–5.

BARNETCH, A. & ILLING, L. V. 1956. The Tamabra Limestone of the Poza Rica oil field, Veracruz, Mexico. *Proc. 20th int. Geol. Congr., Mexico,* 38 pp.

BATHURST, R. G. C. 1975. *Carbonate Sediments and their Diagenesis,* 2nd ed. Elsevier, Amsterdam. 658 pp.

BAY, T. A. 1977. Lower Cretaceous stratigraphic models from Texas and Mexico. *In*: BEBOUT, D. G. & LOUCKS, R. G. (eds) *Cretaceous Carbonates of Texas and Mexico: applications to subsurface exploration. Rep. Bur. econ. Geol. Univ. Texas,* **89**, 12–30.

BEBOUT, D. G. & LOUCKS, R. G. 1974. Stuart City trend, Lower Cretaceous, South Texas. *Rep. Bur. econ. Geol. Univ. Texas,* **78**, 80 pp.

BEIN, A. 1976. Rudistid fringing reefs of Cretaceous shallow carbonate platform of Israel. *Bull. Am. Ass. Petrol. Geol.* **60**, 258–72.

CLOUD, P. 1962. Environment of calcium carbonate deposition west of Andros Island, Bahamas. *Prof. Pap. U.S. geol. Surv.* **350**, 1–138.

COOGAN, A. H., BEBOUT, D. G. & MAGGIO, C. 1972. Depositional environments and geologic history of Golden Lane and Poza Rica trend, Mexico, an alternative view. *Bull. Am. Ass. Petrol. Geol.* **56**, 1419–47.

DICKSON, J. A. D. & COLEMAN, M. L. 1980. Changes in carbon and oxygen isotope composition during limestone diagenesis. *Sedimentology,* **27**, 107–18.

DUNHAM, R. J. 1962. Classification of carbonate rocks according to depositional texture. *In*: HAM, W. E. (ed.) Classification of carbonate rocks. *Mem. Am. Ass. Petrol. Geol.* **1**, 108–21.

—— 1969. Early vadose silt in Townsend mound (reef), New Mexico. *In*: FRIEDMAN, G. M. (ed.) *Depositional Environments in Carbonate Rocks: a symposium. Spec. Publs Soc. econ. Paleont. Miner., Tulsa,* **14**, 182–91.

FÜRSICH, F. T. 1973. *Thalassinoides* and the origin of nodular limestone in the Corallian Beds (Upper Jurassic) of Southern England. *Neues Jb.*

Geol. Paläont. Mh. **3**, 136–56.

GINSBURG, R. N. & JAMES, N. P. 1974. Holocene carbonate sediments of continental shelves. *In*: BURKE, C. A. & DRAKE, C. L. (eds) *The Geology of Continental Margins.* 137–55. Springer-Verlag, New York.

HUDSON, J. D. 1977. Stable isotopes and limestone lithification. *J. geol. Soc. London,* **133**, 637–60.

JORDAN, C. F., CONNALLY, R. C. & VEST, H. A. 1981. Upper Cretaceous carbonates of the Mishrif Formation, Fateh field, Dubai, UAE. *In*: ROEHL, P. O. & CHOQUETTE, P. W. (eds) *Carbonate Petroleum Reservoirs.* Authors' Workshop, Vail, Colorado, 1980.

KAUFFMAN, E. G. & SOUL, N. F. 1974. Structure and evolution of Antillean Cretaceous rudist frameworks. *Verh. Natur. Ges. Basel,* **84**, 399–467.

KENDAL, C. G. ST C. & SCHLAGER, W. 1981. Carbonates and relative changes in sea level. *Mar. Geol.* **44**, 181–212.

KLEMME, D. 1977. Giant gas fields contain less than 1% of world's fields but 75% of reserves. *Oil Gas. J.* **75**, 164.

MCLIMANS, R. K. 1981. Applications of fluid inclusion studies to reservoir diagenesis and petroleum migration: Smackover Formation, US Gulf Coast, and Fateh field, Dubai. *Bull. Am. Ass. Petrol. Geol.* (Abstract), **65**, 957.

MURRIS, R. J. 1980. Middle East: stratigraphic evolution and oil habitat. *Bull. Am. Ass. Petrol. Geol.* **64**, 597–618.

NELSON, R. A. 1981. Significance of fracture sets associated with stylolite zones. *Bull. Am. Ass. Petrol. Geol.* **65**, 2417–25.

PHILIP, J. 1972. Paleoécologie des formation a rudistes du Crétacé Supérieur—l'example du sud-est de la France. *Palaeogeogr. Palaeoclim. Palaeoecol.* **12**, 205–22.

READ, J. F. 1982. Carbonate platforms of passive (extensional) continental margins: types, characteristics and evolution. *Tectonophys.* **81**, 195–212.

TAYLOR, J. C. M. 1977. Sandstones as reservoir rocks. *In*: HOBSON, G. D. (ed.) *Developments in Petroleum Geology (1),* 147–96. Applied Science Publishers, London.

WILSON, J. L. 1975. *Carbonate Facies in Geologic History,* Springer-Verlag, New York. 471 pp.

TREVOR P. BURCHETTE and SELINA R. BRITTON, Sedimentology Branch, The British Petroleum Company plc, Britannic House, Moor Lane, London EC2Y 9BU.

Subject Index